MORPHOLOGICAL AND PHYSICAL CLASSIFICATION OF GALAXIES

ASTROPHYSICS AND SPACE SCIENCE LIBRARY

A SERIES OF BOOKS ON THE RECENT DEVELOPMENTS
OF SPACE SCIENCE AND OF GENERAL GEOPHYSICS AND ASTROPHYSICS
PUBLISHED IN CONNECTION WITH THE JOURNAL
SPACE SCIENCE REVIEWS

VOLUME 178
PROCEEDINGS

MORPHOLOGICAL AND PHYSICAL CLASSIFICATION OF GALAXIES

PROCEEDINGS OF THE FIFTH INTERNATIONAL WORKSHOP
OF THE OSSERVATORIO ASTRONOMICO DI CAPODIMONTE
HELD IN SANT'AGATA SUI DUE GOLFI, ITALY,
SEPTEMBER 3–7, 1990

Edited by

G. LONGO
Osservatorio di Capodimonte, Napoli, Italy
M. CAPACCIOLI
Osservatorio Astronomico, Università di Padova, Italy
and
G. BUSARELLO
Osservatorio di Capodimonte, Napoli, Italy

SPRINGER SCIENCE+BUSINESS MEDIA, B.V.

Library of Congress Cataloging-in-Publication Data

Osservatorio astronomico di Capodimonte. International Workshop (5th : 1990 : Sant'Agata sui Due Golfi, Italy)
Morphological and physical classification of galaxies : proceedings of the Fifth International Workshop of the Osservatorio astronomico di Capodimonte, held in Sant'Agata sui Due Golfi, Italy, September 3-7, 1990 / edited by Giovanni Busarello, Massimo Capaccioli, Giuseppe Longo.
p. cm. -- (Astrophysics and space science library ; v. 178)
Includes bibliographical references and index.
ISBN 978-94-010-5113-2 ISBN 978-94-011-2522-2 (eBook)
DOI 10.1007978-94-011-2522-2
1. Galaxies--Congresses. I. Busarello, G. (Giovanni) II. Capaccioli, M. III. Longo, G. (Giuseppe) IV. Title. V. Series.
QB856.O88 1990
523.1'12--dc20 92-1502

ISBN 978-94-010-5113-2

Printed on acid-free paper

On the afternoon of January 5, 1992, our friend and highly esteemed colleague Jean-Luc Nieto met with a fatal accident while climbing the North face of Pic du Midi in the Pyrenees. Full of sorrow, we all dedicate this volume to his memory.

TABLE OF CONTENTS

EDITORS' FOREWORD xiii

SCIENTIFIC AND LOCAL ORGANIZING COMMITTEE xv

LIST OF PARTICIPANTS xvii

INVITED PAPERS: *GALAXY TYPES*

MORPHOLOGY OF GALAXIES: AN OVERVIEW
R. Buta 1

THE LUMINOSITY FUNCTIONS OF GALAXIES
H. Jerjen, G.A. Tammann, B. Binggeli 17

ELLIPTICAL GALAXIES
M. Franx 23

STRUCTURE AND KINEMATICS OF SPIRALS
P.C. van der Kruit 39

BRIGHTEST CLUSTER MEMBERS
J. Schombert 53

CENTRAL REGIONS OF EARLY—TYPE GALAXIES
J.-L. Nieto 69

STELLAR CONTENT AND CHEMICAL EVOLUTION IN SPHEROIDAL POPULATIONS
M. Rich 85

INVITED PAPERS: *GALAXIAN COMPONENTS*

STELLAR DISKS IN EARLY-TYPE GALAXIES
M. Capaccioli, N. Caon 99

GASEOUS DISKS IN ELLIPTICAL GALAXIES
F. Bertola 115

FORMATION AND MAINTENANCE OF SPIRAL STRUCTURE
E. Athanassoula 127

DARK MATTER IN ELLIPTICAL GALAXIES
T. de Zeeuw 139

DARK MATTER IN LATE–TYPE GALAXIES
S.M. Kent 153

DARK MATTER IN DWARF GALAXIES
C. Pryor 163

ELLIPTICAL GALAXIES IN X–RAYS
L. Ciotti, A. D'Ercole, S. Pellegrini, A. Renzini 179

THE WARM COMPONENT OF THE ISM OF ELLIPTICAL GALAXIES
F. Macchetto, W.B. Sparks 191

THE BAR COMPONENT IN DISK GALAXIES
A. Bosma 207

PECULIAR KINEMATICS
M. Balcells 221

COLD GAS IN ELLIPTICAL GALAXIES
J. van Gorkom 233

COLD GAS IN LATE–TYPE GALAXIES
R. Sancisi 239

ABUNDANCE GRADIENTS AND RELATED PROPERTIES IN GALAXIES
F. Matteucci 245

STRUCTURE AND METALLICITY OF LOW–LUMINOSITY EARLY–TYPE GALAXIES
P. Prugniel, E. Bica, D. Alloin 261

INVITED PAPERS: *ENVIRONMENT AND SYSTEMATICS*

INTERACTIONS: WEAK
F. Combes 265

INTERACTIONS: STRONG
J. Barnes 277

THE MORPHOLOGY OF BINARY GALAXIES
J.W. Sulentic 293

INTRINSIC SHAPES OF ELLIPTICAL GALAXIES
D. Merritt 309

ON UNDERSTANDING THE GLOBAL PROPERTIES OF GALAXIES
D. Burstein 321

GALAXY MANIFOLDS AND GALAXY FORMATION
G.S. Djorgovski 337

SUB-COMPONENTS IN ELLIPTICAL GALAXIES AND THEIR FORMATION
R. Bender 357

MORPHOLOGY AND CLASSIFICATION OF GALAXIES – WHERE DO WE STAND?
I.R. King 371

POSTER PAPERS

DETAILED MORPHOLOGICAL CLASSIFICATION OF EARLY–TYPE GALAXIES IN THE PERSEUS CLUSTER
P. Poulain, J.–L. Nieto, E. Davoust 377

PROPERTIES OF DWARF SPHEROIDALS
G. Djorgovski, R. de Carvalho 379

BRIGHT CORES IN NEARBY SOUTHERN GALAXIES
O.R. Hainaut, B.J. Jarvis 381

SHELL FORMATION AND THICK DISKS IN ELLIPTICAL GALAXIES
R.C. Thomson 385

THE ORIGIN OF STRONG ISOPHOTAL TWISTS IN ELLIPTICAL GALAXIES
J.–L. Nieto, Ch. Lagoute 387

PHOTOMETRIC PROPERTIES OF DISKS IN ELLIPTICAL GALAXIES
C. Scorza, R. Bender 389

ELLIPTICAL GALAXIES WITH DARK MATTER
G. Bertin, R.P. Saglia, M. Stiavelli 391

NUCLEAR DUST–DISCS IN RADIO ELLIPTICALS
W. B. Sparks, F. Macchetto, D. Fraix-Burnet,
R. Bender, M. Capaccioli, J.–L. Nieto 393

EXTENDED Hα FILAMENTS IN ELLIPTICAL GALAXIES
D. Fraix-Burnet, F. Macchetto, W.B. Sparks 395

THE ORIGIN OF THE IONIZED GAS IN S0 GALAXIES
F. Bertola, L.M. Buson, W.W. Zeilinger 397

LINE–STRENGTH GRADIENTS IN S0 GALAXIES
A. Paquet, R. Bender, W. Seifert 399

ULTRAVIOLET SPECTRA OF EARLY TYPE GALAXIES
A. Danks, M. Perez 401

ON THE NATURE OF THE COMPACT GALAXIES IN THE UGC
E. Davoust, S. Considère, Ph. Prugniel, N. Denis 403

MORPHOLOGY OF GALAXIES SURROUNDING COMPACT GROUPS
G.C. Baiesi–Pillastrini, V. Tornatore, G.G.C. Palumbo 405

MEAN SURFACE DENSITY AND MORPHOLOGICAL TYPE IN DISC–GALAXIES
M. Prieto, A. Campos Aguilar, C. García 407

PECULIAR KINEMATICS IN INTERACTING ELLIPTICAL GALAXIES
R. Madejsky 409

MORPHOLOGY AND KINEMATICS OF SPIRAL GALAXIES WITHOUT BAR: THE IMPORTANCE OF THE CENTRAL MASS
G.C. Baiesi–Pillastrini 411

THE CORRELATION BETWEEN VELOCITY GRADIENTS AND ARM CLASSIFICATION IN SPIRAL GALAXIES
A. Biviano, M. Girardi, G. Giuricin, F. Mardirossian, M. Mezzetti 413

OPTICAL LINE/RADIO LUMINOSITY CORRELATION AND THE EFFECTS OF THE INTERSTELLAR MEDIUM
R. Morganti 415

THE LUMINOSITY–DIAMETER RELATION FOR DISK GALAXIES: A SEARCH FOR ENVIRONMENTAL EFFECTS
M. Girardi, A. Biviano, G. Giuricin, F. Mardirossian, M. Mezzetti 419

GLOBAL PROPERTIES IN EARLY TYPE GALAXIES
A. Rifatto 421

MASSES AND MASS–TO–LIGHT RATIOS OF ELLIPTICAL GALAXIES
G. Busarello, G. Longo 423

ON THE PHOTOMETRIC SCALING LAWS FOR EARLY–TYPE GALAXIES
N. Caon, M. Capaccioli 425

PROPERTIES OF BRIGHTEST CLUSTER MEMBERS
S. Djorgovski, R. de Carvalho, I. Shlosman, J. Schombert 427

THE FUNDAMENTAL PLANE FOR EARLY–TYPE GALAXIES
I. Jørgensen, P. Kjærgaard 429

ELLIPTICAL GALAXIES AND THE THICKNESS OF THE FUNDAMENTAL PLANE
F. Simien, P. Prugniel 431

DISSIPATIONLESS FORMATION OF ELLIPTICAL GALAXIES
M. Stiavelli, P. Londrillo, A. Messina 433

THE TWO–COMPONENT COLOUR–STRUCTURE OF THE gE GALAXY NGC 4472
P. Surma, R. Bender, J.–L. Nieto 435

THE VELOCITY DISPERSION ANISOTROPY IN NGC 720
R. van der Marel, P. Cinzano 437

NGC 205 AND THE NATURE OF COMPACT NUCLEI IN DWARF GALAXIES
S. Djorgovski, O. Bendinelli, G. Parmeggiani, F. Zavatti 439

LIGHT AND COLOR DISTRIBUTION IN THE CORE OF M32
R. Michard, J.-L. Nieto 441

NGC 3384 AND ITS "SATURN-LIKE" APPEARANCE
M. Ferrario, M. Capaccioli 443

CIGALE OBSERVATIONS OF NGC 3198
R.L.M. Corradi, J. Boulesteix, A. Bosma,
M. Capaccioli, P. Amram, M. Marcelin 445

MAPPING THE STAR FORMATION IN NEARBY GALAXIES: THE MAGELLANIC IRREGULAR NGC 3109
F. Bresolin, M. Capaccioli, G. Piotto 447

MM OBSERVATIONS OF THE MAGELLANIC CLOUDS FROM ANTARCTICA
P. Calisse, G. Dall'Oglio, M.T. Di Bari, A. Iacoangeli, L. Martinis 449

SEYFERT GALAXIES WITH CLOSE COMPANIONS. A METHOD FOR IDENTIFICATION
P. Rafanelli, P. Marziani 451

GLOBULAR CLUSTERS AS PROBES OF THE GALACTIC POTENTIAL
M. Stiavelli, G. Piotto, M. Capaccioli 455

RELATION BETWEEN THE PROPERTIES OF GLOBULAR CLUSTER SYSTEM AND THE MASS DISTRIBUTION ON THE GALACTIC DISK
V.G. Surdin 457

TOWARDS A PRECISE DEFINITION OF THE MORPHOLOGICAL STRUCTURES OF GALAXIES AS A BASIS FOR THEIR AUTOMATIC CLASSIFICATION
A. Accomazzi, D. Delfini, M. J. Kurtz, P. Mussio 459

FOURIER ANALYSIS OF BARRED GALAXIES: A CONVENIENT ALTERNATIVE TO THE COMPONENT DECOMPOSITION
P. Dubath, B. J. Jarvis, L. Martinet, D. Pfenniger 461

IMPROVEMENTS OF THE SIGNAL-TO-NOISE RATIO BY ADAPTIVE FILTERING
H. Lorenz, G. Richter 465

INDEX OF OBJECTS 469

EDITORS' FOREWORD

The morphological scheme devised by Hubble and followers to classify galaxies has proven over many decades to be quite effective in directing our quest for the fundamental parameters describing the extragalactic manifold. This statement is however far more true for spirals than for ellipticals. Echoing the concluding remarks in Scott Tremaine's summary talk at the Princeton meeting on Structure and Dynamics of Elliptical Galaxies, *"the Hubble classification of spirals is useful because many properties of spirals (gas content, spiral arm morphology, bulge prominence, etc.) all correlate with Hubble time. By contrast, almost nothing correlates with the elliptical Hubble sequence E1 to E7."*

During the last few years much effort has been put into the search for a more meaningful classification of ellipticals than Hubble's. Concomitantly, forwarded by some provocative conjectures by R. Michard, the classical question of whether E galaxies form a physically homogeneous family has been brushed up once more. Results of these and other parallel studies look rather promising and point to suture part of the dichotomy between ellipticals and disk galaxies which had become popular in the early eighties, owing to dynamical arguments. At the same time it appears more and more clear that, besides the usual genetic varieties of galaxies, products of environmental evolution must also be contemplated in building our modern picture of the "reign of galaxies".

The above considerations prompted us to solicit Prof. Mario Rigutti, Director of the Capodimonte Astronomical Observatory (OAC), to devote the 5th OAC Conference at photographing the state of the art of this exciting transition process from the morphological to the physical "era" in classification of galaxies. The documentary role chosen for the meeting called for a structure in just invited and poster papers, with ample space for discussions. Three oral sessions were devised accordingly to the following scheme: Hubble classes, galaxy components, and physical parametrization.

We planned to gather a number of 'experts' in an isolated spot in order to favour interactions (and to discourage tours evasions). A suitable location was found in the hotel *O' Sole Mio* at the outskirts of Sant'Agata sui Due Golfi, a small village placed at 500 meters above sea–level at the tip of the splendid Sorrento peninsula. The answer from the community has been quite good. The attendance to the meeting, from the 2nd to the 6th of September of 1990, was of $\sim$ 100 scientists coming from all over the world.

Practically all the invited speakers provided us with their written contributions, and some of them even within the theoretical deadline. These papers are included here together with the accompanying discussions and the written versions of the posters. They account on the scientific activity; happy hours are instead documented by the photographs with which we have filled up occasional blank space.

This meeting would not have been possible or just equally successful on the organizational ground without the help of Ing. P. Petillo of DIGITAL Equipment Corporation, Mr. F. Serino of OLIVETTI Computers, Dr. F. Bevilacqua of AERITALIA, and Dr. F. Simeoli of the municipality of Massa Lubrense. We also acknowledge the professionality of Mss. G. Iaccarino and T.Ievolella as meeting secretaries, and the "southern" hospitality of M. Insigne, hotel manager, and his family.

Because of some "late deliveries", these proceedings are sent to the printer quite a few months past the first *formal* and *strict* deadline. However, the domain of LaTeX and the dedication of Mss. E. Acampa and A. D'Orsi allowed to cut drastically the technical time for the preparation of the manuscript.

Gianni Busarello, Massimo Capaccioli, and Giuseppe Longo

Naples, December 21, 1991

SCIENTIFIC ORGANIZING COMMITTEE

Ralf Bender, Germany
Francesco Bertola, Italy
David Burstein, USA
Massimo Capaccioli, Italy (Chairman)
Francoise Combes, France
Harold G. Corwin, USA
John Danziger, ESO
Tim de Zeeuw, The Netherlands
George Djorgovski, USA
Ferruccio Macchetto, USA
Jorge Melnick, Chile
Mario Rigutti, Italy (Director OAC)
Piet C. van der Kruit, The Netherlands
Simon White, USA

LOCAL ORGANIZING COMMITTEE

Giovanni Busarello
Giuseppe Cuccaro
Giuseppe Longo (Chairman)

HOTEL O SOLE MIO

LIST OF PARTICIPANTS

ACCOMAZZI Alberto, Harward–Smithsonian C.f.A., Cambridge, USA
AMICO Paola, Dept. of Astronomy, Univ. of Padova, Italy
ARNABOLDI Magda, SISSA, Trieste, Italy
ATHANASSOULA Evangelia, Astronomical Obs., Marseilles, France
BALCELLS Marc, Kapteyn Laboratory, Groningen, The Netherlands
BARNES Joshua, Univ. of Toronto, CITA, Canada
BENDER Ralf, Astronomical Obs., Heidelberg, Germany
BENDINELLI Orazio, Dept. of Astronomy, Univ. of Bologna, Italy
BERTIN Giuseppe, Scuola Normale Superiore, Pisa, Italy
BERTOLA Francesco, Dept. of Astronomy, Univ. of Padova, Italy
BOSMA Albert, Astronomical Obs., Marseilles, France
BURSTEIN David, Dept. of Phys. & Astron., Arizona State Univ., USA
BUSARELLO Giovanni, Astronomical Obs. of Capodimonte, Napoli, Italy
BUSON Lucio, Astronomical Obs., Padova, Italy
BUTA Ronald, Dept. of Phys. & Astron., Univ. of Alabama, USA
CAON Nicola, SISSA, Trieste, Italy
CAPACCIOLI Massimo, Dept. of Astronomy, Univ. of Padova, Italy
CENZON Giovanni, Dept. of Astronomy, Univ. of Padova, Italy
CERIELLO Antonio, Astronomical Obs. of Capodimonte, Napoli, Italy
CINZANO Pierantonio, Dept. of Astronomy, Univ. of Padova, Italy
CIOTTI Luca, Dept. of Astronomy, Univ. of Bologna, Italy
COMBES Francoise, Radioastronomy Obs., Meudon, France
CONSIDÉRE Suzanne, Astronomical Obs., Besançon, France
CORRADI Romano, Dept. of Astronomy, Univ. of Padova, Italy
DANKS Anthony, ST System Corporation, USA
DAVOUST Emmanuel, Obs. of Midi–Pyrénées, Toulouse, France
DELFINI Davide, Dept. of Physics, Univ. of Milano, Italy
DE MARTINO Ferdinando, Astronomical Obs. of Capodimonte, Napoli, Italy
DE ZEEUW Tim, Astronomical Obs., Leiden, The Netherlands
DJORGOVSKI George, CALTECH, Berkeley, USA
D'ONOFRIO Mauro, SISSA, Trieste, Italy
DUBATH Pierre, Astronomical Obs., Geneve, Switzerland
FEOLI Antonio, Dept. of Physics, Univ. of Salerno, Italy
FERRARIO Massimo, Dept. of Astronomy, Univ. of Padova, Italy
FRAIX–BURNET Didier, Space Telescope Science Institute, Baltimore, USA
FRANX Marijn, C.f.A., Optical Infrared Div., Cambridge, USA
FRIEDLI Daniel, Astronomical Obs., Geneve, Switzerland

GIRARDI Marisa, SISSA, Trieste, Italy
GIURICIN Giuliano, Dept. of Astronomy, Univ. of Trieste, Italy
GREGGIO Laura, Astronomical Obs., Bologna, Italy
GRØSBOL Preben, ESO, Germany
GUELIN Michel, IRAM, St. Martin d'Heres, France
HICKSON Paul, University of BC, Canada
JARVIS Brian, ESO, Santiago, Chile
JØRGENSEN Inger, Copenhagen University Obs., Denmark
KENT Stephen, C.f.A., Cambridge, USA
KING Ivan R., Dept.of Astronomy, Univ. California, USA
LONGO Giuseppe, Astronomical Obs. of Capodimonte, Napoli, Italy
MACCHETTO Duccio, Space Telescope Science Institute, Baltimore, USA
MADEJSKY Rainet, Astronomical Obs., Heidelberg, Germany
MARZIANI Paolo, SISSA, Trieste, Italy
MATTEUCCI Francesca, Max Planck Inst. for Astroph., Munich, Germany
MELNICK Jorge, ESO, Santiago, Cile
MERRITT David, Dept. of Phys. & Astron., Rutgers Univ., USA
MEURER Gerhardt, Anglo–Australian Obs., Australia
MORGANTI Raffaella, Inst. of Radioastronomy, Bologna, Italy
NIETO Jean-Luc, Astronomical Obs., Toulouse, France
OOSTERLOO Thomas, ESO, Germany
PALUMBO Giorgio, Dept. of Astronomy, Univ. of Bologna, Italy
PAQUET Alexander, Astronomical Obs., Heidelberg, Germany
PELLEGRINI Silvia, Dept. of Astronomy, Univ. of Bologna, Italy
PERSIC Massimo, Astronomical Obs., Trieste, Italy
PICCIRILLO Lucio, ESA–ESTEC Astrophys. Div., Noordwijk, The Netherlands
PIOTTO Giampaolo, ESO, Dept. of Astronomy, Univ. of Padova, Italy
PIŞMIŞ Paris, UNAM, Mexico City, Mexico
POULAIN Patrick, Astronomical Obs., Toulouse, France
PRIETO Mercedes, Astrophysics Institute of Canarias, Tenerife, Spain
PRUGNIEL Philippe, ESO, Germany
PRYOR Carlton, Dept. of Physics & Astronomy, Rutgers Univ., USA
RAFANELLI Piero, Dept. of Astronomy, Univ. of Padova, Italy
RENZINI Alvio, Dept. of Astronomy, Univ. of Bologna, Italy
RICH Michael, Columbia Astrophys. Lab., Dept. of Astronomy, N.Y., USA
RICHTER Gotthard, Academy of Sciences, Berlin, East Germany
RIFATTO Agatino, Astronomical Obs. of Capodimonte, Napoli, Italy
SALZA Antonio, AERITALIA, Div. Sistemi Spaziali, Italy
SANCISI Renzo, Kapteyn Laboratory, Groningen, The Netherlands
SCORZA Cecilia, Astronomical Obs., Heidelberg, Germany

SCHOMBERT James, Dept. of Astronomy, Univ. of Michigan, USA
SPARKS William, Space Telescope Science Institute, Baltimore, USA
STIAVELLI Massimo, ESO, Germany
SULENTIC Jack, Dept. of Physics & Astronomy, Univ. of Alabama, USA
SURDIN Vladimir, Sternberg Institute, Moscow, USSR
SURMA Peter, Astronomical Obs., Heidelberg, Germany
TAMMANN Gustav, Astronomical Obs., Basel, Switzerland
THOMSON Robert, Institute of Astronomy, Cambridge, England, UK
TYSON Neil, Columbia Astrophys. Lab., Dept. of Astronomy, NY, USA
VAN ALBADA Tijerd, Kapteyn Laboratory, Groningen, The Netherlands
VAN DER KRUIT Piet, Kapteyn Laboratory, Groningen, The Netherlands
VAN DER MAREL Roeland, Astronomical Obs., Leiden, The Netherlands
VAN GORKOM Jaqueline, Dept. of Astronomy, Univ. of Columbia, USA
VAN WOERDEN Hugo, Kapteyn Laboratory, Groningen, The Netherlands
WARMELS Rein, ESO, Germany
ZAVATTI Franco, Dept. of Astronomy, Univ. of Bologna, Italy
ZEILINGER Werner, ESO, Germany
ZWITTER Tomaz, SISSA, Trieste, Italy

MORPHOLOGY OF GALAXIES: AN OVERVIEW

R. Buta
Department of Physics and Astronomy
University of Alabama, USA

ABSTRACT. Morphological classification has been and still is a very useful tool in modern extragalactic astronomy. In this overview, I discuss galaxy morphology with regard to the techniques and problems of classification, as well as recent advances in the field.

1 Introduction

Since Hubble's classification famous 1926 paper outlining his system, understanding galaxy morphology has been an important goal of extragalactic astronomy. For a long time, the subject was little more than descriptive because of a lack of basic data, but today morphology is very much more than a series of type codes and symbols. There is little doubt that morphology was a first logical step in approaching an understanding of galaxies. However, it is reasonable to ask where morphology has led us up to now. This is important to consider at a time when extragalactic astronomy is making great strides on both theoretical and observational fronts.

In this overview I will focus mainly on the fundamentals of morphological classification, both in theory and in practice. The subject is very broad and cannot cover all of the quantitative aspects which have been addressed in recent years. Nevertheless, an attempt is made to review advances in understanding specific types from research during the past 15 years.

2 Morphology: clues to formation and evolution

The principal goal of morphological studies has been to obtain insight into galaxy formation and evolution. Fundamental problems, such as the nature of S0 galaxies, the effects of environment, morphological segregation in clusters, the origin of bars, the driving mechanisms for spiral structure, the possibility of significant secular evolution of structure within a Hubble time, and the underlying factors which determined the various types at the time of galaxy formation, all require accurate knowledge of morphology in order to be addressed reliably. How effective morphology can be in addressing these problems depends on how well the relationships between the various types of galaxies

G. Longo et al. (eds.), Morphological and Physical Classification of Galaxies, 1–16.

are established, and the extent to which follow-up observations and theoretical analyses are carried out.

Morphology has been useful in spite of the fact that classification has been largely a subjective visual exercise. Its techniques, advantages, disadvantages, correlations, etc. have been discussed by many authors. The review by Sandage (1975) covers all of the older references (including those of Hubble, de Vaucouleurs, Holmberg, Morgan, and van den Bergh), while more recent papers include van den Bergh (1976, 1980a), van den Bergh, Pierce, and Tully (1990), Dressler and Sandage (1978), Sandage and Brucato (1979), Dressler (1980a, 1984), Sandage and Binggeli (1984), Sandage (1986), Kormendy (1979, 1982), Kennicutt (1981), Bothun (1982a,b), and Vorontsov-Velyaminov (1987). The problems of classification are fairly well-understood, but this has not prevented astronomers from uncritically accepting the types in catalogues without regard to source, or from blaming the classifiers or the classification systems for the inadequacy of type information for some classes of galaxies. The following five points help to place the problems of galaxy morphology into a general perspective:

1. In galaxy morphology, many distinct structures or "components" are seen: bars, rings, lenses, bulges, disks, spiral patterns, etc. These features presented in various combinations and at various inclinations leave the impression that galaxy morphology consists of an almost "impenetrable thicket of forms" (to borrow a phrase from Stephen Jay Gould, 1985, from an essay on biological taxonomy).
2. Galaxies have a wide range of surface brightnesses, luminosities, and other measured properties. This means that *selection effects are always important.* Details that are needed for classification can be easily missed on inadequate image material.
3. Galaxy structure is by and large *continuous.* In a multi-dimensional classification space, *transition* cases almost always exist between any two distinct and sufficiently common morphologies.
4. The distribution of morphologies in rich clusters is often much narrower than for field galaxies. This implies that environment may be important in determining the galactic form.
5. Galaxies have a large cross section for collisions, mergers, or interactions. This can lead, on one hand, to the possible *evolution of rare or transient galactic forms* (e.g., ring galaxies), but, on another hand, it could also be responsible for some of the more common forms (e.g., E galaxies).

These points highlight the fact that galaxy morphology is a complex problem. The goal of classification is to reduce that complexity somewhat by searching for order within the "utter chaos" (again, to borrow a phrase from S. J. Gould) of a wide range of forms whose relationships to each other may not be obvious. If a classification system eventually sheds light on these relationships, then it could provide the needed physical insight for addressing the ultimate goals of understanding galaxy formation and evolution.

3 Morphological classification: theory

In theory, morphological classification is a fairly simple process. Basic types are defined within the scope of a system of nomenclature which assigns galaxies into "cells" of similar-appearing objects. Using a set of criteria, and a set of standards or prototypes which best illustrate the criteria, an observer attempts to determine the appropriate cell position for any object of interest not included among the standards or prototypes. As long as only a few criteria define a system, and if image material of a similar quality to that which formed the basis of the system is used, then there will be a greater ease of applicability and reproducibility of that system by independent observers. If one later finds correlations between fundamental observables and classifications, then the system could lead to physical insight as has, for example, the Hubble sequence.

The actual assignment of types is more an art than a physical measurement. De Vaucouleurs once told me that his approach to classification is to first identify what a galaxy is *not*. Then one narrows in on the part of the classification continuum where an object may appropriately belong. Classification is fairly straightforward for spirals, where a variety of features (bars, rings, bulge strength) and arm characteristics (resolution, openness) provide a basis for discrimination of types, as noted by Hubble. The classification of S0's depends on distinguishing a "fundamental plane" or envelope surrounding a bright bulge (Sandage, 1961), with the progressive differentiation of disk details, such as a lens, ring-like enhancements or dust lanes, or bars, providing further criteria for early and late S0's. True S0's by definition do not have spiral structure. S0's with obvious lenses, rings, dust lanes, or bars are the least ambiguous and generally are not difficult to classify with good image material. However, non-barred S0's which are early in the S0 sequence, and which show only a trace of a disk or envelope or lens, are very difficult to distinguish from ellipticals and require high quality image material, generally better than the sky surveys.

Although cell morphology is useful, it does have limitations. For instance, it would be a mistake to consider any cell as having a sharp boundary because of point 3 above. Thus a galaxy may not be simply "barred" or "nonbarred", or "ringed or nonringed", but could be "weakly-barred", or have a "broken or partial ring". Some classification systems take this continuity into account better than others (e.g., de Vaucouleurs, 1959 three-dimensional classification volume as opposed to Hubble's revision in Sandage, 1961). Cell morphology is also not the only viable approach to galaxy structure. One could also view galaxies as being composed of a small number of "building blocks" known as distinct components which are assumed to interact. This approach was pioneered by Kormendy (1979; see also Djorgovski, this conference).

4 Morphological classification: in practice

Galaxy classification in practice is difficult for several reasons. First and foremost is that not every galaxy presents a favorable orientation to the line of sight. High inclination makes it difficult to (1) estimate a type consistent with low inclination galaxies, and (2) recognize bars (end-on ones especially), rings, or other disk details. In the case of spirals, the first problem arises because three criteria (bulge-to-disk ratio, the degree of resolution, and the degree of openness of the arms) can be used for typing face-on examples, while only one criterion (bulge-to disk ratio) generally can be used for edge-on examples. This can lead to problems since Hubble's three classification criteria for spirals have been known for a long time to be inconsistent in some galaxies (Sandage, 1961). The second difficulty means that the statistical frequency of important features such as bars and rings will be underestimated in highly inclined galaxies (see, e.g., de Vaucouleurs and Buta, 1980).

Another problem which makes classification difficult is that the image material often used for types (e.g., the small-scale sky surveys) is entirely inadequate for some types. As noted by Sandage and Brucato (1979), *"the classification of E, S0, and early Sa galaxies is often confused"* when the types are based on dense, overexposed plates or paper prints or on underexposed, small-scale plates taken with short focal length telescopes. On overexposed images such as are often found on the SERC IIIa-J southern sky survey films, high surface brightness bars, rings, or lenses are easily missed. Even on Palomar Sky Survey prints, overexposure of high surface brightness galaxies can lead to very misleading classifications; one interesting case, NGC 3928, looks like type E0 on the PSS but appears as a small, tight spiral on a CFHT prime focus plate (van den Bergh, 1980b; see also Taniguchi and Watanabe, 1987).

At the other extreme, underexposed images may cause important faint details to be completely missed. For example, low surface brightness rings, spiral patterns, disks, or other features (e.g., shells) can be difficult to detect on PSS prints or ESO-B films but clearly distinguishable on the SERC and Palomar II surveys or on processed or amplified images. Failure to detect these features could lead to serious misclassification or misinterpretations in some cases, or even to uncataloguing of an object in a diameter-limited survey (Bothun et al., 1987, 1990).

Another problem is that no classification system is perfect enough to encompass all galaxies. Many galaxies cannot be easily fit into one or another system. As emphasized by Sandage and Binggeli (1984), Hubble's classification system encompassed mainly giant, high luminosity galaxies whose forms did not readily extend to much lower luminosities. Galaxies near or outside the fringes of the old classification systems have often been peculiar (colliding or merging systems).

5 Systems of classification

The classification systems in use today are all in some way related to the system described in Hubble's (1926) paper. A detailed review of several of these, in particular Hubble's revised system, de Vaucouleurs' revised Hubble system, van den Bergh's modified Hubble (or DDO) system with luminosity classes, Morgan's spectral form classification (or Yerkes) system, and Vorontsov-Velyaminov's purely descriptive (or MCG) system, has already been provided by Sandage (1975).

The only system which has changed significantly since 1975 is van den Bergh's modified Hubble system. For various reasons, van den Bergh (1976) disagreed with Hubble's placement of S0's in the "transition region" between ellipticals and spirals, and instead proposed placing S0's in a sequence parallel to spirals (called the RDDO system). His sequence of modified Hubble types uses bulge-to-disk ratio as the main classification criterion, while Hubble's final division of spiral types was based principally on the appearance of the arms. Using B/D ratio as the sole stage criterion, van den Bergh identified transition cases between S0's and normal spirals which appeared to be spirals with little star formation in the arms. These were given the term "anemics" and were assumed to be poor in HI (see Bothun and Sullivan, 1980). Van den Bergh, Pierce, and Tully (1990) have recently discussed the application of a further modified RDDO system to CCD images of galaxies.

Revisions to other systems are less drastic. Slight revisions to the Yerkes system are described by Morgan, Kayser, and White (1975), while Sandage and Brucato (1979) discuss refinements to the classification system in Sandage (1961). The ***Revised Shapley-Ames catalogue*** (RSA, Sandage and Tammann, 1987), which is one of the main applications of Hubble's revised system, now recognizes Sd and Sm types as in the de Vaucouleurs revised Hubble system. In addition, what de Vaucouleurs has called I0 and which was assigned a coded numerical stage T = 0 in the ***Second Reference Catalogue of Bright Galaxies*** (RC2, de Vaucouleurs et al., 1976), is now regarded as a special class of objects outside the scope of the Hubble sequence (Sandage and Brucato, 1979; see section 7.3).

6 Morphological classification: recent applications

The most important sources of morphological information have been the large catalogues of galaxies produced since the 1960's. Morphological classifications on the de Vaucouleurs revised Hubble system available prior to 1976 for some 4400 galaxies are summarized in RC2. A standard standard reference for morphology for many years, RC2 will soon be replaced by RC3 (de Vaucouleurs et al., 1991; see below).

Sandage and Tammann in the RSA have applied Hubble's own revision (with modifications) to the 1,246 galaxies in the old Shapley-Ames all-sky survey of bright galaxies, the southern part of which is based in part on a Las Campanas imaging survey described by Dressler and Sandage (1978) and Sandage and Brucato (1979). This is a valuable source for morphology for a relatively well-defined sample. High quality photographs of many of these objects are provided in Sandage and Bedke (1988). Dressler (1980b) also applied the Hubble-Sandage system to thousands of galaxies in 55 rich clusters. A recent application of the Yerkes system was made by Wirth and Gallagher (1980) in the Hydra I and Fornax Clusters based in part on CTIO 4-m prime focus plates. Because of the great recent interest in the so-called "Great Attractor", van den Bergh (1989) gives RDDO classifications for several hundred spirals in that direction.

The Palomar Sky Survey led to some of the largest catalogues of morphological information ever produced. These include the UGC (Nilson, 1973) and the MCG (see Vorontsov-Velyaminov, 1987 and references therein). The more recently produced ESO-B and SERC sky surveys have also proven to be gold-mines for morphology in the zones south of declination $-17°$. Hubble types based on the ESO-B survey are provided for 16,000 galaxies by Lauberts (1982), while Corwin, de Vaucouleurs, and de Vaucouleurs (1985) took advantage of the much finer grained and deeper SERC films and plates to produce the ***Southern Galaxy Catalogue*** (or SGC), which includes, among other things, detailed de Vaucouleurs revised Hubble types and DDO luminosity classes for 5,364 galaxies. This was followed by the ***Extension to the Southern Galaxy Catalogue*** (ESGC, Corwin and Skiff, 1990, in preparation), which provides more accurate type information (based on Palomar I copy plates) in the little studied zone from $-17° < \delta < -2°$.

More specialized catalogues have also been based the SERC charts or their northern equivalent, the Palomar II Sky Survey. For example, the SERC survey has been used for the *Atlas of Southern Peculiar Galaxies and Associations* (Arp and Madore, 1987) and the *Catalogue of Southern Ringed Galaxies* (or CSRG, see Buta, 1986a, 1990a), while the PSS II is being used to compile new lists of northern low surface brightness galaxies (Schombert and Bothun, 1988).

Finally, RC3 continues the tradition of its predecessors RC1 and RC2 in bringing together basic information on many galaxies from a variety of sources, including morphological types. The types are based on 10 sources, but mainly the large databases in RC2, SGC, ESGC, UGC, and CSRG. The overlap among these different sources was used to make triangular comparisons between estimates of the coded numerical de Vaucouleurs stage index, T, which allowed estimates of the mean errors of each type and checks on the reproducibility of the system. In a detailed analysis by S. Mitra first, and later by myself, it was found that scale errors and zero point differences between sources are small and that, on average, the error of an estimate of T (not marked * or ?) for a galaxy having $D_{25} = 2'$ and axis ratio $R_{25}^{-1} = 0.6$ is $\sigma(T) \sim 0.7$. Since the vast majority of types from these sources are based on the small-scale sky surveys, this error is representative of those types and not types based on large scale plates taken with 2.5-5m class telescopes, which should be considerably better. The comparisons confirmed that the revised Hubble system is reproducible at a reasonable level of certainty. Combining all sources, RC3 gives types for 17,775 galaxies and is the largest source of Hubble morphological type information ever produced.

7 Recent advances in morphology and new classes of galaxies

Modern technology has given us high-quality images of more galaxies than ever before. The greater depth of exposures possible, and the all sky coverage due to the PSS, ESO, and SERC surveys, has led to the recognition of new types of galaxies and to more enlightened views of some older classes. In this section I focus on a few of these recent advances.

7.1 Early-type galaxies

There can be little doubt that accurate classification of early-type galaxies requires imaging material of the highest quality. Careful and extensive studies of large numbers of E and S0 galaxies have, over the past 10 years, revealed some of the complexities and inhomogeneities in these two classes of objects.

Several excellent recent reviews of E galaxies already exist (e.g., Nieto, 1988; Kormendy and Djorgovski, 1989, Franx, 1990), as well as a whole I. A. U. Symposium (de Zeeuw, 1987), so I will not list any specific references here. The general consensus of these works, which summarize a great deal of highly focussed research, is that the apparent simplicity of E galaxies is highly deceptive. It is now generally accepted that E's may be characterized more by triaxial intrinsic shapes than oblate shapes, that a significant fraction of E's have dust, that some have formed through merger processes as suggested by boxy isophotes and interleaved shells, that many have weak disks that are hard to detect if face-on, and that many have accreted material since they first formed which manifests itself in unusual HI properties, counter-rotating cores, disks of dust, or polar-type rings. The most interesting aspect of all of this research is just how much can be learned from a concerted and widespread effort on a single type of galaxy.

S0 galaxies have also been studied in great detail, but their relationship to spirals is still controversial. Most interesting has been the detection of neutral gas (e.g., Wardle and Knapp, 1986; van Driel, 1987) and ionized gas (Pogge and Eskridge, 1987) in some S0's. Photometric decompositions (e.g., Simien and de Vaucouleurs, 1986) and bulge studies (Dressler and Sandage, 1983) have favored Hubble's placement of S0's between E's and spirals, rather than in a parallel sequence to spirals, while recent spectroscopic studies (Gregg, 1989) and color analyses (e.g., Bothun and Gregg, 1990) have favored the "burnt-out" spiral theory where S0's are simply spirals that have exhausted their gas supply through astration. However, from a statistical study of S0 luminosities, van den Bergh (1990) has concluded that the S0 class is a ***"repository of physically quite distinct sorts of objects that exhibit only superficial similarities"***, indicating that the ***"various kinds of S0 galaxies might have arrived at their present morphological state along quite different evolutionary tracks."*** This suggests to me that there is still quite a bit more to be learned about the S0 phenomenon in general.

Finally, the properties of the Morgan D and cD classes, the latter usually found in the centers of rich Abell clusters, have been studied in great detail recently by Schombert (1986, 1987, 1988), who also discusses the classification of these objects in terms of morphology and surface brightness profile properties. The understanding now is that cD galaxies form a unique class of objects that may be related to mergers. "BCM's" are reviewed by Schombert (this conference).

7.2 Spiral arm character, multiplicity, and disk resonances

One of the most important recent advances in morphology concerns spiral arm character. Kormendy and Norman (1979) demonstrated that spiral arm morphology depends directly on whether the disk suffers from a global instability, such as a bar, oval, or companion. In the absence of an internal or external non-axisymmetric perturbation, and in the presence of differential rotation, a galaxy will tend to have "flocculent" (or piece-meal) spiral structure, while the presence of perturbations will generally lead to well-defined global or "grand design" spiral patterns in spite of differential rotation. Elmegreen and Elmegreen (1982) proposed a system of "arm classes" to recognize this distinction and everything in between for the purpose of studying density waves in galactic disks. A summary of the physical insights derived from this scheme is given by Elmegreen and Elmegreen (1987).

An important recent use of morphology has been to identify resonance locations in galactic disks, leading to estimates of pattern speeds. In the case of pure spirals, Elmegreen, Elmegreen, and Seiden (1989) and Elmegreen and Elmegreen (1990) have used regularities (gaps, enhancements) in the arms of M 51, M 81, M 100 and NGC 1566 to trace wave resonances and evaluate the modal and stellar dynamical theories of spiral structure. In other galaxies, rings may be the most prominent tracers of specific bar resonances (Schwarz, 1979, 1981; Buta, 1986a,b). The inner and outer Lindblad, inner 4/1, and corotation resonances figure prominently in these studies because the morphological expectations of each of these resonances are fairly well-understood (see

B. G. Elmegreen, 1990 and references therein). Arm multiplicity is also important for density wave studies (B. G. Elmegreen, 1990; D. M. Elmegreen, 1990), and may also shed light on resonance associations and pattern speeds.

The relationship between bars and the form of spiral structure has been an additional important topic. Kormendy (1979) demonstrated that most barred spirals have global spiral structure. Elmegreen and Elmegreen (1982) confirmed this observation and found that 79% of field barred or oval galaxies have grand design patterns. This fraction increased to more than 90% when restricted to binary galaxies.

7.3 Amorphous galaxies

This class was introduced by Sandage and Brucato (1979) to encompass galaxies "*which are not E, S0, or any type of spiral but which have an amorphous appearance to the unresolved light, sometimes with imbedded resolved stars... All members of the class have well-developed early-type absorption spectra spread throughout the disk.*" These galaxies are related to Holmberg's Irr II class and de Vaucouleurs I0 class. A detailed study of a prototype amorphous galaxy, NGC 1800, was made by Gallagher, Hunter, and Knapp (1981). They suggested that the properties of this object favor a system that for some reason has a flatter IMF than normal and has been extremely efficient in forming stars. The color properties and star formation histories of a larger sample of amorphous galaxies are discussed by Gallagher and Hunter (1987).

7.4 Dwarf and low surface brightness galaxies

A detailed photographic study of low-luminosity members of the Virgo Cluster, which *contains galaxies of every known morphological type*, has been presented by Sandage and Binggeli (1984). With this large body of morphological data they propose a refined classification of dwarf galaxies that covers late and early-type galaxies. Particularly interesting is their recognition of a dwarf S0 class (dS0), that is, a class of low surface brightness dwarfs which morphologically are distinguishable from dwarf E's in showing direct evidence for a disk or an inflection in the brightness distribution. The recognition of dwarf S0's, however, does not imply the existence of dwarf spirals of comparable luminosity to the dS0's. Sandage and Binggeli comment that *"there are ... no convincing candidates in the Virgo Cluster for dSa"*, and that *"there is no equivalent in the spirals of the Hubble sequence at faint absolute magnitudes."*

Other interesting findings from this paper are a new class of galaxies which are dwarfs in luminosity but not in size ("huge" Im and dE types), and of variations in the properties of the dwarf ellipticals (e.g., with or without nuclei, see also Kunth et al., 1988). Sandage and Binggeli find that 80% of the Virgo Cluster galaxies fainter than 14th magnitude are dE's while 20% are Sm or Im types. A more detailed study of a subset of dE's in Virgo is given by Impey, Bothun, and Malin (1988).

Perhaps one of the most interesting realizations the past few years is that there exist disk galaxies which are low in surface brightness but which are nevertheless neither dwarfs in luminosity nor in size. Bothun et al. (1987, 1990) discuss two examples discovered so far, Malin 1 and F 568-6, and the implications of their properties with regard to the evolutionary time-scales of disks. These galaxies are examples of massive, low surface brightness disk galaxies.

7.5 Ringed and lensed galaxies

Revisions to the classification of ringed and lensed galaxies in the de Vaucouleurs system are discussed by Kormendy (1979) and Buta (1986a, 1989), the former based on a survey of 121 barred galaxies and the latter on the CSRG discussed above. These cover mostly the recognition of lenses as distinct from rings and account for variations in the morphology of pseudo-outer rings in barred galaxies.

A small number of especially well-defined ringed nonbarred or weakly-barred S0 and S0/a galaxies, NGC 3081, 7187, 7020, and 7702, from the CSRG has been studied by Buta (1990b,c,d, 1991). These galaxies highlight how high contrast rings are often associated with galaxies which are not classified as SB or even SAB. The rings are discussed in terms of resonances and the secular evolution of bars idea proposed by Kormendy (1979).

7.6 Morphology of bars

Barred galaxies have received much attention the past 15 years, but only recently have the properties of the bar form been analyzed in detail. Elmegreen and Elmegreen (1985) have identified two subcategories of bars which are distinguished by their *photometric properties*: "flat" bars, where the intensity along the bar length is roughly uniform; and "exponential" bars, where the intensity along the bar declines exponentially. There is a rough correlation between these bar types and Hubble type: flat bars are prevalent among early-type SB galaxies while exponential bars are prevalent among later type SB galaxies. The Elmegreens discuss these differences in terms of orbit resonances in the disk plane.

Another line of morphological work concerns the shape of bars. The bars of a significant fraction of early-type galaxies appear to be somewhat rectangular in shape (e.g., Athanassoula et al., 1990; Ohta, Hamabe, and Wakamatsu, 1990), while the bars of some late-type SBm spirals can be somewhat triangular (Odewahn, 1989). A particular striking example of a rectangular bar with oval "ears" has been found in the southern outer-ringed galaxy NGC 7098 (Crocker and Buta, 1991, in preparation).

7.7 "Boxy" or "peanut" bulges and "x" galaxies

Boxy or peanut bulge galaxies have been an active subject of study the past 10 years. Major lists of the best cases have been compiled by Jarvis (1986) and Shaw (1987). The phenomenon has been variously explained in terms of cylindrical rotation (see Rowley, 1988, and references therein), edge-on or other preferred views of bars (Combes and Sanders, 1981, Combes et al., 1990), or merger effects (Whitmore and Bell, 1988). The boxy or peanut character is usually very subtle in photographs, although it can be extremely obvious in some cases, as in, for example, NGC 128 (Sandage, 1961; Jarvis, 1989) and IC 4767 (Whitmore and Bell, 1988). With suitable image processing, the latter object also shows an X-shaped structure crossing its inner regions, which has been interpreted by Whitmore and Bell as evidence for a recent merger. The highly inclined, but non-edge-on, S0 galaxy NGC 7020 shows a similar but possibly unrelated feature (Buta, 1990d).

7.8 Ring, polar ring, and "hoag-type" ring galaxies

These rare phenomena have been the subject of a fair amount of recent research. Ring and polar ring galaxies are believed to be produced by collisions or mergers. Recent studies of morphology and statistics of ring galaxies have been made by Few and Madore (1986), Appleton and Struck-Marcell (1987 and references therein), and Arp and Madore (1987). A catalogue of polar rings has been compiled by Whitmore et al. (1990), who also give many of the previous relevant references.

The prototype of "Hoag-type" ring galaxies is Hoag's Object (A1515+2146), which has been studied in great detail recently. This object is remarkable for having a clear ring surrounding a distinct spheroidal galaxy, with no trace whatsoever of a bar. It has been interpreted as a ringed galaxy whose bar dissolved after the ring formed (Brosch, 1985), and as a case where an E galaxy accreted a small companion into something like a polar ring (Schweizer et al., 1987). Wakamatsu (1990) studied another example, NGC 6028, but demonstrated the presence of a small bar in the central region and suggested that at least in this case, bar-driven gas dynamics could explain the presence of the ring. A possible related object is NGC 7187 (Buta, 1990c).

8 Conclusions

This review has only covered a limited portion of a very broad topic, and I apologize for any other important aspects I have left out. However, I hope I have conveyed an adequate impression of what morphology is all about and how it has helped in our understanding of galaxies. In my opinion, the best approach to morphology is to view it as an imprecise science which serves as a means to an end but which is not an end in itself. Morphology cannot generally stand alone and, as pointed out by Dressler (1980a), is not a suitable substitute for physics. However, as pointed out by Kormendy (1979) and Dressler (1980a), morphology can suggest powerful avenues for further research. A purely physical classification is not likely to replace morphological classification for many years.

To answer the question as to where morphology has led us in understanding galaxies, I think it has given us insights by causing us to focus our attention on specific types of objects. Spiral galaxies were the subject of much research in the 1960's and 1970's, while the 1980's could be called the decade of the elliptical galaxies. For the various classes of galaxies, we have sought to measure more basic physical parameters, such as optical and infrared luminosities, colors, surface brightness distributions, HI contents and distributions, CO distributions, rotation curves and mass distributions, diameters and intrinsic flattenings, velocity dispersions, radio continuum fluxes and distributions, and kinematics and dynamics of internal structures, and we have used these physical measurements to piece together what determines morphology and what role the invisible parts of galaxies may play in the overall dynamics. The beauty of morphology is that there are not purely random correlations between many of these parameters and galaxy structure, as is highlighted by the smooth variations of some quantities with position along the Hubble Sequence (see Whitmore, 1984 and Watanabe et al., 1985 for recent applications of principal component analysis). Thus, morphological classification in conjunction with physical measurement has been an important tool in extragalactic astronomy.

The advent of CCD technology, the high quality sky surveys, and the Hubble Space Telescope suggest to me that there is an exciting future for morphological studies. Considering the extensive interest in early-type galaxies among the attendees at this conference, a homogeneous imaging survey of all real or suspected E, S0 galaxies in a well-defined sample (e.g., Sadler, 1984) may be more practical than ever before and may help to alleviate some of the uncertainty in interpreting HI and other properties of these objects. Even more exciting are the prospects from HST, which at some point should allow the study of morphology in very distant galaxies where we may be able to see evidence for secular evolution. This will be important not only for star formation histories, but also for bars, rings, spiral structure, and the distribution of types. For nearer galaxies, CCD's are already providing excellent image material for morphology studies. The efficiency and linearity of these detectors makes it possible to image more galaxies on a large scale and to a deeper limiting surface brightness than ever before. With such high quality images becoming available, more and more attention will be paid to finer details of morphology as theories and observations get more sophisticated. Finally, future large surveys will probably involve computer classification from digital images to a great extent (see Thonnat, 1989; and Delfini, Accomazzi, Kurtz, and Mussio, this conference).

References

Appleton, P.N. and Struck-Marcell, C.: 1987, *Astrophys. J.* **312**, 566.

Arp, H.C. and Madore, B. F.: 1987, *A Catalogue of Southern Peculiar Galaxies and Associations*, Cambridge Univ. Press, Cambridge.

Athanassoula, E., Morin, S., Wozniak, H., Puy, D., Pierce, M.J., Lombard, J., and Bosma, A.: 1990, *Mon. Not. R. Astr. Soc.* **245**, 130.

Bothun, G.: 1982a, *Astrophys. J. Suppl. Ser.* **50**, 39.

Bothun, G.: 1982b, *Publ. Astron. Soc. Pac.* **94**, 774.

Bothun, G. and Gregg, M.: 1990, *Astrophys. J.* **350**, 73.

Bothun, G. and Sullivan, W.T.: 1980, *Astrophys. J.* **242**, 903.

Bothun, G., Impey, C.D., Malin, D.F., and Mould, J.R.: 1987, *Astron. J.* **94**, 23.

Bothun, G.D., Schombert, J.M., Impey, C.D., and Schneider, S.E.: 1990, *Astrophys. J.* **360**, 427.

Brosch, N.: 1985, *Astron. Astrophys.* **153**, 199.

Buta, R.: 1986a, b, *Astrophys. J.* **306**, 768.

Buta, R.: 1989, *The World of Galaxies* (H.G. Corwin and L. Bottinelli, Eds.), 29, Springer, N.Y..

Buta, R.: 1990a, *Dynamics of Galaxies and Molecular Clouds Distribution*, I.A.U. Symposium no. 146 (F. Combes and F. Casoli, Eds.), in press, Kluwer, Dordrecht.

Buta, R.: 1990b, *Astrophys. J.* **351**, 62.

Buta, R.: 1990c, *Astrophys. J.* **354**, 428.

Buta, R.: 1990d, *Astrophys. J.* **356**, 87.

Buta, R.: 1991, *Astrophys. J.* in press.

Corwin, H.G., de Vaucouleurs, A., and de Vaucouleurs, G.: 1985, *Southern Galaxy Catalogue*, Univ. Tex. Mon. Astr. no. 4, Austin. - (=SGC).

Combes, F. and Sanders, R. H.: 1981, *Astron. Astrophys.* **96**, 164.

Combes, F., Debbasch, F., Friedli, D., and Pfenniger, D.: 1990, *Astron. Astrophys.* **233**, 82.

de Vaucouleurs, G.: 1959, *Handbuch der Physik*, **53**, 275.

de Vaucouleurs, G. and Buta, R.: 1980, *Astrophys. J. Suppl. Ser.* **44**, 451.

de Vaucouleurs, G., de Vaucouleurs, A., and Corwin, H.G.: 1976, *Second reference Catalogue of Bright Galaxies*, Univ. Tex. Mon. Astr. no. 2, Austin. - (=RC2).

de Vaucouleurs, G., de Vaucouleurs, A., Corwin, H.G., Buta, R., Paturel, G., and Fouque, P.: 1991, *Third Reference Catalogue of Bright Galaxies*, Springer, N.Y. - (=RC3).

de Zeeuw, T., Ed.: 1987, *The Structure and Dynamics of Elliptical Galaxies*, I.A.U. Symposium no. 127, Reidel, Dordrecht.

Dressler, A.: 1980a, *Astrophys. J.* **236**, 351.

Dressler, A.: 1980b, *Astrophys. J. Suppl. Ser.* **42**, 565.

Dressler, A.: 1984, *Ann. Rev. Astron. Astrophys.* **22**, 185.

Dressler, A. and Sandage, A.: 1978, *Publ. Astron. Soc. Pac.* **90**, 5.

Dressler, A. and Sandage, A.: 1983, *Astrophys. J.* **265**, 664.

Elmegreen, B.G.: 1990, In *Galactic Models* (J.R. Buchler, S.T. Gottesman, and J.H. Hunter, Jr., Eds.), 40, New York Acad. Sci., N.Y..

Elmegreen, D.M.: 1990, In *Galactic Models* (J.R. Buchler, S.T. Gottesman, and J.H. Hunter, Jr., Eds.), 53, New York Acad. Sci., N.Y..

Elmegreen, D.M. and Elmegreen, B.G.: 1982, *Mon. Not. R. Astr. Soc.* **201**, 1021.

Elmegreen, B.G. and Elmegreen, D.M.: 1985, *Astrophys. J.* **288**, 438.

Elmegreen, D.M. and Elmegreen, B.G.: 1987, *Astrophys. J.* **314**, 3.

Elmegreen, B.G. and Elmegreen, D.M.: 1990, *Astrophys. J.* **355**, 52.

Elmegreen, D.M., Elmegreen, B.G., and Seiden, P.E.: 1989, *Astrophys. J.* **343**, 602.

Few, J.M.A. and Madore, B.F.: 1986, *Mon. Not. R. Astr. Soc.* **222**, 673.

Franx, M.: 1990, In *Galactic Models* (J.R. Buchler, S.T. Gottesman, and J.H. Hunter, Jr., Eds.), 67, New York Acad. Sci., N.Y..

Gallagher, J.S., Hunter, D.A., and Knapp, G.R.: 1981, *Astron. J.* **86**, 344.

Gallagher, J.S. and Hunter, D.A.: 1987, *Astron. J.* **94**, 43.

Gould, S.J.: 1985, *The Flamingo's Smile.*

Gregg, M.: 1989, *Astrophys. J.* **337**, 45.

Hubble, E.: 1926, *Astrophys. J.* **64**, 321.

Impey, C.D., Bothun, G.D., and Malin, D.F.: 1988, *Astrophys. J.* **330**, 634.

Jarvis, B.J.: 1986, *Astron. J.* **91**, 65.

Jarvis, B.J.: 1989, *Dynamics and Interactions of Galaxies* (R. Wielen, Ed.), 416, Springer, Berlin.

Kennicutt, R.C.: 1981, *Astron. J.* **86**, 1847.

Kormendy, J.: 1979, *Astrophys. J.* **227**, 714.

Kormendy, J.: 1982, *Morphology and Dynamics of Galaxies*, p. 115, Geneva Obs., Geneva.

Kormendy, J. and Norman, C.: 1979, *Astrophys. J.* **233**, 539.

Kormendy, J. and Djorgovski, S.: 1989, *Ann. Rev. Astron. Astrophys.* **27**, 235.

Kunth, D., Maurogordato, S. and Vigroux, L.: 1988, *Astron. Astrophys.* **204**, 10.

Lauberts, A.: 1982, *The ESO-Uppsala Survey of the ESO (B) Atlas*, European Southern Observatory, Munich.

Morgan, W.W., Kayser, S., and White, R.A.: 1975, *Astrophys. J.* **199**, 545.

Nieto, J.-L.: 1988, *Boletin de la Academia de Ciencias Cordoba*, **58**, 329.

Nilson, P.: 1973, *Uppsala General Catalogue of Galaxies*, Upp. Obs. Ann., Vol. 6. - (=UGC).

Odewahn, S.: 1989, PhD Thesis, University of Texas.

Ohta, K., Hamabe, M., and Wakamatsu. K.I.: 1990, *Astrophys. J.* **357**, 71.

Pogge, R.W. and Eskridge, P.B.: 1987, *Astron. J.* **93**, 291.

Rowley, G.: 1988, *Astrophys. J.* **331**, 124.

Sadler, E.M.: 1984, PhD Thesis, Australian National University.

Sandage, A.: 1961, *The Hubble Atlas of Galaxies*, Carnegie Institution of Washington Publication no. 618, Washington, D.C.

Sandage, A.: 1975, *Galaxies and the Universe, Stars and Stellar Systems Vol. IX* (A. Sandage, M. Sandage, and J. Kristian, Eds.), 1, Chicago.

Sandage, A: 1986, *Astron. Astrophys.* **161**, 89.

Sandage, A. and Bedke, J.: 1988, *Atlas of Galaxies Useful for the Cosmological Distance Scale*, NASA, Washington, D.C.

Sandage, A. and Binggeli, B.: 1984, *Astron. J.* **89**, 919.

Sandage, A. and Brucato, R.: 1979, *Astron. J.* **84**, 472.

Sandage, A. and Tammann, G.A.: 1987, *A Revised Shapley-Ames Catalog of Bright Galaxies*, Carnegie Institution of Washington Publ. no. 635, Washington, D.C. - (=RSA).

Schombert, J.M.: 1986, *Astrophys. J. Suppl. Ser.* **60**, 603.

Schombert, J.M.: 1987, *Astrophys. J. Suppl. Ser.* **64**, 643.

Schombert, J.M.: 1988, *Astrophys. J.* **328**, 475.

Schombert, J.M. and Bothun, G.D.: 1988, *Astron. J.* **95**, 1389.

Schwarz, M.P.: 1979, PhD Thesis, Australian National University.

Schwarz, M.P.: 1981, *Astrophys. J.* **247**, 77.

Schweizer, F., Ford, W.K., Jedrzejewski, R., and Giovanelli, R.: 1987, *Astrophys. J.* **320**, 454.

Shaw, M.: 1987, *Mon. Not. R. Astr. Soc.* **229**, 691.

Simien, F. and de Vaucouleurs, G.: 1986, *Astrophys. J.* **302**, 564.

Taniguchi, Y. and Watanabe, M.: 1987, *Astrophys. J.* **313**, 89.

Thonnat, M.: 1989, *The World of Galaxies* (H.G. Corwin and L. Bottinelli, Eds.), 53, Springer, N.Y..

van den Bergh, S.: 1976, *Astrophys. J.* **206**, 883.

van den Bergh, S.: 1980a, *Phil. Trans. R. Soc. London, Ser. A*, **296**, 380.

van den Bergh, S.: 1980b, *Publ. Astron. Soc. Pac.* **92**, 409.

van den Bergh, S.: 1989, *Astron. J.* **87**, 1556.

van den Bergh, S.: 1990, *Astrophys. J.* **348**, 57.

van den Bergh, S., Pierce, M., and Tully, R.B.: 1990, *Astron. J.* **359**, 4.

van Driel, W.: 1987, PhD thesis, University of Groningen.

Vorontsov-Velyaminov, B.A.: 1987, *Extragalactic Astronomy*, Harwood Publ., New York, N.Y..

Wakamatsu, K.: 1990, *Astrophys. J.* **348**, 448.

Wardle, M. and Knapp, G.R.: 1986, *Astron. J.* **91**, 23.

Watanabe, M., Kodaira, K., and Okamura, S.: 1985, *Astrophys. J.* **292**, 72.

Whitmore, B.C.: 1984, *Astrophys. J.* **278**, 61.

Whitmore, B.C. and Bell, M.: 1988, *Astrophys. J.* **324**, 741.

Whitmore, B.C., Lucas, R., McElroy, D.B., Steiman-Cameron, T.-Y., Sackett, P.D., and Olling, R.P.: 1990, *Astron. J.* **100**, 1489.

Wirth, A. and Gallagher, J.S.: 1980, *Astrophys. J.* **242**, 469.

Discussion

G. Djorgovski - If one insists on a purely morphological classification (which I don't think is necessarily a good idea), it may be more profitable to classify galactic ***subsystems*** rather than entire galaxies. For example, ellipticals/bulges, thick stellar disks, thin disks, etc. Think of them as generalized stellar populations. Their physics and origins are easier to trace and one could use them to "build" galaxies.
R. Buta - Ultimately this is what we would eventually hope to be able to do. We still, however, must rely on catalogues to choose our samples, and morphological classification is an efficient process. Building a physical picture of galaxies need not mean dispensing entirely with morphological classification.

P. van der Kruit - It seemed at a time, when luminosity classes were introduced, that one could estimate a galaxy's absolute magnitude by just looking at it. Of course, now we have the Tully-Fisher relation. What is the current situation on luminosity classes, i.e., how well do these correlate with total luminosity?
R. Buta - In a recent paper, van den Bergh, Pierce, and Tully (1990) demonstrate that luminosity classes based on CCD images correlate better with total luminosity than do RSA luminosity classes. They attribute this to the superior quality of the digital images, and quote an rms dispersion of ~0.68 mag for classes I-IV. The dispersion for RSA classes is closer than 1.0 mag. The method still can't compete with the T-F relation, except perhaps for nearly face-on galaxies.

M. Capaccioli - Your last comment about HST made me think of an aspect of galaxy morphology which has always puzzled me: which is the ideal resolution (in terms of details you can resolve) for the morphologist? Could you quantify a limit at which resolution improvements become counterproductive?
R. Buta - Since Hubble based his classification system on plates taken with 1.5-5m class telescopes, then the typical resolution of those telescopes at prime focus would presumably define the depth and level of detail required for accurate classification on the system. Your point is a good one, however, because too much resolution may be as undesirable as too little. The deep SERC-J plates, for example, show more detail than some images in the Hubble Atlas. This helps to improve the classifications for some galaxies, but may cause problems for others.

D. Burstein - Two comments: (1) I would like to emphasize two further classifications: The Irr II (or I0) classification has lead to studies of interacting systems and has helped the study of merging systems. The S0/a classification appears to be the current "garbage bin" of the Hubble sequence. If many I0's are interacting systems, S0/a's may be these same systems later in their evolution. (2) The advantage of morphology is that it permits galaxies to be selected for study. That is also its disadvantage, as often the morphological classification is taken as defining a class, rather than simply as a starting point to ***help*** define a class of galaxies.
R. Buta - 1. In RC2 the conventional T = 0 class included not only legitimate S0/a galaxies but also de Vaucouleurs' I0 class. At the time, it seemed that I0's fit better in that part of the Hubble sequence. Since RC2, more has been learned about the I0 class

and in RC3 we have not adopted the T = 0 specification. I should point out that the S0/a class is actually well-defined, especially for barred or weakly-barred galaxies. It may be prone to problems, however, if some nonbarred cases that are really peculiar are placed into the class. 2. I agree with your comment.

S. Kent - In response to D. Burstein's question: I suspect that every bin in the Hubble sequence has been referred to in the literature at some time as a "garbage bin."
F. Bertola - There are often cases of galaxies classified either as E or S0 by different authors. Can you suggest the criteria to arrive at a proper classification? With improving classification is the number of elliptical galaxies increasing or decreasing?
R. Buta - These discrepancies are unavoidable for visual classifications. Face-on, non-barred S0's with little structure other than a clear bulge and faint disk are difficult to distinguish from E's visually. How we view the discrepancies depends on whether we accept continuity between these classes, thereby allowing them to have fuzzy cell boundaries, or whether we view the classes as sharply distinct. Rings of dust or stars and lenses help to make the distinction more obvious. I think if we want to be more accurate we need to make well-defined requirements on the luminosity profiles. A comparison between types in RC2 and in SGC suggests that improved classification will decrease the number of "true" E's.

D. Burstein - I know that you and Gérard de Vaucouleurs have experimented with classifying galaxies on the basis of integrated growth curves. Would you care to comment on how well this kind of quantitative morphology works?
R. Buta - Integrated growth curves are sensitive to morphology, as is well known from the type dependence in their shapes illustrated in RC2. In calculating magnitudes from aperture photometry for RC3, we let a program find the best-fitting curve, which gave us a "photometric" type. The correlation between catalogued visual type and photometric type was found to be fairly good, especially for the E and S0 types and Sm, Im types, but more scatter was found for intermediate to late spirals. The curves are sensitive to bulge-to-disk ratio, and are very distinctive for early-type barred galaxies, but my feeling is that the photometric type is not a suitable substitute for a visual type. Growth curves are not a very efficient way of producing types, and data more often than not are so limited that to get a total magnitude, one can do no better than to force fit the standard curve for the visual catalogue type.

S. Kent - More of a comment than a question, one current problem with classification systems is that we need a good catalog of standard types in order for observers to classify additional objects in an objective way. Ideally, we would like a more quantitative measure of morphology, such as growth curves, although, as you pointed out, they are not yet all that useful.
R. Buta - I agree that a good set of standards is essential to apply a system to additional galaxies. The RSA presents a set of excellent photographs illustrating the van den Bergh luminosity class system, while the Hubble Atlas and the recent Sandage and Bedke NASA atlas provide good type standards.

H. van Woerden - As I understand it, the distinction between ellipticals and S0's is

that the latter have a bulge and a disk. Is it not true that the transition from bulge to disk is more easily seen if one has several plates of different depths and gammas?
R. Buta - Yes, S0's are a type of galaxy where long, medium, and short exposure images could improve the classification. In the case of the sky surveys, usage of both the SERC-J and ESO-B atlases for early-type galaxies would be a prudent approach. Neither of those atlases alone would be ideal for such galaxies.

P. Pişmiş - A further provocative question. How much have we taken advantage as regards the physical nature of galaxies, through detailed classification of galaxies? How much have we learned from our classification? I fear that our classification defeats the purpose of a classification.
R. Buta - From our classification we have learned that the spiral form is very common in the universe, that distinct types of galaxies do exist, that environment may be important. As long as we take a guarded approach to morphology and try to identify what is really important, as I think Hubble did, then I think our classification will not defeat its purpose.

M. Balcells - Can or should we include kinematic information into the classification systems?
R. Buta - For those classes of galaxies where kinematic distinctions within the class are well-established and measurable (e.g., E galaxies), then I would say yes.

Alvio Renzini (in the back) looking at Ron Buta and Albert Bosma arguing about ring galaxies.

THE LUMINOSITY FUNCTIONS OF GALAXIES

H. Jerjen[1], G.A. Tammann[1,2], B. Binggeli[1]
[1]*Astronomisches Institut der Universität Basel, Switzerland*
[2]*European Southern Observatory, Garching, Germany*

ABSTRACT. It is pointed out that the different Hubble types of galaxies have different luminosity functions. Because the mix of Hubble types changes with the density of the environment, the overall luminosity function of all types *must* change with the environmental density, i.e. *there is no unique* (Schechter) *luminosity function* of galaxies. For four different type mixes the Schechter parameters M^* and α are given.

1 Introduction

It is known for some time that the Hubble type mix changes with the density of the surrounding field (Dressler, 1980). It is also known that the galaxy luminosity function (hereafter LF) changes with the Hubble type (Sandage et al., 1985). The necessary conclusion is that the LF over all types cannot be the same in different galaxy environments (Binggeli, 1987). Indeed it was demonstrated that the overall LF of the Virgo cluster is distinctly different from a typical field LF (Binggeli et al., 1988). Following the pioneering work of Thomson and Gregory (1980), we present here synthetic overall LFs for four different galaxy environments, ranging from the very dense Coma cluster to the extreme (low-density) field. For a quantitative comparison, the resulting LFs are fitted by the analytic Schechter (1976) function, and the variation of the Schechter parameters M^* and α is discussed. All magnitudes quoted are in the B_T system.

2 Data

The shapes of the type-specific LFs, best known from the Virgo cluster galaxies (Sandage et al., 1985), are quite distinct. E, S0, and spiral galaxies have a turn-over in their LF; they exist only above a certain threshold luminosity. Irregular galaxies are in general fainter than spirals; they have a rather broad LF whose turn-over point is still not established beyond doubt. Dwarf elliptical (dE), on the other hand, have an exponentially increasing LF with ever increasing numbers (to present observational limits) of intrinsically faint objects.

We assume that the *shape* of the LF for a given Hubble type does not change with the environment (Thomson and Gregory, 1980; Binggeli, 1987). This assumption may even-

G. Longo et al. (eds.), Morphological and Physical Classification of Galaxies, 17–22.

tually prove to be an oversimplification, but it is in good agreement with the presently available data (Binggeli et al., 1988; Ferguson and Sandage, 1988; 1991).

Consequently, the LF shapes found for the Virgo Cluster galaxies (Sandage et al., 1985) may be taken as universal ("true"). For computational purposes we have represented these shapes by analytic forms, provided in part already by Sandage et al. (1985). The adopted LF shapes for the five main morphological classes considered here are given in table 1. The rare species of dS0 galaxies is added to the class of dwarf ellipticals. The "irregular" class comprises magellanic types (Im) as well as blue compact dwarfs (BCDs).

Table 1. The adopted universal LF shapes

E	$\varphi(M) \sim \begin{cases} exp[-\frac{1}{2}(\frac{M-\mu}{\sigma_1})^2] & : M < \mu = -17.45^m, \quad \sigma_1 = 2.75^m \\ exp[-\frac{1}{2}(\frac{M-\mu}{\sigma_2})^2] & : M \geq \mu = -17.45^m, \quad \sigma_2 = 1.17^m \end{cases}$
S0	$\varphi(M) \sim exp[-\frac{1}{2}(\frac{M-\mu}{\sigma})^2] \ : \ \mu = -19.05^m, \quad \sigma = 1.50^m$
Spiral	$\varphi(M) \sim exp-\frac{1}{2}(\frac{M-\mu}{\sigma})^2] \ : \ \mu = -18.45^m, \quad \sigma = 1.45^m$
Irregular	$\varphi(M) \sim dex[-0.4(\alpha+1)M]exp[-dex[0.4(M^* - M)]] \ : \ M^* = -17.15^m, \quad \alpha = -1.0$
dE+dS0	$\varphi(M) \sim dex[-0.4(\alpha+1)M]exp[-dex[0.4(M^* - M)]] \ : \ M^* = -17.78^m, \quad \alpha = -1.3$

If one fixes in addition the type mix of a population down to a certain limiting magnitude, it is possible to determine the overall LF (Thomson and Gregory, 1980). It was decided here to synthesize the LFs for four different populations, which are characterised as "Coma", "Virgo", "groups", and "extreme field". The Coma Cluster is the environment of the highest galaxy density known, while the opposite (the "extreme field") is meant to represent the lowest density case. The specific type mixes of the four populations are given in table 2. The numbers quoted are percentages of the different morphological types, counted to a limiting absolute magnitude of $M = -15^m$. For the Coma and Virgo Clusters the data are from Thomson and Gregory (1980) and Sandage et al. (1985), respectively. The "group" sample contains all nearby galaxies that belong to a group, based on an unpublished update of the "10 Mpc Catalog" (Kraan-Korteweg and Tammann, 1979). The numbers for the "extreme field" are hypothetical.

Table 2. Adopted Galaxy Populations (Numbers in percent)

	E	S0	Spiral	Irr	dE+dS0
"Coma"	10	12.5	2.5	0	75
"Virgo"	6	9	25	15	45
"groups"	3	6	62	26	3
"extreme field"	0	0	50	50	0

The individual contributions of different Hubble types to the overall LFs are illustrated for the four cases in the upper panels of figure 1a and figure 1b.

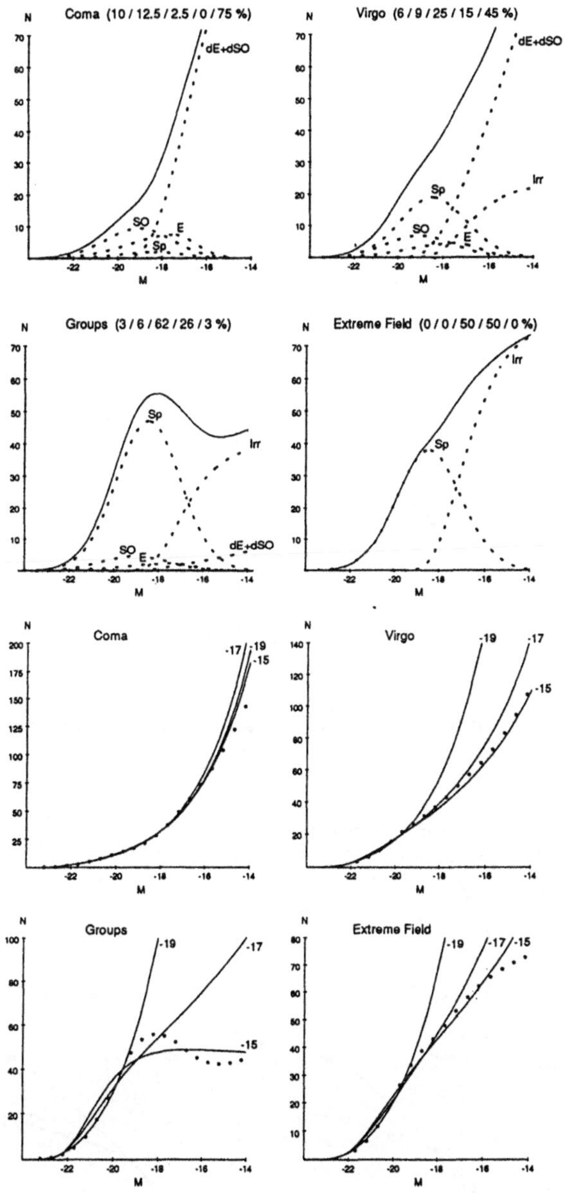

Figure 1a. Differential luminosity functions for four different galaxy environments (Coma Cluster, Virgo Cluster, nearly groups, low-density field; with the percentages of E:S0:Sp:Irr:dE +dS0 given in parentheses). Upper panel: the synthetic overall LFs (full curves) and the individual contributions by the different types (broken curves). Lower panel: Schechter fits for three limiting magnitudes ($M = -15^m, -17^m, -19^m$) to the overall LFs (dotted curves).

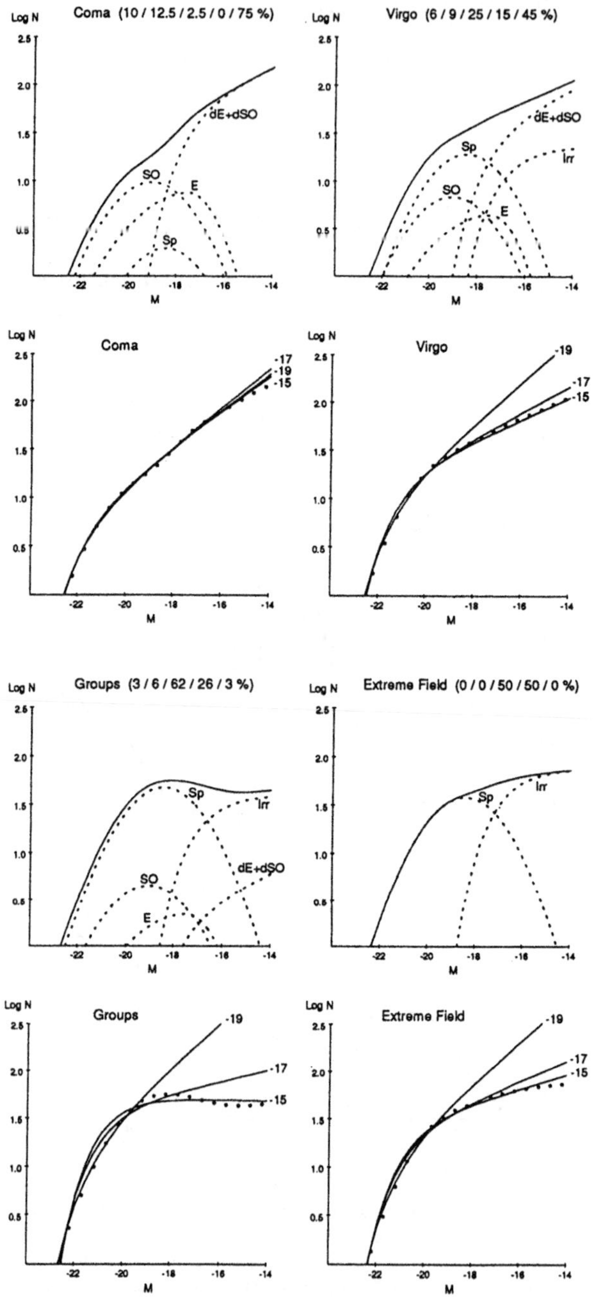

Figure 1b. Same as figure 1a but drawn logarithmically.

3 Schechter fits

A Schechter (1976) function of the form

$$\varphi(M) \sim dex[-0.4(\alpha+1)M]exp[dex[0.4(M^* - M)]]$$

with M^* and α as free parameters, was fitted (if necessary by force), to the overall LFs shown in the upper panels of figure 1. A blind eye had to be turned to the wavy nature of the LFs of the "Coma" and "groups" populations; these waves cannot be reproduced in principle by the Schechter function.

It turned out that the best-fitting values of M^* (the "knee" magnitude) and α (the faint-end slope) depend strongly on the cut-off in absolute magnitude. Therefore these parameters were calculated for three different cut-off luminosities i.e. M_{lim}=-15^m, -17^m, and -19^m.

The resulting values of M^* and α as well as the goodness of the fit χ^2/ν (ν the number of degrees of freedom) are listed in table 3. The fits are illustrated in the lower panels of figure 1b and figure 1 and figure 1b.

Table 3. Best-fitting Schechter parameters

Coma (10/12.5/2.5/0/75 %)				**Virgo** (6/9/25/15/45 %)			
M_{lim}	M^*	α	χ^2/ν	M_{lim}	M^*	α	χ^2/ν
-15	-22.05	-1.47	0.091	-15	-21.52	-1.29	0.107
-17	-22.28	-1.52	0.037	-17	-21.63	-1.34	0.122
-19	-22.16	-1.49	0.029	-19	-22.00	-1.56	0.057
groups (3/6/62/26/3 %)				**extreme field** (0/0/50/50/0 %)			
M_{lim}	M^*	α	χ^2/ν	M_{lim}	M^*	α	χ^2/ν
-15	-21.02	-0.98	1.244	-15	-21.13	-1.19	0.317
-17	-21.25	-1.16	0.828	-17	-21.26	-1.27	0.265
-19	-21.83	-1.59	0.113	-19	-21.64	-1.56	0.081

4 Conclusions

The value of α depends on the presence ("Coma" and "Virgo") or (near) absence ("groups" and "extreme field") of dwarf ellipticals. Clearly in the former case, which corresponds to higher densities, one expects higher values of α. This result is relevant for the long debate on the "true" value of α. It is clear that no single value of α can exist.

The reason why the idea of a universal Schechter function persisted so long is a simple bias towards high luminosities. This is clear from table 3. For a bright magnitude cut-off at $M^* = -19^m$, the Schechter parameters are nearly stable at $M^* \approx -22^m$ and $\alpha \approx -1.5$. But as the LFs are sampled to a deeper level the variance of the type mix according to the environment unfolds full weight, and differences of up to $\Delta M^* \approx 1^m$ and $\Delta\alpha \approx 0.5$ appear.

Even if one compares the M^* values only within a fixed population, it varies up to $0^m.80$ depending on the cut-off luminosity. If one wants to use the "knee" luminosity M^* as a standard candle for relative distance determinations (Dümmler, 1991), one has therefore to be sure to consider only homogeneous populations (i.e. only dense clusters or only "extreme fields") *and* one must assure to include only galaxies down to a uniform cut-off luminosity.

There are weak points in the present analysis, e.g., we would like to have more realistic values for the "extreme field" population. However, that should not distract from the main point, i.e. that the overall LF (and with it the Schechter parameters M^* and α) *does* change for different galaxy populations. There is no universal overall LF, let alone a universal Schechter function. There are only - to first order - universal *type-specific* LFs. The consequences for cosmological studies, in which often a universal Schechter LF is assumed, have yet to be explored.

Financial support from the Swiss National Science Foundation is gratefully acknowledged.

References

Binggeli, B.: 1987. In *Nearby Normal Galaxies*, ed. S. Faber, p.195. New York: Springer.

Binggeli, B., Sandage, A., Tammann, G.A.: 1988, *Ann. Rev. Astron. Astrophys.* **26,** 509.

Dressler, A.: 1980, *Astrophys. J.* **236,** 351.

Dümmler, R.: 1991, *Astron. Astrophys.* in print.

Ferguson, H., Sandage, A.: 1988, *Astron. J.* **96,** 1520.

Ferguson, H., Sandage, A.: 1991, *Astron. J.* **101,** 765.

Kraan-Korteweg, R., Tammann, G.A.: 1979, *Astron. Nachr.* **300,** 181.

Sandage, A., Binggeli, B., Tammann, G.A.: 1985, *Astron. J.* **90,** 1759.

Schechter, P.: 1976, *Astrophys. J.* **203,** 297.

Thomson, L.A., Gregory, S.A.: 1980, *Astrophys. J.* **242,** 1.

Discussion

G. Richter - I'd like to stress tight connection between G. Tammann's arguments and some problems discussed in Buta's paper. R. Buta told us that morphological classification gives cells with unsharp boundaries. If an unsharp boundary lies on a steep part of the frequency distribution of the objects we can be sure to get discordant results anyway, because small differences of the boundaries between different authors influence the results very strongly. Hence, it should be a basic criterium of classification to put boundaries onto flat parts of the frequency distribution (in the best case into a minimum, of course). G. Tammann criterium works so good (and can work at all) only because the classes he used (E and S) obey this criterium.

ELLIPTICAL GALAXIES

M. Franx
Center for Astrophysics
Harvard University, USA

ABSTRACT. An enormous amount of work on ellipticals has been done in the last decade, which has changed our ideas about these simple looking galaxies dramatically. The problems in understanding many of these results are highlighted by analysis of the kinematics of ellipticals. Two parameters are considered: first the rotation parameter v/σ, and second the minor to major axis rotation v_{minor}/v_{major}. The various correlations between v/σ and mass, surface brightness, isophotal shapes, etc., are discussed. The interpretation of these correlations is not yet clear. The minor axis rotation gives constraints on the alignment of the angular momentum, and the intrinsic shapes of elliptical galaxies. These results are compared to predictions for dark halos from hierarchical formation theories. The lack of theoretical predictions for the luminous components is the main reason for our limited understanding of the observational results.

1 Introduction

The ideas about elliptical galaxies have undergone significant change since the early 70's. It was believed previously that ellipticals are old systems, formed early in the history of the universe by a fast monolithic collapse. The galaxies were thought not to have evolved significantly afterwards. Their structure was assumed to be rather simple. In contrast, spiral galaxies have had high star formation rates over a much longer period of time, extending until the present epoch. In this picture, the bulges of spirals formed early, at about the same time as the ellipticals. See also Tinsley (1980) and references therein for a discussion.

Since then, much work has been done in this field, sparked by the discovery that ellipticals rotate slowly (Bertola and Capaccioli, 1975, Illingworth, 1977), and the realization that interactions and secondary evolution may be important in their formation (Toomre, 1977). The field has opened up enormously with observations at many wavelengths (from radio to X-ray), and the use of computers for modeling and simulations. The new material is too extensive for a complete review here, the reader is referred to two recent conference proceedings (de Zeeuw, 1987, and Wielen, 1990) and this conference proceedings for reviews on most relevant topics.

This new work has caused serious doubts as to whether ellipticals formed very quickly in the early history of the universe. A currently popular assumption is that (many)

G. Longo et al. (eds.), Morphological and Physical Classification of Galaxies, 23–38.

ellipticals formed through mergers *up until the present time*. And indeed, the observational evidence that (some) ellipticals have formed through merger is overwhelming (e.g. Schweizer, 1990). On the theoretical side, many of the galaxy formation theories like CDM do predict that galaxies generally build up through the merging or infall of smaller clumps. Major questions still remain unanswered, especially at what epoch most galaxies formed, what their properties were at that time, and what the cause was for the different morphologies of galaxies and the morphology–density relation.

It is unlikely that a full answer to these problems can be obtained from observations of only nearby galaxies. Clearly, results on high redshift galaxies are giving important constraints on galaxy formation, e.g. the faint galaxies counts (Koo, 1986, Tyson, 1988), Butcher-Oemler effect and spectral evolution of cluster galaxies (Butcher and Oemler, 1984; Dressler and Gunn, 1990), and high redshift radio galaxies (Lilly, 1989, Chambers et al., 1990, van Breugel and McCarthy, 1990). Nevertherless, the high spatial resolution and the good S/N that we can achieve on nearby galaxies provides us with unique opportunities to test formation scenarios.

Given the impossibility to review the subject of ellipticals comprehensively in such little space, I focus below on their kinematics. The kinematic properties of ellipticals are among their most distinguishing features, and the last decades have given us many surprises regarding this topic. Furthermore, they are directly related to the formation process and illustrate clearly the problems that we deal with in the interpretation of many observables.

2 Mean rotation of ellipticals

The first measurements of the rotation of elliptical galaxies produced the well-known result that bright ellipticals rotate too slow to explain their flattening by rotation (Bertola and Capaccioli, 1975; Illingworth, 1977; Binney, 1978). Recent data on NGC 1600 by Jedrzejewski and Schechter (1988) illustrate the point dramatically (Fig. 1).

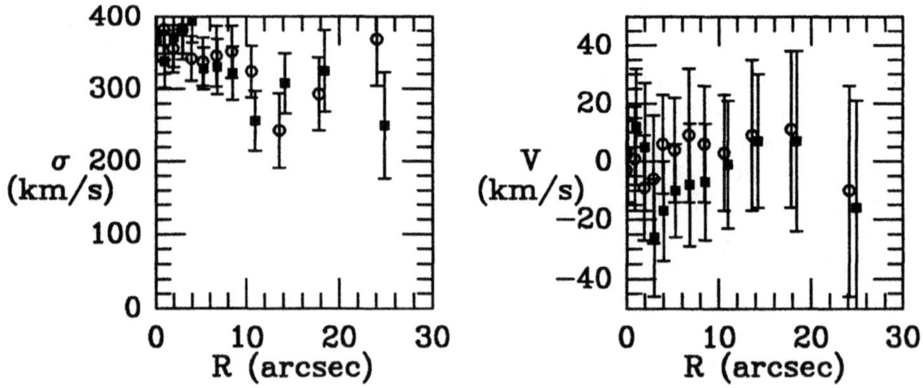

Figure 1. The major axis kinematics of NGC 1600, measured by Jedrzejewski and Schechter (1989). These authors derived a rotation of 1.9 ± 2.3 km/s, compared to the mean velocity dispersion of 300 km/s. The ellipticity of the galaxy is 0.17.

The authors derive a rotational velocity of 1.9 ± 2.3 km/s, and a velocity dispersion of 300 km/s. Hence v/σ, which is an estimate of the rotational support, is smaller than 2% for this galaxy. Of course, projection effects may dominate for one particular galaxy, but generally speaking, bright ellipticals have a mean $v/\sigma \approx 0.2$. The anisotropy parameter $(v/\sigma)^*$ is the ratio of the measured v/σ and v/σ_{ISO}, which is the value of v/σ if the flattening of the galaxy is caused by rotation (Binney, 1978; Davies et al., 1983; hereafter DEFIS). Bright ellipticals have a mean $(v/\sigma)^*$ of 0.4, with a large spread.

The second surprise was the rapid rotation of bulges (Kormendy and Illingworth 1982, Whitmore et al., 1984). The v/σ is larger than 0.5 for most bulges, and their flattening is mostly due to rotation. These measurements are not easy, as the contamination of disk-light can cause spuriously high rotation velocities. Most authors have therefore studied (nearly) edge-on bulges. Until now, no slowly rotating bulges have been found (say, with $v/\sigma < 0.2$). Thus, the bulges have kinematics very different from bright ellipticals.

The apparent contradiction between bulges and ellipticals was better understood with the kinematic measurements of faint ellipticals by DEFIS. These authors found that faint ellipticals, which have roughly the same luminosity as typical bulges, are also fast rotators. The ellipticals with $-20.5 < M_B < -18.5$ have a mean $(v/\sigma)^*$ of 0.9 and a mean v/σ of 0.7. Thus, for $M_B < -18.5$, the rotation v/σ decreases strongly with increasing absolute magnitude (see DEFIS and figures 2b and 2c). The scatter in the relation is large.

The origin of this correlation is not well understood. In simulations of dissipationless hierarchical formation scenarios, no strong correlation is found between the mass of a galaxy and its rotation v/σ (DEFIS, Barnes and Efstathiou, 1987, Quinn and Zurek, 1989). Thus it appears necessary to include baryonic processes like gas dissipation, and star formation in disks, to explain these results.

Several authors have tried to find correlations between the rotation and other structural parameters for a better understanding of the kinematics. Wyse and Jones (1984) found that v/σ correlates with the surface brightness of the galaxy (see figure 2d), the lower the surface brightness the lower the rotation of a galaxy. On the basis of this result, they predicted that the low surface brightness dwarf galaxies would be slow rotators. The effect may be caused in a scenario where the specific angular momentum of a galaxy is constant, and where the rotation speed increases during contraction. If the contraction stops earlier, the rotation will be lower, and the surface brightness will be lower. It is not clear if this can be the whole explanation (Bender and Nieto, 1990). Figure 2e shows the specific angular momentum as a function of mass. If we fit a line to the distribution of points which results from assuming $v_{rot} = \sigma$, then we find that the bright ellipticals lie significantly below that line. Hence their specific angular momenta may be "abnormal" as well.

Later, it was found that the rotation of galaxies is related to the isophotal shapes (Carter, 1987; Bender, 1988a). Detailed CCD photometry allows the detection of the deviations of the isophotes from ellipses. Galaxies with "disky" isophotes (i.e. isophotes with deviations which can be caused by the superposition of a disk on a spheroidal system) generally rotate fast. Boxy galaxies have a large spread in their rotational properties. The interpretation of this effect is not (yet) clear.

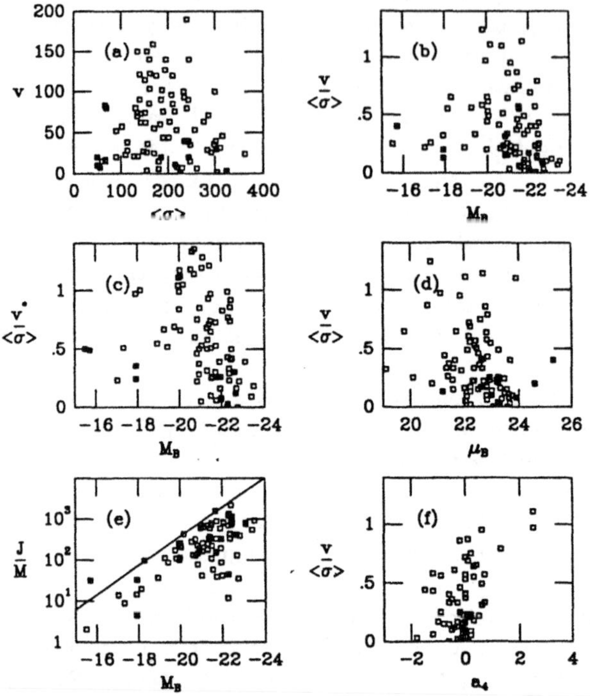

Figure 2. The rotation of elliptical galaxies correlated with other parameters. The rotational velocity is shown against the mean velocity dispersion in (a). The rotational velocities increase with σ for $\sigma < 200$ km/s, and decrease for $\sigma > 200$ km/s. The rotation is correlated with absolute magnitude in (b). (c) shows the anisotropy parameter $(v/\sigma)^*$ as a function of absolute magnitude. Both (b) and (c) show that rotation is relatively unimportant for the high luminosity galaxies ($-M_B > 22$) and low luminosity dwarfs ($-M_B < 18$). The correlation between rotation and surface brightness is show in (d). In (e), the specific angular momentum J/M is plotted versus absolute magnitude. The units of J/M are arbitrary. The line is a fit to the distribution of points when v is set equal to σ. The high luminosity galaxies show a significant scatter below that line. In (f) the rotation is plotted as a function of the isophotal shape a_4, as defined by Bender, 1988a. Disky galaxies ($a_4 > 0$) rotate fast, boxy galaxies ($a_4 < 0$) show a large scatter.

Possibly the simplest explanation is that the "disky" isophotes are caused by a relative luminous disk (contributing on the order of 10% of the light), as indicated by a statistical analysis of the isophotal shapes by Rix and White (1990). These disks may influence the rotation disproportionally, because their velocity dispersion is low. However, other properties like radio emission and X-ray luminosity also correlate with isophotal shapes, and several authors have argued that the isophotal shapes distinguish ellipticals of fundamentally different type (Bender et al., 1989).

The complexity was even increased with the finding of slowly rotating dwarf ellipticals (Held et al., 1990; Carter and Sadler, 1990; Bender and Nieto, 1990). Bender and Nieto observed five dwarf ellipticals with $-M_B < 18.0$. The rotation parameter v/σ is as low as 0.1 for some of these galaxies. The anisotropy parameter $(v/\sigma)^*$ is lower than 0.26 for three out of five. The lower surface brightness galaxies generally rotate slower than the higher surface brightness galaxies, in accordance with the relation for bright ellipticals (Wyse and Jones, 1984). The specific angular momentum of most of these dwarfs falls along the line predicted for galaxies with constant rotational support (see figure 2f). Thus, their specific angular momenta may be considered to be normal, in contrast with the slowly rotating high luminosity ellipticals. It is generally thought that supernova driven winds play an important role in the evolution of dwarf galaxies (e.g. Dekel and Silk, 1986), but it is not clear if they can be the single cause of the low rotation and anisotropy of the dwarfs. A detailed discussion of these issues can be found in Bender and Nieto (1990).

Possibly the last surprise concerns the rotation of galaxies in their outer parts. Mould et al. (1990) have published radial velocities for the globular cluster systems of two Virgo ellipticals, NGC 4472 and M 87. In both cases they measure rotation, $v/\sigma \approx 0.3$. The rotation in NGC 4472 is hardly significant, given the low number of globular clusters measured, but surprisingly enough, it is in the same sense as the rotation of the stars. The rotation of the globular cluster system of M87 is more significant, and it is much higher than the rotation of the stars. The best case for more rapid rotation in the outer parts is Cen A. Ford et al. (1989) and Hui et al. (1991) have measured radial velocities of planetary nebulae out to a radius of 4 r_e. The rotation velocity of the planetary nebulae rises to 100 km/s, whereas the rotation of the "normal" stellar component is only 40 km/s in the inner parts. It will be very interesting to see if further observations of these and other galaxies confirm the conjecture that the outer parts of ellipticals rotate with typical speeds of $v/\sigma \approx 0.3$. If this conjecture is confirmed, it has important consequences: First, the rotation is similar to the rotation of the (few) measured globular cluster systems in spiral galaxies. Second, this rotation agrees well with predictions for the rotation of halos in hierarchical formation scenarios, and hence these formation scenarios may be a good description of the (last) formation stages of these systems. Third, it would suggest that the rotation of the inner parts of the galaxies can be distinctly different, supposedly due to processes occurring when the galaxy is being assembled (e.g. gaseous processes, energy dissipation through cooling etc.). The angular momentum of the inner luminous parts is relatively small compared to the outer parts, and hence we run the risk of studying variations in the details of the kinematics of luminous parts of galaxies, whereas the global kinematics may be much easier to understand. Fourth, it would be very hard to explain rapidly rotating globular cluster systems in the context of the classical picture of a monolithic fast collapse. In such a scenario the specific angular momentum is roughly conserved, and the inner parts would be expected to rotate faster than the outer parts. This fast rotation is indeed observed in bulges, but not in bright ellipticals. This strongly suggests that (at least) an additional mechanism has been at work.

We conclude this section with the remark that it is still difficult to understand the systematics of the rotation of ellipticals. We find many relations, correlations, with a lot of scatter. We lack at this time detailed predictions which can be compared to the observations. In a hierarchical formation scenario, the ellipticals are possibly formed by mergers of preexisting stellar systems. Simulations of such mergers show that the end results may be very similar to bright ellipticals (Barnes, 1988, 1990, 1991). The rotation of these remnants is always low, and hence the rapid rotation of the lower luminosity ellipticals is somewhat more difficult to explain. These simulations did not include gas, and so more general simulations including gas and star formation may solve this problem. In general, one has many free parameters in a merger process, such as all the orbital parameters, the structural parameters of the progenitors (masses, bulge to disk ratio, gas fraction, depth of potential well of the halos, etc.), star formation efficiency, etc. It remains to be seen if the variation of these parameters can reproduce the variation in rotational properties, and what systematics in the progenitors are needed to produce the current systematics between the observables.

3 Minor axis rotation and intrinsic shapes

If elliptical galaxies are triaxial, rotation along the apparent minor axis is naturally expected (Contopoulos, 1956; Kondrat'ev and Ozernoy, 1978; Binney, 1985). It is impossible to derive the intrinsic shape of an individual galaxy from its major and minor axis rotation, but Binney showed how the observed minor axis rotation of a sample of galaxies can be used to constrain their intrinsic shapes. He predicted that apparently round galaxies have the highest probability of significant minor axis rotation.

Shortly afterwards, Davies and Birkinshaw (1986) found strong minor axis rotation in NGC 4261. The rotation along the minor axis is 100 km/s for this galaxy, the major axis rotation is not significant. It was the first case of a galaxy which appeared to rotate around its long axis. In general, minor axis rotation can be due to two effects: First, the apparent minor axis of a triaxial systems does not coincide with the projected intrinsic short axis. Hence if a galaxy rotates around its intrinsic short axis, a residual velocity gradient will be measured along the apparent minor axis. Second, the angular momentum of a triaxial galaxy does not necessarily lie along the intrinsic short axis. Triaxial galaxies without figure rotation have two classes of orbits that can cause internal streaming: the long axis tubes (which cause streaming around the long axis), and the short axis tubes (which cause streaming around the short axis). A selfconsistent model, found by either Schwarzschild's technique (Schwarzschild, 1979), or by analytic means (Dejonghe, 1987) specifies the relative mass contribution of each orbit. Within each tube orbit, stars can rotate in two ways; and the sense of the rotation is a free parameter for the models. Thus, all stars can rotate in the same sense, half the stars can rotate in one sense, and the other half in the opposite sense, and many intermediate constructions are possible. Hence the angular momentum contributed by each orbit can vary from 0 to some maximum (when all the stars rotate in the same sense); and the sign is arbitrary. Thus the total angular momentum components along the x–axis and the z–axis can vary from 0 to some maximum. Consequently, the angular momentum of the galaxy can lie along any direction in the plane containing the short and the long axis.

It might seem counter-intuitive that galaxies exist which rotate around their *long* axis. However, for some particular galaxies, a strong case can be made (Franx, Illingworth, Heckman, 1989; Franx, Illingworth, de Zeeuw, 1991, FIZ). NGC 4406 is a bright Virgo galaxy with a complicated dynamical structure (Fig. 3). The three different groups which observed this galaxies found consistently the same result. In the outer parts, minor axis rotation is observed, and almost no major axis rotation. In the inner parts, strong major axis rotation is observed, and almost no minor axis rotation. Hence the rotation axes of the inner and outer parts project perpendicular to each other. At least one of these systems must have non-zero rotation around the intrinsic long axis. Suppose the outer parts rotate around the intrinsic short axis. Then the line–of–sight must lie in the plane of the intrinsic long and short axis (Binney, 1985, FIZ). As a consequence, the rotation around the long axis and the rotation around the short axis both project to rotation around the apparent long axis. This type of model cannot account for the rotation around the apparent short axis in the center. Thus, we conclude that a model in which the outer parts rotate around the intrinsic long axis is more appropriate. NGC 4365, another Virgo galaxy, has a similar structure. Hence, at least for two galaxies, we have a good reason to believe that they rotate around their long axis in the outer parts. These galaxies appear to have normal photometry and stellar populations, and many more galaxies may be expected to have significant rotation around their long axis.

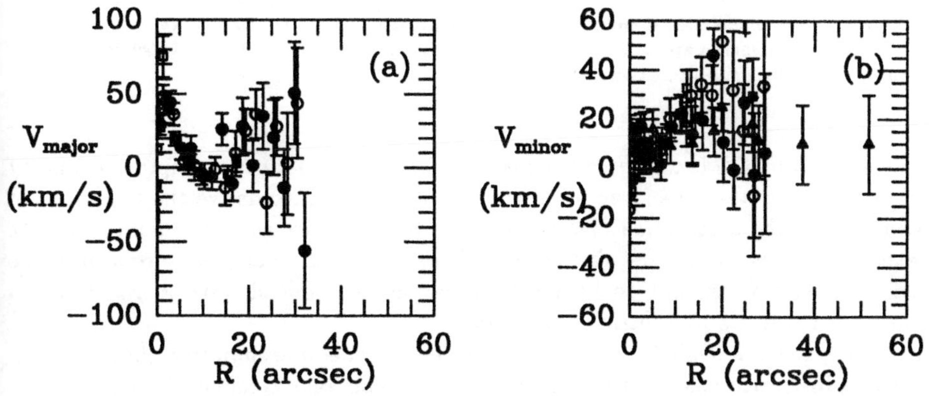

Figure 3. The major and minor axis kinematics of NGC 4406. The data sources are Bender (1988b) and Wagner et al. (1988) (circles), Franx et al. (1989) (squares), and Jedrzejewski and Schechter (1989) (triangles). In the inner 10 arcsec, the galaxies appears to rotate around the apparent major axis. In the interval of 10-20 arcsec, the galaxy appears to rotate around the apparent minor axis. It is argued in the text that the galaxy rotates around intrinsic long axis at this interval.

We can hypothesize that the angular momenta of triaxial galaxies are not constrained by formation mechanism, and lie in an arbitrary direction in the plane of the short and long axis. In this case we expect to see roughly the same number of galaxies with $v_{minor} > v_{major}$ as galaxies with $v_{major} < v_{minor}$.

The minor axis rotation is expressed by the kinematic misalignment angle Ψ, which is defined by $\tan\Psi = v_{minor}/v_{major}$. Galaxies without minor axis rotation have $\Psi = 0$, galaxies without major axis rotation have $\Psi = 90°$. Figure 4 shows a compilation of available data (FIZ). The sources are Davies and Birkinshaw (1988), Franx, Illingworth and Heckman (1989), Jedrzejewski and Schechter (1989). Clearly a significant number has minor axis rotation, but in general, the histogram of Ψ peaks towards low values. Evidently, the intrinsic angular momentum is reasonably well aligned with the intrinsic short axis, otherwise we would not see the peak at $0 < \Psi < 10°$. This good alignment of the angular momentum is not required by the equilibrium dynamics, and is probably due to the formation process.

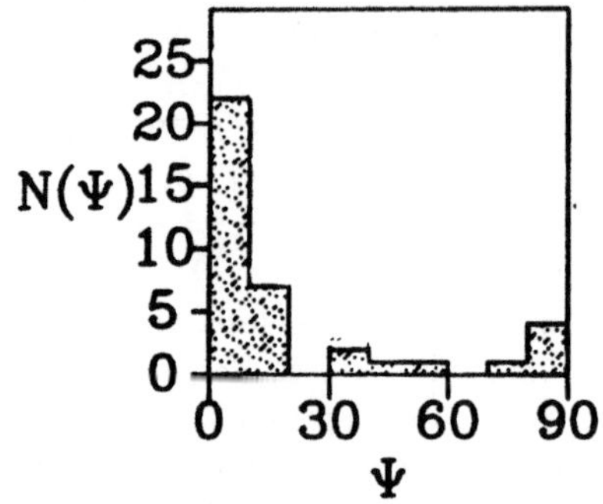

Figure 4. The distribution of the apparent kinematic misalignment angle Ψ. Galaxies without minor axis rotation have $\Psi = 0$, galaxies without major axis rotation have $\Psi = 90°$. Note the peak of galaxies at small apparent misalignments. This implies that the intrinsic [Aic angular momenta are relatively well aligned with the intrinsic short axes, which is not required by the equilibrium dynamics.

The histogram of Ψ can be used to constrain the intrinsic shapes of ellipticals. Binney (1985) showed how the kinematic misalignment can be used to derive the triaxiality of ellipticals ***under the assumption that the angular momenta are perfectly aligned with the intrinsic short axes.*** From the discussion above, it is clear that this approach is too limited, as galaxies can and do have significant intrinsic misalignments ψ_{int} between their angular momenta and intrinsic short axis.

FIZ analyzed the available data under the more general assumption that both triaxiality and intrinsic misalignment contribute to the apparent kinematic misalignments. The problem is degenerate, as there are three intrinsic parameters (triaxiality, flattening, and intrinsic misalignment), and two observed parameters (apparent flattening and apparent misalignment). The problem can be simplified somewhat in the following way: The apparent position angle of a galaxy Γ_{minor} is a function of the two axial ratios b/a, c/a, where a, b, c are the semiaxis lengths with $a > b > c$, and the two viewing angles ϕ, θ: $\Gamma_{minor} = \Gamma_{minor}(b/a, c/a, \phi, \theta)$. It turns out that the dependence on b/a and c/a can be expressed through one parameter T, which we call the triaxiality parameter

$$T = \frac{1 - (b/a)^2}{1 - (c/a)^2}. \tag{1}$$

For oblate galaxies, $T = 0$, and for prolate galaxies $T = 1$. More general triaxial galaxies have $0 < T < 1$. Franx (1988) and FIZ showed that $\Gamma_{minor} = \Gamma_{minor}(T, \phi, \theta)$. Thus, all galaxies with the same value of T have the same position angle, if the viewing angles are the same. This relation is not self-evident, but follows from the special properties of the projections of ellipsoids. The first implication is that we can redefine the cause of position angle twists: if the principal axes of the isodensity surfaces are aligned, a position angle twist is caused by a radial change in T, and vice versa. A change in ellipticity is not required to accompany a position angle twist, as b/a and c/a can conspire to give constant apparent ellipticity but changing position angle. Furthermore, a triaxial galaxy

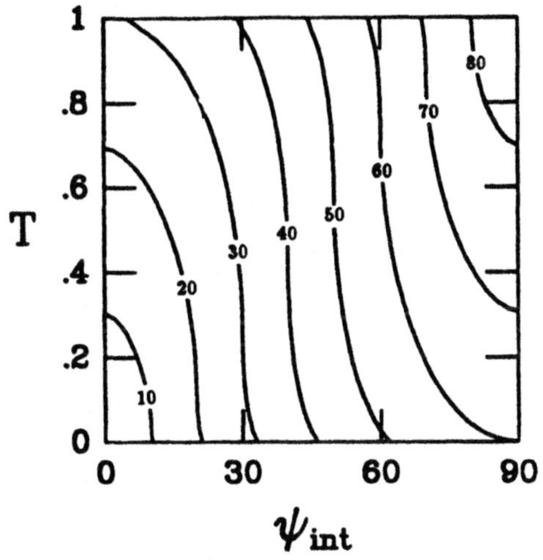

Figure 5. Contours of constant mean apparent misalignment $\langle\Psi\rangle$ as a function of intrinsic misalignment ψ_{int} and triaxiality T. The mean apparent misalignment is the misalignment for a galaxy averaged over all viewing angles. It increases with increasing ψ_{int} and increasing T. Thus, if the observed $\langle\Psi\rangle$ is low, upper limits can be obtained on $\langle T\rangle$ and $\langle\psi_{int}\rangle$. Note the $\langle\Psi\rangle$ is only weakly dependent on T for intermediate values of ψ_{int}.

can have changing c/a as a function of radius, but constant T, and thus can have constant apparent position angle, but changing apparent flattening. The only conclusion that we can derive from the absence of an isophotal twist is that T is constant, it can be 0 (oblate shape), 1 (prolate shape), or any intermediate value (triaxial shape).

We can now simplify the modeling: the apparent misalignment Ψ is a function of the intrinsic misalignment ψ_{int} and the triaxiality parameter T, and the viewing angles. We can calculate the probability distribution $p(\Psi)d\Psi$ of observing a misalignment between Ψ and $\Psi + d\Psi$ for random viewing angles (see FIZ). These probability distributions are generally non-zero for all Ψ, and show a singularity at $\Psi = \psi_{int}$. We introduce the mean apparent misalignment $\langle\Psi\rangle$, which is the first moment of $p(\Psi)$. It is the value of Ψ averaged over all viewing angles. The dependence of $\langle\Psi\rangle$ on ψ_{int}, T is shown in figure 5. The mean apparent misalignment is 0 for models with (ψ_{int}, T)=(0,0), and increases to 90° at (ψ_{int}, T)=(90°,1). The mean apparent misalignment increases always when T increases or when Ψ increases. It is also clear that $\langle\Psi\rangle$ does not depend strongly on T for intermediate values of ψ_{int} ($20° < \psi_{int} < 70°$), but is strongly dependent on T outside that range. Thus, for small observed $\langle\Psi\rangle$, we cannot obtain unique values for the mean triaxiality $\langle T\rangle$ and the mean intrinsic misalignment $\langle\psi_{int}\rangle$, but we can put upper limits to their values.

We have modeled the available data by fitting to the histogram of apparent misalignments. We assumed that galaxies occupy a line in the (ψ_{int}, T)-plane, calculated the probability distribution $p(\Psi)$ for each point along that line, and inverted the observed histogram $N(\Psi)$ to derive the distribution of galaxies along the line in the (ψ_{int}, T)-plane. With help of Lucy's method we obtained 2 solutions for each set of models: a solution after 5 iterations which is smooth, and a solution after 1000 iterations which is very much peaked. The solutions are shown in figure 6. In figure 6a, we have assumed that the intrinsic misalignment is zero for all galaxies.

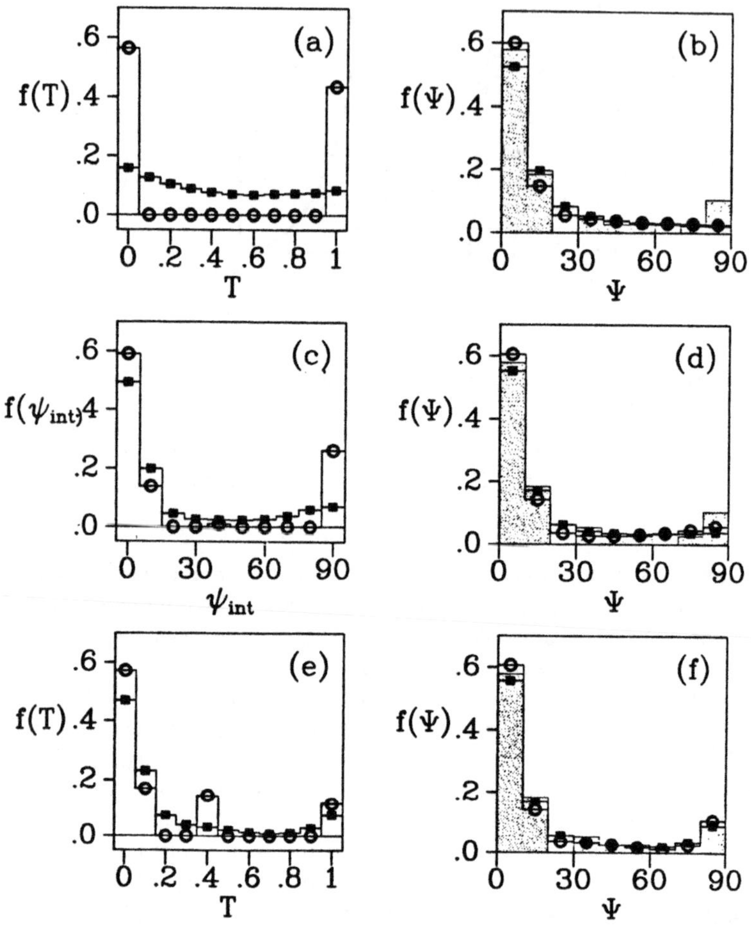

Figure 6. Three types of solutions for the intrinsic shapes and intrinsic misalignments of elliptical galaxies. The models plotted in (a) have no intrinsic misalignment. The squares show the smooth solution after 5 iterations, the circles show the peaked solution after 1000 iterations. Both fit the observed histogram of Ψ equally well. The models in (c) have no triaxiality ($T = 0$), but intrinsic misalignment. The squares and circles fit the observed histogram well (d). The models shown in (e) have $\psi_{int} = T \times 90^\circ$. These are truly intermediate between the models of (a) and (c). The apparent minor axis rotation is caused by both intrinsic misalignment and triaxiality. This set of models can fit the little peak at $\Psi = 90^\circ$ best (f).

The first solution is fairly regular and extends from oblate shapes to prolate shapes. The other solution is strongly peaked, and has only oblate shapes and prolate shapes. Both fit the data equally well (Fig. 6b). None are able the reproduce the small (and hardly significant) peak at $\Psi = 90°$.

The next set of solutions are oblate ($T = 0$), and have non-zero intrinsic misalignments. These models are not entirely realistic, as oblate galaxies cannot have intrinsic misalignments, but they illustrate very well what the effects are of intrinsic misalignment. They are also representative of triaxial models with intrinsic misalignments, but with very small T. The two solutions show both peak at $\psi_{int} = 0$, decline rapidly, and have a low fraction of galaxies with intermediate and high intrinsic misalignments.

The third set of solutions have $\psi_{int} = 90° \times T$. For these models the misalignment increases with increasing triaxiality, which may be expected from models based on distribution functions (FIZ). A smooth and peaked solution is shown in figure 6e. Both solutions peak at $\psi_{int} = T = 0$, decline to low values at intermediate values of T, and show a secondary peak at $\psi_{int} = 90°, T = 1$. They reproduce the observed histogram very well, including the little peak at $\Psi = 90°$.

The solutions shown here are certainly not the only solutions possible, there is no special reason that galaxies should lie along a line in the (ψ_{int}, T)–plane. However, they do suggest certain systematic properties which all solutions might have to satisfy: First of all, despite the very different character of the two solutions show in figure 6a, they have almost the same mean triaxiality. $\langle T \rangle$ is 0.43 and 0.44 for both solutions. Thus the distribution in T is not well constrained, but the mean triaxiality is! The models in figure 6c have mean intrinsic misalignments of $\langle \psi_{int} \rangle$=21° and 25°. The fraction of galaxies with very low intrinsic misalignments ($\psi_{int} < 10°$) is higher than 50% for both models. For the models of figure 6f, the apparent misalignment is caused by both intrinsic misalignment and triaxiality. The mean triaxiality is 0.2, and the mean intrinsic misalignment is 18° for both models. Comparing these values to the previous models, we note that the apparent misalignment is mostly due to the intrinsic misalignment, and not the triaxiality of these galaxies (as the mean intrinsic misalignment is similar to the mean intrinsic misalignment of the oblate models with intrinsic misalignments, but the mean triaxiality is a factor of two lower than for the models without intrinsic misalignments). By this type of modeling, we can clearly put upper limits on both the mean triaxiality and the mean intrinsic misalignment. In general, a higher mean misalignment implies a lower mean triaxiality. The current models satisfy $\langle T \rangle < 0.45$ and $\langle \psi_{int} \rangle < 25°$. A bigger sample is needed to verify these numbers. Further more, the sample should be made complete in apparent flattening. The current sample is strongly biased towards round galaxies (FIZ).

4 Predictions for intrinsic shapes and angular momenta

The results described in the previous section can provide valuable constraints on the formation mechanism of ellipticals. In order to draw any conclusions, we need theoretical predictions from formation scenarios. At this moment, no simulations or analytic predictions exist for the shapes and angular momenta of the luminous parts of ellipticals. There are predictions for halos of galaxies, mostly based on hierarchical formation scenarios (including, and extending beyond CDM-type models). Bardeen et al. (1986) gave analytical predictions for the shapes of peaks in smoothed Gaussian random fields. Their approach ignores the dynamical effects that occur when clumps merge to form halos, so they can be thought of a characterization of the initial conditions before virialization of halos. They found that the distribution of shapes is relatively independent of the spectrum of the fluctuations. In figure 7 we show the predictions for the distribution of T which apply to most realistic power spectra (FIZ). Note that all the distributions peak at large triaxiality. The lowest mean triaxiality is 0.6. These values are significantly higher than the mean values derived from the data. Numerical simulations by Frenk et al. (1988) also showed a tendency for triaxial-prolate halos, but these results are still uncertain because of the low number of particles per halo.

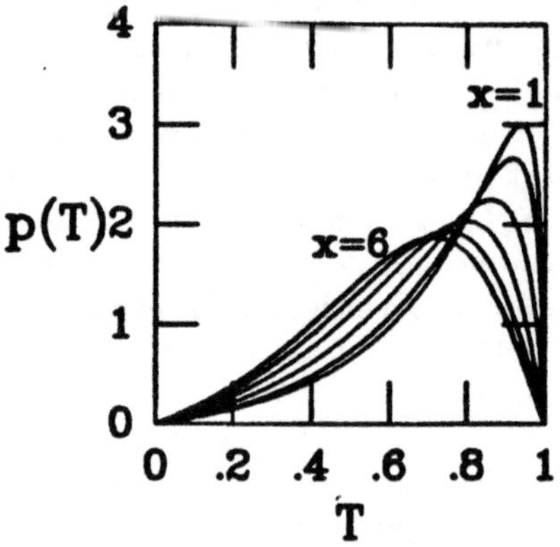

Figure 7. The predicted distribution of T for peaks in smoothed Gaussian random fields. This distribution can be thought typical of the shapes of the initial conditions of halos before collapse. Non-linear effects are ignored in this type of analysis. The shapes depend only on the parameter x, which is a measure of the width of the peaks, and which lies between 1 and 6 for most relevant power spectra. Note that all curves predict rather high values of T.

Several authors have analyzed the intrinsic misalignment of the angular momentum vectors with the symmetry axes of the halos (Barnes and Efstathiou, 1988; Quinn and Zurek, 1988). These authors found alignments of the angular momenta, but not as tight as required by the data.

The above comparison is very preliminary, as the simulations are being improved in resolution, and as gaseous processes are being taken into account. However, on the whole, the current simulations predict higher apparent misalignments than observed. This may be due to one of several effects: Gradients may exists in the shapes of the halos, and the inner parts are more axisymmetric-oblate; possibly the resolution of the simulations is not yet sufficient, and the halos may become more axisymmetric if very high resolution is used. We note that Miralda-Escudé and Schwarzschild (1989) found that small core radii change the orbital structure of galaxies significantly, and it may be difficult to construct selfconsistent halos with small cores and triaxialities around 0.5. The current simulations lack the resolution to find these effects. Thirdly, the gaseous processes which have caused the separation between luminous and dark material may have influenced the shapes on the resulting luminous structures significantly. As an example, cool gas tends to settle in a disk. In a triaxial halo, the disk will be extended in a direction perpendicular to the elongation of the halo in the disk-plane, and effectively the triaxiality of the potential will be decreased.

It is too early now to decide which of these effects are relevant, but we can expect some answers in the near future.

5 Conclusion

We have reviewed the rotational properties of elliptical galaxies. Both the statistics of the rotational amplitude and the minor to major axis rotation reflect the conditions at the formation of ellipticals, and are not caused by the laws of stellar dynamics alone. We still lack detailed theoretical models which can explain all the observations. These difficulties are probably caused by our limited understanding of the relevant processes acting during formation, especially the hydrodynamics, star formation, etc.. Predictions for the halos may be easier to make than for the luminous parts of galaxies. One of the main observational uncertainties is the question whether ellipticals have dark halos. The hot X-ray gas, globular cluster systems, and planetary nebula systems may provide us with answers in the near future. The globulars and planetary nebulae also give us information on the kinematics at large radii. These may provide additional tests of galaxy formation scenarios.

Many relevant topics were ignored in this review, such as the luminosity function, the stellar populations, the scaling relations for Ellipticals. Other papers in these proceedings discuss these matters in more detail.

The author hopes that a synthesis of the material that is becoming available at this time, together with the information that we are obtaining from intermediate and high redshift galaxies, will allow for a significant improvement of our understanding of the formation of galaxies.

Acknowledgements

It is a pleasure to thank the organizers of the conference for a pleasant and informative workshop in a beautiful environment, and for their financial support. Much of the work reported here was done in collaboration with Tim de Zeeuw and Garth Illingworth. The Harvard Society of Fellows and the Center for Astrophysics supported me during the course of this work.

References

Bardeen, J.M., Bond, J.R., Kaiser, N., Szalay, A.S.: 1986, *Astrophys. J.* **304**, 15.

Barnes, J.E.: 1988, *Astrophys. J.* **331**, 699.

Barnes, J.E.: 1990, In *Dynamics and Interactions of Galaxies*, ed R. Wielen, (Berlin: Springer–Verlag), p. 186.

Barnes, J.E.: 1991, this conference proceedings.

Barnes, J.E., Efstathiou, G.: 1987, *Astrophys. J.* **319**, 575.

Bender, R., Nieto, J.L.: 1990, *Astron. Astroph.* **239**, 97.

Bender, R.: 1988a, *Astron. Astroph. Letters* **193**, L7.

Bender, R.: 1988b, *Astron. Astroph. Letters* **202**, L5.

Bender, R., Surma, P., Döbereiner, S., Möllenhof, C., Madejsky, R.: 1989, *Astron. Astroph.* **217**, 35.

Bertola, F, Capaccioli, M.: 1975, *Astrophys. J.* **200**, 439.

Binney, J.J.: 1978, *Astron. Astrophys. Suppl. Ser.* **183**, 501.

Binney, J.J.: 1985, *Astron. Astrophys. Suppl. Ser.* **212**, 767.

Butcher, H., Oemler, A.O.: 1984, *Astrophys. J.* **285**, 426.

Carter, D.: 1987, *Astrophys. J.* **312**, 514.

Carter, D., Sadler, E.M.: 1990, *Astron. Astrophys. Suppl. Ser.* **245**, 12.

Chambers, K.C., Miley, G.K., can Breugel, W.J.M.: 1990, *Astrophys. J.* **363**, 21.

Contopoulos, G.: 1956, *Zs. Astroph.* **39**, 126.

Davies, R.L., Efstathiou, G., Fall, S.M., Illingworth, G.D., Schechter, P.L.: 1983,, *Astrophys. J.* **266**, 41, DEFIS.

Davies, R.L., Birkinshaw, M.: 1986, *Astrophys. J. Letters* **303**, L45.

Davies, R.L., Birkinshaw, M.: 1988, *Astrophys. J. Suppl. Ser.* **68**, 409.

de Zeeuw, P.T.: 1987, IAU Symposium 127: *Structure and Dynamics of Elliptical Galaxies*, (Dordrecht: Reidel).

Dejonghe, H.: 1987, in IAU Symposium 127: *Structure and Dynamics of Elliptical Galaxies*, Ed. P.T. de Zeeuw, (Dordrecht: Reidel), p. 495.

Dekel, A., Silk, J.: 1986, *Astrophys. J.* **303**, 39.

Dressler, A., Gunn, J.E.: 1990, in ASP conf. ser. vol. 10 *Evolution of the Universe of Galaxies*, Ed. R. G. Kron, 200.

Ford, H.C., Ciardullo, R., Jacoby, G.H., Hui, X.: 1989, in IAU Symposium 131, *Planetary Nebulae*, Ed. S. Torres–Peimbert, 335. (Dordrecht: Reidel)..

Franx, M.: 1988, *Astron. Astrophys. Suppl. Ser.* **231**, 285.

Franx, M., Illingworth, G.D., Heckman, T.M.: 1989, *Astrophys. J.* **344**, 613.

Franx, M., Illingworth, G.D, de Zeeuw, P.T.: 1991, *Astrophys. J.* in press (=FIZ).

Frenk, C.S., White, S.D.M., Davis, M., Efstathiou, G.: 1988, *Astrophys. J.* **327**, 507.

Held, E.V., Mould, J.R., de Zeeuw, P.T.: 1990, *Astron. J.* **100**, 415.

Hui, X., Ford, H.C., Freeman, K.C.: 1991, in preparation.

Illingworth, G.D.: 1977, *Astrophys. J.* **218**, L43.

Jedrzejewski, R.I., Schechter, P.L.: 1989, *Astron. J.* **98**, 147.

Kondrat'ev, B.P., Ozernoy, L.M.: 1979, *Sov. Astron. Lett.* **5**, 37.

Koo, D.C.: 1986, *Astrophys. J.* **311**, 651.

Kormendy, J., Illingworth, G.D.: 1982, *Astrophys. J.* **256**, 460.

Lilly, S. J.: 1989, *Astrophys. J.* **340**, 77.

Miralda–Escudé, J., Schwarzschild, M.: 1989, *Astrophys. J.* **339**, 752.

Mould, J.R., Oke, J.B., de Zeeuw, P.T., Nemec, J.M.: 1990, *Astron. J.* **99**, 1823.

Quinn, P.J., Zurek, W.H.: 1988, *Astrophys. J.* **331**, 1.

Rix, H., White, S.D.M.: 1990, *Astrophys. J.* **362**, 52.

Schwarzschild, M.: 1979, *Astrophys. J.* **232**, 236.

Schweizer, F.: 1990, In *Dynamics and Interactions of Galaxies*, Ed. R. Wielen (Berlin: Springer–Verlag), 60.

Tinsley, B.M.: 1980, *Fund. Cosmic Physics* **5**, 287.

Toomre, A.: 1977, In *The Evolution of Galaxies and Stellar Populations*, eds B.M. Tinsley & R.B. Larson, 401–416. (New Haven: Yale Univ. Observatory)..

Tyson, J.A.: 1988, *Astron. J.* **96**, 1.

van Breugel, W. J. M., McCarthy, P. J.: 1990, in ASP conf. ser. vol. 10 *Evolution of the Universe of Galaxies*, Ed. R. G. Kron, 359.

Wagner, S.J., Bender, R., Möllenhof, C.: 1988, *Astron. Astroph. Letters* **195**, L5.
Whitmore, B.C., Rubin, V.C., Ford, W.K.: 1984, *Astrophys. J.* **287**, 66.A
Wielen, R.: 1990, *Dynamics and Interactions of Galaxies*, (Berlin: Springer–Verlag).
Wyse, R.F., Jones, B.J.T.: 1984, *Astrophys. J.* **286**, 88.

Discussion

D. Burstein - I am part of a group studying elliptical galaxies at distances 8000–12000 km/s, in an effort to obtain $D_n - \sigma$ distances. As such, we feel justified in obtaining luminosity profiles using circular apertures (to obtain D_n). I am surprised by our finding that many of our galaxies ($\sim 200 - 300$ out of 500) are fit very well by an $r^{1/4}$ law, and more are fit by an $r^{1/4}$ law + low luminosity exponential disk.
M. Frank - The similarity of the intensity profiles of ellipticals is generally thought to be the result of violent dissipationless formation. Mergers and collapse simulation are effective in producing $r^{1/4}$ laws.

V. Surdin - Do you know about some cases of differences between shapes of elliptical galaxies and shapes of globular cluster subsystems?
M. Franx - Now I do not know of data that address this.
I. King - Only in radial distribution.

I. King - You said a great deal about dynamics but did not discuss populations.
M. Franx - I did not discuss many other properties of elliptical; the field is too wide for a general review in 25 minutes.

P. Pişmiş - Did I hear you say that elliptical galaxies are older than the spiral galaxies? I am not in agreement with this statements.
M. Franx - This is not what I said. The general issue when galaxies formed is not yet resolved.

M. Balcells - You seem to suggest the globular cluster system may be the strongest similarity between spirals and ellipticals. Aren't bulges the strongest similarity?
M. Franx - The point is, that if globular cluster systems are confirmed to have v/σ of $\sim 0.2 - 0.3$ in luminous ellipticals, ,they are similar to the systems in spirals. This raises the possibility that their formation is very similar.

B. Jarvis - Do you know of any well flattened bulges of disk galaxies with low v/σ?
M. Franx - No, I don't know of any low v/σ bulges. It would be interesting to try to find such systems.

J.–L. Nieto - I compiled a sample of 60 elongated elliptical galaxies from our own observations and literature data. The idea behind it was to have a sample free as much as possible from projection effects. All the correlations with isophotal shapes (see Bender et al., 1989) appear much more clearly than with a sample of E's with no selection. It appears notably than the kinematical behaviour of all boxy irregular E's is anisotropic ($(v/\sigma)^* < 0.7$) while all disky E's are rotationally flattened.

G. Busarello - In the relationships of kinematical properties with other quantities, you use of course projected (integrated along the line of sight) $v(r)$ and $\sigma^2(r)$, which are usually different from the spatial (intrinsic) ones. The effect of deprojecting v and σ is that the v/σ ratio generally increases. The effect on some correlations is that they may also disappear.
I think that this effect must be taken into account before trying physical interpretations of the correlations.
M. Franx - I have not seen your paper, so I cannot comment directly. The effect of deprojection is largest for the most rapidly rotating galaxies; this might affect the correlations. I do not see how a galaxy like NGC 1600 can be deprojected to rotate, except if the angular momentum is chosen to lie in the line of sight.

Marijn Franx addressing the audience.

STRUCTURE AND KINEMATICS OF SPIRALS

P.C. van der Kruit
Kapteyn Astronomical Institute
Groningen, The Netherlands

ABSTRACT. In this review I discuss various aspects of the large scale structure and dynamics of spiral galaxies. It is first pointed out that spirals basically exist of two discrete components, which may indicate that there have been two epochs of star formation. In spite of the different light distributions the two galaxies NGC 891 and 7418 have very similar rotation curves, which argues against maximum disk models and revised gravity.

Discussion of photometric and kinematic data on spiral disks show: *(1)* Disks are exponential with an "edge" at about 4.5 scalelengths and have a constant thickness with radius. *(2)* The observed central surface brightness has a preferred value, probably also in the near-IR. *(3)* There are no systematic radial color gradients. *(4)* Stellar kinematics indicates that (M/L) and Toomre's Q are constant with radius. *(5)* Actual measurements show $(M/L) \approx 6 \pm 2$ in B *(old disk population)* and $Q_* \approx 1.5 - 2$. *(6)* The stellar velocity dispersion correlates with luminosity, such that Q and (M/L) are also constant from galaxy to galaxy. *(7)* The exponential nature of disks and the "edge" can be explained by a collapse of the disk with detailed conservation of angular momentum. *(8)* The constant scaleheight follows from a dynamical heating of the disk to a value for Q somewhat larger than 1 at all radii. *(9)* Freeman's law follows if all galaxies have the same ratio of dark to luminous matter. *(10)* The question of whether or not disks are optically thick has not yet been answered satisfactorily.

1 Introduction

In this review I will concentrate on aspects of the measurement of light distributions and kinematics of disks of spiral galaxies. The purpose will be to summarize what is known about the variation of properties within and among disks. Before starting this I wish to make three remarks of a more general nature.

(1). An important feature for understanding the formation of spiral galaxies is the well-known fact that such galaxies consist basically of two **discrete components**. This is most easily seen in surface brightness distributions of edge-on galaxies. These two components will be called *"disk"* and *"bulge"*. Good examples of this are NGC891 and NGC7814, which also are two extreme examples in that NGC891 is disk-dominated (the old-disk population contains 82% of the light in old stars) and NGC7814 bulge-dominated (the bulge contains 93% of the light). In NGC891 (van der Kruit and Searle, 1981b) a flat disk can be subtracted from the observed light distribution and a bulge with a distinctly different flattening remains. In NGC7814 (van der Kruit and Searle, 1982b; Wainscoat

G. Longo et al. (eds.), Morphological and Physical Classification of Galaxies, 39–52.

et al., 1990) the procedure is to subtract an $R^{1/4}$-bulge of moderate flattening and then a very flat disk remains. In our Galaxy at least there is mounting evidence for a "thick disk" with age, flattening, kinematics and abundance intermediate between disk and halo, but this contains only about 10% of the light in the old disk population. It may be associated with the epoch of violent star formation, when the gas remaining after population II formation settled in the disk and therefore probably does not contradict this basic picture of two discrete components.

(2). Van der Kruit and Searle (1982b) argued from this that there are also two **discrete epochs of star formation**. Population II stars formed before the galaxy collapsed and the bulge evolved dissipationlessly, while disks formed dissipationally. One epoch of star formation occured before and the other after virialisation of the bulge and the collapse of the disk.

(3). The two galaxies NGC891 and NGC7814 can also be used to draw an interesting set of conclusions. In spite of their completely different distributions and central concentrations of luminous material, their **rotation curves are indistinguishable**, both rising within 3 kpc or so to a value of about 225 km s^{-1} and remaining flat thereafter out to at least 20 kpc. This indicates that the gravitational field is dominated by the dark component and that disks are less massive than inferred in so-called *"maximum disk"* models, where the disk contribution to the rotation curve is maximized such that it essentially provides the full amplitude of the rotation curve in the inner regions. Also this observation is strong evidence against models involving revised Newtonian gravity, since no completely different mass distributions can be expected to provide the same rotation curve.

2 Photometry of disks

I will now briefly summarize the general properties of the distribution of light in disks of spiral galaxies.

(1). From the early work of de Vaucouleurs (1959) and the classical paper by Freeman (1970) it is a well-known result that the radial distribution of surface brightness in disks can be described by an exponential function

$$L(R) = L(0)\exp(-R/h). \tag{1}$$

Here and in the following I will denote surface brightness in linear units ($L_\odot$ pc^{-2}) as L and in logarithmic units (mag $arcsec^{-2}$) as μ.

(2). Freeman (1970) also found the interesting property that the (inclination corrected) extrapolated central surface brightness of disks appeared restricted to a small range. This property, sometimes referred to as *"Freeman's law"*, has been a matter of much debate. Many workers in the field, most notably Disney (1976), suspected that this result derived from observational selection, when observers selected galaxies for their surface photometry. Faint surface brightness systems are difficult to see against the sky background and bright ones have generally a small angular extent and these properties make both kinds of systems less suitable for surface photometry. It has only recently been possible to select complete samples and study the question properly. Disney and Phillipps (1983) (but see also Davies, 1990) have developed the necessary methods to allow for selection effects in samples that are selected to for example a minimum angular diameter at a uniform surface brightness. This formalism was used in van der Kruit (1987) in analysing a complete sample of 51 galaxies with a major axis diameter in excess of 2 arcmin at about 26.5 B-mag arcsec^{-2}. It was used in two forms, depending on whether distances were or were not used (they were not available for the whole sample). In the last case the visibilities were normalised to unity and only general conclusions can be drawn concerning the effects of sample selection. Disney and collaborators have objected against this use, but I maintain that this does give useful general information, while the distance dependent analysis is entirely correct according to their prescriptions.

When the sample was divided into galaxies of type Sc or earlier ("non-dwarfs") and those of later type ("dwarfs") a small range was confirmed. In about B-band the mean and r.m.s. scatters were respectively 21.5 ± 0.4 and 22.5 ± 0.3 mag arcsec^{-2}. Disney and collaborators have obtained complete samples by fully scanning deep Schmidt plates. In an area centered on the Fornax cluster, Phillipps et al. (1987) find that normal galaxies form a uniform background sample and that their central surface brightness indeed peaks at about 21.7 B-mag arcsec^{-2}. Lower surface brightness galaxies clearly exist, but are concentrated to the cluster and apparently part of it. Davies (1990) on the other hand claims that the first conclusion does not hold up, when effects of bulges are taken into account, but this is sensitive to the assumption that the bulges continue as $R^{1/4}$ laws to large radii and that these there again exceed the surface brightness from the disks. All studies agree that low surface brightness systems appear to occur predominantly in clusters and much less in the field.

(3). The scalelength appears independent of wavelength. This can be seen for example in the data of Wevers et al. (1986), where no systematic color gradients were seen in detailed surface photometry. Although some individual galaxies may become somewhat bluer or redder with galactocentric radius, there is no systematic trend in the sample. Westerhof and van der Kruit (unpublished) have repeated the van der Kruit (1986) sample from IIIa-F plates (somewhere between V and R). The ratio of the scalelengths is 1.07 ± 0.13. Again the central surface brightness has small ranges (20.63 $\pm$ 0.49 for non-dwarfs and 21.99 $\pm$ 0.44 for dwarfs). Elmegreen and Elmegreen (1984) give scalelengths in B and I and the ratio is 1.16 ± 0.47. Their sample has some very extreme differences in the two colors and when these are deleted the ratio becomes 1.08 $\pm$ 0.29. Giovanardi and Hunt (1988) even go to the near-infrared K-band and find a small range in central surface brightness and no statistically different scalelengths from the optical. The next few years will see a dramatic increase in near-infrared surface photometry of galaxies with the availability of large two-dimensional array detectors.

(4). At about 4 to 5 scalelengths there is a sharp drop in the light distribution of disks. This is most obvious in edge-on spirals (van der Kruit, 1979; van der Kruit and Searle, 1981a, 1982a), but can also be seen in less inclined systems (van der Kruit, 1988). It is evident from the smaller separation in isophotes at large radii, but does not show up in azimuthally averaged profiles due to departures from exact circular symmetry in disks.

(5). Edge-on galaxies allow the study of the vertical distribution of luminosity and therefore the three-dimensional distribution of the old disk population. The young population I resides too close to the plane to be studied due to absorption effects. Van der

Kruit and Searle (1981a, 1982a) proposed to use the description of the locally isothermal sheet with a scaleheight independent of galactocentric radius, so that the total light distribution becomes

$$L(R,z) = L(0,0)\exp\left(\frac{R}{h}\right)\,\mathrm{sech}^2\left(\frac{z}{z_0}\right), \tag{2}$$

for $R \leq R_{\max}$. This is an exponential at larger z. Wainscoat et al. (1989) have found in the near infrared a sharper increase at low z. This is to be expected, because there the stellar mix consists of a range of generations with each a different velocity dispersion. They propose to use an exponential for the vertical distribution. Van der Kruit (1988) concludes that this is probably too extreme and proposes as an intermediate solution to drop the exponent 2 from the sech-function. This gives a quite reasonable variation of velocity dispersion with z near the plane. Above the dust lane all three function are essentially the same. Wainscoat and van der Kruit (in preparation) have performed K-band photometry of a few edge-on systems and in NGC891 and NGC4565 there is a sharper increase than in the isothermal case.

At this point a comment on the effects of absorption is in order. In this matter we should always remember the ideal case in which there is a thin dust layer located centrally in a stellar disk. If this layer is optically thick, we always miss half the light and there is no effect on the behaviour of surface brightness as a function of inclination (except of course the geometrical effect) and the colors are also unaffected. By going to the near-infrared we may see the effect of the dust layer becoming more transparent in the integrated colors. Disney et al. (1989) have argued from simple models that observations are at least consistent with disks being optically thick. Valentijn (1990) furthermore used a sample of 9381 Sa to Irr galaxies and found that the *observed* central surface brightness showed no dependence upon inclination. This means that even the expected geometrical effect is absent. From this he concluded that disks are opaque. In fact, Burstein (1980, unpublished) has pointed to similar behavior among RC2 galaxies.

At present no detailed comparison of inferred surface brightness from Valentijn's data and independent (preferably CCD) photometry has been carried out and it is surprising that he finds that bulges make such a small difference between his various methods for early-types in his sample. Furthermore his result is puzzling, in that an absence of dependence of observed surface brightness on inclination can only occur, when the stellar disk and the dust layer have comparable thickness, which is strongly contradicted by observations in the optical and near-infrared of edge-on galaxies. Finally, reddened globular clusters are known to be seen through the disk of M31, while there are examples where we see galaxies through disks of foreground spirals. Good examples are NGC 45 (an apparently distant edge-on spiral is seen projected near a spiral arm) and NGC 450 (see Rubin and Ford, 1983), where UGC 807 (systemic velocity 11,587 km s^{-1}) is seen projected onto the disk of the foreground galaxy (systemic velocity 1863 km s^{-1}).

It is true that so-called "starburst" galaxies are known in which the IRAS flux is comparable to or may even considerably exceed the optical flux. However, for the majority of spirals the ratio is much smaller than unity. For the sample of (by IRAS) *detected* spirals in the Revised Shapley - Ames Catalogue, $\log(L_{\mathrm{FIR}}/L_{\mathrm{B}}^{0}) = -.33 \pm .39$ (van den Broek, 1990) and this value does *not* support the case of Disney et al., that IRAS fluxes show evidence of large amounts of light being absorbed by dust and reradiated in the infrared. The average inferred fraction of absorbed light from IRAS data is only about one-third. Although parts of disks are known that are optically thin (the solar neighborhood) or thick (dust on the inside of spiral arms), the effects on azimuthally averaged radial light profiles are not fully understood. At present Valentijn's case is not conclusive and leads to unphysical geometries for the dust.

3 Stellar kinematics

The constant thickness of the stellar disks (remember that this refers only to the *old disk population*) with galactocentric distance implies for a constant mass-to-light ratio M/L that the *vertical* velocity dispersion of these old disk stars should decrease outward from the center. If the integrated surface density is exponential then this decrease has the form

$$\langle V_z^2 \rangle_*^{1/2} \propto [\sigma(R) z_0]^{1/2} \propto \exp\left(-\frac{R}{2h}\right). \tag{3}$$

This has been observationally tested now in a number of galaxies (e.g. van der Kruit and Freeman, 1986; Bottema, 1990 and references therein). The fact that this dependence has indeed been found then implies that the M/L of the old disk population is indeed in first approximation independent of galactocentric distance in galactic disks. This is of course consistent with, but not demanded by, the observed absence of colour gradients.

A different dependence can in principle be expected for the *radial* velocity dispersion of these stars, namely when the constant Q, which relates to Toomre's (1964) criterion for local stability, is also independent of radius. This parameter is

$$Q = \frac{\langle V_R^2 \rangle_*^{1/2} \kappa}{3.36 G\sigma}. \tag{4}$$

Here κ is the epicyclic frequency. For a flat rotation curve the condition Q is constant then gives

$$\langle V_R^2 \rangle_*^{1/2} \propto R \exp\left(-\frac{R}{h}\right). \tag{5}$$

This dependence seems rather different from the one for the vertical velocity dispersion above and therefore to imply that the axis ratio of the velocity ellipsoid varies with radius. In the first place it should be noted that the velocity dispersions can be measured only over the radial extent $R \approx (1-2)h$ and over this limited range in R the difference is too small to measure. To investigate this further, let us assume that both the vertical and the radial velocity dispersion vary with radius with an exponential with twice the scalelength of the light, in other words such that the thickness *and* the axis ratio of the velocity ellipsoid remain constant and then see in what way Toomre's Q depends on R. In order to do this let us assume some relevant values that are reasonable for the disk of our Galaxy:

$$h = 5 \text{ kpc}, \tag{6a}$$

$$\sigma_* = 425 \exp(-R/h) \text{ M}_\odot \text{ pc}^{-2}, \tag{6b}$$

$$\langle V_R^2 \rangle_*^{1/2} = 100 \exp(-R/2h) \text{ km s}^{-1}, \tag{6c}$$

$$\sigma_{\rm HI} = 30 \exp(-R/2h)\ {\rm M}_\odot\ {\rm pc}^{-2}, \tag{6d}$$

$$\langle V^2 \rangle_{\rm HI}^{1/2} = 8\ {\rm km\ s}^{-1}, \tag{6e}$$

$$V_{\rm rot} = 225\ {\rm km\ s}^{-1}. \tag{6f}$$

The dependence of the HI surface density on R follows from the inference of the sample in Wevers et al. that the scalelength of HI is about 1.8 times that of the light. The values for the HI surface density take account of the contribution from helium and approximately of that of molecular hydrogen. Now calculate from this the value of Q inferred by using only the stellar parameters (this will be denoted as Q_*) and that when the effects of the HI on the surface density and the effective total velocity dispersion are taken into account (denoted $Q_{\rm total}$). The following table then obtains.

R	R/h	σ_*	$\sigma_{\rm HI}$	$\langle V_{\rm R}^2 \rangle_*^{1/2}$	Q_*	$Q_{\rm total}$
5	1	156	18.2	61	1.71	1.45
7.5	1.5	94.8	14.2	47	1.46	1.19
10	2	57.5	11.0	37	1.41	1.09
12.5	2.5	34.9	8.6	29	1.45	1.05
15	3	21.2	6.7	22	1.55	1.05
17.5	3.5	12.8	5.2	17.4	1.71	1.07
20	4	7.8	4.1	13.5	1.92	1.11
22.5	4.5	4.7	3.2	10.5	2.19	1.19
25	5	2.9	2.5	8.2	2.53	1.34

The column with Q_* shows that the assumed variation of the velocity dispersion on radius (which for a velocity ellipsoid of constant axis ratio provides a disk with constant thickness) corresponds to a major increase of Q with radius. However, when we take the effects of the gas into account the effective value of Q becomes much less dependent on R. The table shows that for this numerical example, which should be typical for spiral disks, the condition for a constant thickness disk actually corresponds to one in which the *effective* Q is rather constant. It can therefore be hypothesized that disks do heat up under secular effects (such as encounters with massive interstellar concentrations or spiral arms, etc.) until an equilibrium value of $Q_{\rm total}$ is obtained somewhat larger than unity. Then stability is assured. That situation corresponds to an increase in the *apparent* value for Q, when only the stellar parameters are used to estimate it. So, the constant thickness of disks may simply be the result of a secular heating of the disk with a constant axis ratio velocity ellipsoid until a value of $Q_{\rm total}$ similar at all radii is reached.

Actual measurements of the value for the old disk M/L from velocity dispersions and thicknesses of both stellar disks and of HI gas-layers gives 6 ± 2 (in solar B-units) and for Q_* about $1.5 - 2$. This is discussed in detail in van der Kruit (1990).

Bottema (1988 and private communication) has found that a fiducial value of the stellar velocity dispersion correlates closely with the integrated luminosity of the old disk and with the amplitude of the rotation curve (the two are equivalent through the Fisher - Tully relation). For this he compares the observed radial velocity dispersion at one optical scalelength with the total luminosity for inclined systems. He then adds face-on galaxies, where only the vertical velocity dispersion can be measured, by plotting the (extrapolated) value at the center. If the axis ratio of the velocity ellipsoid is as in the solar neighborhood, then these two should be comparable if these decrease with the double optical scalelength. His current updated diagram contains 13 disks (including our Galaxy from Lewis and Freeman, 1989). There is a clear relation, which is fitted by

$$\langle V_{\rm R}^2 \rangle_{\rm h}^{1/2} \propto L^{0.35 \pm 0.15}. \tag{7}$$

This can be easily explained in qualitative terms. Take an exponential disk with constant scaleheight, constant axis ratio of the velocity ellipsoid and a flat rotation curve. Then

$$\langle V_{\rm R}^2 \rangle_{\rm h}^{1/2} \propto \mu_0 \left(\frac{M}{L} \right) \frac{Qh}{V_{\rm rot}}. \tag{8}$$

The total luminosity is

$$L \propto \mu_0 h^2, \tag{9}$$

with μ_0 constant according to Freeman's law. The Fisher - Tully relation gives approximately

$$L \propto V_{\rm rot}^4. \tag{10}$$

Then

$$\langle V_{\rm R}^2 \rangle_{\rm h}^{1/2} \propto \left(\frac{M}{L} \right) Q L^{1/4}. \tag{11}$$

This indeed is roughly what has been observed, if M/L and Q (or at least their product) are constant between galaxies. Actual measurements of M/L and Q do indeed give similar values from galaxy to galaxy.

So the conclusion of all this is that there are indications that μ_0, Q and M/L (for the old disk) are constant from galaxy to galaxy and that the last two are also constant within disks as a function of galactocentric distance. This is remarkable and should be a strong constraint on models for disk formation and subsequent evolution. Note that a constant M/L for an old stellar component implies a constant initial mass function over the relevant stellar mass range.

4 Galaxy formation and the origin of the exponential disk

I will now discuss a simple scenario for the formation of disk galaxies. In the first place I recall the basic two-component structure of disk galaxies in their light distributions and the inference that there are two discrete epochs of star formation (see above). In NGC 7814, van der Kruit and Searle (1982b) did observe a color gradient in the spheroid and noted that the isochromes and the isophotes have similar flattening. Now models have been presented, where bulge formation proceeds by a dissipational process, in which stars form out of a relatively slowly contracting gas sphere, which is continually enriched (e.g. Larson, 1976). It seems a general property of such models, especially when rotation is included, that the isochromes are considerably flatter than the isophotes, while the inner ones are flatter than those further out. This is so because on this hypothesis a galaxy is a superposition of a chronological sequence of stellar generations in which flattening and metal abundance both increase with time of birth. The observed color gradients suggest that the binding energy of the stars is correlated with metallicity.

Van Albada (1982) has performed n-body simulations of the dissipationless collapse process of galaxy formation. He found that the resulting distributions in collapsed systems with a variety of irregular initial conditions could be fit excellently to the $R^{1/4}$-law over a region between radii containing 10 and 99% of the total mass, provided that the collapse factor is large. A second important finding was that the binding energy of the particles before and after collapse was correlated. This means that any structure in the protogalaxy, such as an abundance gradient would survive statistically in the process of dissipationless collapse and violent relaxation. This study thus indicates that bulges with their observed luminosity distributions and abundance gradients can indeed be produced without invoking dissipational processes.

Peebles (1969) has studied the possibility that protogalaxies acquire angular momentum from tidal torques from their neighbors in the early universe. He found that the amount of angular momentum predicted by this mechanism could be described by a single dimensionless parameter

$$\lambda = J\,|E|^{1/2}\,G^{-1}\,M^{-5/2}, \tag{12}$$

where J is the total angular momentum, E the total energy and M the total mass. Peebles estimated that λ would be about 0.08, while numerical experiments have shown it to be about 0.07 with a standard deviation of about 0.03. A problem was that the estimated amount of angular momentum that results from this mechanism was far too little to explain the observed rotation. This was addressed by Fall and Efstathiou (1980). They found that the requirement that tidal torques provide the angular momentum observed translates into the requirement that the mean collapse factors $R_{\rm m}/h$ (the ratio of the radius of the protogalactic cloud and the disk scalelength) are about 20 and the mean halo-to-disk mass ratio's of order 10.

Mestel (1963) had made an interesting inference from the then known rotation curve and mass distribution of the disk of the Galaxy. He calculated the distribution function of specific angular momentum and noted that it was rather similar to that of a sphere with a uniform density distribution and with uniform (angular) rotation speed. He then proposed that galaxies collapsed with detailed conservation of angular momentum and started from uniform, uniformly rotating spheres. Gunn (1982) used this hypothesis to show that if such a sphere settles in the force field of a dark halo with a flat rotation curve a roughly exponential disk results.

I will now take the following approach in my discussion of a possible scenario of disk galaxy formation (see also van der Kruit, 1987, 1990). Assume that protogalaxies form in the early universe and acquire angular momentum from tidal torques from neighbors and that these protogalaxies can roughly be described as uniform, uniformly rotating spheres. Then let the dark matter quickly settle (dissipationlessly) in a distribution that resembles an isothermal sphere. This means that the dark halos have at large R a density distribution proportional to R^{-2} and provide a potential field that at larger radii corresponds to a flat rotation curve. Early on also some stars form, especially in the inner regions, which also settle quickly thereafter in the inner bulge. These bulges will have a small amount of rotation, as is observed. If indeed this star formation is concentrated towards the central area's of the protogalaxy, it is mostly self-gravitating and will collapse as van Albada simulated in a density distribution that closely follows that of the $R^{1/4}$-law. The amount of mass that is turned into stars at this stage determines the Hubble type of the galaxy that is forming. It may have to do with details of the angular momentum distribution in the protogalaxy. An abundance gradient, that will naturally develop in the protogalaxy, will survive the collapse and result in the abundance structure observed in the bulges.

As Fall and Efstathiou, I will assume that the dark halo has the same angular momentum distribution as the protogalaxy and this means that at all positions in the protogalaxy there is an equal ratio of dark to luminous matter. The distribution of specific angular momentum (h_s) in the Mestel sphere is given by $M(h_s)/M$, which is the fraction of matter with specific angular momentum $\leq h_s$:

$$\frac{M(h_s)}{M} = 1 - \left(1 - \frac{h_s}{h_{\max}}\right)^{3/2} . \tag{13}$$

The protogalactic sphere has a density ρ_o and therefore radius $R_m = (3M/4\rho_o)^{1/3}$. The gravitational potential energy is $\Omega = -3GM^2/5R_m$ and the total angular momentum $J = 2/5Mh_{\max}$. Now suppose that at the time of the start of the collapse, when the protogalaxy detaches itself from the expanding universe and reverses its expansion into contraction, its total energy is essentially gravitational: $|E| = |\Omega|$. Then we find from Peebles' λ that

$$h_{\max} = 3.23G^{1/2}\lambda M^{1/2}R_m^{1/2} . \tag{14}$$

Now assume that a mass fraction $(1 - \Gamma)$ is in the form of dark matter. If it settles in a roughly isothermal sphere with radius R_H, its gravitational energy after collapse is in good approximation

$$\Omega_H = -\frac{GM_H^2}{R_H} = -G(1-\Gamma)^2\frac{M^2}{R_H} \tag{15}$$

and according to the virial theorem its total energy

$$E_H = \frac{\Omega}{2} = -G(1-\Gamma)^2\frac{M^2}{2R_H} . \tag{16}$$

The total energy of the protohalo is under the assumptions made of equal distribution of specific angular momentum equal to $(1 - \Gamma)$ times that of the protogalaxy or

$$E_H = -\frac{3}{5}(1-\Gamma)\,G\frac{M^2}{R_m} . \tag{17}$$

Since the dark halo collapses without dissipation of energy these two energies need to be equal and we get

$$R_H = \frac{5}{6}(1-\Gamma)\,R_m \tag{18}$$

and the asymptotic circular speed in the rotation curve is (this is of course not the rotation of the halo itself)

$$V_m^2 = \frac{GM}{R_H} = \frac{6}{5}\frac{G}{1-\Gamma}\frac{M}{R_m} . \tag{19}$$

Now let us look at the remaining material, which is gas that will eventually settle in the disk. Its mass is ΓM and its specific angular momentum distribution that of the Mestel sphere. Let this settle in a flat disk under conservation of specific angular momentum in the potential field of the dark halo. As discussed in more detail in van der Kruit (1987, 1990) it can be shown that, if we let such material settle in the force field corresponding to a flat rotation curve under detailed conservation of angular momentum, we end up with a surface density distribution that is very close to exponential and truncates at 4.5 scalelengths. This truncation corresponds to the maximum value of the specific angular momontum in the Mestel sphere ($h_{\rm max}$).

The gas that settles in the disk will quickly start to form stars. The final process of disk formation will certainly be accompanied by cloud collisions and enhanced star formation. This initial generation of disk stars will be expected to have large random motions (due to the scattering of stellar orbits by the many gas concentrations present and the quickly changing gravitational field of the collapsing disk) and intermediate metallicities. It must then settle in a thicker distribution. I suggest that this is now evident in the "thick disk" population and maybe also in Zinn's (1985) subsystem of disk globular clusters. The subsequent star formation will form the old disk population and will then naturally display the edges at 4 to 5 scalelengths.

Recently models have been proposed in which there are significant viscous effects in the disks, that redistribute the angular momentum (e.g. Yoshii and Somer-Larsen, 1989). Although such models appear to be able to explain the exponential nature of disks and conserve an abundance gradient in the gas, these have no natural way of explaining the edges or to predict at what radii these should occur.

This leaves the question of the HI gas that is observed beyond the optical edges. We can only assume that this is gas that fell in at a later stage and has been falling in from larger initial radii than the extent of the protogalaxy. There is then no requirement for it to settle in the same plane as the disk. This would explain the larger angular momentum than present in the stellar disk and also why the observed warps usually start at about the optical edges. Actually, there is no reason, why this process would not also provide gas at smaller radii, which would fit in with the observation that the HI surface density shows no feature at the optical edge.

The gas with mass ΓM of the protogalaxy will settle dissipationally in a flat, exponential disk with a scalelength $h = h_{\rm max}/4.5V_{\rm m}$. In the following I will for simplicity ignore self-gravity, but this does not affect the conclusion concerning the formation of an exponential disk. The assumption roughly corresponds to the situation, where the dark halo everywhere dominates the rotation curve. The resulting rotation curve will be approximately flat at the level of the asymptotic value $V_{\rm m}$ for the dark halo. With the equations above for $h_{\rm max}$ and $V_{\rm m}$ we can then write (using $\lambda = 0.07$)

$$h = 0.046 \frac{R_{\rm m}}{(1-\Gamma)^{1/2}}. \tag{20}$$

So, as Fall and Efstathiou we do expect collapse factors of order 20 or so, if Γ is small. The central surface density of the disk is

$$\sigma_\circ = 197 \frac{\Gamma}{1-\Gamma} \rho_\circ^{2/3} M^{1/3}. \tag{21}$$

The initial density ρ_o is the density of the protogalaxy when it detaches itself from the rest of the universe. If galaxies form at about the same time, this density may not vary by large factors. Indeed in the discussion of Fall (1979) on density perturbations for hierarchical clustering, the density spectrum in a critical universe $\Delta\rho/\rho$ is proportional to a low negative power of M and therefore the dependence of σ_o on ρ_o and M almost disappears. If Γ is constant for all galaxies, we recover Freeman's law of constant central surface brightness and even find an indication that dwarf systems should on average be somewhat fainter than spirals of types Sa to Sc (which in general are non-dwarfs). This says that everywhere in the early universe we should have had equal ratio's of dark and luminous matter and that this still applies between present-day galaxies.

The luminosity of the disk is

$$L_{\rm disk} \propto (L/M)\Gamma^2(1-\Gamma)\frac{V_{\rm m}^4}{\mu_o}. \tag{22}$$

So, for constant M/L, Γ and central surface brightness μ_o we get the Tully-Fisher relation.

It should be stressed that the scenario given here is very schematic; yet it qualitatively explains the general features that we know about spiral galaxies.

5 Conclusions

Observations show the following general characteristics of galactic disks:

- Disks are exponential with an "edge" at about 4.5 scalelengths and have a constant thickness with radius.
- The observed central surface brightness has a preferred value, probably also in the near-IR.
- There are no systematic radial color gradients in disks.
- (M/L) for the old disk population and Q (corrected for effects of the gas) are constant with radius.
- Actual measurements show $(M/L) \approx 6 \pm 2$ in B *(old disk population)* and $Q_* \approx 1.5 - 2$.
- The stellar velocity dispersion correlates with luminosity, such that Q and (M/L) are also constant from galaxy to galaxy.

Possible explanations and deductions from this are:

- The exponential nature of disks and the "edge" can be explained by a collapse of the disk with detailed conservation of angular momentum.
- The constant scaleheight follows from a dynamical heating of the disk to a value of Q somewhat larger than 1 at all radii.
- Freeman's law follows if all galaxies have the same ratio of dark to luminous matter.
- The question of whether or not disks are optically thick has not yet been satisfactorily answered.

References

Albada, T.S. van: 1982, *Mon. Not. R. Astr. Soc.* **201,** 939.

Bottema, R.: 1988, *Astron. Astrophys.* **197,** 105.

Bottema, R.: 1990, *Astron. Astrophys.* **233,** 372.

Broek, A.C. van den: 1990, *A Study of Extreme IRAS Galaxies*, Ph.D. Thesis, Univ. of Amsterdam, p.153.

Davies, J.I.: 1990, *Mon. Not. R. Astr. Soc.* **224,** 8.

Disney, M.J., Davies, J., Phillipps, S.: 1989, *Mon. Not. R. Astr. Soc.* **239,** 939.

Disney, M.J., Phillipps, S.:1983, *Mon. Not. R. Astr. Soc.* **205,** 1253.

Elmegreen, D.M., Elmegreen, B.G.: 1984, *Astrophys. J. Suppl. Ser.* **54,** 127.

Fall, S.M.: 1979, *Rev. Mod. Phys.* **51,** 21.

Fall, S.M., Efstathiou, G.: 1980, *Mon. Not. R. Astr. Soc.* **193,** 811.

Freeman, K.C.: 1970, *Astrophys. J.* **160,** 811.

Giovanardi, C., Hunt, L.K.: 1988, *Astron. J.* **95,** 408.

Kruit, P.C. van der: 1979, *Astron. Astrophys. Suppl. Ser.* **38,** 15.

Kruit, P.C. van der: 1987, *Astron. Astrophys.* **173,** 59.

Kruit, P.C. van der: 1988, *Astron. Astrophys.* **192,** 117.

Kruit, P.C. van der: 1990, in: *The Milky Way as a Galaxy*, 19th Saas-fee course by G. Gilmore, I.R. King and P.C. van der Kruit (Geneva Observatory and University Science Books).

Kruit, P.C. van der, Freeman, K.C.: 1986, *Astrophys. J.* **303,** 556.

Kruit, P.C. van der, Searle, L.: 1981a, *Astron. Astrophys.* **95,** 105.

Kruit, P.C. van der, Searle, L.: 1981b, *Astron. Astrophys.* **95,** 116.

Kruit, P.C. van der, Searle, L.: 1982a, *Astron. Astrophys.* **110,** 61.

Kruit, P.C. van der, Searle, L.: 1982b, *Astron. Astrophys.* **110,** 79.

Larson, R.B.: 1976, *Mon. Not. R. Astr. Soc.* **176,** 31.

Lewis, J.R., Freeman, K.C.: 1989, *Astron. J.* **97,** 139.

Mestel, L.: 1963, *Mon. Not. R. Astr. Soc.* **126,** 553.

Peebles, P.J.E.: 1969, *Astron. Astrophys.* **11,** 377.

Phillipps, S., Disney, M.J., Kibblewhite, E.J., Cawson, M.G.M.: 1987, *Mon. Not. R. Astr. Soc.* **229,** 505.

Rubin, V.C., Ford, W.K.: 1983, *Astrophys. J.* **271,** 556.

Toomre, A.: 1964, *Astrophys. J.* bf 139, 1217.

Valentijn, E.A.: 1990, *Nature* **346,** 153.

Vaucouleurs, G. de: 1959, *Handbuch der Physik* **53,** 511.

Wainscoat, R.J., Freeman, K. and Hyland, A.H.: 1990, *Astrophys. J.* **348,** 85.

Wevers, B.H.M.R., Kruit, P.C. van der, Allen, R.J.: 1986, *Astron. Astrophys. Suppl. Ser.* **66,** 505.

Yoshii, Y., Sommer-Larsen, J.: 1989, *Mon. Not. R. Astr. Soc.* **236,** 779.

Zinn, R.: 1985, *Astrophys. J.* **293,** 424.

Discussion

D. Burstein - As I will show in my talk on Thursday, E. Valentijn is correct in saying that spirals have optically-thick disks, but for the wrong reasons.

P. van der Kruit - I eagerly await your talk.

D. Burstein - Optically-thick means on average, not necessarily uniformly. Just as an exponential disk of a spiral is derived from an azimuthal-averaged light distribution, optical-depth should also be averaged azimuthally. Patchy dust distribution would permit galaxies to have regions of low extinction. We live in such a region of low extinction in our own galaxy. If we didn't, I doubt whether this kind of meeting could be held!
P. van der Kruit - There is no question that optically thin *and* optically thick parts of disks exist. From looking at galaxies through disks such as NGC 450 it is possible to see what part of the disks will be optically thick, and such systems (also NGC 45) suggest that it is a small fraction. Whatever is true though, Valentijn's absence of correlation of central surface brightness demands a dust layer as thick as the stellar disk.

S. Kent - Although the vertical distribution of stars in disks is often assumed to be of a $sech^2$ form, there have been claims made that the distribution is actually closer to a pure exponential form. What are your views on the subject?
P. van der Kruit - The $sech^2$ was introduced to describe old disk populations in edge-on galaxies above the dust lane where is was noted that in the solar neighborhood stars of age a few Gyr or more all have roughly the same velocity dispersion. The $sech^2$ corresponds to the isothermal sheet and it was clear from the start that it would not hold at small z; also note that it was a model for the old disk population only, and therefore to describe the mass distribution of a uniform population. At low z there must be excess light from younger populations (ignoring the dust) and due to extra gravity also the old disk population must be more peaked. I concluded in a recent paper (1988) that something in between an exponential and the $sech^2$ would be appropriate. This is consistent with the data we have at K-band.

M. Capaccioli - Piet, you notice that, if μ_0, M/L, and V_r are roughly costant for all spirals, then mass and angular momentum will depend on one parameter only, the disk scale length, for instance. Isn't this embarrassing?
P. van der Kruit - This is true, but it is not really V_r that is constant between galaxies, but rather Q. It seems to me that in a first approximation the total mass is a parameter that will determine most of the resulting properties. However, in the early universe the mass of the protogalaxy is not the only parameter, as for each mass there will be a range of dimensions (or equivalently density). In the picture of tidal origin of angular momentum there will again be a range of angular momentum for each mass. So there will be total mass as one parameter which will be reflected in the disk scale length, but this scale length will also depend on the total angular momentum. So I doubt that all this means that spirals are strictly a one-parameter family

E. Athanassoula - You told us that the Q of the stars is constant within each galaxy and from galaxy to galaxy. Since the gas percentage varies significantly from galaxy to galaxy, should we not infer that the mean Q (stars and gas) should change significantly from one galaxy to the other? In that case is there any reason to believe that the mean Q (stars and gas) you derived for our Galaxy is representative for disc galaxies?
P. van der Kruit - I used the galaxy really as an example to show that the real Q is likely to be roughly constant when that derived from the stellar component is not. Also the derived values I quoted are all derived around one to two scale length from the center

where little gas is present. So it is a stellar value that should be corrected for gas, but at these radii the correction itself and certainly its variation between galaxies is smaller that the uncertainty or range 1.5–2.0 that I quoted.

F. Matteucci - It looks a bit puzzling to me that spiral galaxies which show clear abundance gradients do not have color gradients. On the other hand, elliptical galaxies show color gradients and these are interpreted as abundance gradients. Do you have any comment about that?

P. van der Kruit - There are two points: in the first place the abundance gradient are seen in the gas! As all models of chemical evolution show the *mean* stellar abundance is always close to the (effective) yield and in spite of the observed gaseous gradient a much smaller one is predicted in the stars. The gradients in disks are already small in the gas (when translated into colors) and will be even smaller in integrated starlight. The second point is that disk light is a mixture of ages and the color is more sensitive to age distributions than to mean abundance. It is only in systems with no or very little young stars, such as bulges or ellipticals that color can be used to estimate (or guess) abundances.

Piet Van der Kruit chasing a missing transparency.

BRIGHTEST CLUSTER MEMBERS

J. M. Schombert
Department of Astronomy
University of Michigan, USA

ABSTRACT. A review of the structural properties of BCM's in poor and rich clusters is presented in order to define the morphological types gE, D and cD in ellipticals. Although the difference between bright field ellipticals and BCM's is subtle (primarily an enlargement in characteristic radius for BCM's) it is quantifiable on the fundamental plane for ellipticals. Various indirect arguments, plus direct comparison to N-body simulations, demonstrate that these structural differences are the result of a past history of early mergers plus late accretions of lesser clusters members. Although present day tidal stripping is evident in some clusters, cD envelopes must have their origin before cluster virialization.

1 Introduction

The purpose of this conference is to debate the usefulness of the morphological classification of galaxies and whether it should be replaced by a scheme based on the physical properties of galaxy types. In either case, one example of where morphological classification has been very successful is in the understanding of first-ranked ellipticals in clusters, i.e. the brightest cluster members (BCM's). In this incidence, I define successful in the sense that the morphological classification of BCM's alerted the astronomical community of their special nature and, with further study, led to a substantial contribution to our understanding of dynamical evolution. In this review I will summarize the morphological classes of BCM's, outline the structural meaning of these classes, particularly in differentiating the interior versus halo properties of BCM's, and demonstrate that the interior properties of BCM's are due to a history of mergers, while the extended envelopes are primordial in origin.

BCM's are extreme examples of a class of very homogeneous objects, elliptical galaxies. They are extreme in their luminosities ($L > 10^{11} L_\odot$, $H_o = 100$ km sec^{-1} Mpc^{-1} assumed throughout this review), being the brightest objects in the Universe that emit light strictly from stellar photospheres. They are extreme in their absolute sizes with some cD galaxies measuring over 1.0 Mpc in radius (Oemler, 1976, Schombert, 1984). They are also extreme in the environment in which they reside, the cores of rich clusters. In modern astrophysics it is generally perceived that special objects imply a special process of formation or evolution, and the process most often invoked is one of growth of BCM's from the cannibalism of lesser galaxies by dynamical friction. Note that there

G. Longo et al. (eds.), Morphological and Physical Classification of Galaxies, 53–68.

are certain dynamical reasons to believe this effect would *not* be strong in present-day clusters since, although the cross sections for galaxy interactions are high in the cores of a rich cluster, those encounters are fast and not typical of the bound, merging orbits (see Merritt, 1985).

2 BCM morphological types

The most extensive morphological classification of BCM's is from the catalogs of Struble and Rood (1984). Within these catalogs BCM's are represented in a variety of forms; gE (giant elliptical), B (binary galaxy), B_p (peculiar binary, usually relating to a common envelope), cD (a gE with evidence of a diffuse and extended halo) and cD_n (a "nested" cD of several multiple nuclei imbedded in the primary's core). Sketches of several examples are shown in figure 1, taken from Struble and Rood (1984).

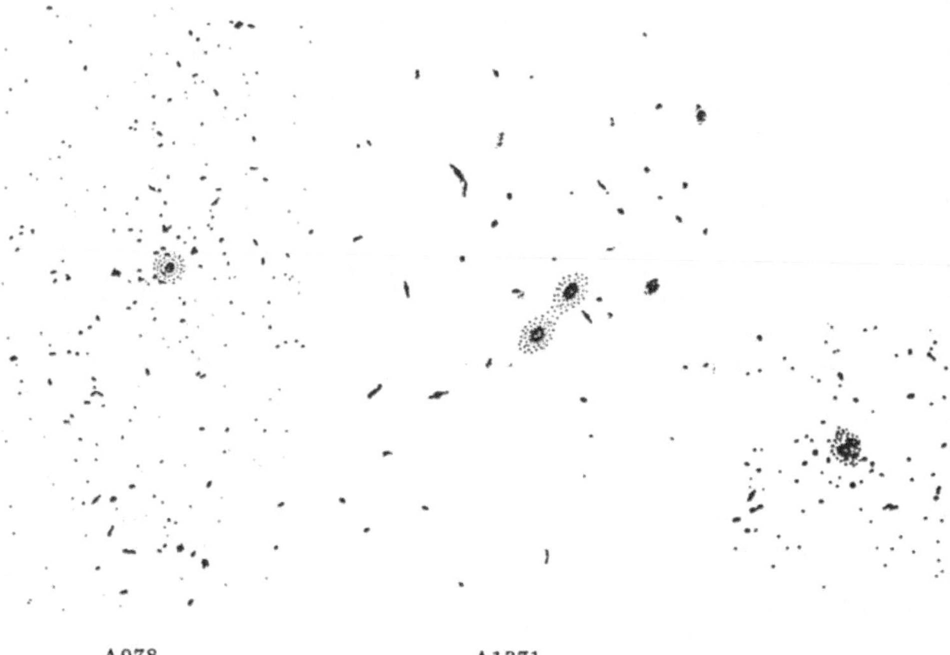

Figure 1. Examples of BCM's from Struble and Rood (1984), A978 is a flat cluster (F-type) with a D galaxy, A1371 is a B_p cluster and A1927 is an example of a nested cD galaxy.

Although BCM's come in a variety of flavors, a majority are of the plain cD or the multi-nuclei cD variety where the embedded companions are sufficiently small enough to be ignored for structural analysis. For the rest of this review, I will concentrate on the parent body or primary mass concentration of BCM's, however, one should not forget that a significant fraction of BCM's are of the dumbbell variety (e.g. A400) where the companion is of comparable size and luminosity. An excellent study of the binary BCM's was performed by Valentijn and Casertano (1988).

It is instructive to remember that classification of BCM's is usually performed by visual examination of wide-field photographic plates (e.g glass copies of the Palomar Sky Survey), involving the material from a non-linear detector using a logarithmic device. This places a high weight, in the classification, on the mid-regions of an elliptical's halo. The centers are usually too high in density with little contrast to the eye and the faint, outer envelopes are below eye detection. Thus, it is not too surprising, as detailed in the following section, that a BCM's classification is strongly influenced by the slope of the radial luminosity distribution at 5 to 10% of sky brightness (Malumuth, 1983).

3 Classification by surface photometry

Upon examining luminosity distributions, it is found that BCM's are generally distributed into three morphological types; gE, D and cD. A normal elliptical is represented by a smooth manifold of surface photometric profiles. On the low mass end of the sequence these profiles are exponential in shape (Binggeli, Sandage and Tarenghi, 1984). On the high mass end these profiles become approximately $r^{1/4}$ shaped. The transition from exponential to power-law is smooth and an interested reader can view template profiles for ellipticals in figure 1 of Schombert (1987). At the extreme end in luminosity of the normal elliptical sequence are, by my nomenclature, giant ellipticals. They are a natural extension of the elliptical sequence, being roughly $r^{1/4}$ in appearance, and lie on the same fundamental plane of structural and kinematic properties (Djorgovski and Davis 1987) as normal ellipticals. I simply delineate them from other ellipticals by their size ($R \approx 75$ to 100 kpc) and magnitude ($M_V < -22$). On the other hand, the D class galaxies, similar in luminosity, size and mean surface brightness as gE's, display special properties with respect to normal ellipticals on the fundamental plane. For example, their surface brightness profiles are more extended at a characteristic radius than normal ellipticals causing a shallower profile slope (see below). The cD class is recognized by D-like interiors plus a large, low surface brightness ($\overline{\mu}$=26 to 27 mag arcsec^{-2}) envelope (Oemler, 1976). All three types are shown in figure 2 along with a intermediate luminosity elliptical. The manifold of profiles for gE and D type galaxies overlap; however, the brightest D galaxies are more luminous than the brightest gE types.

Individual examples of surface brightness profiles of all these morphological classes can be found in figure 2 of Schombert (1987). Most notable of these examples are some "classical" cD galaxies, such as the BCM in A2029, which, although being a very large D galaxy, does not have the extended envelopes that I associate with the cD class. Extreme cD examples are seen with the BCM's in A1413 ($L_{env} = 7 \times 10^{11} L_\odot$) and the southern cluster Shapley 8 ($L_{env} = 2 \times 10^{12} L_\odot$). Homology merger theory (Hausman and Ostriker, 1978) predicts that cD galaxies should have depressed central surface brightnesses; however, as noted by Oemler (1976), most D and cD galaxies have high central surface brightnesses (central refers to the inner 2 kpc, not a core value). Nonetheless, there does exist a subsample of cD galaxies with depressed central surface brightness (e.g., A85 and NGC 6166 in Schombert, 1987). These objects are rare, but it is interesting to note that most are associated with emission lines, IRAS emission and other evidence of recent star formation in their cores.

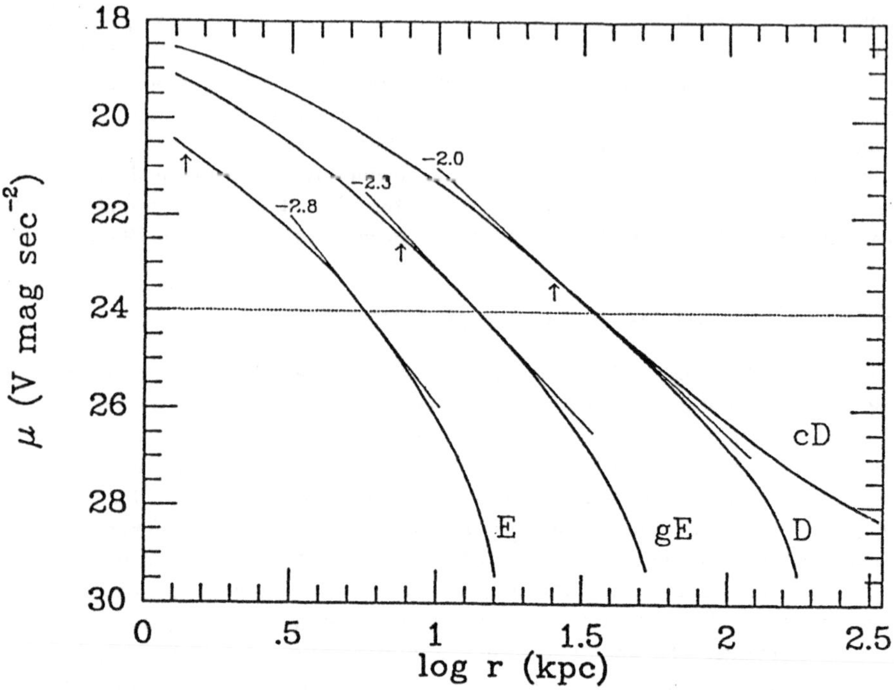

Figure 2. Surface brightness profiles for the four classes of BCM's, normal elliptical, giant elliptical, D and cD. The dotted line indicates the region of a profile most strongly weighted in visual classification.

Several characteristics are common to BCM's evidenced through comparison of their surface brightness types. The first is that almost all BCM's are of the D or cD class. This is true for poor clusters (Thuan and Romanishin, 1981) as well as rich clusters (Malumuth, 1983, Schombert, 1987). A second point is that cD envelopes are not detected by eye (most being below 1% of sky brightness). The "diffuseness" seen for BCM's is not the extended cD envelope, but rather the slope of the profile around the 24 V mag arcsec^{-2} level. This is an important distinction since I reserve the cD class for objects with extended envelopes only, a matter for deep surface photometry, and that the "diffuseness" is a characteristic independent of the existence of a cD envelope. This is seen in figure 3, where the morphological type by surface photometry is compared to the morphological type as determined by visual classification of Struble and Rood (1984). What Struble and Rood would call a cD galaxy is equally divided into the D and cD classes meaning their definition of cD had nothing to do with extended envelopes, but rather depended on profile slope. Fortunately, as I will prove in the following section, this shallow profile is a merger signature and, therefore, the cD classification by Struble and Rood still has merit with respect to the galaxy type. Also, very few gE's are called cD by Struble and Rood. Sometimes, as in NGC 6034, cD galaxies are misidentified as S0 galaxies, probably due to a high elongation combined with a shallow profile which simulates the appearance of a disk. Lastly, not all D or cD galaxies are BCM's; however, no D or cD type galaxies have been found in the field despite an extended search of bright field ellipticals from redshift surveys (Schombert, 1987) and all are positioned on local cluster density enhancements (Beers and Geller, 1983).

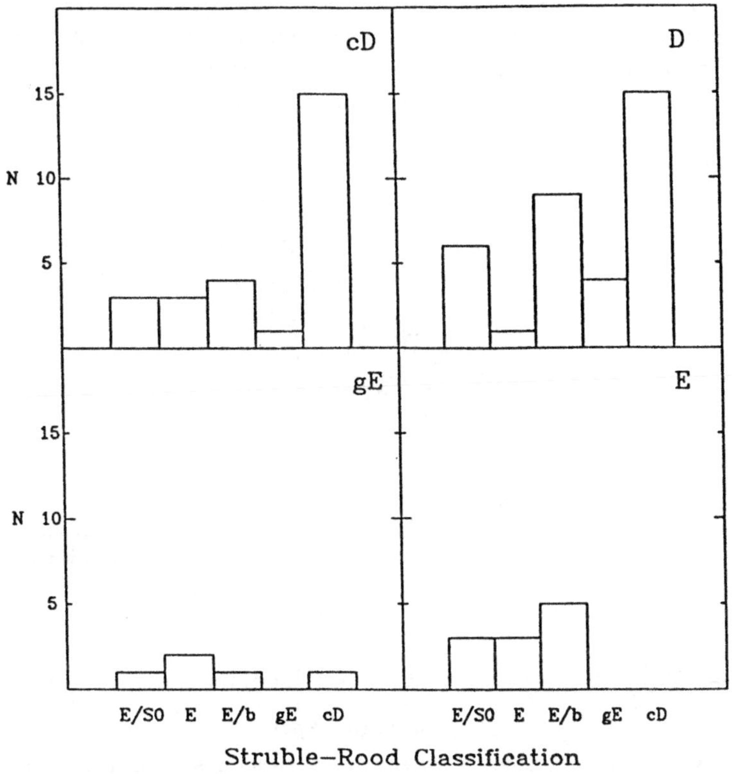

Figure 3. Comparison of morphology from Struble and Rood versus classification based on surface photometry profiles as shown in figure 1.

4 Are BCM's an extension of the normal elliptical sequence?

In the next two sections I will address two questions. One concerns the interior properties of BCM's and the second concerns the origin of cD halos. The first question comes down to the problem of whether BCM's are a simple extension of the elliptical sequence, in terms of luminosity and structural properties, or whether they are special objects whose formation or evolution is due to some external factor, such as mergers or accretions. At first glance, the naive answer to this question is that BCM's have many properties in common with normal ellipticals. The profiles of D galaxies and the interiors of cD galaxies are basically $r^{1/4}$ in shape and there is no discontinuity in luminosity from gE and D type galaxies. The diffuse appearance of BCM's, as compared to gE's, is a difficult attribute to quantify. On the other hand, BCM's exhibit luminosity characteristics that imply a two population interpretation (Schneider, Gunn and Hoessel, 1983, Bhavsar, 1989) as well as external differences from other galaxies, such as their location on density enhancements beyond a mere morphology-density relationship (Beers and Geller, 1983).

A crude technique, but still enlightening, is a test of three structural properties: mass (aperture luminosity), scale length (characteristic radius) and density (surface brightness), three of the four segments of the fundamental plane. These three diagrams are reproduced in figures 4, 5 and 6 from the work of Schombert (1987).

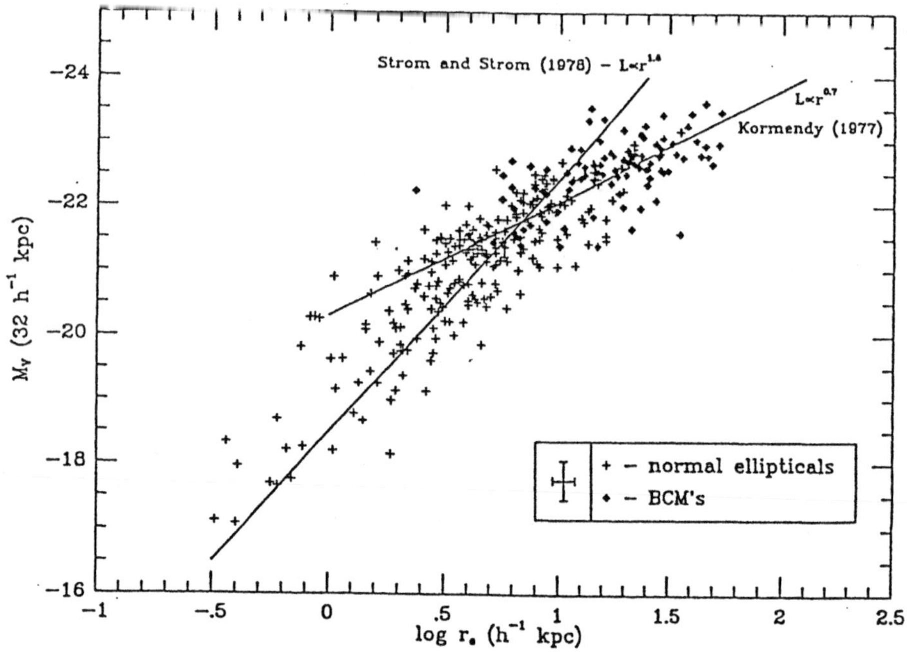

Figure 4. Aperture magnitude versus effective radius for normal ellipticals and BCM's. The two lines are relationships taken from early work on ellipticals.

Figure 4 displays a plot of aperture magnitude (M_{16kpc}) versus a characteristic scale length, in this case the half-light radius, r_e, from $r^{1/4}$ fits. Although BCM's appear to follow a different relationship from normal ellipticals, closer inspection of this diagram shows that the bright field ellipticals also deviate from the $L \propto r^{1.6}$ line above $M_V = -21.5$. I interpret this to mean that the kinematic properties of bright ellipticals (rotators below $M_V = -21.5$, and velocity supported above) are reflected in the structural relationship of mass and scale length. In figure 5, mass versus density is represented by magnitude and an inner metric surface brightness (μ at the 2 kpc radius). Similarly in this diagram, BCM's are not strongly deviant, although the majority appear overly bright for their inner luminosity densities. On the other hand, the special nature of BCM's can clearly be seen in a plot of density (effective surface brightness, μ_e) versus scale length (r_e again) in figure 6. Here 87% of the BCM's lie above the normal sequence, where even the brightest gE galaxies follow the relationship established by Kormendy (1980). I consider this diagram to be the strongest evidence for structural difference between BCM's and normal ellipticals. And, I interpret this difference to be that BCM's are enlarged (have larger effective radii) for a given surface brightness (rather than higher luminosity density for a given radii) which is also consistent with a "diffuse" appearance from morphology. Thus, D and cD galaxies are *not* in the same sequence as the elliptical template profiles from Schombert (1987), although they are very similar with BCM's being slightly shallower in profile slope than gE of the same scale length or magnitude. Lastly, a recent analysis of the fourth leg of the fundamental plane (velocity dispersion) for BCM's and field ellipticals (Djorgovski et al., 1990), from Malumuth and Kirshner (1985) data, confirms the above conclusions finding that BCM's follow different scaling relations.

I have argued in previous papers that these differences on the fundamental plane are exactly what is predicted from N-body merger and accretion simulations. I am particularly influenced by the series of papers by Farouki et al., (1983) for equal mass mergers and hierarchical accretions in which they publish surface density profiles for their remnant objects after each stage of mergers. Sifting through these models and others in the literature (Villumsen, 1982, White, 1982, May and van Albada, 1984) has led me to propose the following scenario for BCM formation and evolution. At the time of galaxy formation there were a range of protogalactic lumps from dwarf sized to 1 or 2 L_* sized bodies formed by a dissipative collapse processes (see Larson, 1990 for a review of these galaxy building processes). Immediately after this early epoch, all objects which will become bright ellipticals (either field or cluster members) are formed as the product of hierarchical mergers between these early lumps (White and Rees, 1978, Tinsley and Larson, 1979). One argument for this particular origin to all bright ellipticals is shown in figure 4 where the transition from exponential to $r^{1/4}$ shaped profiles occurs at the same point where there is a break in the M_{16kpc} vs r_e diagram, figure 4. May and van Albada (1984) argue that a $r^{1/4}$ profile is a natural result from merging and the simulations of Farouki et al., also note the tendency for an $r^{1/4}$ profile to form in the remnant. However, to maintain the mass-metallicity relation and the old stellar population observed today, this merging must be at or before the epoch of first star formation. After this early merger epoch, the field and cluster population diverge as the cluster ellipticals continue the process of cannibalizing their neighbors. The field ellipticals, on the other hand, have no neighbors from which to continue this process and, thus, promotes the expectation that cluster ellipticals will depart from the field population only in the degree of dynamical evolution. At this stage, seed cluster ellipticals (ones which will become the BCM's) are located in small subclusters which will later coalesce to form a rich cluster. Within this subgroup, the relative velocities are low and dynamical friction speeds the accretion process. However, unlike hierarchical mergers which led to equalpartition of internal energy and loss of identity of the two original components, later accreting objects will, in all probability, be smaller in mass. Their fate is such that, as they fall into the BCM their outer envelopes are disrupted, placing their stellar material in orbit near the effective radii of BCM's. The remaining small, tightly bound core survives to reach the center of the BCM and increase its central surface brightness (White, 1982). After a crossing time (the orbital period at the effective radius), the cannibalized material redistributes into a smooth core and halo $r^{1/4}$ appearance. The result is that the BCM grows mostly in size due to the fact that the accreting material is located in the outer regions (diffuseness).

There is little change in the core surface brightness, relate to other bright ellipticals of the *same* size or aperture magnitude, while a $r^{1/4}$ shape is maintained.

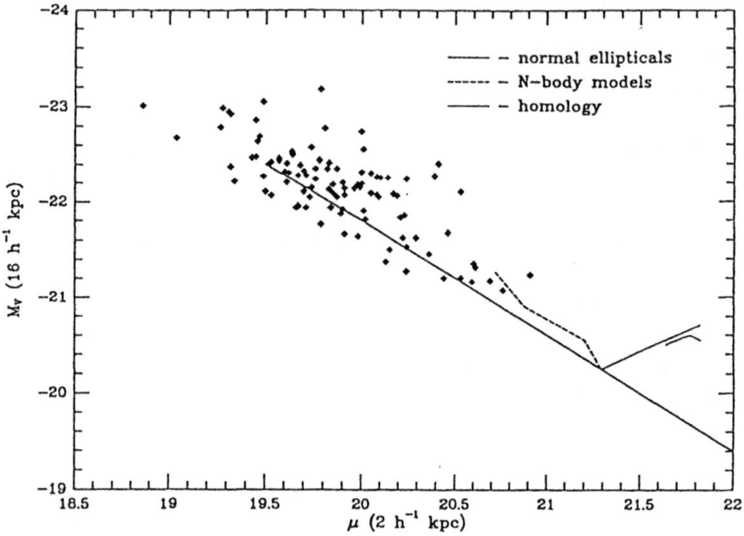

Figure 5. Aperture magnitude versus surface brightness for BCM's. BCM's tend to scatter above the relationship for normal ellipticals.

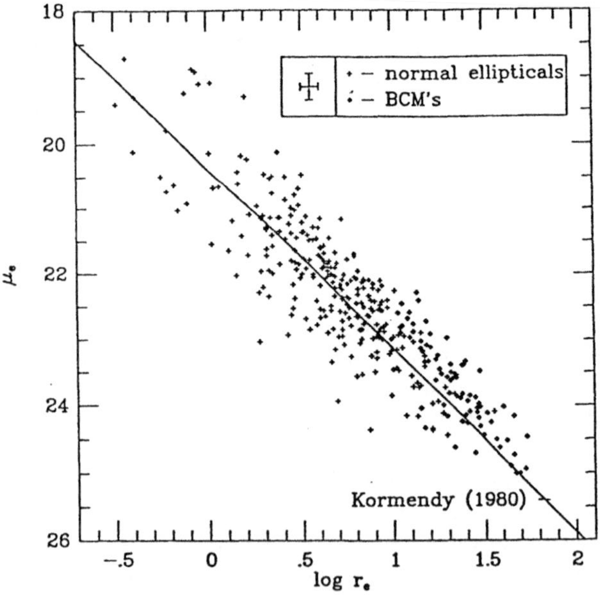

Figure 6. Surface brightness versus effective radius for normal ellipticals and BCM's. Most BCM's lie to the right of the normal elliptical relationship as defined by Kormendy (1980).

An interesting special case is if the victim galaxy is not centrally concentrated (e.g. a disk galaxy). In this scenario, the BCM consumes the object yet no core exists to increase the central density of the BCM. The result is a low central surface brightness BCM with dissipating gas collecting in the core, which would be visible as emission gas, if heated by infall, or compressed to ignite star formation. These are precisely the characteristics exhibited by A85 and NGC 6166, our two best examples of low central surface brightness cD's.

5 Are cD envelopes primordial in origin?

The shape of cD envelopes is most easily understood by a simple two component model of a galaxy potential on top of a more extended cluster potential indicating that envelopes surrounding cD galaxies are structurally independent of the underlying galaxy. Figure 7 shows one attempt to model the envelope of A2670 (Schombert, 1988).

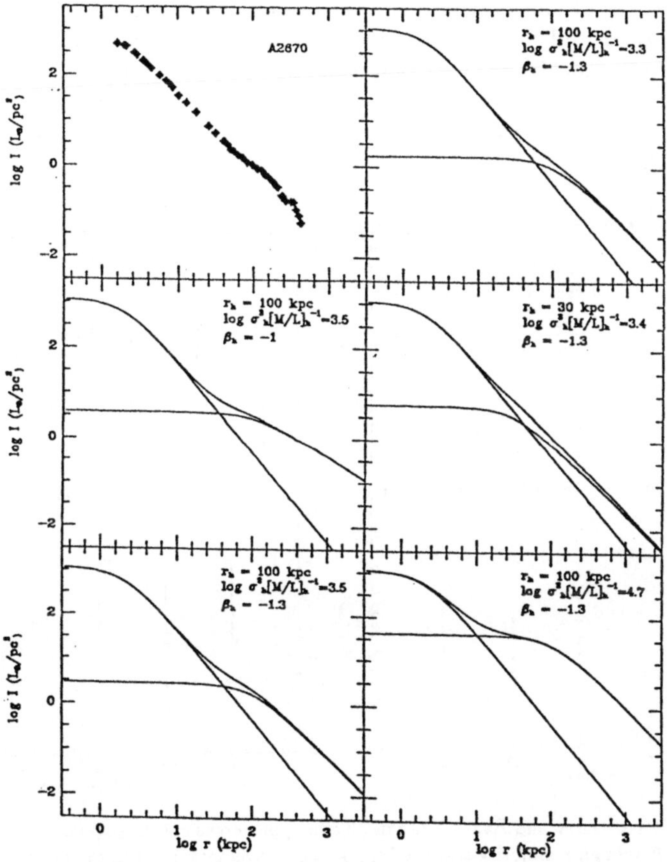

Figure 7. A two component cD envelope model for A2670. Various values of halo radius and velocity dispersion, as well as halo M/L and power-law slope β, are shown.

The major limitations to this method are an unresolved cluster core value, r_h, and lack of velocity dispersion data on cD envelopes. However, adequate models are produced from typical cluster core radii and velocity dispersions, lending support to the idea that cD envelopes are a "sea" of stars bound not to the underlying galaxy, but to the cluster potential as a whole. The underlying galaxy is simply at rest with respect to the cluster's potential and benefits from its unique position with the adhesion of the surrounding cluster light as a cD envelope. This argues that we should decouple the properties of cD envelopes from the underlying galaxy. The discovery of cD envelopes either independent of an underlying galaxy (cluster light) or with non-concentric isophotes (such as the starpile in A545, Struble, 1988) supports this idea.

Since cD envelopes are cluster-sized entities, it would not be surprising to find correlations between envelope properties, such as occurrence and luminosity, and global cluster properties, such as cluster morphological type. However, figure 8 shows that this is not as pronounced as expected. Both the occurrence and luminosity of cD envelopes are shown with respect to cluster Rood-Sastry type (a measure of the dynamical state of the cluster, where cD and B type cluster are assumed to be highly evolved and L/F/I clusters are irregular and in a pre-collapse state) and Bautz-Morgan type (a measure of the dominance of the BCM, also reflecting the dynamical state of a cluster). Although there is a tendency for more and brighter cD envelopes to occur in advanced cluster types, there are still significant numbers of envelopes in unevolved clusters with comparable brightnesses.

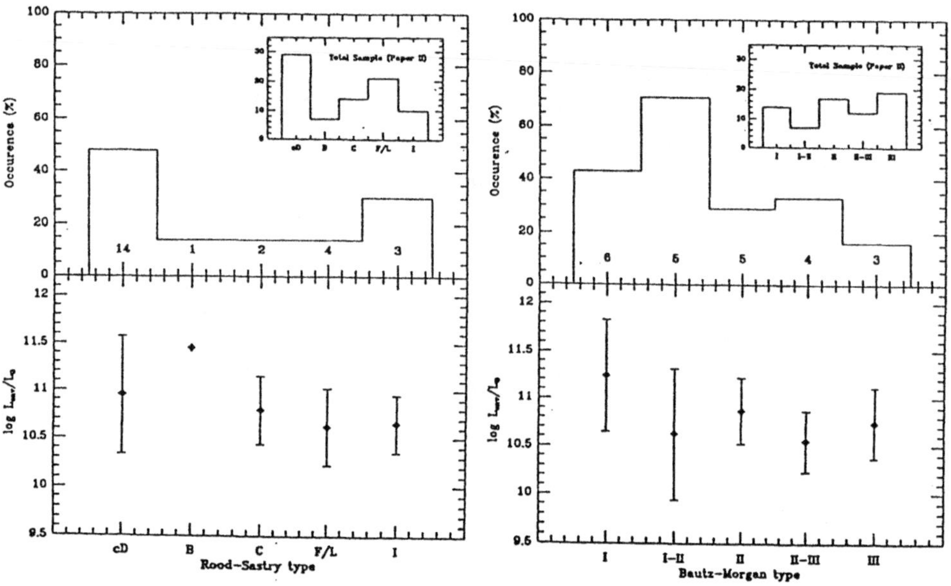

Figure 8. Rood-Sastry and Bautz-Morgan cluster type versus cD envelope occurrence and luminosity. Although a majority and the brightest cD envelopes occur in the evolved cluster types, a significant number exist in unevolved clusters.

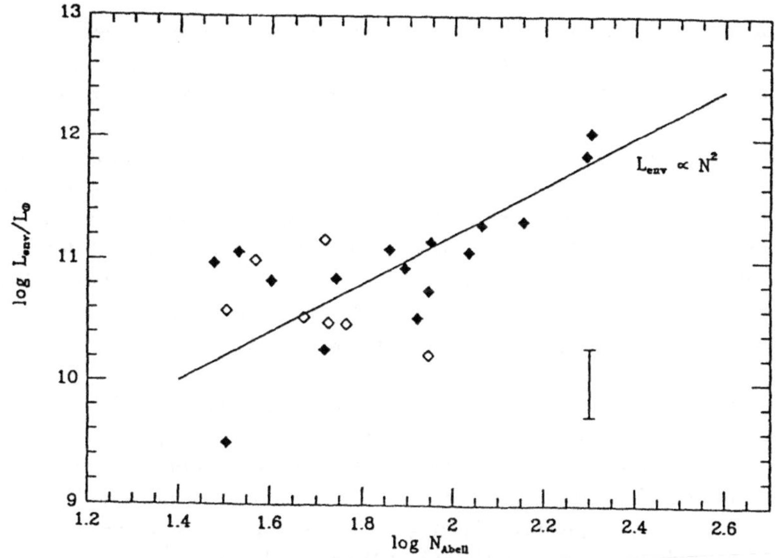

Figure 9. Cluster richness versus cD envelope luminosity.

Comparing the physical properties of clusters to cD envelopes is even less edifying; the correlations are weak to non-existent. In figures 9 and 10, the cD envelope luminosity is plotted against two cluster properties, richness and total X-ray luminosity. If the stripping of cluster galaxies by the mean tidal field is the source of light for cD envelopes, then one would expect a direct correspondence between number of cluster members and envelope luminosity. Figure 9 shows there is a relationship (a correlation coefficient of 0.72), similar to the $L_{env} \propto N^2$ trend predicted by Malumuth and Richstone (1984), although weak. An equivalent weak relation exist between X-ray and envelope luminosity as shown in figure 10.

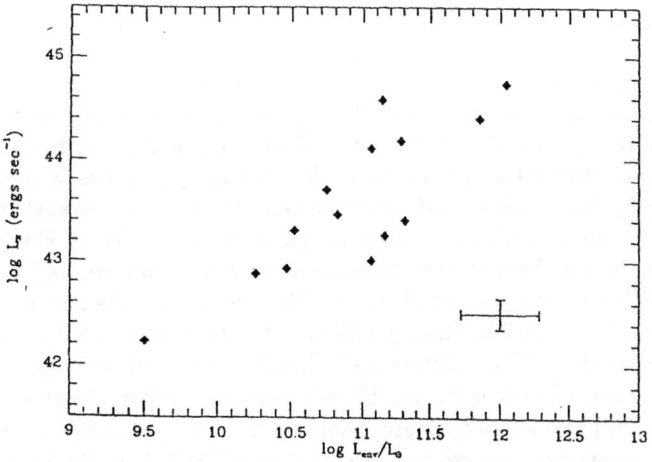

Figure 10. X-ray luminosity versus cD envelope luminosity.

Additional evidence as to the origin of cD envelopes comes from inspection of their broadband colors (i.e. a measure of the dominant stellar population). Although color information at the surface brightnesses typical for cD envelopes is extremely difficult, requiring long integrations with CCD's that have excellent flattening characteristics, I have examined several cD galaxies with various filter schemes and have concluded that no color gradients are visible through the transition from galaxy to envelope light, and the mean colors are the same as an old elliptical-like population. This is somewhat surprising in view of the fact that Bothun and Schombert (1990) found evidence for truncated galaxies in the present-day cD clusters. Even if most of these truncated galaxies are ellipticals, one would expect, from the color-magnitude relation for galaxies ($B-V=0.9$ for low luminosity ellipticals) plus any blue disk material stripped from spiral, a minimal change of 0.1 to 0.2 $B-V$ gradients.

It is difficult to draw a strong conclusion of dynamical origin to cD envelopes while lacking clear cluster correlations. Combined with the stellar population information, I believe that the formation of cD envelopes was an early process, from one of two scenarios: (1) the envelopes have their origin in tidal stripping of proto-galactic material of subcluster members or (2) as Merritt suggests, the envelope is a remnant of the special location of the BCM; all galaxies initially had large envelopes, but were stripped by the cluster mean tidal field *except* the BCM's. This latter view is attractive since then the parallel evolution of mergers for the BCM's interior and stripping for envelopes both reflect, to some degree the richness and mass concentration of the initial subcluster and, thus, are partially reflecting in present-day global cluster correlations. However, this scenario would also predict that all BCM's should be cD's and does not explain where all the excess material from early tidal stripping is today (e.g., not in the form of stars). In addition, Bothun and Schombert (1990) have shown strong evidence of ongoing dynamical friction and tidal stripping in present day clusters. The amount of material available may be insufficient to form cD envelopes, which would further support an earlier epoch for envelope formation.

6 Summary

The morphological classification of BCM's, when augmented with quantitative surface photometry, leads to the conclusion that BCM's are special objects, beyond being mere extrapolations of the luminosity function. Morphology paved the way for an examination of these objects and has led to an interpretation of their properties within the context of dynamical evolution, mergers and cannibalism. When the structural information is combined with kinematics of the underlying galaxies, dynamics of cluster cores and color information about the component stellar populations, we are persuaded that the evolution of BCM's has two primary elements. The first is that the interior properties of BCM's are determined by a long history of hierarchical merging (i.e. the cannibalism of smaller cluster members). The "diffuseness" of BCM's and their enlarged characteristic radii are all signatures of past mergers. These conclusions define the class of D galaxies and are drawn from parameters that are independent of the cD envelopes found around some BCM's. The second process in the evolution of BCM's is the occurrence of cD envelopes, a feature in BCM's which is invisible to the eye and, thus, morphological classification. All evidence points to the origin of cD envelopes as primordial, before cluster collapse, at the same epoch of star formation that produced the populations in galaxies. Whether all galaxies had cD envelopes to be later stripped off by the cluster mean tidal field, or whether they developed from stripping other cluster members in local subgroups before cluster collapse, remains unknown.

References

Beers, T.C. and Geller, M.J.: 1983, *Astrophys. J.*, **274**, 491.
Bhavsar, S.P.: 1989, *Astrophys. J.*, **338**, 718.
Binggeli, B., Sandage, A. and Tarenghi, M.: 1984, *Astron. J.*, **89**, 64.
Bothun, G.D. and Schombert, J.M.: 1990, *Astrophys. J.*, **360**, 436.
Djorgovski, S.G., de Carvalho, R.R., Shlosman, I. and Schombert, J.: 1990, preprint.
Djorgovski, S. and Davis, M.: 1987, *Astrophys. J.*, **313**, 59.
Farouki, R.T., Shapiro, S.L. and Duncan, M.J.: 1983, *Astrophys. J.*, **265**, 597.
Hausman, M.A. and Ostriker, J.P.: 1978, *Astrophys. J.*, **224**, 320.
Kormendy, J.: 1980, in *ESO Workshop on Two-dimensional Photometry*, p. 191.
Larson, R.B.: 1990, *Publ. Astron. Soc. Pac.*, **102**, 709.
Malumuth, E.M.: 1983, Ph.D. Thesis, University of Michigan.
Malumuth, E.M. and Kirshner, R.P.: 1985, *Astrophys. J.*, **291**, 8.
May, A. and van Albada, T.S.: 1984, *Mon. Not. R. Astr. Soc.*, **209**, 15.
Merritt, D.: 1985, *Astrophys. J.*, **289**, 18.
Oemler, A.: 1976, *Astrophys. J.*, **209**, 693.
Porter, A.: 1988, Ph.D. Thesis, California Institute of Technology.
Malumuth, E.M. and Richstone, D.O.: 1984, *Astrophys. J.*, **276**, 413.
Schneider, D.P., Gunn, J.E. and Hoessel, J.G.: 1983, *Astrophys. J.*, **268**, 476.
Schombert, J.M.: 1984, Ph.D. Thesis, Yale University.
Schombert, J.M.: 1987, *Astrophys. J. Suppl. Ser.*, **64**, 643.
Schombert, J.M.: 1988, *Astrophys. J.*, **328**, 475.
Struble, M.F.: 1990, preprint.
Struble, M.F. and Rood, H.J.: 1984, *Astron. J.*, **89**, 1487.
Thuan, T.X. and Romanishin, W.: 1981, *Astrophys. J.*, **248**, 439.
Tinsley, B.M. and Larson, R.B.: 1979, *Astrophys. J.*, **221**, 554.
Valentijn, E.A. and Casertano, S.: 1988, *Astron. Astrophys.*, **206**, 27.
Villumsen, J.V.: 1982, Ph.D. Thesis, Yale University.
White, S.D.M.: 1982, *Saas-Fee Lectures*, p. 291.
White, S.D.M. and Rees, M.J.: 1978, *Mon. Not. R. Astr. Soc.*, **183**, 341.

Discussion

D. Burstein - Many of your E galaxy luminosity profiles show deviations from an $r^{1/4}$ law at faint surface brightnesses. Can you prove that such deviations *cannot* be due to sky subtraction? Sky subtraction is a problem we have to deal with at the 1-2% level in the data on E galaxies being taken for our own survey. It is a straightforward test to make.

J.M. Schombert - Of course I cannot *prove* that the deviations at faint surface photometry levels are indeed real. The behavior of elliptical profiles at the 27 to 28 mag arcsec^{-2} level is in the realm of 0.5% of sky brightness and are very sensitive to proper sky subtraction as you noted. However, there is a great deal of overlap in surface photometry in the literature; i.e. same galaxy from different telescopes and observers. A comparison shows that there is very good agreement down to 28 mag arcsec^{-2}. In other words, faint surface photometry is repeatable. I believe everyone in this field has to be very careful and very honest about their techniques. The burden of proof here is for someone to show that there is a legitimate reason to believe that everyone has been systematically wrong in their sky subtraction.

P. Surma - At Heidelberg Observatory we have performed a study of deep surface photometry of E's, where we also looked at the galaxies in Kormendy's 'tidal class' galaxy sample. What we have found is that the occurrence of so-called 'tidal halos' is *strongly* coupled with the problem of accurate sky subtraction. M. Capaccioli has been emphasizing the importance of sky subtraction in order to reach the faint outer regions of E's, S0's and also cD's. It is evident that you need a sky subtraction accurate to <<1% if you want to reach brightness levels of 24 mag arcsec^{-2}, and this is the region you need to reach in order to see cD envelopes! I wonder how much of the discussion about cD's is influenced by these problems.
J.M. Schombert - First, I strongly disagree with your statement that better than 1% sky is need to obtain the 24 arcsec^{-2} photometry level. A sky of 1% (in the Johnson V band) corresponds to 26 not 24 mag arcsec^{-2}. The exact shape of the outer halos of ellipticals is certainly open to debate and new work; however, CCD's have achieved 0.1% flattening, on average, for many years now. And many observers, with different telescopes and detectors, using different techniques and software, have time and time again proven the repeatability of faint surface photometry. I cannot accept the claim that everyone has performed their sky subtraction incorrectly for 20 years. cD envelopes have been known for 15 years and *cannot* be the result of sky errors.

S. Kent - D and gE galaxies have identically shaped profiles (i.e. $r^{1/4}$). Hence, they can only differ in the value of μ_e at r_e. Is your "diffuse" characteristic of D galaxies another way of saying that they have lower μ_e?
J.M. Schombert - No, D galaxies are noted by their shallower $r^{1/4}$ fits. This places them larger and brighter in the $\mu_e - r_e$ diagram (see figure 6). I prefer to interpret this as enlarged r_e as predicted by N-body simulations.

R. Buta - What are the three-dimensional shapes of cD's, particularly the envelopes?
J.M. Schombert - Alan Porter's thesis (Caltech, 1988) addresses this problem more fully. His answer is that cD isophotes become more flattened with radius, presumably shifting from the galaxy potential to the cluster potential. However, it is an excellent future project, with the advent of reimaging cameras, to study the behavior of cD envelopes with respect to twisting and central shifts.

G. Djorgovski - Two more bits of information come from Porter's thesis. First, BCM's show systematic ellipticity gradients, in the sense of becoming flatter outwards, whereas the normal ellipticals show no such trend. Second, the envelopes tend to be well aligned with the parent clusters (the Binggeli effect). To me, this also suggests that the origin of envelopes is at the epoch of cluster formation. The BCM galaxy and the cD envelope may be independent entities, sharing the same special location (i.e. the bottom of the cluster potential well).
I. King - I didn't understand your argument that the extended envelopes must be primordial. Can you distinguish against the case of a merger of galaxies that already

consisted only of old stars? In answering, please note that the merging process favors massive galaxies, which have colors like those of cD envelopes. Also, with regard to Djorgovski's arguments about the alignment of cD envelopes with the major axis of their cluster, this could be simply due to the greater velocities of galaxies in this direction.
J.M. Schombert - If tidal stripping is the origin of cD envelopes, then a majority of their luminosity must be taken from low luminosity members of the cluster. The mass-metallicity relation predicts that this material will be bluer than the stellar population of the cD parent body. Although only a handful of cD envelopes have been studied by myself (in Johnson B and V plus a Strömgren narrow band system) all show very flat color profiles. The merging of bright ellipticals can maintain a red cD interior, but the outer envelope is going to be influenced by lower luminosity E's and, of course, any disk galaxies in the cluster. I agree that the elongation of the cD envelope could be due to the orbits of galaxies being stripped if that is the mechanism of formation. However, a more likely explanation is that the cD envelope is reflecting the shape of the cluster potential in which it was born which, if galaxies are on strongly radial orbits, is the same statement.

J. Melnick - I disagree with your statement that there is no correlation between cluster parameters and cD envelopes. I see in your plots a rather strong correlation which, of course, eventually disappears if one eliminates critical data (such as the brightest envelopes). My question is, is there any physical or statistical arguments to eliminate the points that give a strong correlation?
J.M. Schombert - I agree that the extreme points in these diagrams are the "best" cD galaxies, but a simple correlation test gives only marginal significance. My statement is not that the correlations do not exist, but that they are surprising weak, and, therefore, have been partially erased during cluster collapse.

D. Merritt - Is there any evidence that the presence of a cD galaxy correlates with the dynamical properties of a cluster, such as its velocity dispersion?
J.M. Schombert - A comparison of cD envelope properties with cluster dynamical properties (velocity dispersion, Bautz-Morgan type, etc...) is an exploration into the world of bad statistics. The correlations exist, but they are only marginally significant, strongly weighted by the very few brightest cD's. I argue that this merely reflects the early cluster state into the present epoch, and that cD envelopes are linked to the local conditions (i.e. its original subcluster) at birth as you predicted in 1985.

D. Merritt - Does this fact argue for a primordial origin of the cD's, as well as for the diffuse envelopes?
D. Schombert - There is only a weak correlation between the underlying galaxy luminosity and the cD envelope luminosity. Therefore, I don't believe we can conclude that the merging that produced the parent galaxy was only primordial. In fact, bound populations exist today with dynamical friction timescales of only 5×10^9 yrs, so the process of dynamical evolution must still be ongoing. My interpretation of the parent to envelope correlations is that parallel dynamical formation processes are being echoed from early epochs (i.e. mergers versus tidal stripping)

J.–L. Nieto - Can you comment on the flat central profiles of BCM's? Are they intrinsically flat or is your material resolution limited?

J.M. Schombert - Due to the large variety of surface photometry for the sample; some from CCD's, others from large scale photographic plates, I cut the inner limit at 2 kpc and, therefore, can say very little on the core properties of BCM's.

A cluster of astronomers, merging in Pompei. On the left of the pillar Jack Sulentic in deep meditation.

CENTRAL REGIONS OF EARLY–TYPE GALAXIES

J.-L. Nieto[1,2]
[1]*Observatoire Midi-Pyrénées, Toulouse, France*
[2]*Canada-France-Hawaii Telescope Corporation, USA*

1 Introduction

The central regions of galaxies are magical 'dumping places' where the archeologist-astronomer may uncover fascinating aspects of the formation and subsequent evolution of galaxies. Insights can be gained into the nature of supermassive black holes, the last bursts of star formation, recurrent nuclear activity, violent relaxation, as well as unmistakable signatures of merging processes. The usual terminology christened the central part of the stellar distribution of galaxies that defines the form of their central potential as the 'core'. Within the core, there may exist another body, directly seen so far only in our Galaxy and the most nearby galaxies, M 31 and M 32 (and possibly a few others): the 'nucleus'. These nuclei (when they exist) have a scale of $\simeq$ 10 pc, a scale unresolved from the ground at the distance of, e.g. the Virgo Cluster, while cores, according to the usual terminology, may extend up to 1 kpc.

This terminology was defined when King (1962) proposed an empirical density law (the 'King law') that fits with great success the luminosity profiles of globular clusters and the dwarf elliptical galaxies. This law is essentially defined by two parameters: the 'core radius' and the central surface brightness (a third parameter, called the 'tidal radius', concerns the outer parts and will not be discussed hereafter). The 'core radius', introduced as a scale factor, corresponds to the radius at which the central intensity has dropped by a factor of 2. This empirical law, that was also thought to be applicable to the profiles of bright galaxies (King, 1966), found a strong theoretical support with dynamical models (the so-called 'King models'): in these, the central regions when fitted by a King law, have a near-isothermal behavior that could result from violent relaxation (King, 1966). Currently, the term 'core' is used generically to define the central region (even when the central profile does not fit King's empirical law) and, unfortunately, seems to imply that this region is isothermal or near isothermal, as the centers of globular clusters, where the King models apply beautifully. However, there is nothing at the moment that leads one to believe that the cores of globular clusters and bright galaxies have had similar histories and evolution. In fact, besides dynamical relaxation times being widely different, recent observational results show direct evidence against such a suggestion.

G. Longo et al. (eds.), Morphological and Physical Classification of Galaxies, 69–83.

Previous observational review on galaxian cores may be found in Kormendy (1982, 1987a,b) Kormendy and Djorgovski (1989) and Lauer (1989). More specifically, counter-rotating cores have been discussed by Illingworth and Franx (1989), Bender (1990); see also Barnes, Bender, Franx, Balcells (this conference). In this paper, we present a brief historical and critical review of our knowledge of the observational properties of the central regions of (early-type) galaxies, essentially classical E's and SO's, from the turning point when it became clear to many that seeing effects have a major incidence on the observed properties of these regions.

2 Effects of seeing on central profiles

"The centers [of elliptical galaxies] possess extremely bright central pips. How sharp these centers are, how starlike, is impossible to say at the moment, for atmospheric effects smear the central pip into an apparent disk" (Hoyle, 1965).

The possible effects of seeing on the appearance of the central regions of elliptical galaxies has been pointed out many times in the past but were first quantitatively evaluated during the 70's. For a brief review of early references, see Schweizer (1979). Several studies addressed this question, but only on a few individual prototypical galaxies: M87 (Young et al., 1978, de Vaucouleurs and Nieto, 1979), NGC 3379 (de Vaucouleurs and Capaccioli, 1979), NGC 6251 and the two brightest Coma galaxies (Young et al., 1979), NGC 4486B (Nieto, 1980). A few other objects were also studied with the balloon-borne telescopes: NGC 4151 (Danielson et al., 1986), M 31 (Light, et al., 1974) similar data on M 32 were unfortunately not reduced, for lack of a Point-Spread-Function (PSF) for the camera that was used for that galaxy (King, private communication).

Because of the small sample available, no clear broad view of the physical parameters of the central regions could be presented. This situation changed with the (photographic) photometric study of 16 E's and one S0 by King (1978) for which he gave the experimental PSF (expressed with the dispersion of a Gaussian profile fitting best the PSF, called hereafter $\sigma*$). However, the effects of seeing on central profiles were still not clearly perceived.

Re-analyzing King's data, Schweizer (1979) strongly emphasized that the low central surface brightness of giant ellipticals (relative to that of Local Group galaxies) is certainly due to smearing of intrinsic light distribution by seeing that generates *artificial cores* in otherwise peaked galaxies; to reinforce this claim, he adopted $r^{1/4}$ laws (that are indeed strongly centrally peaked) fitting best the outer parts, convolved them with realistic PSFs and reproduced the flat-topped profiles of E's.

However, the brightest galaxies of the sample appeared 'resolved' in the sense that they showed an apparent core radius significantly broader than that obtained from the seeing-convolved $r^{1/4}$ law. Schweizer found then that the observed central profiles of (almost) all galaxies could indeed be fitted by King profiles, much flatter in the center than $r^{1/4}$ laws, but with an intrinsic core radius $r_{c,int}$ much smaller than the apparent core radius ($r_{c,app}$, with obvious relations linking $r_{c,app}$, $r_{c,int}$, and $\sigma*$). This enlightening paper indicated that we had little access, with available seeing, to the real physical properties of the central regions.

However, and conservatively, because of the central behavior of the brightest galaxies, it has been preferred to assume that all central regions of galaxies could be fitted with King profiles (fainter galaxies being suspected of having smaller cores) and, with a few exceptions, all galaxies were claimed to be 'isothermal' (see, e.g., Kormendy, 1982). In a sense, the lack of an alternative theoretical model/scenario combined with the theoretical expectation from the violent relaxation process and the considerably attractive consequence of the isothermal hypothesis, easy derivation of the central M/L (King and Minkowski, 1972; Schechter, 1980; Richstone and Tremaine, 1986) and of the phase space density (Carlberg, 1986), were certainly good and understandable reasons to assume that observations were correctly described by this convenient theoretical model.

In this view (Kormendy, 1982), the only problem remaining would be the correct derivation $r_{c,int}$, the intrinsic core radius, from $r_{c,app}$, the apparent core radius, and $\sigma*$. (This deconvolution problem is not too difficult since the real shape of the galaxy profile is known.) Non-isothermal cores (namely central regions not fitted with a King profile) would thus either (i) reveal the presence of a nuclear black hole as in M87 but also NGC 3379 (de Vaucouleurs and Capaccioli, 1979; Nieto and Vidal, 1984) or (ii) indicate the signature of a merger, as in NGC 1316 (Hausman and Ostriker, 1981; Schweizer, 1981).

Two major advances permitted progress to be made in our knowledge of the observational properties of cores, but without departing much from the above views. The first advance was the use of CCDs, which produced a higher photometric accuracy, free of the many problems plaguing photographic material (see, e.g., Nieto, 1983). Lauer (1985a, b), working with CCD data having spatial resolutions similar to those of King (1978), was able to 'resolve' a few additional galaxies and, using a deconvolution procedure, to estimate the intrinsic core radii for all the galaxies in his sample.

The second advance came from the gain in spatial resolution available at the CFHT. Kormendy (1985a, b) obtained CCD frames with mean resolution $\sigma^* = 0.34$ arcsec, almost two times better than Lauer's material, that allowed him to resolve the central part of additional galaxies. He also compared his core radii (seeing-corrected, assuming that cores follow King profiles) to Lauer's deconvolved values, and concluded (see Kormendy, 1987a, b) that the degree to which an image 'resolves' a given galaxy can be correctly described in terms of the ratio $r_{c,app}/\sigma^*$ (introduced by Schweizer, 1979). If this parameter is smaller than 3, the galaxy is not really resolved, seeing corrections are large and the subsequent estimate of the intrinsic core radius is erroneous or largely dependent on the assumption made to model the real central profile. If the ratio is larger than 5, the galaxy is well-resolved, the seeing correction is modest and the value of the intrinsic core radius can be quite safely estimated, largely free of assumptions. Situations with $5 > r_{c,app}/\sigma* > 3$ are intermediate.

The above discussion shows that the strongly peaked central profiles of many galaxies resist deconvolution procedures. Indeed, Lauer (1985b) found spurious features not confirmed by the higher-resolution data of Kormendy; similarly, Bendinelli et al. (1982) found a nucleus in M 32 from average resolution data, which was not confirmed by data at the better CFHT resolution (Kormendy, 1989; Michard and Nieto, 1991). It is likely that the spatial resolution expected from Hubble Space Telescope has discouraged efforts towards deriving reliable deconvolution procedures and that the problems encountered by HST will foster a strong of such studies.

3 Supermassive black holes?

A first practical application of the derivation of reliable central profiles is to search for the expected signatures of supermassive black holes. Historically, the first case was that of M87, based on both photometric (Young et al., 1978) and spectroscopic (Sargent et al., 1978) data. These results were challenged by several authors: on the theoretical side, Duncan and Wheeler (1979), Binney and Mamon (1982) (see also Simien, 1980) advocated instead that a well-chosen anisotropic behavior would fit equally well the data with no center mass excess (namely, a constant M/L); on an observational basis, Dressler (1980), Illingworth (1980) and Kormendy (1985a) showed that the central properties of M87 were not exceptional with respect to those of other galaxies. Recent theoretical developments (e.g., Merritt, 1987) and observational results (Dressler and Richstone, 1990) left the situation unsettled. An excellent critical detailed discussion can be found in Binney and Petit (1989; see also Dejonghe, 1989).

Additional detections of supermassive black holes were claimed in galaxies nearer than M87, but showing no peculiar sign of activity: M 31 (Dressler and Richstone, 1988; Kormendy, 1988a) and M 32 (Tonry, 1984, 1987, Dressler and Richstone, 1988), NGC 4594, the Sombrero galaxy (Kormendy, 1988a) and NGC 3115 (Kormendy and Richstone, in preparation). In all cases, the objects were studied with slit spectroscopy along several directions, and the evidence for a central high M/L comes from a very steep gradient of the rotation velocity across the center in the major axis direction (often neglecting information along the perpendicular axis) as well as a steep rise in the velocity dispersion profile. Modelling leads to extremely concentrated mass distribution: $10^8 M_\odot$ within $r = 2$ pc in M 31, $3 - 10 \times 10^6 M_\odot$ within 1 pc in M 32, etc.

A few (possibly important) provisons are worth being mentioned:

(i) Although the different authors reach similar conclusions (a very high M/L ratio), their rationales sometimes contradict each other, which creates some confusion to the picture: for example, Kormendy (1988a) identifies, after bulge subtraction, the central structure of M 31 as a disk, while Dressler and Richstone (1988), with data of similar quality, fail to substantiate this claim.

(ii) Most of the modelling has been done assuming spherical or spheroidal models. However, Gerhard (1989, 1990) argues that the nuclear disk, the bulge, and the outer disk of M 31 must all be in the same equatorial plane; therefore, to reproduce the observed isophotal twists, the bulge must be prolate-triaxial and the nuclear bar, seen nearly end-on. The large apparent rotation velocity would be due to large non-circular streaming velocities in the bar and the nuclear M/L would remain reasonably low. Similar arguments could apply to M 32 and the Sombrero (Gerahard, 1989).

(iii) A high M/L does not necessarily mean a supermassive black hole. A likely alternative is a dense stellar cluster that, for instance, could results from core collapse; the cores of M 31 and M 32 are dense enough for this (see Dressler and Richstone, 1988; Kormendy, 1989; Michard and Nieto, 1991).

(iv) Another point is more technical: slit spectroscopy is not a technique well adapted to high spatial resolution, notably when several directions are used for 2D mapping. Even a small error in centering the slit is disastrous when high velocity gradients are involved. For example, as Gerhard (1989) correctly noticed, the minor-axis rotation in the Sombrero galaxy (disregarded by Kormendy) may not be a centering error, but instead may have a physical significance. The case of M 31 may be even more complex since the photometric and kinematic centers do not coincide, and in a way which does not seem to be explained by dust. Obviously, two-dimensional spectroscopy with an integral field spectrograph is the only way to settle these questions.

4 Systematics of cores

From his CCD high resolution imaging survey of a large sample of early-type galaxies (see sect. 2), Kormendy (1985a) obtained several interesting results. Except for the very brightest E's whose central profiles are found to be well described with King profiles (and are thus interpreted as being 'consistent with projected isothermal models'), virtually all elliptical galaxies depart from King profiles such that the departures increase as total luminosity decreases. Bulges of disk galaxies also show strong departures from a King profile and rise steeply toward the center; as in M 31, several bulges contain extra nuclei superposed on large cores.

Kormendy (1985b) further correlated the *intrinsic* core parameter (core radius and central surface brightness calculated from the *apparent* core parameters, the seeing correction being made assuming a King profile) with other parameters, namely luminosity and central velocity dispersion. (For galaxies departing from a King profile, seeing correction are obviously too small and give only an upper limit to core radii and a fainter limit to surface brightness.) The main result is that core parameters, like global parameters, correlate with each other and with total luminosity: the most luminous galaxies have the largest core radii, lowest surface brightnesses and highest central velocity dispersions. Also, the intrinsic core radius and the central surface brightness correlate with the 'central' velocity dispersion. These correlations are universal in the sense that they hold from Brightest Cluster Members (BCMs) to M 32, namely over $\simeq$ 8 magnitude range. Similarly, the M/L ratio (computed assuming that cores are isothermal) correlates nicely with luminosity.

These results have two applications. One is the derivation of phase-space densities as related to the problem of galaxy formation. Since phase-space densities can only decrease during violent relaxation, Kormendy (1989) concluded that the high central phase-space densities (derived from the isothermal hypothesis, following Carlberg, 1986) and small core radii in low-luminosity E's and bulges imply that they formed essentially by dissipative collapse, whether triggered by a merger or not (see also Nieto, 1990).

He further suggested, based on the 'cooling diagram' for cores, that low-luminosity E's formed with more dissipation than high-luminosity E's. The second important result (Kormendy, 1985b, 1987a, b) is the strong differences between the core properties of dwarf spheroidals and classical E's; this suggests that these two families experienced two different formation processes.

Because of the extreme difficulty of this observational study and the great astrophysical importance of these results, it is important to be aware of some sources of uncertainties that may affect their significance. One is that a few galaxies (e.g., NGC 1316) may depart from the correlations. Second, for galaxies whose central profiles depart from a King profile, only upper limits to core radii and fainter limits to central surface brightnesses can be derived. Another problem is the fact that cores are not isothermal, and by far (see sect. 7): therefore, even if an intrinsic core radius (namely, the half widths at half maximum of the intrinsic profiles) and a corresponding surface brightness can be measured, it is probably risky to assume that these two quantities reflect the physical state of an isothermal core, and to consequently use them for deriving other astrophysical quantities such as central M/L and phase-space density. The following discussions will permit making these considerations more precise.

5 Decoupled and counter-rotating cores

"Observations in recent years have shown that the cores of E's are not just the isotropic central regions of violently relaxed stellar systems, but in general have a non trivial structure" (Bender, 1990).

It has been realized recently that galaxies accrete significant amounts of material. Therefore, if ellipticals are the product of dissipationless hierarchical merging, the most stable parts of their progenitors should survive as substructures in the center of the merger remnant. Farouki et al., (1983), for instance, attempted to derive from N-body simulations the central profile of a galaxy resulting from hierarchical merging. A pioneering observational study on this matter is that of Efstathiou et al. (1982) on NGC 5813, often since called the 'core-within-a-core galaxy'. Indeed, they suggested that the central profile of this galaxy could be fitted by a (large) core containing a core of smaller size and higher surface brightness. The inner core rotates more rapidly than the outer one and has a smaller velocity dispersion. These results prompted Kormendy (1984) to propose the following basic elements of a scenario based on merging and leading to these observed features: if the ratio of the core radii is within a factor 6–10, the secondary would survive the primary tidal field and settle in its central part, so that the remnant would have, at its center, a higher surface brightness, a lower velocity dispersion, a larger rotation velocity and bluer colors; these are all properties of small E's.

Several related observational results were published at almost the same time by Franx and Illingworth (1988), Jedrzejewski and Schechter (1988) and Bender (1988). Bender (1988, 1990) and Surma and Bender (1990) obtained several interesting results that deserve special mention:
(a) out of seven low-rotation galaxies, four have a rotating or counterrotating core;
(b) the ellipticity profiles of three of these cores show specific features (sudden change of slope) at the radius where the kinematical decoupling occurs;
(c) in one of them (NGC 5322), the core appears to contain a rotating disk (based on both 'pointed' isophotes and a decomposition of the absorption line profile);
(d) the central luminosity profiles of three galaxies which show kinematical decoupling exhibit a clear departure from an $r^{1/4}$ law at exactly the distance where the inner kinematical decoupling appears. This result raises the question of a physical definition for the core of a galaxy (see sect. 8).

Following Bender (1990), several scenarios (within the general merger scenario) can be envisaged, that lead to counter-rotating/decoupled cores:

1. Accretion of a gas-rich companion. The counter-rotating gas (as seen throughout many ellipticals; see, e.g., Bertola et al., 1990) dissipatively settles into the central region, leading to a decoupled core.

2. Merging of two spirals, as advocated by Schweizer (1990). The example is NGC 7252, still at an early stage of the merger and where the gas, seen counter-rotating inwards, is expected to form a counter-rotating core after dissipation.

3. Multiple merging in compact groups (see, e.g., Barnes, 1990)

4. Merging of a luminous elliptical with a low-luminosity object. This is the scenario advocated by Kormendy (1984); the counter-rotating core is the relic of roughly opposite spin axes in the progenitors. Balcells and Quinn (1990; see also Balcells, Barnes, this conference) have carried out N-body simulations of such encounters. They confirmed several of Kormendy's expectations, except that the spin of the secondary core settling at the center of the primary is of *orbital* origin, always aligned with the orbital angular momentum. Counter-rotation appears only when the merger is *retrograde* at the core, the particles of *both* the secondary and the primary are counter-rotating with respect to the main body of the remnant/primary. The peculiar kinematics is thus not the result of a superposition of two galaxies, but instead the consequence of a *physical transformation* during the merger.

The detection rate of counter-rotating cores, after a spectacular start (Bender's rate of 4 out of 7 slowly rotating ellipticals let us envisage a total percentage among E's as high as 30%), became stable: only 15 possible cases are known, 8 of them being well established. At least two reasons may explain this stagnation. One is physical: the high dispersion velocity of the systems prevents measurements of small-amplitude rotations; the second one is more technical: the rotation axis of the core may not necessarily be aligned with the slit axis adopted for the spectroscopic work (major or minor axis). Further possible cases may be found in the obvious present mergers (see, e.g., Balcells and Stanford, 1990), which seem destined to become ellipticals.

6 More systematics of cores

This section is in tune with the topic of the conference, as it relates to the physical classes of early-type systems. A series of work has recently shown that elliptical galaxies can be divided into two overlapping physical classes whose specific properties are well-correlated with the type of deviation from isophotal shapes, as defined by the $\cos 4\theta$ term, a_4, in the Fourier series describing the isophotal contour. While E's with 'pointed' isophotes ($a_4 > 0$, hereafter called disk E's) resemble S0 galaxies in the sense that they are oblate, rotationally flattened systems with little or no radio emission, boxy ($a_4 < 0$) or irregular galaxies are intrinsically brighter than S0's and disk E's and show all kinds of kinematical behavior (see Bender, this conference). Clearly, these classes appear to define two different evolutionary tracks and, thus, a correlation of core properties (believed to be representative of the past history of the galaxy) with isophotal classes must also be expected. Nieto, Bender and Surma (in preparation) have therefore studied the central luminosity profiles of a large sample of early-type galaxies (essentially E's) from high-resolution CCD images. The preliminary results show an interesting correlation between $r_{c,app}/\sigma*$ and isophotal class (Fig. 1); while no single disk galaxy is resolved ($r_{c,app}/\sigma* < 5$), a fair proportion of boxy and boxy/irregular E's is resolved. This result does not seem to be due to a luminosity effect, that is to say a consequence of two well-known correlations: (i) between core radius and luminosity (Sect. 4) and (ii) between isophotal class and luminosity (Nieto, 1991 and Bender, this conference). Indeed, in the range where the two isophotal classes overlap, the correlation between $r_{c,app}/\sigma*$ and isophotal class still pertains (although with a lesser degree of significance).

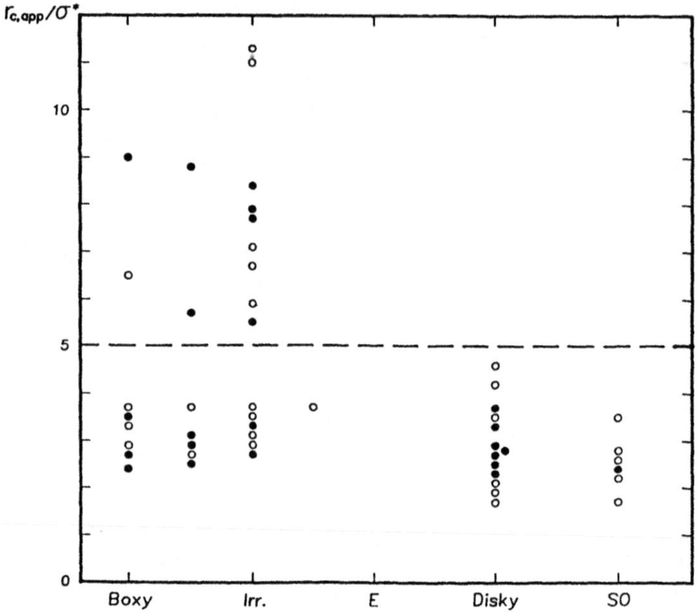

Figure 1. The $r_{c,app}/\sigma*$ values for all galaxies of a large sample of early-type galaxies (from Nieto, Bender and Surma, in preparation) are plotted against isophotal class (either boxy, irregular, disky or S0), with intermediate steps in the classification, the few E galaxies being not plotted for clarity. *Solid circles* stand for galaxies whose luminosities fall within the common luminosity range of boxy and disk E's (others are plotted with *open circles*). All disk E's appear to be unresolved systems, since $r_{c,app}/\sigma* < 5$ while a fair percentage of boxy and irregular E's is resolved. This result probably cannot be attributed to a luminosity effect (see the text).

Interestingly enough, these results add further evidence to the physical similarity between disk E's and S0's; indeed, they suggest that disk E's, like S0's (see sect. 4), have higher central phase-space densities and therefore must form with a higher degree of dissipation than boxy and irregular E's. In addition, since either the more massive galaxies or the boxy/irregular ones are suspected to owe their present structure to merging processes (Binney and Petrou, 1985; Nieto and Bender, 1989), it also suggests that merging processes tend to produce flat cores. This conclusion is very consistent with BCMs having very flat cores (Kormendy, 1985a). It gives further support to the dichotomy hypothesis (Bender et al., 1989a and Bender, this conference) and its interpretation in terms of merging/accretion processes while disk E's would be more S0-like and thus have formed in a more dissipative way (Nieto and Bender, 1989).

7 The shape of cores

The morphological characteristics of the kinematically decoupled cores put forward by Bender (1988, 1990; sect. 5) and a a few cases of disky cores (Nieto et al., 1991a) prompted Nieto et al., (1991b) to investigate the isophotal shapes of the central regions of a sample of 75 E's using high-resolution CCD images. The results are:

1. About one third (25) of ellipticals shows features in their ellipticity, P.A. and a_4 profiles, produced by unexpected changes in their slopes; these features are indicative of separate entities in their centers.

2. 15 (plus a few uncertain cases) out of these 25 separate entities show pointed isophotes indicative of disks which, from statistical considerations of projection effects, suggests that they *all* harbor a small disk.

3. In all (5) galaxies with kinematically decoupled cores, these separate disky entities coincide in radius with the entities kinematically detected.

4. All but one well-resolved ($r_{c,app}/\sigma* > 5$) galaxies (of a total of 14) show either a separate disky substructure (9) or, if not, a large amount of dust (4).

5. The other galaxies showing these separate bodies are of various kinds (S0-like, disky, boxy and boxy low-mass).

Several consideration suggest that the number of galaxies containing a decoupled core would certainly increase with better angular resolution. In addition, there is no clear correlation between the apparent core radius of the galaxy, $r_{c,app}$, and the radius at which the decoupling, either photometric or kinematical, r_{dec}, appears. In at least one case (Fig. 2), the core radius is found to be slightly larger than the radius of decoupling, which means that the 'resolved core' contains a separate disky substructure (in all other cases, $r_{dec} > r_{c,app}$).

Nieto et al. conclude from the clear identification of kinematical and geometric decoupling that, in galaxies showing both phenomena, these certainly have the same origin. Consequently, a scenario based on merging or accretion processes (responsible for a transfer of mass toward the center from the outside) is likely for a large percentage of E galaxies. However, if this scenario is valid for the most massive galaxies or those whose structure most likely results from merging processes, the large variety of E galaxies harboring a separate body in their center suggests that, in some ellipticals, such structures may result from an *internal* transfer of mass.

A natural question is therefore to ask whether the presence of a central separate body has an impact on, or may be detected by, the color behavior. Franx and Illingworth (1988) found, in the central region of IC 1459, a color gradient very typical of that for ellipticals and conclude that the core population in this galaxy is not different from that of the cores of other E's. An instructive way of discriminating colors within the same galaxy is to use both high-resolution data and large-scale low-resolution data. A first result in this sense has been obtained by Surma, Bender and Nieto (this conference) for NGC 4472, where the two-component structure is clearly seen on the color profile, reflecting the inhomogeneous formation history of this galaxy.

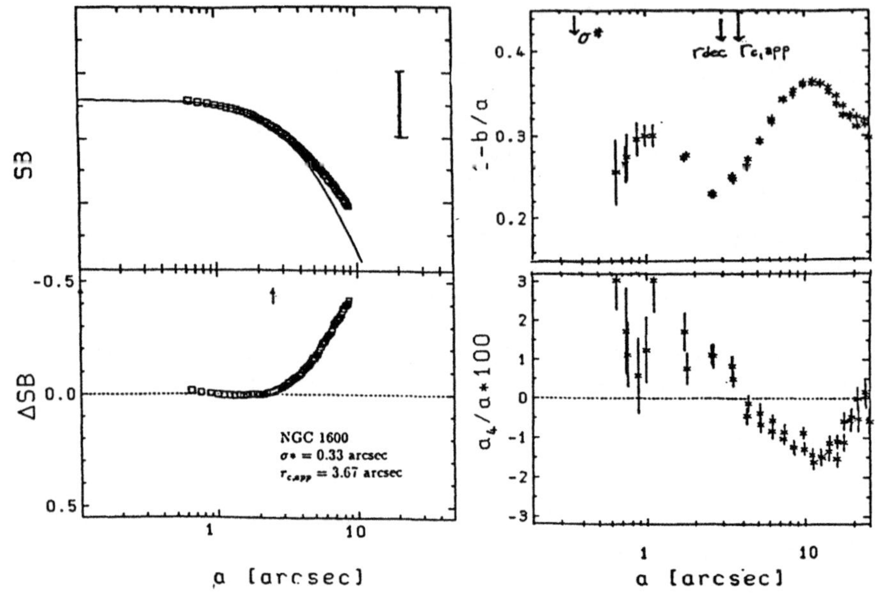

Figure 2. NGC 1600 is the best resolved galaxy of the sample of Nieto et al. (1991b), with $r_{c,app}/\sigma* \simeq 11$. Its central profile in the R band is shown in (a), where the *solid line* defines the King profile fitting best the central points within the largest possible radius range (defined by an arrow); $r_{c,app}$ is derived from this fit; the *vertical bar* defines 1 mag. The differences (Obs-Model) presented in (b) show that the range where the fit could be acceptable (< 0.10 mag) hardly reaches the core radius and that the departure is as high as $\simeq 0.3$ mag already at $\simeq 2r_{c,app}$. The ellipticity and the a_4 profiles from the composite of two R images are presented in (c) and (d). These geometrical data should be taken with caution within $2\sigma* \simeq 0.6$ arcsec. Within $r_{c,app} \simeq r_{c,int}$ a hump in the ellipticity profile and a positive a_4/a, reaching $\simeq 3\%$, appear in the central 2.5 arcsec and indicate the presence of a central disky substructure. In some other galaxies, the central substructure shows also an isophotal twist with respect to the rest of the galaxy.

8 Concluding remarks

The concept of 'core' seems to imply that the picture everyone has in mind of the evolution of the cores of globular clusters may also apply (at least to a first approximation) to the central regions of galaxies, to the extent that the centrally relaxed region eventually collapses into an entity such as a supermassive black hole (Kormendy, 1987c). However, cores are farther from an isothermal state than the extensive use of the related photometric parameters suggests. It is important to note that the 'isothermal' hypothesis (and related developments) is based only on the photometric profiles and relies upon two steps: a) the quality of the fit with a King profile and b) the King profile being univocally interpreted in terms of isothermal or, rather, near-isothermal structure. In fact, the photometric profiles of even the best resolved galaxies do not fit a King profile further out than the core radius (see figure 2). The isothermal approach is thus no longer valid, now that geometrical and kinematical considerations show anisotropy of the velocity residuals as well as kinematical and geometrical decoupling, possibly even within the region defined by the core radius. Therefore, the quantities derived as a direct consequence of the isothermal hypothesis, such as central M/L or central phase-space densities, may be incorrect by a large amount. Furthermore, even if the core radius certainly is an important parameter for defining the overall shape of the central potential, describing cores obviously requires introducing additional parameters, including the radius of decoupling. Similarly, the present definition of the 'core' probably needs to be revised also: it should probably be the central separate substructure, in which case a physically meaningful 'core radius' is the radius where decoupling appears.

If isothermal cores do not give a good physical description of the physics of central regions of galaxies, what is the meaning of the two important results based on core parameters (discussed in sect. 4), and can we still use the 'core radius' as a reliable scaling length? In any event, the first result, namely the correlations involving core properties, retains a deep physical meaning in that the most massive/luminous galaxies have the lowest central luminosity gradients and the lowest central stellar densities. However, we may wonder whether the second result, the distinction between E's and diffuse dwarfs (dE's), could possibly result from the use of quantities (core parameters) of uncertain physical significance. There are three reasons for being cautions: (i) central regions of diffuse dwarfs (whose photometric profiles can indeed be fitted with King profiles) also have been found to have anisotropic velocity residuals, indicating that these systems are not isothermal (Bender and Nieto, 1990; Carter and Sadler, 1990; Bender et al., 1991); (ii) there is no clear kinematical distinction between the brightest dwarfs and the faintest E's (Bender and Nieto, 1990); and (iii) global photometric parameters fail to show the E-dE distinction as clearly as the core parameters (see, e.g., Binggeli et al., 1984; Caldweel and Bothun, 1987; Nieto, 1988, 1990). It seems that the question of the E-dE distinction has been definitely settled, at least on the photometric side, by Binggeli and Cameron (1991) who showed that the photometric profiles in the inner regions of E or dE galaxies come in two distinct types, independent of any model and of any morphological classification.

In the above discussion, many questions remain unsettled. New advances require investigating in detail the nature, origin and physical properties of the central substructure, observing with a better resolution a few galaxies, overcoming the uncertainties attached to slit spectroscopy so that 'supermassive black holes' can be discussed on a safe ground and, maybe also, finding a scale length of sounder physical significance than the core radius. To answer these questions, if we except HST, two recent instrumental developments at CFHT, that already have given encouraging first results, may prove useful: the image stabilizing High Resolution Camera (McClure et al., 1989), and the two-dimensional spectrograph, TIGER (Bacon et al., 1988). The former has proven to be able to provide images with $\sigma* = 0.14$ arcsec (Nieto et al., 1990), thus allowing a gain by almost a factor of two in resolution with respect to classical imaging. The latter has already produced spectacular results on various astrophysical objects (Adam et al., 1989; Pecontal et al., 1990) and great advances can be envisaged based on its particular ability of providing two-dimensional spectra at sub arcsecond spatial resolution.

Acknowledgements

This review owes very much to daily conversations with several colleagues: R. Bender, C. Möllenhoff and P. Surma at Heidelberg, and R. Bacon and G. Monnet at CFHT. Some of the original data presented here were obtained with the help of J. Arnaud and R. McClure. I particularly thank B. Binggeli for enlightening discussions on dwarf spheroidals and I. King for many discussions on the subject and critical remarks on an earlier version of the manuscript. T. Davidge, J. Glaspey and R. Racine are also gratefully acknowledged for careful readings and critical comments. Part of this review was prepared while the author was guest at the Ladessternwarte (Heidelberg) with support from the Deutsche Forschunggemeinschaft (SFB 328).

References

Adam, G., Bacon, R., Courtès, G., Georgelin, Y., Monnet, G., Pécontal, E.: 1989, *Astron. Astrophys.* **208**, L15.

Bacon, R., Adam, G., Baranne, A., Courtès, G., Dubet, D., Dubois, J.P., Georgelin, Y., Monnet, G., Pécontal, E., Urios, J.: 1988, *ESO Conference on VLTs and their Instrumentation.*

Balcells, M., Quinn, P.J.: 1990, *Astrophys. J.* **361**, 381.

Balcells, M. Stanford, S.A.: 1990, *Astrophys. J.* **362**, 443.

Barnes, J., 1990, in *Dynamics and Interactions of Galaxies*, Ed. R. Wielen, Springer Verlag, Heidelberg, p. 186.

Bender, R., 1988, *Astron. Astrophys.* **202**, L5.

Bender, R., 1990, in *Dynamics and Interactions of Galaxies*, Ed. R. Wielen, Springer Verlag Heidelberg, p. 232.

Bender, R, Döbereiner, S., Möllenhoff, C.: 1988, *Astron. Astrophys. Suppl. Ser.* **74**, 385.

Bender, R., Nieto, J.-L.: 1990, *Astron. Astrophys.*, in press.

Bender, R., Paquet, A., Nieto, J.-L.: 1991, submitted.

Bender, R., Surma, P., Döbereiner, S., Möllenhoff, C, Madejski, R.: 1989, *Astron. Astrophys.* **217**, 35.

Bendinelli, O., Parmeggiani, G., Zavatti, F.: 1982, *Astrophys. Space Sci.* **83**, 239.

Bertola, F., Bettoni, D., Buson, L.M., Zeilinger, W.W.: 1990 in *Dynamics and Interactions of Galaxies*, Ed. R. Wielen, Springer Verlag Heidelberg, p. 249.

Binggeli, B. Cameron, L.M.: 1990, submitted.

Binggeli, B., Sandage, A.R., Tammann, G.: 1984, *Astron. J.* **89**, 64.

Binney, J., Mamon, G.: 1982, *Mon. Not. R. Astr. Soc.* **200**, 361.

Binney, J., Petit, J.-P.: 1988, in *Dynamics of Dense Stellar Systems*, Ed. D. Merritt, Cambridge University Press, p. 43.

Caldwell, N., Bothun, G.D.: 1987, *Astron. J.* **94**, 1126.

Carlberg, R.G.: 1986, *Astrophys. J.* **310**, 593.

Carter, D., Sadler, E.: 1990, *Mon. Not. R. Astr. Soc.* **245**, 12P.

Dejonghe, H.: in *Dynamics of Dense Stellar Systems*, Ed. D. Merritt, Cambridge University Press, p. 69.

Dressler, A.: 1980, *Astrophys. J.* **240**, L11.

Dressler, A., Richstone, D.O.: 1988, *Astrophys. J.* **324**, 701.

Dressler, A., Richstone, D.O.: 1990, *Astrophys. J.* **348**, 120.

Duncan, M.J., Wheeler, J.C.: 1980, *Astrophys. J.* **237**, L27.

Efstathiou, G., Ellis, R.S., Carter, D.: 1982, *Mon. Not. R. Astr. Soc.* **201**, 975.

Farouki, R.T., Shapiro, S.L., Duncan, M.J.: 1983, Astrophys. J. **265**, 597.

Franx, M., Illingworth, G.: 1988, *Astrophys. J.* **327**, L55.

Gerhard, O.E.: 1988, in *Dynamics of Dense Stellar Systems*, Ed. D. Merritt, Cambridge University Press, p. 61.

Gerhard, O.E.: 1990, *Mon. Not. R. Astr. Soc.*, in press.

Hausman, M.A., Ostriker, J.P.: 1978, *Astrophys. J.* **224**, 320.

Hoyle, F.: 1965, *Nature* **208**, 111.

Illingworth, G.: 1980, in *ESO Workshop on Two Dimensional Photometry*, Ed. P. Crane and K. Kjär,(Geneva: ESO), p. 299.

Illingworth, G., Franx, M.: 1988, in *Dynamics of dense Stellar Systems*, Ed. D. Merritt, Cambridge University Press, p. 13.

Jedrzejewski, R.I., Schechter, P.L.: 1988, *Astrophys. J.* **330**, L87.

King, I.R.: 1962, *Astron. J.* **67**, 471.

King, I.R.: 1966, *Astron. J.* **71**, 64.

King, I.R.: 1978, *Astrophys. J.* **222**, 1.

King, I.R., Minkowski, R.: 1972, in IAU Symposium 44 *Internal Galaxies and Quasi Stellar Objects*, Ed. D.S. Evans (Dordrecht: Reidel), p. 87.

Kormendy, J.: 1982, in *Morphology and Dynamics of Galaxies*, 12th Advanced Course of the Swiss Society of Astronomy and Astrophysics, Ed. L. Martinet and M. Mayor (Sauverny: Geneva Observatory).

Kormendy, J.: 1984, *Astrophys. J.* **287**, 577.

Kormendy, J.: 1985a, *Astrophys. J.* **292**, L9.

Kormendy, J.: 1985b, *Astrophys. J.* **295**, 73.

Kormendy, J.: 1987a, in *Structure and Dynamics of Elliptical Galaxies*, IAU Symposium 127, Ed. T. de Zeeuw, Dordrecht: Reidel, p. 17.

Kormendy, J.: 1987b, in *Nearly Normal Galaxies*, Ed. S.M. Faber, New-York: Springer Verlag, p. 163.

Kormendy, J.: 1987c, in *Dark Matter in the Universe*, IAU Symposium 117, Ed. J. Kormendy, G.R. Knapp, Dordrecht: Reidel, p. 139.

Kormendy, J.: 1988a, *Astrophys. J.* **325**, 128.

Kormendy, J.: 1988b, *Astrophys. J.* **335**, 40.

Kormendy, J.: 1989, *Astrophys. J.* **342**, L63.

Kormendy, J. Djorgovski, S.: 1989, *Ann. Rev. Astron. Astrophys.* **27**, 235.

Lauer, T.R.: 1985a, *Astrophys. J. Suppl. Ser.* **57**, 473.

Lauer, T.R.: 1985b, *Astrophys. J.* **292**, 104.

Lauer, T.R.: 1988, in *Dynamics of Dense Stellar Systems*, Ed. D. Merritt, Cambridge University Press, p. 3.

Light, E.S., Danielson, R.E., Schwarzschild, M: 1974, *Astrophys. J.* **194**, 257.

McClure, R., Grundmann, W.A., Rambold, W.N., Fletcher, J.M., Richardson, *Astron. Soc. Pac.* **101**, 1156.

Merritt, D.: 1987, *Astrophys. J.* **319**, 55.

Michard, R., Nieto, J.-L.: 1991, in press.

Nieto, J.-L.: 1980, in *ESO Workshop on Two Dimensional Photometry*, Ed. P. Crane and K. Kjär,(Geneva: ESO).

Nieto, J.-L.: 1988, in *2da Reunion de Astronomia Extragalactica*, Academia Nacional de Ciencias de Cordoba, p. 239.

Nieto, J.-L.: 1990, Heidelberg in *Dynamics and Interactions of Galaxies*, Ed. R. Wielen, Springer Verlag Heidelberg, p. 258.

Nieto, J.-L.: 1983, *Astron. Astrophys. Suppl. Ser.* **53**, 383.

Nieto, J.-L., 1991, *Astron. Astrophys.*, in press.

Nieto, J.-L., Bender, R.: 1989, *Astron. Astrophys.* **215**, 266.

Nieto, J.-L., Bender, R., Arnaud, J., Surma, P: 1991, submitted.

Nieto, J.-L., McClure, R., Fletcher, J.M., Arnaud, J., Bacon, R., Bender, R., Comte, G., Poulain, P.: 1990, *Astron. Astrophys.* **235**, L17.

Nieto, J.-L., Poulain, P., Davoust, E., Rosenblatt, P.: 1991, *Astron. Astrophys. Suppl. Ser.*, in press.

Nieto, J.-L., Vidal, J.-L.: 1984, *Mon. Not. R. Astr. Soc.* **209**, 21P.

Pécontal, E., Adam, G., Bacon, R., Courtès, G., Georgelin, Y., Monnet, G.: 1990, *Astron. Astrophys.* **232**, 331.

Richstone, D.O., Tremaine, S.: 1986, *Astron. J.* **92**, 72.

Sargent, W.L.W., Young, P.J., Boksenberg, A., Shortridge, K., Lynds, C.R. Hartwick, F.D.A.: 1978, *Astrophys. J.* **221**, 731.

Schechter, P.L.: 1980, *Astron. J.* **85**, 801.

Simien, F.: 1980, Thèse d'Etat, Université de Provence.

Surma, P., Bender, R.: 1990, Heidelberg in *Dynamics and Interactions of Galaxies*, Ed. R. Wielen, Springer Verlag Heidelberg, p. 277.

Schweizer, F.: 1979, *Astrophys. J.* **233**, 23.

Schweizer, F.: 1981, *Astrophys. J.* **246**, 722 .

Schweizer, F.:1990, Heidelberg in *Dynamics and Interactions of Galaxies*, Ed. R. Wielen, Springer Verlag Heidelberg, p. 60.

Tonry, J.L: 1984, *Astrophys. J.* **283**, L27.

Tonry, J.L: 1987, *Astrophys. J.* **322**, 632.

de Vaucouleurs, G., Capaccioli, M.: 1979, *Astrophys. J. Suppl. Ser.* **40**, 699.

de Vaucouleurs, G., Nieto, J.-L.: 1979, *Astrophys. J.* **230**, 697.

Williams, T.B.: 1981, *Astrophys. J.* **244**, 458.

Young, P.J., Westphal, J.A., Kristian, J., Wilson, C.P., Landauer, F.P.: 1978, *Astrophys. J.* **221** , 721.

Young, P.J., Sargent, W.L.W., Kristian, J., Westphal, J.A.: 1979, *Astrophys. J.* **234**, 76.

Discussion

S.G. Djorgovski - What is known about the metallicity and color behavior in kinematically decoupled cores?
J.-L. Nieto - We have not looked at this aspect in detail yet because the different PSFs of the B and R images frequently observed in our images prevent one from giving an immediate answer, except for those galaxies having flat cores. In NGC 4472 for instance, the comparison of these data with large-scale data (Surma, Bender, Nieto, this workshop) show a strong reddening inwards, which suggests a metallicity effect. We have all data necessary to answer this question in a near future.

D. Burstein - Todd Boroson, who is not here, has done a careful study of the line strength gradient in NGC 4406, in both the visual and near-IR (to 1 μm). He finds a

;rong gradient in *both* sodium features (Na 5898 and Na 8190), in the sense of decreasing utward. It will be interesting to compare his results to the kinematical and photometric udies that you and Marijn have described.

V.B. Sparks - I wonder if the effects of dust could be more significant that you implied. lthough I think the conclusion would be the same (mergers), it seems that the examples f NGC 5813 and IC 1459 show irregular dust (see poster). Could the photometric roperties be significantly changed (twists, a_4, core radii)?

.-L. Nieto - In principle, yes. However, for most of the galaxies in our sample, we have btained images at two wavelengths, generally B and R. Therefore, dust can be detected rom the only comparison of the geometrical parameters derived from our images (this not always simple because the resolution of the images is poorer in B than in R). For hose galaxies where dust is detected and not too strong, we used the R image. The emaining galaxies (numbering very few) where dust prevents any analysis were simply iscarded. Note that in some of our highest resolution images, the effect of dust can be orrected (Nieto et al.: 1990, Astron. Astrophys. **235**, L17).

Jean-Luc Nieto and George Djorgovski in the poster room.

STELLAR CONTENT AND CHEMICAL EVOLUTION IN SPHEROIDAL POPULATIONS

R.M. Rich
Department of Astronomy
Columbia University, USA

ABSTRACT. Intense study of the stellar population of the nuclear bulge during the last decade has provided important insights into the populations of other bulges and ellipticals. Due to the wide range of abundances in these populations, one expects to find a complex evolved population, with late M giants and RR Lyrae stars coexisting in the same volume. In the galactic bulge, the abundance distribution can be fit by the closed box model; the discovery of correlations between color and local escape velocity in ellipticals suggests that the one-zone model with a wind could describe the abundance distributions in most spheroidal populations. There is growing evidence for an age range as well. Luminous M giants have been found in the bulge of the Milky Way and most recently in the bulge of M31. We consider the possibility that bulges may be formed in starbursts that do not necessarily have to occur at early times.

1 Introduction

In the broadest morphological sense, it is easy to classify galaxies as either "disks" or "spheroids" or some combination of the two. Looking at a deep photograph of an edge-on spiral such as NGC 4565, it is easy to imagine the bulge in the middle as being closely related to, or precisely identical to, an isolated elliptical galaxy in the field. On the other hand, bulges are often strongly boxy or even peanut-shaped. And unlike elliptical galaxies, it generally appears that their shapes are supported by rapid rotation.

Within the last decade, advances in astronomical instrumentation have given us the unprecedented opportunity to study the detailed properties of stars in the bulge of the Milky Way. Images from the IRAS and COBE satellite infrared surveys reveal our galactic bulge as a distinct, boxy stellar system that resembles distant bulges. Therefore, as a consequence of investigating our galactic bulge we are in a better position to discuss the stellar content in all spheroidal populations. Study of our bulge teaches us that the idealized condition of population synthesis models are not appropriate. Real spheroidal populations have a range in abundance, and perhaps in age. In real spheroidal populations, the horizontal branch and asymptotic giant branch (AGB) stars make an important contribution to the integrated light of the population. If we are to learn from the integrated light of distant bulges and ellipticals, we must build models that are faithful to reality.

G. Longo et al. (eds.), Morphological and Physical Classification of Galaxies, 85–98.

Baade (1944) resolved the bulge of M31 into stars, and was able to show that the color-magnitude diagram bore closer resemblance to globular clusters than to actively star-forming regions of the disk. He subsequently discovered RR Lyraes in the bulge of the Milky Way (Baade, 1951), strengthening that conceptual link. Introductory textbooks in astronomy still describe spheroidal populations as being old and metal poor, like globular clusters.

Baade succeeded at identifying a portion of spheroidal populations with the familiar globular clusters. Study of their integrated light immediately told another story. Baum (1959) found elliptical galaxies too much redder than globular clusters. Morgan and Mayall (1957) found the integrated light of spheroidal populations to have strong CN λ3883Å absorption, like K giants. And Nassau and Blanco (1958) found huge numbers of late M giants concentrated toward the galactic center. Well before 1960, there was strong evidence for high metallicities and even abundance spreads. Indeed, Baade (1963) understood as early as 1958 that the bulge of M31 has an abundance range, and contains relatively old stars of high metallicity. He suggested that the process of "enrichment" (read chemical evolution) proceeded rapidly in the M31 bulge, leaving a relatively old stellar population with a wide range in abundance. Indeed, after a few significant detours, modern thinking has come around to this viewpoint.

Johnson (1966) obtained photometry of ellipticals from 3500 Å to 3.4 μm, and used his stellar library to perform a first-order population synthesis. In addition to the obvious K giants, success required addition of A0V and late M giants. Johnson discovered the two classical problems (Renzini, 1986) of galaxy population synthesis presuming old stars-the "warm star deficit" (model population have too little flux below 4000 Å) and the "cool star deficit" (models fall too low beyond 1 μm). Johnson believed that Arp's (1965) photometry of the galactic bulge in Baade's Window demonstrated the integrated light of such a stellar population to be giant-dominated. Fitting a synthesis model to observations at 3–4 μm required the addition of M0III and even M8III giants to the model population. Indeed, these were precisely Nassau and Blanco's late M giants. (Whitford, 1975 shows Johnson's spectral energy distribution).

Following the broad-band work, scanners confirmed early photographic spectroscopy (Morgan and Osterbrock, 1969) showing strong absorption lines of MgH and Na D in galaxy populations. As Spinrad and Peimbert (1975) review, the discovery of large virial masses (our present day dark matter problem) led to the development of dwarf-enriched models for galaxy populations. This new idea simultaneously explained the strong Na and MgH in ellipticals (these lines are stronger in dwarfs) but also provided an obvious candidate for baryonic dark matter (low mass dwarfs). Spinrad and Taylor (1969) proposed that galaxies and some stars might have abundances well above solar. Photographic high dispersion spectroscopy seemed to disprove the idea of metal rich stars; as we shall see, the idea of extreme metallicity has survived (cf. Whitford and Rich, 1983; Taylor and Johnson, 1987) but the light of galaxies is dominated by giants and not dwarfs.

Decisive evidence for the giant-dominated nature of spheroidal populations came from infrared photometry: galaxies have strong CO 2.3 μm absorption (Frogel et al., 1978) while dwarfs have virtually no CO (Frogel, 1971). Johnson's late M giants were rediscovered as a unique component of elliptical galaxies using the 1.87 μm H_2O feature

(Aaronson et al., 1978).

Detailed study and synthesis of spheroidal populations receives inspiration both from the prospect of discovering hidden young and intermediate-age populations, as well as interpreting spectra of high redshift galaxies (O'Connell, 1968, 1988). Until recently, practitioners of population synthesis ignored the complication that even simple closed-box models of chemical evolution produce populations of long-lived stars *with a wide abundance range* (see sec. 3 below). More sophisticated efforts now take chemical evolution and abundance ranges into account; Arimoto and Yoshii (1987) account for the abundance ranges due to the simple model of chemical evolution, and successfully fit Johnson's (1966) venerable galaxy photometry. (The above is a good case of this generation's astronomers being interested in *more* than just the previous generation's finding charts).

What is not contained in Arimoto and Yoshii's photometric models (or anyone else's) is a realistic AGB population, and in particular, any proper treatment for the AGB and mass-loss produced by an actual intermediate-age metal rich stellar population. Terndrup (1990) proposes to address this by developing "bulge-based" synthesis models. I have pushed study of spheroidal populations to the evolved giants near the nucleus of M31 (Rich et al., 1989; Rich and Mould, 1991); these efforts begin to show us how diverse spheroidal populations may actually be.

In the following section, I discuss the characteristics of stars in the bulge of our galaxy and M31; the distinctive properties of these populations are due entirely to the wide abundance range and presence of some very metal rich stars. In section 3, I consider how the simple one-zone model of chemical evolution could give rise to the bulge abundance range and spatial gradient. Finally, I consider how these new ideas may be used to discover intermediate-age metal rich populations.

2 Evolved stellar content of bulges

Development of modern observatories at Cerro Tololo and Las Campanas initiated the modern period of discovery in the study of bulge populations. While Morgan and Osterbrock (1969) found strong CN absorption in the bulge of the Milky Way, Whitford (1978) used a modern photoelectric spectrum scanner with careful sky subtraction to show that the integrated light of the bulge near Baade's Window is very similar to that of distant bulges and ellipticals. V. Blanco, M. McCarthy, and B. Blanco trained the 4-m and its grism on fields in the galactic bulge, discovering thousands of new M giants. Whitford and Blanco (1979) obtained R and I plates to reveal the giant branch with enough precision to select an unbiased sample of K giants, with which to study the abundance distribution of long-lived stars (Whitford and Rich, 1983).

The discovery of both RR Lyrae stars and late M giants proved the existence of a wide abundance range. Arp's (1965) photographic photometry, albeit hampered by severe crowding, also found a wide abundance range in the bulge. Figure 1 illustrated the effect of this upon the stellar content of the bulge (and presumably other spheroidal population). Even metal rich galactic globular clusters have luminous AGB stars, some of which are Miras. In addition to the discovery of numerous late M giants in the bulge (see McCarthy, 1990 and Blanco, 1988), we have OH/IR and numerous Miras (Whitelock et al., 1991).

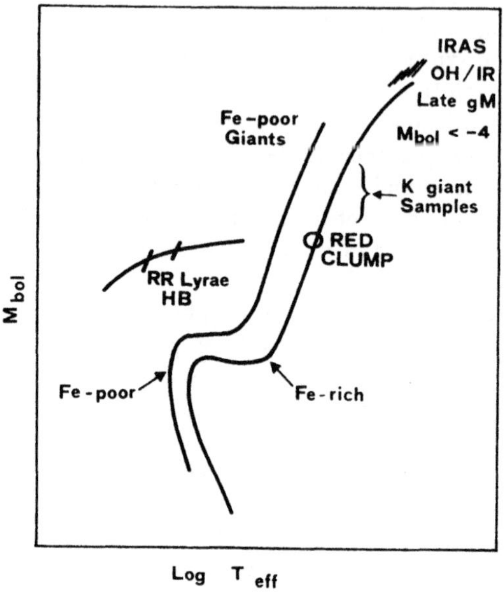

Figure 1. The wide abundance range in the galactic bulge causes an unusual mix of evolved stars. The progenitors of the late M giants, Miras, OH/IR and IRAS stars are metal rich. On the other hand, the RR Lyraes and HB stars are evolved from the metal poor population (certain to be very old). The red "clump" giants (see also figure 4) are evolved from the metal rich population.

The OH/IR stars are so dust-shrouded that they emit virtually all of their energy longward to 2 μm. Indeed, many "normal" elliptical galaxies have 10 μm excesses (Impey and Wynn-Williams, 1986; Jura et al., 1987). Luminous elliptical galaxies may have even brighter AGB stars than are found in the galactic bulge. Figure 2 shows that the integrated light of the galactic bulge has the strong lines characteristic of elliptical galaxies, but is generally not as strong-lined. It is interesting that the integrated light of the bulge has weak CN λ3889 bands compared to the ellipticals. Rich (1988) points out that individual K giants are weak in CN. Like the galaxies, the integrated bulge light has a strong Hβ even though no conspicuous young turnoff is present in the SgrI or Baade's Window bulge fields (Rich, 1985; Whitford, 1986; Terndrup, 1988). While strong Hβ in galaxies has been taken to be evidence for the presence of some young population, it may rather be due to the entire giant branch being elevated in temperature. Infrared photometry of Blanco's giants by Frogel and Whitford (1987) and Frogel (1984) shows that the giant branch lies bluer in (J-K) than expects, fitting securely between the M92 and M67 globular cluster giant branches. Terndrup (1988) finds that the giant branch is blue with optical BVRI photometry. The strong Hβ in the integrated K giant light (Terndrup, 1990) also confirms that the bulge giant branch is remarkably blue. Subsequent infrared photometry by Frogel et al. (1990) confirms this for bulge giants found at a wide range of galactocentric distances.

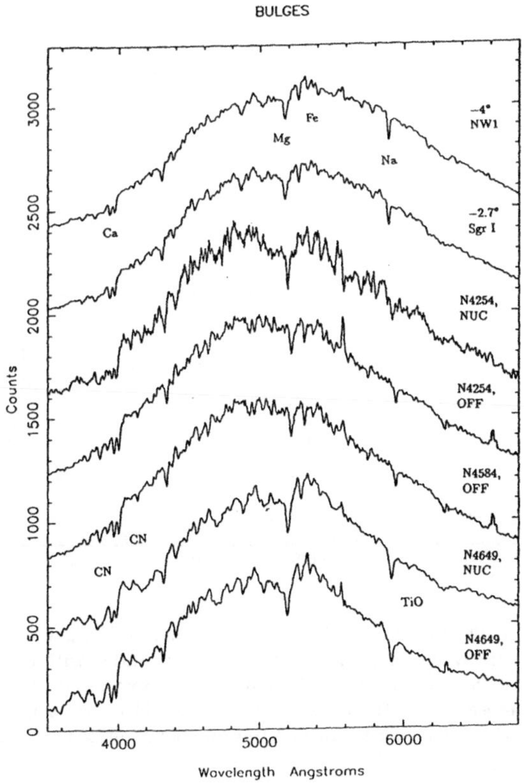

Figure 2. Integrated spectra of patches of the galactic bulge near Baade's Window (b = -4 deg) and near Sgr I ($b = -2.6$ deg); these compared with the integrated spectra of other galaxies. Notice that the line strength does not change between the two latitudes in the inner bulge. Hβ is strong in the bulge fields, but CN and the HK break are noticeably weak relative to the nuclei.

2.1 Ages and abundances in the bulge of the galaxy

The abundance distribution of K giants probably reflects that of the "long-lived" stars with great fidelity. Rich (1988) shows that galactic bulge K giants range from 0.1 to above 5 times the solar metal abundance (Figure 3). Subsequently, there has been intensive study of the M giants (likely to be evolved from metal rich stars) and of the RR Lyraes (likely to be old and metal poor). Not surprisingly, Sharples et al. (1990) find the M giants to be metal rich (+0.3 dex) while Walker and Terndrup (1991) show that the RR Lyraes are nearly all -1 dex. In studies of the M giant properties as a function of galactocentric distance, Frogel et al. (1990) and Terndrup et al. (1990) see little evidence for a gradient in abundance in that stellar population. Equally disturbing is that the bulge giant branch has an extremely narrow range in temperature. Giants in M31 may share this peculiarity (Rich and Mould, 1991).

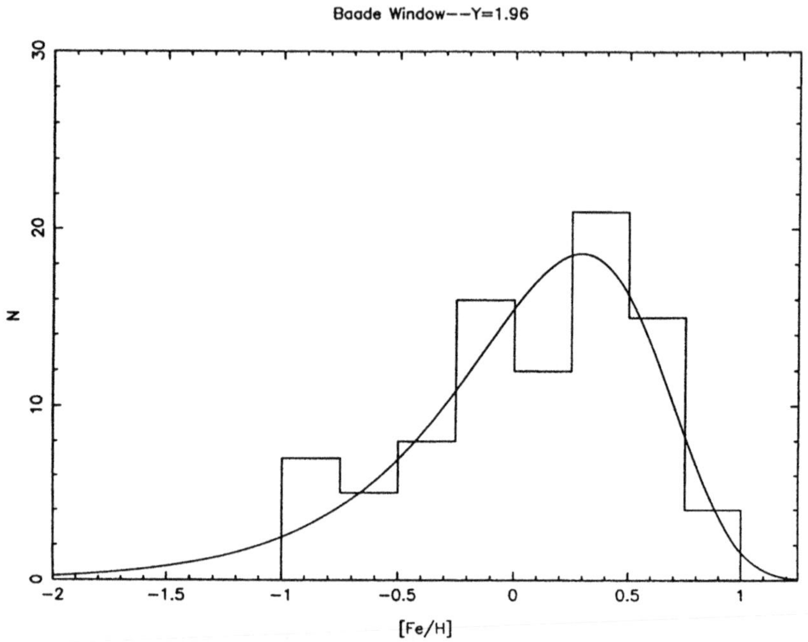

Figure 3. Abundance distribution for 88 K giants in Baade's Window from Rich (1988). Theoretical curve represents the simple model of chemical evolution with $y = 2.0$ (see text). If the evolution were dominated by infall, there would be fewer metal poor stars. If wind outflow dominated the evolution, the theoretical curve would be the same but shifted to lower abundance. If the wind catastrophically terminated evolution, the metal rich end would be sharply cut off.

Presumably the RR Lyraes and M giants evolved from the extrema of the metallicity distribution. Figure 4 illustrates a typical bulge color-magnitude diagram; notice that the red clump is a prominent feature.

While turnoff photometry shows that fields more distant than 1 kpc are older than 5 Gyr (Terndrup, 1988), the crowding, age range, and spatial thickness of the bulge conspire to prevent us from setting an age precise enough to determine whether the bulge is slightly younger than, or older than the galactic globular clusters. We may infer age differences indirectly using the evolved stellar population. The presence of very long period (800 day) Miras and a luminous AGB termination lend support to the idea that the metal rich population may be younger than the halo (cf. Wood and Bessel, 1983; Sharples et al., 1990). Figures 5 and 6 illustrate the remarkably luminous AGB stars in the M31 bulge. The giant branch terminates at $M_{bol} = -5$, and shows little sign of the break found by Frogel and Whitford (1987) at $M_{bol} = -4$. Rich et al. (1989) find that these stars are M giants like those found in the galactic bulge. Renzini (1990) proposes that blue stragglers may be the progenitors of many of the most luminous stars. However, Renzini also shows that low mass metal rich stars can climb to lofty luminosities only if

He remains near solar values in the metal rich stars (it is hard to see how $\Delta Y/\Delta Z = 0$).

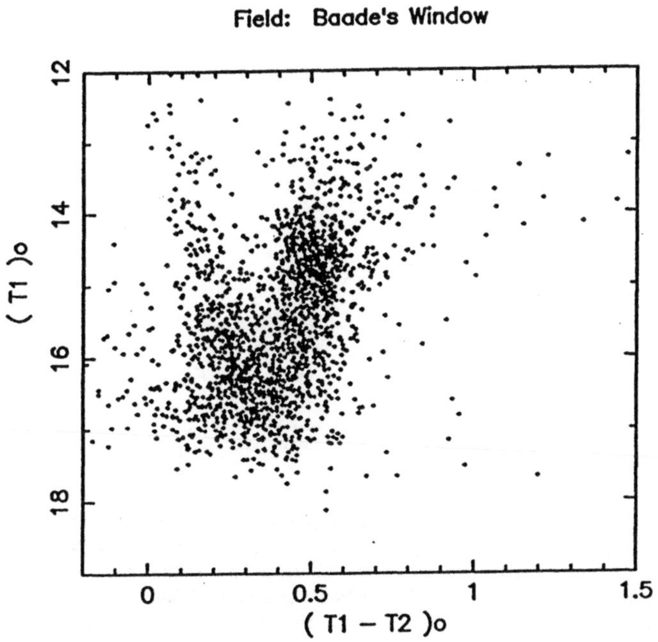

Figure 4. Color magnitude diagram of the giant branch in Baad's Window, using the Washington photometric system. T1 and T2 are equivalent to R and I. Notice the prominence of the clump, and the red giants trailing off to red colors. Blue sequence is due to foreground disk stars (from Tyson 1991).

The new ΔS measurements of bulge RR Lyraes shows that, in contrast to the solar neighborhood, there are no RR Lyraes near solar metallicity (Walker and Terndrup, 1991). Because RR Lyraes only arise in old populations (cf. Graham and Nemec, 1983) it is intriguing to speculate that the metal rich stars in the bulge are too young to make RR Lyraes. If the metal rich bulge is slightly young, then the blue giant branch and high AGB luminosities can be understood.

The general agreement between the infrared photometry of giants in the Milky Way and M31 permits some general conclusions to be drawn. Although the bulge AGB stars are luminous, they are more than a magnitude fainter than their true intermediate-age disk counterparts. Population models now have a clear example of a true bulge population.

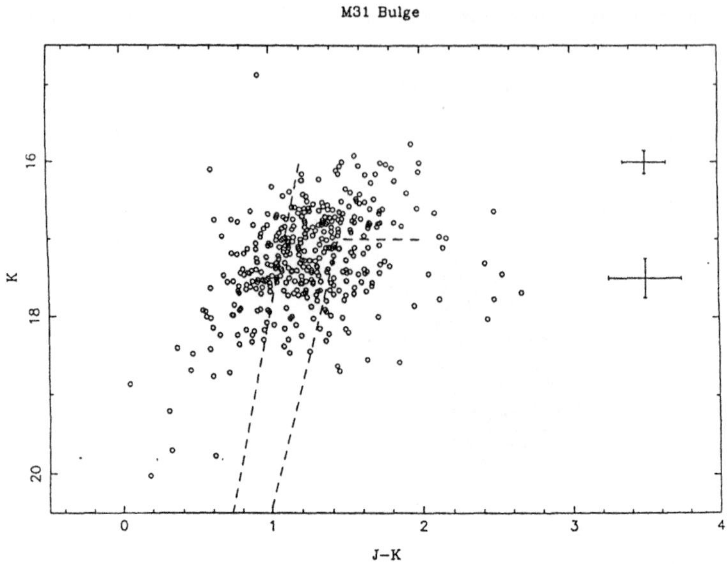

Figure 5. Color-magnitude diagram of luminous giants 3 arcmin SE of the nucleus of M31, obtained with an infrared imaging array (Rich and Mould, 1991). The photometric errors are larger than 0.15 magnitude in each color. Fiducial lines indicate the extent in color and luminosity of the giant branch in the bulge of the Milky Way. Notice that the high luminosity of the giants is seen clearly. Uncertainties in bolometric corrections are larger than the photometric errors.

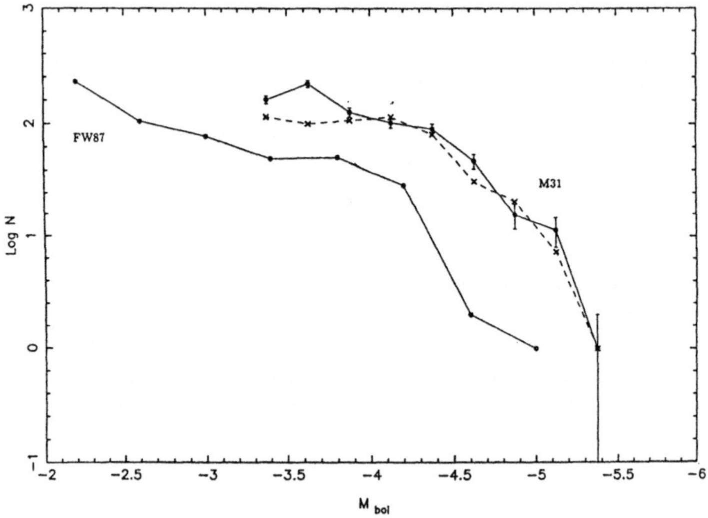

Figure 6. Luminosity functions for the M31 bulge, corrected for incompleteness due to crowding, compared with Frogen and Whitford's (1987) photometry in Baade's Window. The dashed connecting line is based on stars with both J and K colors, and the solid line is the full sample. The bulge luminosity function is below, lacking error flags. Milky Way stars were optically selected, and observed with a single-channel InSb photometer; M31 bulge sample is based on stars detected in the infrared K band.

2.2 Spheroidal populations in distant galaxies

As figure 2 illustrates, the bulge is metal poor relative to its giant elliptical cousins. Because starbursts fade and redden so quickly, distant intermediate-age bulge populations may leave only subtle traces. One such trace would be a large number of circumstellar shells resulting from the AGB evolution of metal rich stars. Impey and Wynn-Williams (1986) report 10 μm excesses in a number of galaxies. The 10 μm emission may be the last trace of a starburst or merger event that occurred more than a few Gyr ago.

The notion of age ranges in bulges and ellipticals may strike some as startling, yet many cases of blue bulges are known. NGC 5102 (cf. Hibbard and Rich, 1990) and NGC 4736 both have the blue colors and strong Balmer lines characteristic of young populations. Many Sc nuclei (such as the stellar nucleus of M33) have strong Balmer lines. O'Connell has long emphasized the need to seriously consider the possibility of age ranges even in populations that appear to be old.

3 Chemical evolution and origin

The bulge's wide abundance distribution fits the one zone of chemical evolution, with a yield of twice solar. As Rich (1990a, b) points out, there is no lack of metal poor stars, as we find in the solar neighborhood. The reader may review figure 3, the distribution in Baade's Window. The first order modifications to the zone model permit gas flows (cf. Pagel and Patchett, 1975; Searle and Zinn, 1978, and Mould, 1984). If inflow of enriched gas is important (as might be the case in a galactic nucleus) the resulting abundance distribution is narrow and peaks at high abundance. If steady outflow of gas dominates the chemical evolution, then the abundance distribution retains the characteristic shape of $e^{-z/y}$, but the effective yield y (mean metal abundance) declines. Indeed, early loss of gas from spheroidal systems has long been postulated to explain low metallicity.

Larson (1974) proposed the idea that star formation might cease when the energy input by SNe equals the binding energy of a galaxy, resulting in a catastrophic wind. The effect of such a sudden, rather than steady wind is to sharply truncate the abundance distribution at high abundances without necessarily reducing the effective yield. N. Tyson and I are determining the abundance distribution in the outer bulge of the Milky Way to determine whether the stellar abundance distribution better fits the steady or catastrophic wind picture. Note that the numerical models of Arimoto and Yoshii have truncated abundance distributions due to the onset of a catastrophic wind when run for the case of low mass ($< 10^{10} M_{\odot}$) galaxies.

If winds are important in chemical evolution, then the local depth of the potential well should be the main factor controlling the abundance of a stellar population. Indeed, Franx and Illingworth (1990) claim that color is correlated with local escape velocity in elliptical galaxies. If wind (outflow) control chemical evolution, then we can roughly predict the abundance distribution at any point in spheroidal population as having the form $e^{-z/\langle z \rangle}$, where the effective yield is just the mean abundance. Also, these models produce color gradients without the need for dissipation and metal rich gas flows toward the nucleus. If the simple model with outflow explains chemical evolution of spheroidal populations, then we must see abundance spreads, even in the nuclei.

It may be premature to rule out dissipation, however, since bulges are found to have rotation supported shapes (Kormendy and Illingworth, 1982) and very steep color gradients (Wirth, 1981). Rich (1990b) finds that metal rich bulge giants have a lower velocity dispersion than metal poor bulge giants. In the Milky Way bulge, Minniti (1991) reports that the metal rich bulge population is supported by rotation, but the metal poor component is not. One must note that these differences in kinematics can be explained as due to observation of different populations rather than due collapse and enrichment in a single event.

Recently, Blitz and Spergel (1991) report that the bulge of the Milky Way may be bar-shaped. Indeed, the many known cases of box/peanut bulge morphology may actually be edge-on bars. Combes et al. (1990) show that stellar disks are unstable to forming thick bars which have the requisite box/peanut morphology when viewed edge-on. Rupen's (1991) observations of non-circular gas motions in the nucleus of NGC 4565 (a classic boxy bulge) lend further support to this idea. A bar might form when infall of gas results in a massive starburst. In such a starburst, all the requirements of the Simple Model of chemical evolution would be satisfied, independent of whether the event takes place shortly after the Big Bang. (Note that Ir luminous galaxies with $\approx 10^{10} M_\odot$ of molecular gas in the nucleus are known; Sanders et al., 1986). Metal rich stellar populations with abundance spreads therefore result from the expected chemical evolution of gas merger/starburst events deep in a galaxy's potential well. The resulting population, viewed a few Gyr later, would have an abundance spread, high metallicity, and a full AGB. However, it would be slightly younger than the halo population. It is interesting to speculate that the Milky Way bulge, and perhaps other bulges, formed in this way.

The stellar content and abundance of the bulge are both inconsistent with a series of recent mergers of Magellanic galaxies, unless the hypothetical system were nearly pure gas (Rich, 1989). The bulge contains no luminous carbon stars, a certain signature of any intermediate-age population. Low mass systems lose their gas before chemical evolution makes metal rich stars; such stars must be formed in a deep potential well. Finally, Ostriker (1988) points out that infall of a stellar system would have dramatically heated the disk during its many crossing of the galactic plane. Therefore, it appears unlikely that the bulge could have accumulated from stellar systems like globular clusters.

4 Summary

In this workshop, we learn that ellipticals commonly have anomalies we would have though remarkable a decade ago: dust lanes, signs of disturbance, molecular gas, and disk structures. In light of these revelations, we must open our mind to the possibility of age spreads in spheroidal populations, and unusual formation scenarios.

On the other hand, we see the glimmer of a binding link between the formation history and resulting stellar content of spheroidal populations: the importance of winds in chemical evolution. Thus while we may not be able to measure the shape of the abundance distribution in distant galaxies, we may have a sturdy theoretical basis to predicts its form. Detailed study of the Milky Way bulge has given us means to predicts the abundance distribution and evolved stellar content of distant bulges and ellipticals.

We may be close to attaining deep insight into the formation and evolution of these systems.

References

Aaronson, M., Cohen, J., Mould, J., Malkan, M.: 1978, *Astrophys. J.* **223**, 824.

Arimoto, N., Yoshii, Y.: 1987, *Astron. Astrophys.* **173**, 23.

Arp, H.C.: 1965, *Astrophys. J.* **141**, 45.

Baade, W: 1944, *Astrophys. J.* **100**, 44.

Baade, W.: 1951, *Publ. Obs. U. Mich.* **10**, 7.

Baade, W.: 1963, in *The Evolution of Galaxies and Stellar Populations*, Harvard University Press: Cambridge, p. 256.

Baum, W.A.: 1959, *Publ. Astron. Soc. Pac.* **71**, 106.

Blanco, V.M.: 1988, *Astron. J.* **95**, 1400.

Blitz, L., Spergel, D.: 1991, *Astrophys. J.* **370**, 205.

Combes et al.: 1990, *Astron. J.* **86**, 981.

Franx, M., Illingworth, G: 1990, *Astrophys. J. Letters* **359**, L41.

Frogel, J.A., Persson, S.E., Aaronson, M., Matthees, T.: 1978, *Astrophys. J.* **220**, 75.

Frogel, J.A., Whitford, A.E.: 1987, *Astrophys. J.* **320**, 199.

Frogel, J.A., Whitford, A.E., Rich, R.M.: 1984, *Astron. J.* **89**, 1536.

Frogel, J.A., et al.: 1990, *Astrophys. J.* **353**, 494.

Frogel, J.A.: 1971, Ph.D. Thesis, Caltech.

Graham, J.A., Nemec, J.M.: 1983, in *IAU Symp. 108: Structure and Evolution of the Magellanic Clouds*, eds. van der Bergh and K.S. De Boer, p. 37.

Hibbard, J.E., Rich, R.M.: 1990, in *Galactic Bulges*, ESO: Garching, p. 76.

Impey, C.D., Wynn-Williams: 1986, *Astrophys. J.* **356**, 62.

Johnson, H.L.: 1966, *Astrophys. J.* **143**, 187.

Jura, M. et al.: 1987, *Astrophys. J.* **312**, L11.

Kormendy, J., Illingworth, G.: 1982, *Astrophys. J.* **256**, 456.

Larson, R.: 1974, *Mon. Not. R. Astr. Soc.* **166**, 585.

Minniti, D.: 1991, Private Communication.

McCarthy, M.F., S.J., Blanco, V.M.: 1990, in *Galactic Bulges* ESO: Garching, p. 11.

Morgan, W.W., Mayall, N.U.: 1957, *Publ. Astron. Soc. Pac.* **69**, 291.

Morgan, W.W., Osterbrock, D.E.: 1969, *Astron. J.* **74**, 515.

Mould, J.: 1984, *Publ. Astron. Soc. Pac.* **96**, 773.

Nassau, J.J., Blanco, V.M.: 1958, *Astrophys. J.* **128**, 46.

O'Connell, R.W.: 1986, in *Stellar Populations* , eds. C. Norman, A. Renzini, M. Tosi, Cambridge: Cambridge Univ. Press.

O'Connell, R.W.: 1988, in *The Edwin Hubble Symp.*, ASP conf. series.

Pagel, B.E.J., Patchett, B.E.: 1975, *Mon. Not. R. Astr. Soc.* **172**, 13.

Renzini, A., Greggio, L.: 1990, in *Galactic Bulges*, ESO: Garching p. 47.

Renzini, A.: 1986, in *Stellar Populations*, Cambridge: Cambridge Univ. Press, p. 456.

Rich, R.M.: 1965, *Mem. Soc. Astron. It.* **56**, 23.

Rich, R.M.: 1988, *Astron. J.* **95**, 828.

Rich, R.M. et al.: 1989, *Astrophys. J. Letters* **341**, L51.

Rich, R.M.: 1990a, in *Galactic Bulges*, ESO: Garching, p. 65.

Rich, R.M., Mould, J: 1991, *Astron. J.* **101**, 1286.

Rupen, M.: 1990, Ph.D Thesis, Princeton University.

Sanders et al.: 1986, *Astrophys. J. Letters* **305**, L45.

Searle, L., Zinn, R.: 1978, *Astrophys. J.* **225**, 357.

Shaples, R.A. et at.: 1991, *Mon. Not. R. Astr. Soc.* **246**, 54.

Spinrad, H., Peimebrt, M.: 1975, in *Stars and Stellar Systems*, vol. IX, A. and M. Sandage, eds., Chicago, p. 37.

Taylor, B.J., Johnson, S.B.: 1987, *Astron. J.* **322**, 930.

Terndrup, D.: 1990, in *Galactic Bulges*, ESO: Garching, p. 199.

Terndrup, D.M.: 1988, *Astron. J.* **96**, 884.

Terndrup, D.M. et al.: 1990, *Astrophys. J.* **357**, 453.

Tyson, N.D.: 1991, Ph.D. Thesis, Columbia University.

Walker, A., Terndrup, D.: 1991, *Astrophys. J.* in press.

Whitelock, P.A. et al.: 1991, *Mon. Not. R. Astr. Soc.* **248**, 276.

Whitford, A.E., Blanco, V.M.: 1979, Bull. AAS **11**, 675.

Whitford, A.E.: 1975, in *Stars and Stellar Systems*, vol. IX, A. and M. Sandage, eds., Chicago, p.159.

Whitford, A.E., Rich, R.M.: 1983, *Astrophys. J.* **274**, 723.

Whitford, A.E.: 1978, *Astrophys. J.* **226**, 777.

Wirth, A.: 1981, *Astron. J.* **86**, 981.

Wood, P., Bessell, M.: 1983, *Astrophys. J.* **265**, 748.

Discussion

F. Matteucci - How do you define dissipative as opposed to non-dissipative formation?
R.M. Rich - I would define dissipative collapse as a galaxy formation scenario in which the kinetic energy of gas cloud motions is dissipated in cloud-cloud collisions, ultimately resulting in a flattened structure rotating with gas clouds falling toward the center. A wind model requires no dissipation to establish a color gradient.

P. van der Kruit - When Searle and I found the color gradients in the bulges of NGC 891 and NGC 7814 we did not regard this as evidence for dissipational collapse. Also the observed similarity of isochromes and isophotes is inconsistent with (for example) Larson's models where the isochromes are much flatter. Suppose that the gradient is set up before virialization of the bulge, because the inner parts of the proto-galaxy, where pop II is forming, can hold on longer to their gas. If followed by a dissipationless collapse through violent relaxation the gradient may be conserved. The numerical work of Tjeerd van Albada shows that in violent collapse there is a statistical relation between binding energy of the stars before and after collapse.
R.M. Rich - The value of observing abundance distribution and kinematics is that we can distinguish between these alternatives.

D. Burstein - In a paper in press (Schweizer et al., *Astrophys. J. Letters*) it is shown that a morphological/kinematical criterion of galaxy interaction (developed by Schweizer

and Seitzer) correlates with deviations of absorption line-strengths and absolute magnitude (or central velocity dispersion). The sense of deviation is that Hβ, CN and Mg indices is that greater evidence of interaction correlates with weaker Mg and CN but strong Hβ; what one expects from recent star formation added to an old stellar population. The evidence indicates that differences in age and differences in metallicity contribute approximately equally to absorption line strength variation.
R.M. Rich - This can probably be better sorted out if we include the UV and 10 μm IR. Intermediate age AGB stars have to lose substantial mass; hence their shells will make excess 10 μm emissions.

A. Renzini - Radial gradient may perhaps survive merging events, but this leaves the question open over how they were formed in the first place. Also, in the merging scenario I find hard to see how the luminosity-metallicity correlation can come to be established. Does anyone of those believing ellipticals are made up of merging spirals who has an explanation of these two questions?
R.M. Rich - If mergers deliver primarily gas to the nucleus, and that gas subsequently undergoes a starburst and *in situ* chemical evolution, then both color gradients and luminosity-metallicity correlations can arise naturally. Both effects would be caused by winds, and would depend on the depth of the potential well. The work of Franx and Illingworth suggests just such a wind-driven link.

Michael Rich "holding the screen".

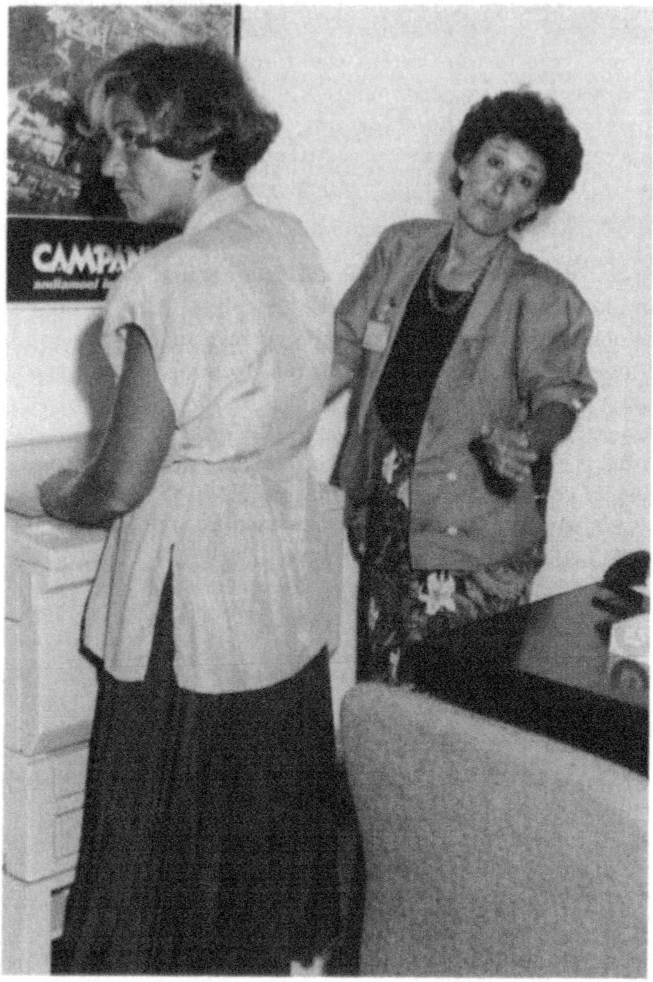

On the pretty side of the LOC: from right to left, Ms. Teresa Ievolella and Ms. Pina Iaccarino.

STELLAR DISKS IN EARLY-TYPE GALAXIES

M. Capaccioli[1] and N. Caon[2]
[1]*Department of Astronomy, University of Padova, Italy*
[2]*International School for Advanced Study, Trieste, Italy*

ABSTRACT. Correlation between isophotal and physical parameters suggests that elliptical galaxies are divided into two main classes: the 'boxy' Es, which are likely the result of a strong environmental evolution (mergers), and the 'disky' Es, which represent the continuation of the S0 sequence toward vanishing disk-to-bulge ratios. The latter class is quite common. The detection of disks, which is of the order of 40%, depends strongly on disk-to-bulge ratio and on inclination angle; statistical considerations indicate that about 50% of disks go undetected.
Disks in E and S0 galaxies are well described by exponentials with peak brightness clustering around the same value found for spirals (μ_0 = 21.65 B-mag arcsec^{-2}). It is still questioned whether the increase of the scale-height with radius is a property common to the whole class or holds only for particular cases.
Even if faint, stellar disks can affect both kinematical and geometrical measurements, so that an appropriate allowance must be made in performing dynamical investigations (e.g., the $(\varepsilon, v/\sigma)$ test) or models.
Disks are effective diagnostics of the galaxian structure: rotation curves of disky galaxies can give information on the radial trend of the mass-to-light ratio, and, under reasonable assumptions, can be used to infer the intrinsic shape of the spheroidal component.

1 Introduction

The discovery that elliptical galaxies possess a kinematical behavior quite different from that of lenticulars and spirals (Bertola and Capaccioli, 1975, Illingworth, 1977) — at variance with the theoretical views in vogue at the time (Gott, 1975, Wilson, 1975) — inspired Bertola and Capaccioli (1978) to speculate on the existence of a physical 'dichotomy' between *disk* (S0 and S) and *diskless* (E) galaxies; a perception anticipated in the pioneering work of van Houten (1961). Soon after, new kinematical studies of bright as well as faint E galaxies, and the comparison of the latter with bulges of S0s and spirals, made it apparent that the dichotomy was actually internal to the class E. It was in fact possible to catalog ellipticals into two *kinematical* families: one dominated by random motions and made up of the most luminous galaxies, and another containing fainter objects characterized by a fair balance between random and ordered motions, just as in bulges of S0 and S galaxies (Davies et al., 1983; but see Bender and Nieto, 1990, for the dwarf Es).

G. Longo et al. (eds.), Morphological and Physical Classification of Galaxies, 99–114.

The next logical step was the re–arrangement of some ellipticals within the S0 class. The idea that quite a few E galaxies might be actually misclassified S0s was first suggested by Michard (1984), who pointed out the striking similarity of the ellipticity profiles of the isophotes of some flat Es with those typical of S0s close to edge–on. Further support came from the occasional discovery of faint disks in otherwise *bona fide* Es, such as NGC 4125 or NGC 4697 (Capaccioli, 1987; Carter, 1987), from considerations and experiments on the 'detectability' of faint disks embedded in bright bulges (Capaccioli, Held, and Nieto, 1987; van den Bergh, 1989; Rix and White, 1990), and from the need of understanding the systematic failures of all tests aimed at deriving the intrinsic shape of ellipticals (Capaccioli, 1987).

In the meantime, the wider use of the Fourier analysis introduced by Cawson (1983) to parametrize the deviations of the actual galaxy isophotes from their best–fitting ellipses (e.g., Lauer, 1985; Bender and Möllenhoff, 1987) led to define two classes of ellipticals according to the sign of the 4 th order Fourier coefficient a_4 (Bender, 1987): galaxies with positive a_4 have 'pointed' or 'lemon shaped' isophotes, while objects with negative a_4 have 'boxy' isophotes. The key fact is that isophotal morphology and kinematical properties appear well correlated to each other, and both seem correlated with other global properties, such as absolute magnitude, X–ray and radio luminosity (Bender et al., 1989a, Bender, 1990). In particular, the strong similarity between 'pointed' Es and lenticular galaxies suggests that the former are 'disky' galaxies representing the extreme case of S0s with vanishing faint disks in the continuation of the disk/bulge sequence (Capaccioli, 1990; Capaccioli and Longo, 1990).

In turn, the characteristic features of 'boxy' Es such as their systematically stronger 'activity' in the radio, UV, X–ray, and Hα regions, the higher luminosity and lower specific angular momentum, and the significantly larger frequency of counter–rotating cores with respect to 'disky' Es, have induced consideration of the possibility that they are merger products (Nieto, 1989); a possibility that should not be abused, as thoroughly discussed by van den Bergh (1990) and directly shown by the emblematic case of NGC 5252 (Fig. 1).

The whole body of new conjectures about 'boxy' and 'disky' ellipticals resurrects and strengthens the former idea of a dichotomy between disk and diskless galaxies, suggesting that it represents the fundamental distinction between genetic varieties on the one side (disk galaxies, which develop along a sequence primarily described by the value of the specific angular momentum; Capaccioli and Vietri, 1989), and on the other side the results of a strong environmental evolution ('boxy' mergers).

The plan of this paper is to review in which way disks can be detected in ellipticals and bulge–dominated lenticulars, and how they can be measured. We will also summarize their photometrical and kinematical properties and, through some examples, report on the use of faint disks for investigating the global structure of the host galaxy (both Es and S0s).

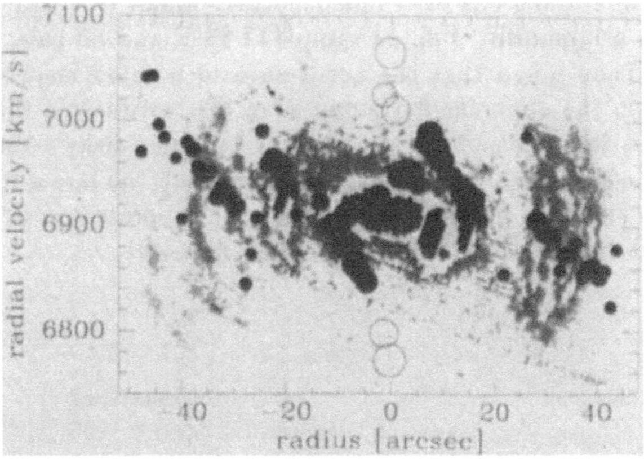

Figure 1. Image of the extended [OIII] λ 5007 emission in the Seyfert II galaxy NGC 5252 and, overimposed, the corresponding radial velocity curve for a cross-section at P.A.= 170° (Capaccioli, Cappellaro, and Held, 1991). The size of the circles is proportional to the line intensity; open symbols indicate those cases where two velocity components coexist at the same location. The remarkable symmetry between the two sides over a range $\Delta R \geq 20$ kpc challanges the interpretation of the extranuclear gas component in terms of a merging event (in spite of its provocative morphology).

2 The 'disky' ellipticals

While it seems by now positive that 'disky' Es represent a large fraction of the total population of E galaxies, the actual percentage — as well as its dependence on environment — is still uncertain. So far, Fourier analysis applied to moderately large samples of objects gives that $\sim 30\% - 40\%$ of all galaxies classified E in the main morphological catalogues exhibit isophotes with a 'pointed' aspect (Bender, 1987). It is also found that about the same percentage of Es display 'boxy' isophotes, with the rest showing irregular isophotes or coexistence of boxiness with diskiness. An ambitious project aimed at performing the isophotal analysis on 300 ellipticals is now under way at ESO as one of the so-called Key Programs with a substantial allocation of telescope time (Bender et al., 1989b).

If diskiness *always* implies that a disk-like structure — possibly small and faint — is harbored inside the dominant bulge[1], then we shall investigate how the statistics of detections is affected by parameters such as the inclination angle of the galaxy principal plane on the line of sight, the disk-to-bulge luminosity ratio, the ratio of the scale-lengths of disk and bulge, and the intrinsic flattening of the bulge. The role played by inclination is described by the diagram in which the apparent ellipticity is correlated with the Fourier coefficient a_4 (Bender et al., 1989a). Not only diskiness but also boxiness is recognized only when $\varepsilon \gtrsim 0.3$; for $\varepsilon \lesssim 0.3$ the value of a_4 is always so low as to prevent a reliable classification of the isophotal shape.

[1]This might not necessarily be the case. As a by-product of n-body calculations related to galaxy formation problems, Stiavelli, Londrillo, and Messina (1991) find that the same spheroidal galaxy may present either a moderate boxiness or some diskiness depending on the view angles.

The same limit of $\varepsilon = 0.3$ has been independently found by Capaccioli, Caon, and Rampazzo (1990) in a luminosity–limited sample of 33 E and S0 galaxies belonging to the Virgo cluster. They noted that the occurrence of pointed isophotes in Es has a similar dependence on the apparent flattening as in S0s, estimating that about 50% of the elliptical galaxies in their sample are actually of type 'disky'; they are generally fainter and more elongated than the other Es. If significant, the larger fraction of 'disky' Es than in Bender's (1987) is possibly due to the higher percentage of faint ellipticals, confirming the correlation of a_4 with luminosity (see figure 2).

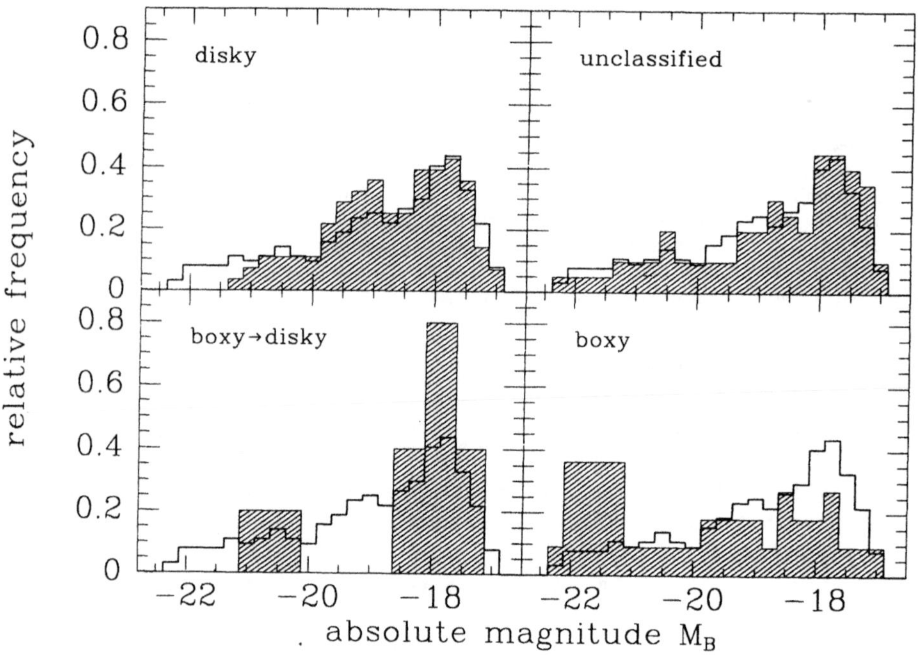

Figure 2. The solid line plots the relative luminosity function for a sample of E and ordinary S0 galaxies brighter than $M_B = -17.3$ mag, belonging to the Virgo and to the Fornax clusters. Data are from Caon et al., (1990), Caon (1991), and Trevisani (1991).
Galaxies are also divided in three groups according to their isophotal shape (filled histograms): pure disky (28 objects), disky with boxy bulges (boxy→disky; 5), pure boxy (11). We also show the remaining 20 galaxies, mostly faint, which could not be classified.

The detectability of a disk as a function of inclination, disk–to–bulge ratio etc., has been investigated by Rix and White (1990) through simple galaxy models consisting of two idealized components: a $r^{1/4}$ spheroidal bulge and an exponential disk, with bulge–to–disk ratio and scale–length ratio chosen within appropriate ranges of values. The positive detection of the disk was set by two criteria:

a) the major axis profiles deviate from the $r^{1/4}$ law by more than 0.1 mag, and/or

b) the a_4 coefficient is larger than 1% of the isophotal major axis a.

The parameters controlling the ability of discovering the disk turned out to be the inclination angle and the bulge–to–disk ratio, while the scale–length ratio and the intrinsic flattening of the spheroid were found to be of secondary importance. For inclination angles larger than $\sim 50°$, there is no appreciable effect on a_4, whatever the bulge/disk ratio is, and significant deviations from the $r^{1/4}$ law can be seen only when $L_{disk}/L_{bulge} > 0.25$.

In summary, Rix and White estimated that the detection probability for disks is about 50%, a figure which would imply that the great majority of Es are 'disky'. Since 'disky' Es have in general lower luminosities than 'boxy' Es, magnitude limited samples are biased toward 'boxy' objects, which explains the high percentage of the latter in such samples.

3 Measuring the disk

Several methods have been so far devised and used to decompose a disk galaxy (S0 or spiral) into its photometric components. Since most methods work well only if the disk–to–bulge ratios are not too small, specific strategies have been conceived to tackle the more difficult problem of disentangling small disks from much brighter bulges, as is the case in 'disky' Es.

We outline here the decomposition methods known to us, listing their basic assumptions and input/output data, and briefly reporting on their methodology.

Kormendy (1977) method

Assumptions: bulge and disk light profiles are idealized by a $r^{1/4}$ and an exponential law respectively:

$$\mu(r) = \mu_e + 8.33 \times \left[(r/r_e)^{1/4} - 1\right]$$

$$\mu(r) = \mu_0 + 1.09 \times \alpha\, r$$

Input data: major axis or equivalent luminosity profile.

Output data: scale parameters for the adopted models, e.g., effective surface brightness and radius, μ_e and r_e, for the bulge, and peak brightness and scale–length, μ_0 and α^{-1}, for the disk. They are used to compute bulge and disk total magnitudes and the disk–to–bulge ratio.

Method: iterative. The procedure, originally devised by de Vaucouleurs (1959) and van Houten (1961), works well when neither bulge or disk dominate. Firstly, the observed surface brightness is plotted against the linear radius to identify by eye, and then fit by a straight line, the range dominated by the exponential disk. The guessed disk profile is extrapolated to the entire radial range and subtracted from the global light profile. The result is fitted by a $r^{1/4}$ formula which is in turn subtracted from the global profile. The difference is again fitted by an exponential, and the above loop is repeated until some convergence criterium is fulfilled. Alternatively — and more objectively — some non-linear fit algorithm is applied to the whole radial domain and solved directly for the bulge and disk parameters. It is worth noticing that, for very small disk–to–bulge ratios departures of bulge light profiles from $r^{1/4}$ (Capaccioli, 1989) may become important in retrieving faint disks.

Kent (1986) method

Assumptions: bulge and disk each have elliptical isophotes of constant flattening (but different for the two components). No *a priori* knowledge of, or assumption on, the radial run of the surface brightness is required.

Input data: luminosity profiles along both the major and the minor axis, $I_{maj}(r)$ and $I_{min}(r)$, and fixed major–to–minor axis ratios for bulge and disk, f_B and f_D (estimated from the run of the ellipticity with radius).

Output data: light profiles of the bulge and the disk separately, $I_B(r)$ and $I_D(r)$.

Method: using the above notation it is

$$I_{maj}(r) = I_B(r) + I_D(r)$$
$$I_{min}(r) = I_B(f_B \times r) + I_D(f_D \times r) \qquad (1)$$

Defining the '*observable*' $D(r)$ as

$$D(r) = I_{min}(r) - I_{maj}(f_D \times r) \qquad (2)$$

one gets

$$I_B(r) = D(r/f_B) + I_B\Big[(f_D/f_B) \times r\Big] \qquad (3)$$

Provided $f_D > f_B$, this equation is used to determine iteratively the bulge profile. In the first iteration I_B is set equal to D and substituted into the right-hand side of eqn. 3 to compute an improved version of I_B. Thé process is repeated until convergence is achieved. I_D is then determined using one of eqns. 1.

Kodaira et al., (1986) method

Assumptions: as in Kormendy's method, bulge and disk profiles are interpolated by the $r^{1/4}$ and exponential formulae.

Input data: the so–called generalized profile and three global parameters, D_{26} (diameter at $\mu_B = 26$ mag arcsec^{-2}), SB (mean surface brightness within the $\mu = 26$ isophote), and $X1(P)$ (mean concentration index).

Output data: scale parameters of bulge and disk light profiles.

Method: a grid of models with different combinations of the above scale parameters is generated, and for each of them the 3 global parameters are computed. Then 10 models are selected for which the distance D between model and observed parameters

$$D^2 = \Big[\Delta \log (D_{26}/\sigma_D)\Big]^2 + \Big[\Delta SB/\sigma_{SB}\Big]^2 + \Big[\Delta X1(P)/\sigma_X\Big]^2$$

is minimum. Lastly, a search is made among these models to identify one giving the best fit to the observed generalized profile.

Capaccioli et al., (1987) method

Assumptions: the bulge possesses purely elliptical isophotes and the disk is affecting only a narrow strip of the galaxy image about the major axis (i.e., it must be nearly edge–on). No *a priori* knowledge of the dependence on radius of the light profiles of the two components is required.

Input data: two–dimensional (2–D) light distribution of the galaxy.

Output data: 2–D light distributions for disk and bulge separately.

Method: firstly, the flattening profile of the bulge is obtained by fitting ellipses to a set of galaxy isophotes cleaned of those parts about the major axis 'judged' to be altered by the disk. The geometrical information (ellipticity and twisting) is then coupled to the complete minor–axis light profile — which, except near the center, is assumed to be unaffected by the disk — to build a 2–D photometric model of the bulge. Subtracting this from the original image of the galaxy one obtains the 2–D light distribution of the disk. In the case of moderate inclination, such estimate of the disk luminosity profile may be used as a first guess in an iterative procedure aimed at accounting for the presence of some disk light along the minor axis also.

Simien and Michard (1990) method 1

Assumptions: there is a one-to-one correspondence between the deviation of an observed isophote from the best–fitting ellipse, and the local luminosity of the disk, assumed to be substantially fainter than the bulge.

Input data: the total 2–D light distribution and the disk flattening profile, $q_D(r)$, estimated/guessed from the total flattening profile.

Output data: bulge and disk 2–D light distributions.

Method: the relative deviations from the best–fitting ellipses are converted into brightness differences through the gradient of the luminosity profile and assigned to the disk. A model of the disk is then computed and subtracted from the original image and the isophote shape analysis is carried out again. The process is iterated, adding up the disk values to those of the preceding steps, until convergence is reached. The residual image then gives the light distribution of the bulge.

Simien and Michard (1990) method 2

Assumptions: bulge isophotes are elliptical. The bulge flattening q_B must be smaller than q_D anywhere.

Input data: major and minor axis light profiles, $I_{maj}(r)$ and $I_{min}(r)$, and the flattening profiles of bulge and disk, $q_B(r)$ and $q_D(r)$, estimated/guessed from the flattening profile of the galaxy isophotes.

Output data: bulge and disk major–axis light profiles, $I_B(r)$ and $I_D(r)$.

Method: it is

$$I_{maj}(r) = I_B(r) + I_D(r)$$
$$I_{min}\Big[r \times q_B(r)\Big] = I_B(r) + I_D\Big[u_1(r)\Big] \tag{4}$$

with $u_1(r)$ such that $u_1(r) \times q_D\Big[u_1(r)\Big] = r \times q_B(r)$. Defining a set of N functions u_i $(i > 1)$ by the relation $u_i(r) = u_1\Big[u_{i-1}(r)\Big]$, it can be shown that:

$$I_B(r) = \sum_{i=1}^{N} \Big\{ I_{min}\Big[u_i(r) \times q_D(u_i[r])\Big] - I_{maj}\Big[u_i(r)\Big]\Big\} + \epsilon$$

where $\epsilon = I_B\Big[u_N(r)\Big]$ is a residual which can be neglected when N is large enough. The disk luminosity is then given by the first of eqns. 4. Several trials can be made with different runs of $q_D(r)$ and $q_B(r)$ in order to seek the best–fit.

Shaw and Gilmore (1990) method

Assumptions: the bulge has elliptical isophotes and a $r^{1/4}$ light profile. The disk is exponential along the major axis, while the orthogonal cross–sections follow a sech^2 law (van der Kruit and Searle 1982).

Input data: 2–D light distribution.

Output data: scale parameters of the components, namely:
1) μ_e, r_e, and $q = b/a$ for the $r^{1/4}$ bulge, and
2) μ_0, α^{-1}, and h_z for the exponential/sech^2 disk.

Method: an iterative non–linear least–squares algorithm is used, which minimizes the sum of the squares of the residuals, or the reduced χ^2, between the observed two–dimensional luminosity distribution of the galaxy and that of the model.

Scorza and Bender (1990) method

Assumptions: bulge isophotes are perfectly elliptical and the disk light profile is exponential.

Input data: the 2–D light distribution.

Output data: scale parameters of the model profile, inclination, and center coordinates of the disk, and position angle of the photometric line of nodes.

Method: 2–D exponential disk model is subtracted from the galaxy image, each time exposing the difference to the isophotal shape analysis. The best model is the one producing a residual image (bulge) with perfectly elliptical isophotes (Fourier coefficients nearly null anywhere).

4 Properties of stellar disks in Es and S0s

Studies of the luminosity profiles of the disks of S0 and E galaxies have shown that they can be well described by exponential laws, with scale–length varying between 1 and 4 kpc, while the central surface brightness values cluster around $\mu_0 = 21.65$ B–mag arcsec^{-2}, just like disks of spiral galaxies (Fig. 3; but see also Scorza and Bender, 1990, for a remarkably discrepant case). The practical constancy of μ_0 and of the circular velocity in the host galaxy — which must be the rotational velocity in the cold disk — seems to indicate that the total mass and angular momentum of a disk depend on just one parameter: the disk's scale length. This fact, and the large range in bulge–to–disk ratio values, may help constraining the mechanisms with which galaxy components acquire angular momentum (Capaccioli and Vietri, 1989).

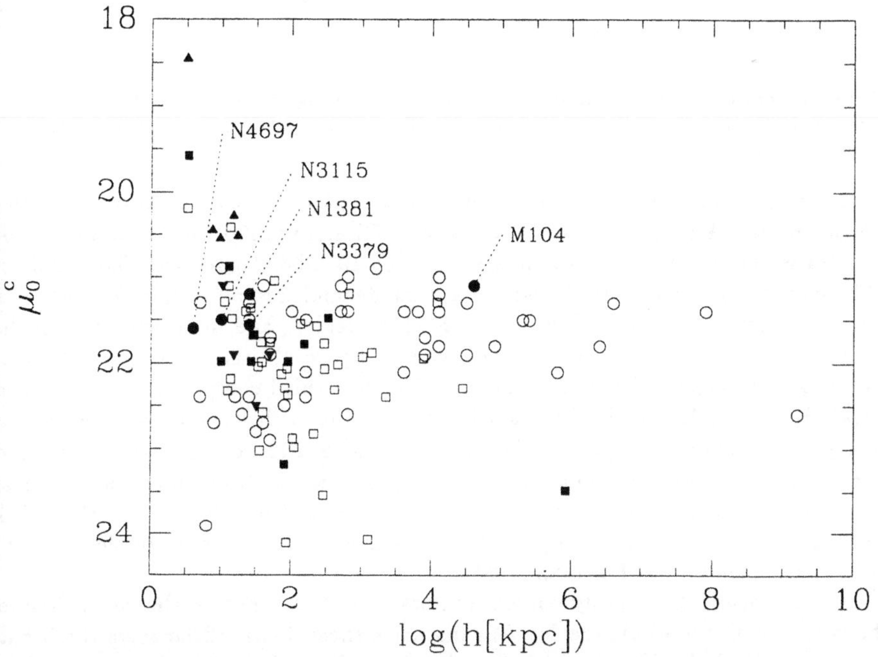

Figure 3. Distribution of galaxian disks in the central surface brightness (B–band) versus scale–length plane. Open symbols represent spiral galaxies from van der Kruit (1987; circles) and from D'Onofrio (1991; squares). Filled squares are early–type galaxies from Simien and Michard (1990), triangles are low–luminosity ellipticals from Trevisani (1991), while inverted triangles are S0s from Caon (1989). Filled circles represent M104 (Burkhead 1986), NGC 1381 (de Carvalho and Da Costa 1987), NGC 3115 and NGC 3379 (Capaccioli et al., 1987, 1991b), and NGC 4697 (Capaccioli, 1987).

Still debated is the color of disks, compared to that of bulges. While most S0s have disks which are bluer than their bulges by 0.1 mag in $U - V$, in NGC 3115 this difference is about 0.5 mag (Silva et al., 1989), to be compared with the typical values for spirals, 0.5 – 0.75 mag (Caldwell, 1983). This suggests that the disk of NGC 3115 has an age intermediate between the disks of typical S0s and those of spirals.

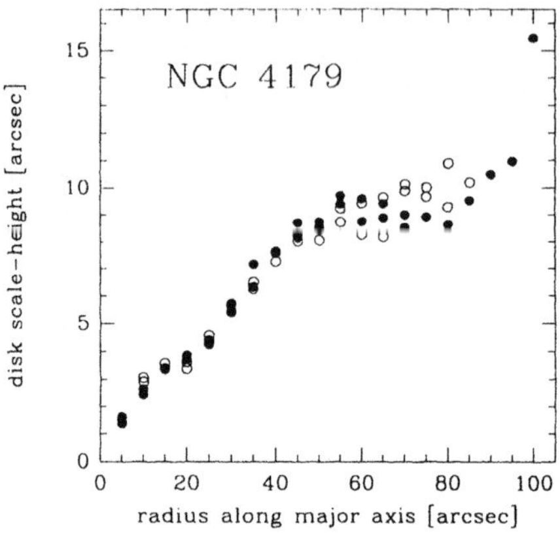

Figure 4. Radial run of the scale-height of the disk of the edge-on S0 galaxy NGC 4179 (Capaccioli, Fasano, and Held, 1991). Open and filled symbols correspond to the R and B bands respectively.

As for the thickness of disks, there are controversial results on the trend of the scale-height with radius. While, for instance, Shaw and Gilmore (1990) find that the scale-height h_z turns out to be constant in a sample of 10 'disky' galaxies, Capaccioli et al., (1987) show that in NGC 3115 there is a clear dependence on radius: h_z increases sharply in the inner part, levels off in an intermediate region, and increases again in the outer part (see another example of this behavior in figure 4). The question is further complicated by the suspected presence, in some S0s, of a third component, the so called 'thick disk' (Burstein, 1979). While a thick disk (that is, a disk whose scale-height is significantly larger than that typical of ordinary disks, $\simeq$ 200 pc) has been recognized is a few cases (e.g., Jensen and Thuan, 1982), its presence has been ruled out for most of edge-on lenticulars studied so far. In any case, it is not clear whether the thick disk must be regarded as a real new component or it is merely the response of the bulge to the disk's gravitational field (Freeman, 1983).

At least for 'disky' Es, the dynamical importance of the disk is almost negligible, given the very low disk-to-bulge ratios. For instance, these disks indeed seem too 'light' to be the cause of the bell-shaped trend of the isophotal flattening of the dominant bulges (Capaccioli, 1987, Caon 1991; see also figure 5). However, even faint disks may affect the kinematical measurements in a systematic way. When applied to multi-component systems, the methods for measuring rotation and velocity dispersion, such as the Fourier quotient and the Cross Correlation technique, give more weight to the narrow-lined, thus more rapidly rotating component (see, e.g., Whitmore, 1980, and Franx and Illingworth, 1988). This, in turn, leads to overestimate the ratio v/σ (e.g., Busarello, Longo, and Feoli, 1990). On the other hand, also the ellipticity measurements are affected by the disk, so that appropriate allowance must be made for these effects in order to investigate the $(\varepsilon, v/\sigma)$ plane, as discussed by Nieto, Capaccioli, and Held (1988).

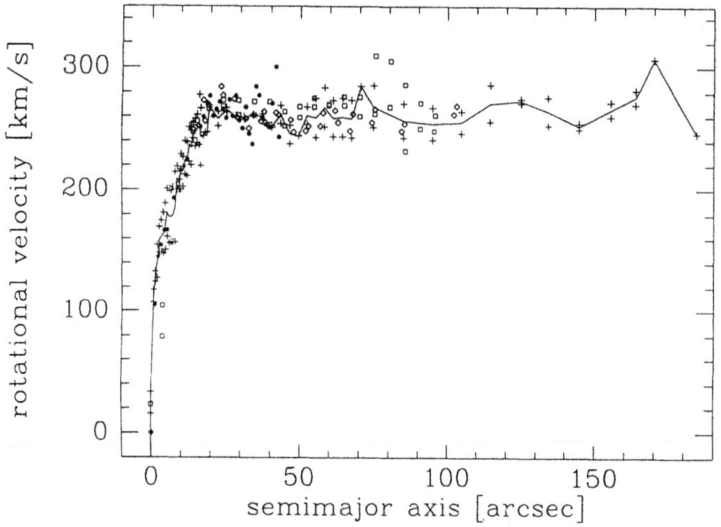

Figure 5. Rotation curve of NGC 3115 (Capaccioli et al., 1991a). The rotational velocity is practically constant from $a \simeq 20"$, where the light of the edge–on disk still dominates over that of the bulge, to $3'$, equivalent to 2 effective radii. This behavior is inconsistent with a constant mass–to–light ratio over the whole radial range covered by these observations.

5 Stellar disks and galaxian properties

Stellar disks have proven to be effective diagnostics of the galaxian structure. For instance the radial trend of the scale–height of the disk of the standard edge–on S0 NGC 3115 has been used by Capaccioli, Vietri, and Held (1988) to compute the ratio M/L of the disk itself under the assumption that the flare–up of the scale–height of the (exponential) disk marks the radius at which external gravity begins to dominate over self–gravity. The fact that the mass–to–light ratio applying to a flat component, $M/L_{disk} = 7$, is almost identical to the global M/L within the same radial range of about 5 kpc (or $\sim 1r_e$), was taken as a negative indication of a significant contribution to the potential field by a dark halo. On the other hand, by deep spectroscopy along the major axis of the same galaxy, Capaccioli et al., (1991a) were able to show that the rotational velocity of the bulge of NGC 3115 remains flat out to $r \sim 2r_e$, a fact implying at least $M/L = 14$ and thus a significant growth in the contribution by dark matter in passing from 1 to $2r_e$ (Fig. 6).

Tests on the intrinsic shape of E galaxies have shown that there is a marked difference between 'boxy' and 'disky' Es. While for the former the deficiency of apparently round galaxies suggests a triaxial shape, for the latter the apparent distribution of the ellipticity is fairly consistent with the oblate shape (Fasano, 1991). It must be said, however, that all our present knowledge about this problem rests on indirect evidence, and that the 'reasonable' assumptions of our statistical tests may fail. An example is offered by an analysis of the correlation between effective parameters of bulge light profiles, as it is in the revisitation by Capaccioli and Caon (1991) and in figure 7, which incorporates new data (Trevisani, 1991).

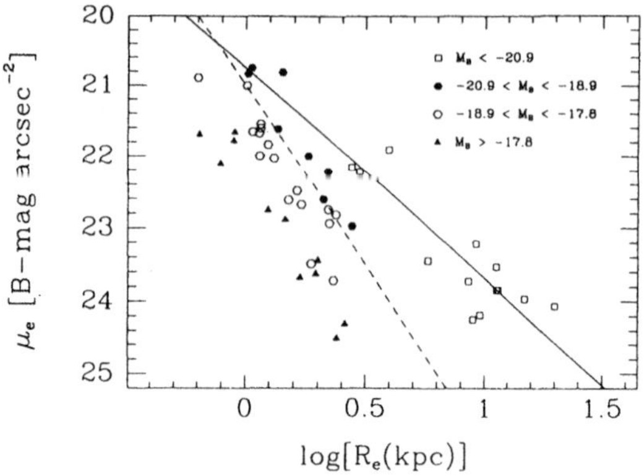

Figure 6. Relation between the effective radius R_e [kpc] and the corresponding surface brightness μ_e for a luminosity limited sample of 52 E and S0 galaxies belonging to the Virgo Cluster (data from Capaccioli and Caon, 1991 and Trevisani, 1991). Galaxies are grouped in four luminosity classes, designated by different symbols. The solid line is the relation given by Hamabe and Kormendy (1987), reduced to the same distance and photometric band of the plotted Virgo data. The locus of the points for galaxies with fixed luminosity and obeying the same scaling law $B_T = \mu_e - 5\log R_e + \text{const.}$, is shown as a dashed line.

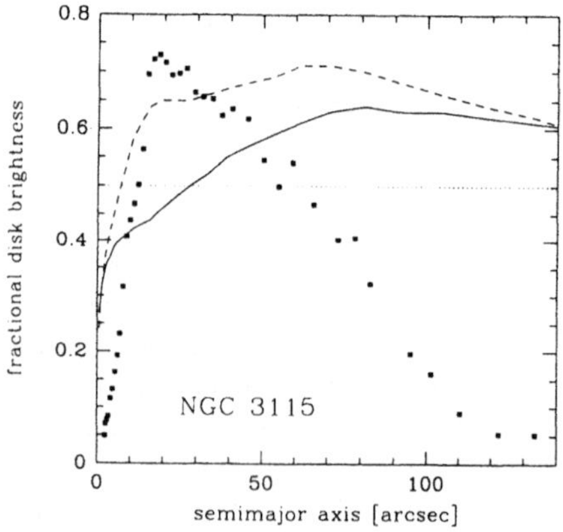

Figure 7. The filled squares represent the radial trend of the surface brightness of the disk of NGC 3115 along the major axis, normalized to the total surface brightness. The dashed and solid lines are the ellipticity profiles of the isophotes of the galaxy image and of the disk–subtracted (≡ bulge) model (Capaccioli et al., 1991a).

It is found that the effective surface brightness μ_e covers an interval of at least 2 magnitudes for bulges of the same total luminosity and with light profiles all following roughly the same law. This spread is mainly intrinsic, since we expect inclination to alter the effective surface brightness of spheroidal galaxies by at most a factor of 2 going from face to edge on view, with obvious consequences on the photometric shape tests.

The presence of even faint disks may help solving the problem of bulge shapes by studying individual galaxies rather than statistical samples. The reason is that very likely a stellar disk provides us with the inclination to the line–of–sight of one on the two principal planes containing the intermediate axis of a generic triaxial bulge (Heisler, Merrit, and Schwarzschild, 1982). Using this fact and postulating a morphological similarity between NGC 3115 and NGC 3379, an E galaxy suspected to host a small disk, Capaccioli et al., (1991b) show that the structure common to these two galaxies is an inner almost oblate spheroid ($b/a = 0.9$, $c/a = 0.5$) whose symmetry plane contains the long and intermediate axes of a moderately triaxial halo with $b/a = 0.75$, and $c/a = 0.5$.

References

Bender, R.: 1987, Ph.D. Thesis, Univ. of Heidelberg.

Bender, R.: 1990, *Dynamics and Interactions of Galaxies*, ed. R. Wielen (Heidelberg: Springer–Verlag), p. 232.

Bender, R., Möllenhoff, K.: 1987, *Astron. Astrophys.* **177**, 71.

Bender, R., Nieto, J.-L.: 1990, *Astron. Astrophys.* **239**, 97.

Bender, R., Surma, P., Döbereiner, S., Möllenhoff, C., Madejsky, R.: 1989a, *Astron. Astrophys.* **217**, 35.

Bender, R., Capaccioli, M., Macchetto, F., Nieto, J.–L.: 1989b, *The Messenger* **55**, 6.

Bertola, F., Capaccioli, M.: 1975, *Astrophys. J.* **200**, 439.

Bertola, F., Capaccioli, M.: 1978, *Astrophys. J.* **219**, L95.

Burkhead, M.S.: 1986, *Astron. J.* **91**, 777.

Burstein, D.: 1979, *Astrophys. J.* **234**, 829.

Busarello, G., Longo, G., Feoli, A.: 1990, *Bulges of Galaxies*, (La Serena), eds. B. Jarvis and D.M. Terndrup (Garching: ESO Workshop Publ.), p. 269.

Caldwell, N.: 1983, *Astrophys. J.* **268**, 90.

Caon, N.: 1989, Laureato Thesis, Univ. of Padova.

Caon, N.: 1991, Master Thesis, ISAS (Trieste).

Caon, N., Capaccioli, M., Rampazzo, R.: 1990, *Astron. Astrophys. Suppl. Ser.* **86**, 429.

Capaccioli, M.: 1987, IAU Symp. 127, *Structure and Dynamics of Elliptical Galaxies*, ed. T. de Zeeuw (Dordrecht: Reidel), p. 47.

Capaccioli, M.: 1989, *Le Monde des Galaxies*, (Paris), eds. H.G. Corwin Jr. and L. Bottinelli (Berlin: Springer–Verlag), p. 208.

Capaccioli, M.: 1990, *Bulges of Galaxies*, (La Serena), eds. B. Jarvis and D.M. Terndrup (Garching: ESO Workshop Publ.), p. 231.

Capaccioli, M., Caon, N.: 1991, *Mon. Not. R. Astr. Soc.* **248**, 523.

Capaccioli, M., Longo, G.: 1990, *Windows on Galaxies*, (Erice), eds. G. Fabbiano, J. Gallagher, and A. Renzini (Dordrecht: Kluwer), p. 23.

Capaccioli, M., Vietri, M.: 1989, *Origin and Structure of Galaxies*, Procs. of the Yellow Mountain Summer School on Relativistic Astrophysics (Cina), ed. Fang Li Zi (Singapore: World Scientific), p. 150.

Capaccioli, M., Caon, N., Rampazzo, R.: 1990, *Mon. Not. R. Astr. Soc.* **242**, 24p.

Capaccioli, M., Cappellaro, E., Held, E.V.: 1991, Procs. of the *6th OAC International Workshop, Stellar Jets and Bipolar Outflows*, (Capri Island), in press.

Capaccioli, M., Fasano, G., Held, E.V.: 1991, in preparation.

Capaccioli, M., Held E.V., Nieto J.–L.: 1987, *Astron. J.* **94**, 1519.

Capaccioli, M., Vietri, M., Held, E.V.: 1988, *Mon. Not. R. Astr. Soc.* **234**, 335.

Capaccioli, M., Cappellaro, E., Held, E.V., Vietri, M.: 1991a, *Astrophys. J.*, submitted.

Capaccioli, M., Vietri, M., Held, E.V., Lorenz, H. : 1991b, *Astrophys. J.* **371**, 535.

Carter, D.: 1987, *Astrophys. J.* **312**, 514.

Cawson, M.C.: 1983, Ph.D. Thesis, Univ. of Cambridge.

Davies, R.L., Efstathiou, G., Fall, S.M., Illingworth, G., Schechter, P.L.: 1983, *Astrophys. J.* **266**, 41.

de Carvalho, R.R., Da Costa, L.N.: 1987, *Astron. Astrophys.* **171**, 66.

de Vaucouleurs G. : 1959, *Handbuch der Phys.* **53**, 331.

D'Onofrio, M.: 1991, Ph.D. Thesis, ISAS (Trieste).

Fasano, G.: 1991, *Mon. Not. R. Astr. Soc.* **249**, 208.

Franx, M., Illingworth, G.: 1988, *Astrophys. J.* **327**, L55.

Freeman, K.C.: 1983, IAU Symposium n. 106, *The Milky Way Galaxy*, ed. H. van Woerden (Dordrecht: Reidel).

Gott, J.R.: 1975, *Astrophys. J.* **201**, 296.

Hamabe, M., Kormendy, J.: 1987, IAU Symp. 127, *Structure and Dynamics of Elliptical Galaxies*, ed. T. de Zeeuw (Dordrecht: Reidel), p. 379.

Heisler, J., Merrit, D., Schwarzschild, M.: 1982, *Astrophys. J.* **258**, 490.

Illingworth, G.: 1977, *Astrophys. J.* **218**, L43.

Jensen, E.B., Thuan, T.X.: 1982, *Astrophys. J. Suppl. Ser.* **50**, 421.

Kent, S.M.: 1986, *Astron. J.* **91**, 1301.

Kodaira, K., Watanabe, M., Okamura, S.: 1986, *Astrophys. J. Suppl. Ser.* **60**, 475.

Kormendy, J.: 1977, *Astrophys. J.* **217**, 406.

Lauer, T.R.: 1985, *Astrophys. J. Suppl. Ser.* **57**, 473.

Michard, R.: 1984, *Astron. Astrophys.* **140**, L39.

Nieto, J.-L.: 1989, *Second Extragalactic Astronomy Regional Meeting*, (Cordoba: Academia de Ciencias), p. 239.

Nieto, J.-L., Capaccioli, M., Held, E.V.: 1988, *Astron. Astrophys.* **195**, L1.

Rix, H.-W., White, S.D.M.: 1990, *Astrophys. J.* **362**, 52.

Scorza, C., Bender, R.: 1990, *Astron. Astrophys.* **235**, 49.

Shaw, M.A., Gilmore, G.: 1990, *Mon. Not. R. Astr. Soc.* **242**, 59.

Silva, D.R., Boroson, T.A., Thompson, J.B., Jedrzejewski, R.: 1989, *Astron. J.* **98**, 131.

Simien, F., Michard, R.: 1990, *Astron. Astrophys.* **227**, 11.

Stiavelli, M., Londrillo, P., Messina, A.: 1991, *Mon. Not. R. Astr. Soc.*, in press.

Trevisani, S.: 1991, Laureato Thesis, Univ. of Padova.

van den Bergh, S.: 1989, *Publ. Astron. Soc. Pac.* **101**, 1072.

van den Bergh, S.: 1990, *Dynamics and Interactions of Galaxies*, ed. R. Wielen (Heidelberg: Springer–Verlag), p. 492.

van der Kruit, P.C.: 1987, *Astron. Astrophys.* **173**, 59.

van der Kruit, P.C., Searle, L.: 1982, *Astron. Astrophys.* **110**, 79.

van Houten, C.J.: 1961, *Bull. Astron. Netherlands* **16**, 1.

Whitmore, B.C.: 1980, *Astrophys. J.* **242**, 53.

Wilson, C.P.: 1975, *Astron. J.* **80**, 175.

Discussion

B. Jarvis - You say that the presence of a disk does not affect the flattening of a bulge. Well a few years ago I published a paper where I computed self-consistent bulge models with a superimposed flat potential field, similar to what you would expect for a disk. I found that indeed the flat disk potential DID produce an observable flattening of the bulge. Moreover, fitting my models to data showed that better fits to the photometry and kinematics were obtained by including a disk (which produced a larger bulge flattening).
M. Capaccioli - Well, I only say that, on my data, the typical bell–shaped flattening profile of disky Es and inclined S0s has its maximum well outside the radial range where the disk is photometrically significant (see my figure 5). Also, the value of this maximum does not depend on the bulge–to–disk ratio. On this ground my guess is that, with light disks such as mine, the effect you speak of contributes only marginally to the flattening intrinsic to the (diskless) bulge. But I admit I haven't checked it with dynamical calculations.

S. Djorgovski - Can you make at least a preliminary estimate what is the typical or a median D/T light ratio? Can you already tell whether the point distribution of D/T light ratios for E's and $S0$'s will be bimodal?
M. Capaccioli - I am afraid that I am presently unable to answer both your questions. Data are still insufficient. We have to wait for the completion of the specific ESO Key Program.

I. King - This is a question that has bothered me for years. When you make a component separation, you consider the light to be additive. But gravitation is a non-linear process; and if you physically add a gravitationally self-consistent bulge and a gravitationally, self-consistent disk, the result will not be the same as that of merely adding the light.
M. Capaccioli - My answer to your profound comment is that self–gravity need not be an 'isotropic' property. For instance, the faint disks occurring in bulge dominated S0s retain their identity because of their high flattening. In fact, up to some radius they are self–gravitating in z, even if, radially, they are dominated by the bulge. What strikes me is that, even so, they remain 'exponential'.

D. Burstein - Ivan King's question regarding the photometric separation of disks and bulges is important, and one that should be answered by each of us that does this kind of work. On my part, I note that I do not understand the physical origin of either bulges or disks, nor why these physical components have forms of luminosity distributions that can be so well-defined (i.e., $R^{1/4}$ law + exponential). I would prefer *not to assume* a physical origin for these components, but rather empirically determine those properties of galaxies that I can, as accurately as I can. I fell we are still in a 'fact-finding' mode, in which many interpretations of data are possible, and no interpretation is certain. For example, the visible part of a galaxy is really self-gravitating?

I. King - I think that there may be a single resolution of this superposition paradox. It rests on the proverb that God takes case of little children and idiots and astronomers. It may well be that component separation works because one component's strongly dominant in one region and the other component strongly dominant in another region. Perhaps the difficult transition region is too small to bother us seriously.

R. Thomson - Leading on from I. King period: "You have described the flaring up of disks in ellipticals when the potential is dominated by the halo". Working with A. Wright (Parkes) we have developed a model for shell galaxies wherein shells are formed by a weak interaction (but not merger) in a thick disk associated with the primary elliptical galaxy. This suggests that disks in massive ellipticals (i.e., dynamically cold stars) will always have a thick structure, as required by our model.

Massimo Capaccioli trying to convince Gustav Tammann to come and listen to his talk.

GASEOUS DISKS IN ELLIPTICAL GALAXIES

F. Bertola
Department of Astronomy
University of Padova, Italy

1 Introduction

Elliptical galaxies with dust lanes (Bertola and Galletta, 1978) are interpreted as the result of the acquisition of gaseous and dusty material. This material forms a disk by settling into the symmetry plane of the elliptical galaxy, if oblate or prolate (Tubbs, 1980; Simonson, 1982) or into one of the two permitted planes (defined by the major and intermediate axis and by the minor and intermediate axis) if the galaxy is triaxial and not tumbling (Steiman-Cameron, 1984). An inspection of the structure of the dust lanes in the ellipticals of the selected list published by Bertola (1987) indicates that we are looking at these disks almost edge-on, at an angle not exceeding 10°. This situation is the most favourable one for detecting the dust lane, since the region absorbed divides the galaxy into two parts which are absorption free. As soon as the viewing angle increases the contrast diminishes and the effect of the dust becomes less appreciable. In this case the best way to detect the acquired disk is to look at the elliptical galaxy through an interference filter centered on an emission line emitted by the gas component. Given the relatively high number of ellipticals with dust lanes (Sadler and Gerhard, 1985) it is therefore expected to detect acquired gaseous disks in the majority of ellipticals with emission lines. The purpose of this paper is to define and study the properties of these disks.

2 Geometrical and kinematical considerations

When a gaseous-dusty disk lying in the equatorial plane of a prolate elliptical or the principal plane perpendicular to the major axis of a triaxial one is seen on edge, the elliptical galaxy is crossed by a dust lane along the minor axis. If, on the contrary, the galaxy is oblate or triaxial with the material on the principal plane perpendicular to the minor axis, an elliptical galaxy with a dust lane along the major axis is seen. This is illustrated in the upper part of figures 1 and 2 for the triaxial case when the line of sight coincides with the intermediate and minor axes and with an axis at 45° with them (for a

G. Longo et al. (eds.), Morphological and Physical Classification of Galaxies, 115–126.

minor axis dust lane elliptical) and with intermediate, major and 45° axes (for a major axis dust lane elliptical).

The lower part of figures 1 and 2 shows cases analogous to those above, the only difference being that the gaseous disk, assumed circular, is seen at an angle of 45°. The circular assumption is reasonable even if the gas is subject to a triaxial potential, since the outer closed orbits tend to be circular.

An interesting phenomenon appears, namely a geometrical decoupling of the gaseous and stellar component in most cases. The gas is aligned with the stars only when the disk lies in the plane perpendicular to the minor axis and the line of sight is contained in the plane yz. The first case of such a geometrical decoupling was noticed thirty years ago by Osterbrock (1960) in NGC 4278.

Four geometrical parameters, namely the apparent ellipticity and major axis position angle of the stellar and gaseous components, can be measured from images obtained in the light of Hα/[NII] and the adjacent continuum. These set constraints on the intrinsic shape of the (triaxial) galaxy. In the plane $b/a, c/a$ they define a restricted area for each of the two permitted planes for the gas. The same is true for the plane of the two viewing angles of the triaxial galaxy, θ and ϕ (where θ is the angle of the plane defined by the major and intermediate axes with the line of sight and ϕ is the azimuth of the intermediate axis with respect to the line of nodes), where the geometrical parameters set constraints (Bertola et al., 1991a).

It should be mentioned that the axial ratios and the viewing angles can be directly determined from an accurate knowledge of the velocity field alone, as has been done for the bulge of NGC 4845 by Gerhard, Vietri and Kent (1989) using the high precision velocity field derived by Bertola, Rubin and Zeilinger (1989).

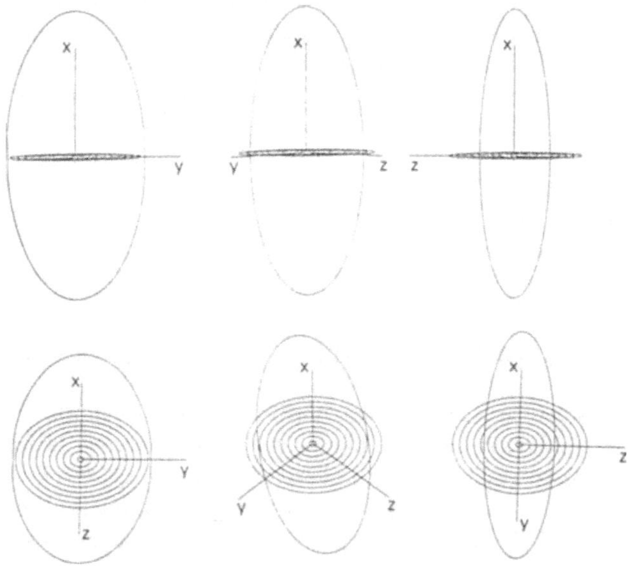

Figure 1. Appearance of a gaseous disk lying on the plane defined by the minor and intermediate axis of a triaxial elliptical galaxy when seen on edge (top) and at 45° (bottom).

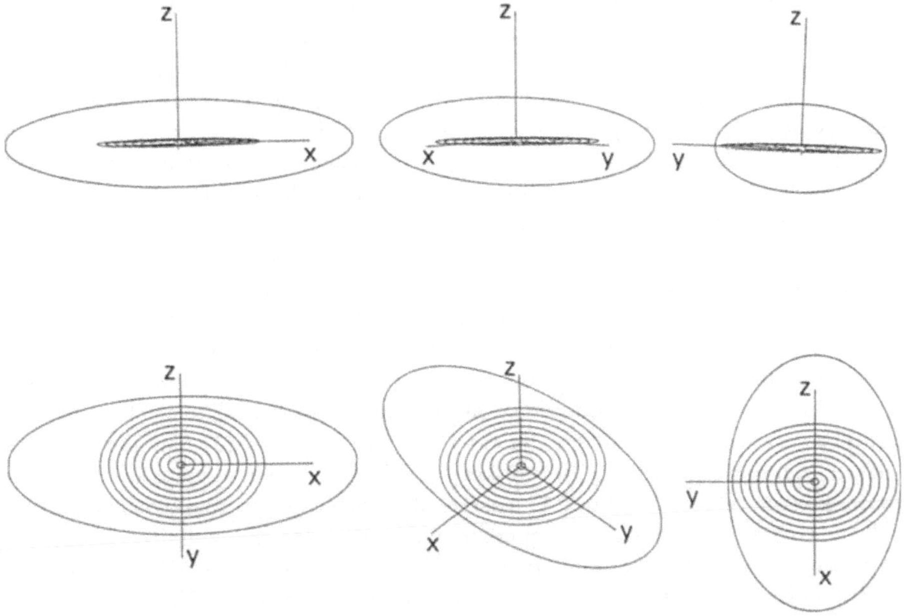

Figure 2. Appearance of a gaseous disk lying on the plane defined by the major and intermediate axis of triaxial elliptical galaxy when sen on edge (top) and at 45° (bottom).

The kinematical study of dust lane ellipticals (which, as mentioned earlier, have their gas disk close to edge–on) shows that in the minor axis ones the angular momentum of the gas is orthogonal to that of the stars (e.g. Bertola, Galletta and Zeilinger, 1985). On the other hand for ellipticals with a dust lane along the major axis the two angular momenta are either parallel or antiparallel (Bertola and Bettoni, 1988; Bertola, Buson and Zeilinger, 1988). In all cases the rotation curve of the gas indicates ordered motions. In the more general case where the gas disk is seen at arbitrary inclination (i.e. the line of sight is not in the plane of the disk), the projected angular momenta of gas and stars will appear decoupled at all angles. This decoupling has been pointed out by several authors (Bertola et al., 1984).

From the above considerations it appears that geometrical and kinematical decoupling and ordered motions (not necessarily circular motions) are strong signatures of the presence of a gaseous disk in elliptical galaxies.

3 Study of a sample of elliptical galaxies with gaseous disks

In order to study the properties of elliptical galaxies with gaseous disks in a systematic way, a well–defined sample of elliptical galaxies with emission line images has been selected. As pointed out by Baum, Heckman and van Breugel (1988), the morphology of the narrow band images can be either regular (flattened or round shape) or filamentary, showing arcs, strings and other peculiar features. From the kinematical point of view they also distinguish between rotators and non rotators, the latter being far less numerous. Therefore in selecting the galaxies particular care has been used in judging the morphological aspect of the gaseous structure. The sample galaxies are included in table 1, which lists elliptical galaxies with available Hα imaging. They were selected from the papers quoted in the sources, with a condition that they were classified as ellipticals in at least one of the following catalogues: RSA (Sandage and Tammann, 1981), RC 2 (De Vaucouleurs et al., 1976), UGC (Nilson, 1973), ESO/Uppsala (Lauberts, 1982). Only galaxies brighter than B$\sim$15 magnitude were included. Only three galaxies (NGC 4406, NGC 4486, NGC 4696) in table 1 exhibit filamentary structures, while the remaining twenty can be considered as *bona fide* ellipticals with gaseous disks. For these galaxies the size of the emitting region, the axial ratios and the orientations of the stellar and gaseous structure have been measured, when not given explicitly in the original sources. When the galaxy is also a radio source, the position angle of its axis is reported. Hα fluxes and the respective luminosities are included where available. The redshifts are from the Catalogue of Principal Galaxies (Paturel et al, 1989) and the distances are generally computed assuming $H_\circ$=50 km/sec/Mpc. In order to visualize the properties of the samples reported in table 1, a series of histograms has been constructed. Figure 3 illustrates the distribution of apparent axial ratios for the stellar component of the galaxies in our sample. For comparison the same distribution recently derived by Fasano and Vio (1990) for a general sample of ellipticals is also plotted. The good agreement of the two distributions tells us that the sample of gaseous disk ellipticals is representative of ellipticals in general. Figure 4 shows the distribution of sizes of the emission line regions as measured at the outermost isophotes. In most cases the emitting region is smaller than 10 kpc, with a median value of 4.5 kpc.

Figure 5 shows the distribution of axial ratios of the emitting regions, again measured at the outer isophotes. There are no objects in which the disk, assumed circular, is seen at an angle less then 20°. This is expected, since at lower inclinations the ellipticals become dust lane ones. There is also a lack of objects seen face on, which might be ascribed to the low surface brightness of the emitting region and the difficulty of detecting it against a stellar continuum. Figure 6 illustrates the distribution of the twisting angles of the stellar and gaseous components, which represents the evidence that a gaseous disk is located in one of the two permitted planes of the triaxial galaxy. Such a distribution $f(t)$ can be used to obtain information on the distribution function $\phi(\frac{b}{a},\frac{c}{a})$ of the intrinsic shapes of the elliptical galaxies, in a way similar to that used by Bertola, Vietri and Zeilinger (1991) in deriving the axial ratios of bulges from the misalignment of the disk and bulge major axes. A procedure to obtain $\phi(\frac{b}{a},\frac{c}{a})$ is here outlined. Let P_{yz} and P_{yx} be the fraction of the total number of galaxies in the sample with the gas in the plane yz and yx respectively; $f_{yz}(t)$ and $f_{yx}(t)$ the corresponding distribution function of twistings. We can write

$$f_{yz}(t) = \int\int K_{yz}\left(t|\frac{b}{a},\frac{c}{a}\right)\phi\left(\frac{b}{a},\frac{c}{a}\right)d\left(\frac{b}{a}\right)d\left(\frac{c}{a}\right) \quad (1)$$

Table 1. ELLIPTICAL GALAXIES WITH Hα IMAGING

Object	Dist. Mpc	V km/s	b/a stars	b/a gas	P.A. stars	P.A. gas	ΔP.A. gas–stars	Size arcsec	Size kpc	P.A. radio	ΔP.A. gas–radio	Flux 10^{-14}	Lum. 10^{40}	Sources
N484	101	5062	0.73	0.65	88	90	2	7.0	3.4	–	–	–	–	1
N547	111	5542	0.80	0.78	110	79	3	10.5	5.7	10	69	2.9	4.2	2,7
N708	100	4999	0.60	0.74	46	163	63	26	12.6	66	83	7.0	8.4	2,7
N745	116	5798	0.70	0.73	47	45	2	7.0	3.9	–	–	–	–	1
N1052	30	1495	0.65	0.80	114	53	61	13	1.9	63	10	81	8.7	4,6
U2748	181	9057	0.56	0.69	151	102	49	5.5	4.8	58	44	2.8	10.9	2,7
N1395	32	1591	0.80	0.45	105	45	60	30	4.6	–	–	–	–	1
N1453	78	3888	0.82	0.70	34	153	61	7	2.6	–	–	–	–	1
N1600	93	4631	0.73	0.65	10	170	20	40	18	181	11	0.6	0.6	3,6
N2768	29	1429	0.47	0.95	94	170	76	10	1.4	–	–	7.5	0.7	4
N2974	37	1846	0.60	0.55	40	50	10	30	5.3	–	–	–	–	1
N3862	128	6397	0.85	0.85	152	82	70	6.3	3.9	30	52	1.3	2.6	2,7
N3894	67	3337	0.56	0.46	19	23	4	14	4.5	–	–	7.3	3.9	4
N3962	32	1616	0.8	0.6	10	50	40	56	8.7	–	–	–	–	1
N4125	30	1499	0.51	0.43	84	98	14	26	3.8	–	–	16	1.7	4
N4261	42	2098	0.81	0.75	158	0	22	9	1.8	88	88	9.6	2.0	4,6
N4278	16	640	0.82	0.63	20	40	20	22	1.8	152	68	–	–	5,6
N4374	20	877	0.85	0.50	127	77	50	24.5	2.4	175	82	12	0.6	2,7
N4406	20	-326	0.74	–	116	–	–	–	–	–	–	5.7	0.3	3
N4486	20	1205	0.93	–	155	–	–	198	19.1	112	48	80	3.8	2,7
N4552	20	237	0.91	0.7	112	90	22	40	3.9	–	–	4.0	0.2	3
N4589	43	2161	0.79	0.3	88	175	87	14	2.9	–	–	4.7	1.0	4
N4636	20	908	0.78	0.55	145	120	25	20	1.9	33	87	–	–	1,6
N4696	54	2723	0.83	–	115	150	30	7.9	100	–	–	–	–	1
N4782	77	3847	0.8	0.6	165	142	23	6.9	2.6	177	35	2.6	1.8	2,7
N5077	54	2692	0.64	0.41	173	86	18	18	4.7	–	–	–	–	1
N5846	36	1782	0.88	0.58	56	14	42	60	10.5	–	–	16	2.5	3
U9799	206	10308	0.64	0.5	33	78	45	23	23	5	23	2.6	13.2	2,7
N6868	56	2819	0.77	0.35	76	30	46	22	6	–	–	–	–	1
E234-21	108	5400	0.70	0.65	175	145	30	12	6.2	–	–	–	–	1
N7097	51	2534	0.70	0.43	27	0	27	16	4	–	–	–	–	1
E406-30	34	1701	0.75	0.45	45	25	20	38	6.3	–	–	–	–	1

Sources: 1 : Bertola et al., 1991b. 2 : Baum et al., 1988. 3 : Di Serego Alighieri, Trinchieri, 1991. 4 : Kim, 1989. 5 : Demoulin Ulrich et al., 1984. 6 : Birkinshaw and Davies, 1985. 7 : Baum et al., 1989a.

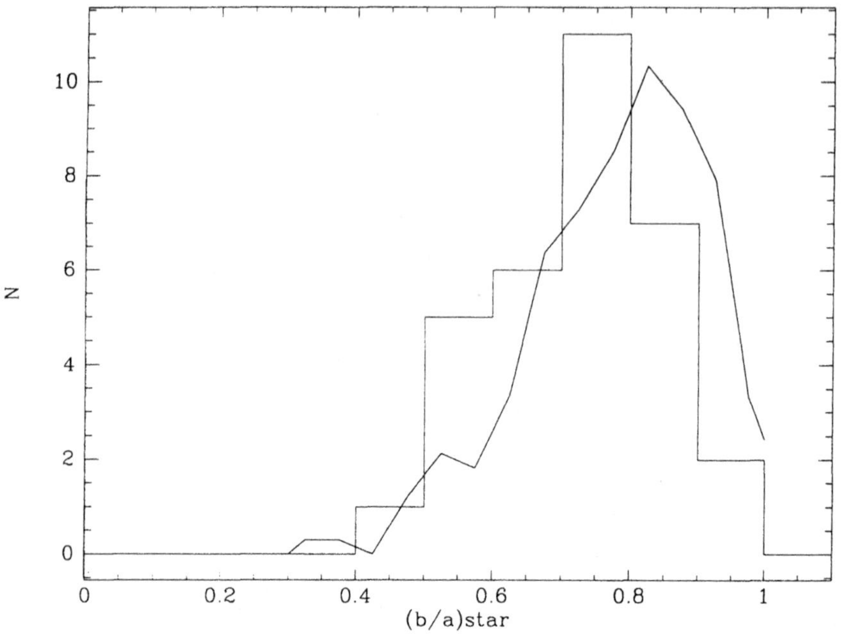

Figure 3. The distribution of the apparent axial ratios of the elliptical galaxies with gaseous disk as compared with that of the general sample (Fasano and Vio, 1991) properly normalized.

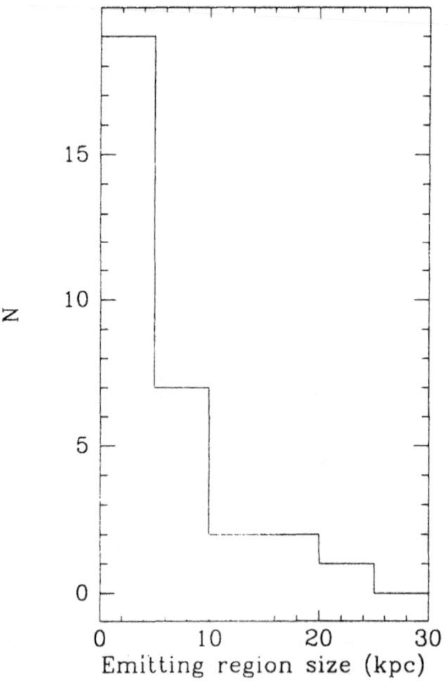

Figure 4. The distribution of the Hα emitting region sizes.

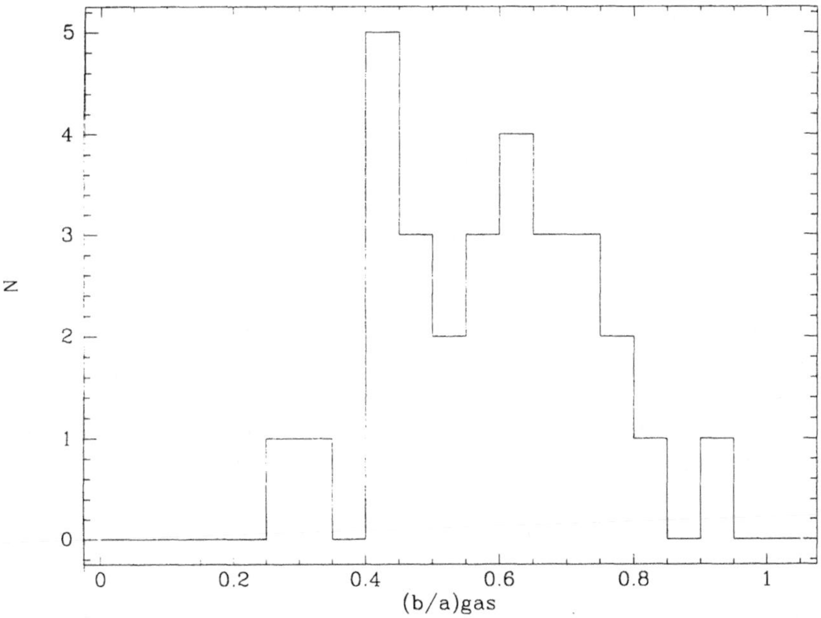

Figure 5. The distribution of the apparent axial ratios of the gaseous disks.

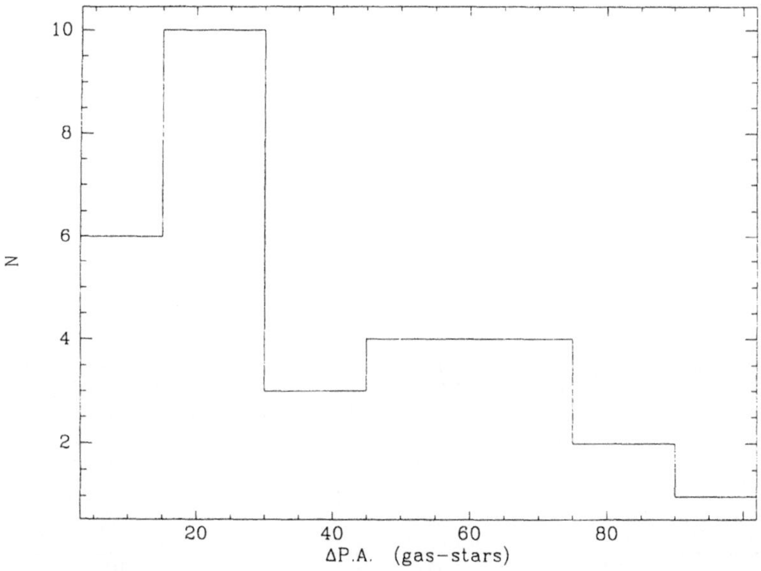

Figure 6. The distribution of the angles between the major axes of the stellar and gaseous component.

where K_{yz} is the distribution of twistings for a given $\frac{b}{a}$ and $\frac{c}{a}$. An analogous expression can be written for $f_{yx}(t)$. On the other hand

$$P_{yz} = \int\int P\left(\frac{b}{a},\frac{c}{a}\right)\phi\left(\frac{b}{a},\frac{c}{a}\right)d\left(\frac{b}{a}\right)d\left(\frac{c}{a}\right) \tag{2}$$

where

$$P\left(\frac{b}{a},\frac{c}{a}\right) = \frac{2}{\pi}\arcsin\left[\left(1-\frac{b^2}{a^2}\right)/\left(1-\frac{c^2}{a^2}\right)\right]^{\frac{1}{2}} \tag{3}$$

is the probability for the gas to settle on the yz plane according to dynamical considerations (Steiman-Cameron, 1984). Finally one obtains

$$f(t) = P_{yx}f_{yx}(t) + P_{yz}f_{yz}(t) = P_{yx}f_{yx} + (1 - P_{yx})f_{yz}(t). \tag{4}$$

If this equation can be properly inverted it is possible to derive the distribution function of the axial ratios $\phi(\frac{b}{a},\frac{c}{a})$.

Figure 7 gives the distribution of the twisting angle (misalignment) between the major axis of the gaseous disks and the axis of the associated radio source, when known.

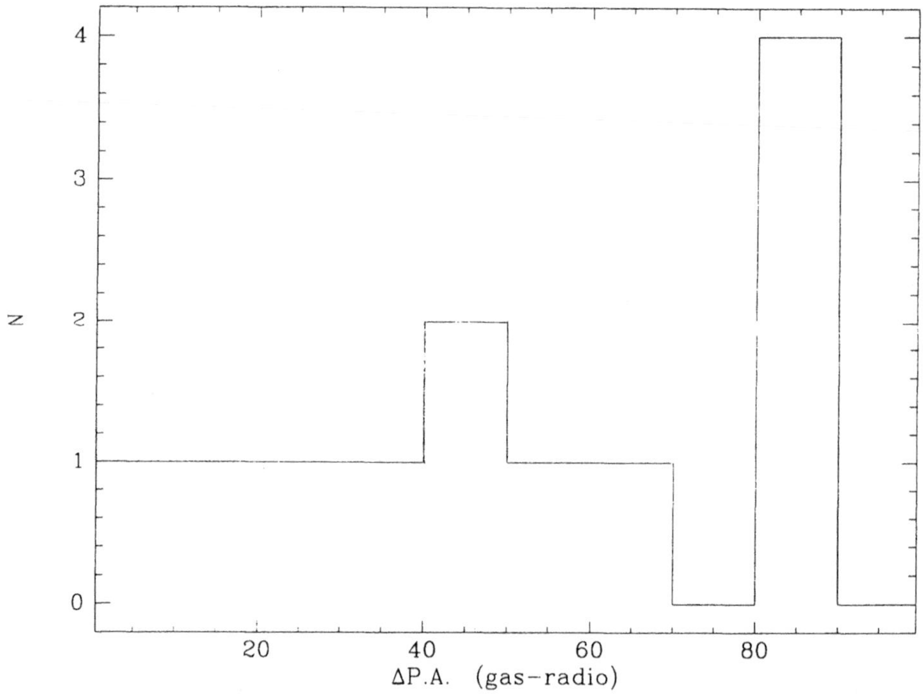

Figure 7. The distribution of the angles between the major axis of the gaseous disks and the axis of the associated radio source.

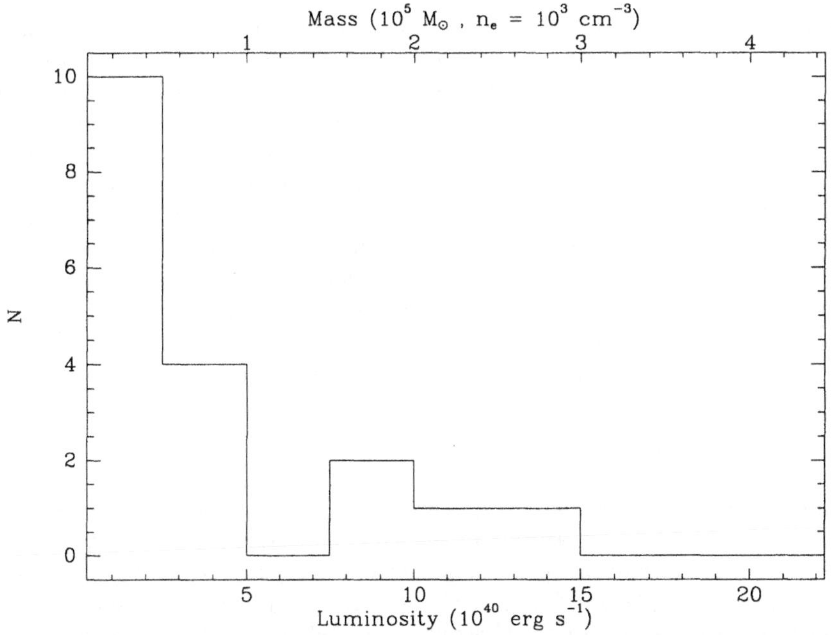

Figure 8. The distribution of the Hα luminosities of the gaseous disks. Masses, computed on the assumption $n_e = 10^3 cm^{-3}$, are also plotted.

The histogram shows a marked tendency for the radio source to be elongated perpendicular to the major axis of the gaseous component. This is analogous to the well known result that radio sources tend to be elongated perpendicular to the dust lane in dust lane ellipticals (Kotanyi and Ekers, 1979; Bertola et al., 1988b), and is again consistent with a continuity between dusty and gaseous disk ellipticals. No preferred values were found for the misalignment between radio and stellar axes.

Figure 8 shows the distribution of Hα plus [NII] luminosity for those galaxies with fluxes available. The mass of ionized material, also shown in the upper scale of figure 8, has been computed assuming case B recombination (Osterbrock, 1974) and an electron density $n_e = 10^3 \mathrm{cm}^{-3}$. The mass of ionized gas is in the range 10^5–10^6 M$_\odot$. Similar values are estimated for the mass of the dust in dust lane ellipticals (Zeilinger, Bertola and Galletta, 1990), while the estimates of the mass of neutral hydrogen in ellipticals are two orders of magnitude larger (Knapp, 1987).

4 Gaseous disks as probes of the gravitational field of elliptical galaxies

While the usual way to derive the rotation curves in spiral galaxies is by observing the neutral or ionized gas, most ellipticals have so far been observed kinematically by looking at their stellar content. However a gaseous disk, when present, provides a very straightforward way to study the mass distribution. Unfortunately the simplified assumption that the gas moves in a spherical or axisymmetric potential gives inconsistent results and often yields mass to light ratios which decrease rapidly towards the center (Caldwell et al., 1986; Kim 1989). If the gas is assumed to move in the potential of a triaxial galaxy whose axial ratios have been determined as described in section II (as has been done by Bertola et al., 1991a; Amico et al., 1991 using the models produced by de Zeeuw and Franx, 1989), it is possible to use the gas velocity field to derive a density distribution for the elliptical galaxy. Usually two solutions are possible, since gas may lie in one or in the other of the two permitted planes. In the sample of galaxies so far studied the two solutions differ only marginally. The main result obtained for NGC 1052, 3998, 4278, 5077, 5266 and 7097 is that the mass to B light ratio M/L_B tends to be constant from the center to the outermost (5–7 kpc) observed points in the gaseous disk, with an average value $M/L_B \sim 4$. This is roughly half the mean value determined with several methods using stellar dynamics, i.e. in a way which is completely independent.

5 Conclusions

This paper argues that dust lane ellipticals and ellipticals with regularly shaped line emission regions are different aspects of the same phenomenon, namely the result of a gaseous-dusty disk lying in the principal plane of an axisymmetric stellar body or in one of the two permitted planes of a triaxial one. Although gas from stellar mass loss is expected to be seen as ionized gas (Bertola et al., 1984), the phenomenology described in this paper indicates that this process tends to be hidden by the acquisition one. In fact several pieces of evidence, kinematical and morphological, point to these disks being acquired from outside. Given the high percentage of ellipticals with dust lanes (Sadler and Gerhard, 1985) and emission lines (Phillips et al., 1986) together with the fact that ordered structures and motions are common (Baum et al., 1988) we have to conclude that gaseous disks are common phenomena in ellipticals and therefore that acquisition processes, although involving small amounts of material with respect to the mass of the galaxy, play a frequent role in the evolution of elliptical galaxies.
The same picture holds for S0 galaxies, where gas in the equatorial plane has been found both rotating and counterrotating (Bertola et al., 1990). S0s with polar rings can be considered as analogous to ellipticals with gas in the plane defined by the minor and intermediate axes. Similar phenomena could also occur in spirals, but they remain unrevealed in a direct way because of the presence of preexisting gas.
We therefore conclude that the acquisition of material is a common event in all types of galaxies.

Acknowledgements

I wish to thank G. Cenzon for setting up the data base, G. Galletta for help in preparing figures 1 and 2 and R. Vio for discussions leading to the statistical treatment outlined in section 3. I also benefited from several discussions with collaborators in the ESO Key Program within which some of the data discussed in this paper were obtained, namely G. Bertin, L. Buson, J. Danziger, H. Dejonghe, T. de Zeeuw, E. Sadler, R. Saglia, M. Vietri and W. Zeilinger. Elaine Sadler kindly read this manuscript and contributed several comments.

References

Amico P., Bertola F., Pizzella A., de Zeeuw T.: 1991, in preparation.

Baum S., Heckman T.: 1989a, *Astrophys. J.* **336**, 681.

Baum S., Heckman T.: 1989b, *Astrophys. J.* **336**, 702.

Baum S., Heckman T., Bridle A., van Breugel W., Miley G.: 1988, *Astrophys. J.* **68**, 643.

Bertola F.: 1987, in *Structure and Dynamics of Elliptical Galaxies*, ed. T. de Zeeuw, I.A.U. Symposium N° 127, p. 135.

Bertola F., Bettoni D.: 1988, *Astrophys. J.* **329**, 102.

Bertola F., Galletta G.: 1978, *Astrophys. J. Letters* **226**, LL115.

Bertola F., Buson L.M., Zeilinger W.W.: 1988a, *Nature* **335**, 705.

Bertola F., Galletta G., Zeilinger W.W.: 1985, *Astrophys. J. Letters* **292**, LL51.

Bertola F., Rubin V., Zeilinger W.W.: 1989, *Astrophys. J. Letters* **345**, LL29.

Bertola F., Vietri M., Zeilinger W.W.: 1991c, *Astrophys. J. Letters*, in press.

Bertola F., Bettoni D., Buson L.M., Zeilinger W.W.: 1990, in *Dynamics and Interactions of Galaxies*, Ed. R. Wielen, Springer–Verlag Berlin, Heidelberg.

Bertola F., Bettoni D., Danziger J., Sadler E., Sparke L., de Zeeuw T.: 1991a, *Astrophys. J.* in press.

Bertola F., Bettoni D., Rusconi L., Sedmak G.: 1984, *Astron. J.* **89**, 356.

Bertola F., Buson L.M., Danziger J., Sadler E., de Zeeuw T., Zeilinger W.W.: 1991b, in preparation.

Bertola F., Galletta G., Kotanyi C.G., Zeilinger W.W.: 1988b, *Mon. Not. R. Astr. Soc.* **234**, 733.

Birkinshaw M., Davies R.L.: 1985, *Astrophys. J.* **291**, 32.

Caldwell N., Kirshner R.P., Richstone D.O.: 1986, *Astrophys. J.* **305**, 136.

Demoulin–Ulrich M.H., Butcher H.R., Boksenberg A.: 1984, *Astrophys. J.* **285**, 527.

Fasano G., Vio R.: 1991, *M.N.R.A.S.*, in press.

Gerhard O.E., Vietri M., Kent S.M.: 1989, *Astrophys. J. Letters* **345**, LL33.

Kim D.W.: 1989, *Astrophys. J.* **346**, 653.

Knapp G.: 1987, in *Structure and Dynamics of Elliptical Galaxies*, ed. T. de Zeeuw, I.A.U. Symposium N° 127, p. 145

Kotanyi C.G., Ekers R.D.: 1979, *Astron. Astrophys.* **73**, L1.

Lauberts A.: 1982, *The ESO/Uppsala Survey of the ESO Atlas*, (ESO Garching).

Nilson P.: 1973, *Uppsala General Catalogue of Galaxies.*

Osterbrock D.E.: 1960, *Astrophys. J.* **132**, 325.

Osterbrock D.E.: 1974, *Astrophysics of Gaseous Nebulae*, W.H.Freeman and Co. San Francisco.

Paturel G., Fouqué P., Bottinelli L., Gouguenheim L.:1989, *Catalogue of Principal Galaxies*, Lyon.

Phillips M.M., Jenkins C.R., Dopita M.A., Sadler E., Binette L.: 1986, *Astron. J.* **91**, 1062.

Sadler E., Gerhard O.E.: 1985, *Mon. Not. R. Astr. Soc.* **214**, 177.

Sandage A., Tammann G.A.: 1981, *A Revised Shapley-Ames catalog of bright galaxies*, Carnegie Inst. of Washington Pub. 635, Washington, D.C..

di Serego Alighieri S., Trinchieri G.: 1991, *Astron. J.* submitted.

Simonson G.F.: 1982, *Ph. D. dissertation*, Yale University.

Steiman–Cameron T.Y.: 1984, *Ph. D. dissertation*, Indiana University.

Tubbs A.D.: 1980, *Astrophys. J.* **241**, 969.

de Vaucouleurs G., de Vaucouleurs A., Corwin H.: 1976, *Second Reference Catalogue of Bright Galaxies*, Univ. of Texas Press, Austin.

de Zeeuw T., Franx M.: 1989, *Astrophys. J.* **343**, 617.

Zeilinger W.W., Bertola F., Galletta G.: 1990, in *Dusty Objects in the Universe*, eds. E. Bussoletti and A.A. Vittone (Kluwer, Netherlands) p. 227.

Discussion

A. Renzini – Di Serego and Trinchieri found the Hα emission to be centrally peaked (with no signs of either disks or filaments) in a fairly large fraction of the objects they have observed. What is this fraction within your own sample of early type galaxies?
F. Bertola - My sample is constituted by elliptical galaxies characterized by the presence of a gaseous disk and includes some of the Di Serego and Trinchieri objects.

B. Jarvis - You showed the results from Heckman on M87 as an example of a galaxy with a chaotic velocity field. Well I have made kinematic observations of an elongated feature, colinear with the nucleus of Hα + [NII] emission and find an ordered velocity gradient along this feature. This feature is very reminiscent of a jet since it is one sided, highly elongated and exactly aligned with the nucleus. In M87 at least and maybe in other galaxies not all of the gas need to be chaotically ordered.

S. Djorgovski - Is there any correlation between the presence and properties of the ionized gas with the X-ray emission? Also can you tell anything about the ionizing mechanism for the gas?
F. Bertola - X–ray fluxes are available for only 10 galaxies in our sample and no correlation with Hα data has been noticed at present. The UV light giving origin to the "rising branch" in the far UV spectrum of ellipticals could be responsible for the ionization of the gas.

J. Barnes - The relative numbers of major versus minor axis dust lanes tells us something about the shapes of elliptical galaxies (e. g. prolate galaxies support only minor axis lanes; oblate support only major lanes). What is the ratio of major o minor axis dust lanes in your sample?
F. Bertola - In the list of bona–fide elliptical galaxies with dust lanes I presented at the Princeton Meeting on the Structure and Dynamics of Elliptical Galaxies (1987), minor axis dust lane ellipticals largely outnumber major axis ones (3 to 1) thus indicating a tendency toward prolateness rather than oblateness.

I. King - Pursuant to Djorgovski's question, is there any correlation between the presence of a gas disk and the property of being "UV – hot"?
F. Bertola - Recent IUE spectra of NGC 5077, one of the ellipticals with a gaseous disk, show an energy distribution typical of the presence of star formation.

FORMATION AND MAINTENANCE OF SPIRAL STRUCTURE

E. Athanassoula
Observatoire de Marseille
Marseille, France

ABSTRACT. I briefly review different mechanisms for forming and maintaining spirals, such as modes, edges and grooves, driving by bars, tidal interactions and local responses. Some of their observational consequences, and in particular the existence of a leading component in nearby galaxies, are discussed.

1 Introduction

The problem of the formation and maintenance of spirals has been with us for some time. If spiral arms were material structures, they would wind up because of differential rotation. Thus density waves were introduced as a solution to this winding dilemma (Lindblad, 1959 and references therein; Lin and Shu, 1964 and 1966). However this remedy brought soon new headaches, since Toomre (1969) showed that a packet of tightly wound waves will propagate radially from corotation towards the Lindblad resonances where it will be absorbed. Thus it became obvious that it is necessary to continuously replenish, drive or stimulate the spirals. I will here discuss briefly five possible ways for doing so, giving more emphasis to recent developments. For more thorough and complete reviews see e.g. Toomre (1977) or Athanassoula (1984).

2 Modes due to feedback cycles and amplification at corotation

The work of Toomre (1969) and Mark (1974) showed clearly that a wave should be shielded from the inner Lindblad resonance (hereafter ILR) in order to avoid being absorbed. This can be achieved either if its pattern speed is higher than $(\Omega - \kappa/2)_{max}$, or if the velocity dispersion increases abruptly inwards at a radius which is larger than that of the ILR. Then any wave travelling inwards would be reflected on this "barrier" before reaching the lethal ILR. In either case the damping from the ILR would be confined to the very small contribution of stars on very excentric orbits. Yet this is not the only source of energy loss. The dissipation of the gas, present not only at the Lindblad resonances, is an important damping factor (Kalnajs, 1972; Roberts and Shu, 1972). Thus mode cycles should necessarily comprise an amplifier in order to keep the spiral going.

G. Longo et al. (eds.), Morphological and Physical Classification of Galaxies, 127–138.

Two amplification schemes have been so far proposed. In the WASER (Wave Amplification via Stimulated Emission of Radiation; Toomre, 1969, Mark, 1976a, b, c, 1977) an outgoing long trailing wave stimulates the emission of two short trailing waves. One of them propagates outwards towards the outer Lindblad resonance (hereafter OLR) while the other propagates inwards and is reflected at the Q barrier into a long trailing wave which propagates towards corotation and closes the cycle. The amplification thus achieved is very small, and negligible in the case of hot discs.

The second one, the SWING (Toomre, 1981), draws on the amplification of a wave when swinging from leading to trailing (Goldreich and Lynden-Bell, 1965, Julian and Toomre, 1966, Toomre, 1981). The closure of this cycle, providing fresh leading waves, can be made by a reflection of the trailing wave at a sharp edge, or at the center of the galaxy (Bardeen, 1975). The SWING involves regimes in which the standard WKBJ approximation (which assumes that the wave is tightly wound) is in trouble. Its analytical treatment is thus non trivial and has been explicitly worked out so far only in the case of the shearing sheet (Pellat et al., 1990). The SWING amplifier is very powerful and can give important amplification rates. Thus several modes based on it have been found so far (e.g. Zang, 1976; Toomre, 1981; Bertin et al., 1989).

Since the fastest growing mode dominates the evolution, that is the relevant one for comparing with observations. There has been a lot of pessimism about whether modes could explain the observed grand design spirals, chiefly because of their very high growth rates. These could be lowered with the help of a halo (however this solution is limited if one does not want to privilege spirals of higher number of arms m), by increasing the stellar velocity dispersions or because of nonlinear effects. Furthermore it should not be forgotten that all these mode calculations do not include, not even schematically, the dissipation by the gaseous component. Yet a balance could be reached between the energy dissipated in the shocks and the energy obtained from the amplification.

3 Edges and their effects

Edges or sharp gradients can induce spiral structure. This can be understood as due to two basic ingredients, the nonaxisymmetric instabilities of isolated thin rings (Maxwell, 1856; Lindblad, 1958; Papaloizou and Lin, 1989; Sellwood and Kahn, 1990) and the response from the disc (Julian and Toomre, 1966). The instability makes clumps which rotate with roughly the same angular velocity as the material in the ring. These clumps may drive hefty spiral responses, similar to the responses to the giant molecular clouds discussed in section 6, provided the disc mass and velocity dispersion allow large swing amplification rates. The number of arms favoured is a compromise between the two ingredients. Indeed ring instabilities have higher growth rates at high m values (Papaloizou and Lin, 1989) while the disc responds best for m = 2 or 3 for realistic disc parameters. (The maximum disc amplification that can be wrung from a given disc depends on $X = r\kappa^2/2\pi G m\mu$, and for $\Gamma = -\mathrm{dln}\ \Omega/\mathrm{dln}\ r = 1$, has a maximum around X = 1.5 (Toomre, 1981)).

Toomre (1981, 1989) in a stability analysis of the Gaussian disc found amidst the other modes an edge mode which he called D mode. This has a corotation resonance near the edge of the disc, as expected from the above discussion. Toomre demonstrated on the one hand the dependance of such modes on the abruptness of the edge by experimenting with different outer density cut-offs of the exponential and of the V = constant disc, and on the other their independence from the feedback loop by freezing the center of the Gaussian disc. Edge modes with similar properties can be easily found in numerical simulations (Athanassoula unpublished). Edges in observed galactic discs have been found photometrically (van der Kruit and Searle, 1981). Whether they are sharp enough to provoke edge modes and how the sharpness is maintained, still remains to be worked out.

Sellwood and Kahn (1990, see also Sellwood, 1989 and 1990) invoke a groove not in the density but in the angular momentum distribution. Of course the two would be the same for the case of cold discs (Lovelace and Hohlfeld, 1978). Both sides of the groove will form clumps, like in the case of an edge. However if the phase of the clumps on the two opposite sides are different they will attract each other in pairs, thus exchanging their angular momenta, as is nicely illustrated and discussed by Sellwood and Kahn. Their numerical simulations further showed that the compromise between the two main ingredients of this instability (instabilities in thin rings and the response of the disc) give, in warm discs, higher growth rates for m = 2. One still has to worry about the reason for forming such grooves in the angular momentum distribution. Sellwood and Lin (1989) saw such grooves form naturally in their simulations of low mass Keplerian discs. Each groove caused an instability with corotation near the radius of the groove. After the instability had grown to a spiral of sizeable amplitude it scattered particles off its Lindblad resonance thus creating new grooves there, which would in turn create new instabilities etc. (The whole process could be closely related to the emission of waves at resonances as discussed by Sygnet et al., 1988). Unfortunately observational tests of this theory seem to be exceedingly difficult. It is thus essential to check the existence of the instability cycle found by Sellwood and Lin in realistic hot discs (since grooves must not only form but also have sufficiently abrupt edges for them to be effective) in order to assess the relevance of this interesting mechanism for real galaxies.

Fridman (1989), inspired from shallow water experiments (e.g. Fridman et al., 1987), discussed the case where the rotation velocity, the density and the sound speed showed a discontinuity at the same radius. For the idealised case where these three quantities can be considered constant, both before and after the radius of the discontinuity, and in the limit of very small or very large Mach numbers allowing for analytical solutions, he showed that such discontinuities could indeed generate instabilities. The formation and persistence of such triple discontinuities in real galaxies has not been so far demonstrated. Nevertheless, at least as far as the rotational velocity is concerned, it is possible to check for such gradients in the observations.

4 Driving by bars

Several simulations of non-self consistent gaseous responses to bar-like or oval potentials have been made (Sanders and Huntley, 1976, Sanders, 1977, Sanders and Tubbs, 1980; Van Albada, 1985; etc.) showing clearly that the response is spiral. Thus a spiral forcing is not a necessity for obtaining a spiral response. A self consistent, but linear, calculation (Athanassoula, 1980) showed that gaseous spirals are more tightly wound than stellar ones and that the thin gaseous arms are embedded in the broad stellar ones for most of their extent. Furthermore the potential of the spiral arm provides a considerable part of the total forcing and therefore should not be neglected in the calculations. In all the above cases either the pattern speed of the bar had been assumed from the onset to be equal to that of the driven spiral (e.g. Athanassoula, 1980) or the self consistency was neglected (Sanders and Huntley, 1976; Sanders, 1977; Sanders and Tubbs, 1980, etc) in which case the equality of the pattern speeds followed naturally. However this need not always be the case.

In several N-body simulations (e.g. Sellwood, 1985) more than one mode have been observed of which the one with the lowest pattern speed has an ILR, and, according to what was discussed in section 2, should not have existed. Tagger et al. (1987) and Sygnet et al. (1988) noted that in all these cases the region of the offensive ILR was the same as that of corotation of the faster rotating mode. They thus explained the existence of the slow wave by the nonlinear coupling of the two modes between them and with their beat wave. This coupling, although constrained only to the narrow region around the two resonances, is particularly strong because of the resonance overlap. This would thus lead to the possibility of having spirals with a different pattern speed than that of the bar that drives them. Sygnet et al. (1988) discussed the implication of this for real galaxies, noting that this might account for the fact that in certain barred spirals the dust lanes are observed on the inner (concave) side of the spiral arm. However they were reluctant to believe that this could happen for galaxies like NGC 1300, where the arms join very smoothly to the bar, because different pattern speeds will result in random relative orientation of the two components. However that pessimism was unwarranted since the spiral may continually break from the bar and reconnect to it, as shown in the simulations of Sellwood and Sparke (1988).

5 Tidal spirals

In order to understand tidal spirals we have to consider two extreme cases. That of satellites on circular orbits and that of unbound companions.

a) Satellites on circular orbits provide a periodic forcing. It is thus reasonable to expect the pattern speed of the excited spiral to be equal to that of the driving and resonances to be of primary importance. If the satellite is direct then the m = 2 component of its forcing will conspire with the m = 2 ILR and a two armed spiral wave will be excited. On the other hand if the satellite is retrograde then the only possible resonance is the m = 1 ILR. Thus an m = 1 wave will be excited, which is furthermore leading (Athanassoula, 1978). Since this is a very unorthodox spiral structure several numerical simulations have tested it (Noguchi and Ishibashi, 1986; Thomasson et al., 1989) and have shown good agreement with the theoretical predictions.

b) In the case of unbound passages it is not clear whether the spiral will have the same pattern speed as the companion at its pericenter or not. In order to test this I made a number of N body simulations using J. Sellwood's N body code (Sellwood, 1981). More information on these can be found in a forthcoming paper and in Athanassoula (1990). Figure 1 of Athanassoula (1990) shows a simulation of the driving of a spiral by a companion in what can be loosely termed a "stable" galaxy: one which if unperturbed would show no spiral features - either in the density plots or by Fourier analysis - for a time considerably longer than that of the simulation. Note that the driven spiral is transient and its propagation towards and absorption at the ILR agree very well with the analytical calculations of Toomre (1969). I measured its pattern speed in three different ways and found it compatible with that of the companion at pericenter. As a second experiment I tried forcing an "unstable" disc. A spiral was again excited but its fate was quite different from that of the previous example. It did not propagate towards the ILR to be absorbed there but rather evolved into a bar. The pattern speed of the spiral was the same as that of the bar it evolved in to, but it was not compatible with that of the companion at pericenter. Instead it seemed to be compatible with the pattern speed of one of the modes that would develop in the same disc if this were to evolve unperturbed, as I could show by a further numerical simulation. Thus the main effect of the companion in this case was to excite, much before its due time, an instability that would develop anyway. These two different behaviours being now clear I wondered what would happen in intermediate cases. I thus submitted a disc, with properties intermediate between those of the "stable" and "unstable" discs of the above examples, to a passage of an external companion. A spiral was forced which after a while died out, very much like that of the "stable" case. After that followed some dull time, during which not much could be seen in the simulation, and then a new spiral formed, grew and evolved into a bar. The pattern speed of the first spiral episode was compatible with that of the companion at pericenter, while the second spiral and the ensuing bar had a different pattern speed, presumably that of the perturbation that would evolve in the undriven disc.

From all the examples described above, and many more corroborative simulations, it is clear that companion driven spirals are short lived, not exceeding a life time of the order of a few times 10^9 years. On the other hand nothing prohibits a later encounter to form a new spiral which will in turn die out, etc. This could continue only as long as the heating of the disc due to the consecutive spirals is balanced by the cooling due to star formation from fresh gas.

A spiral driven by a satellite will also evolve and eventually die out. This is due to the fact that the radius of the satellite's orbit will change because of the dynamical friction and of exchange of energy and angular momentum between the satellite and stars around the ILR of the parent disc (Tremaine, 1981; Quinn and Goodman, 1986). However the time scale involved may well be longer than in the case of a companion, although detailed simulations are necessary to give a quantitative estimate.

6 Ragged spirals and local responses

Most spirals do not have a "grand design" structure as M51 or M81, but show a more ragged structure. Attention to such "swirling hotch-potch of pieces of spiral arms" was focussed by Goldreich and Lynden-Bell (1965) and Julian and Toomre (1966).

Julian and Toomre worked out the response of a disc when forced by an orbiting mass or clump, like a giant molecular cloud. The wake thus formed is a trailing spiral with an excess mass much larger than that in the driving clump and is a driven steady density wave. It can be easily visualised by following the trajectories of particles orbiting near the driven clump. The inclination of the spiral depends on the shearing rate in the sheet.

Toomre (1990) discussed an N-body realisation of such a shearing sheet, where each of the particles can be considered at the same time as driving and being driven. The impression one gets from the evolution of this simulation is that of a "kaleidoscope of spiral features". These features can be understood and explained as the superposition of the wakes of the individual particles.

7 An observational search for the leading component

Histogram of the angle differences between two successive maxima in the m = 2 component for all galaxies in the sample discussed here. A necessary ingredient for the SWING feedback cycle discussed in section 2 is the existence of a leading component. It is thus important to establish whether such a component exists in real galaxies. It may be easily spotted on isodensity contours of a given multiplicity m. Figure 1b shows the result obtained by adding on the spiral of figure 1a its mirror image (i.e. leading if that of figure 1a is trailing) at 30% of its amplitude. The superposition of the trailing and the leading component shows a very characteristic interference pattern with maxima roughly 90° apart. Thus one promising way of looking for leading components is to search for such maxima using photometry of nearby spirals. Elmegreen et al. (1989) mention the existence of maxima in M 81 and M 100. In M 81 they are approximately 120° apart and in M 100 approximately 100° and 140°. No other relevant information could be found in the literature.

I thus tried to look for such maxima along the spiral arms in the data obtained by Considère and Athanassoula (1988) from the Fourier decomposition of the light distribution in 16 nearby disc galaxies. Four out of those galaxies were either flocculent or ringed and technical problems prohibited analysing the data for a fifth one. I was thus left with a sample of 11 galaxies with some spiral structure. I found the maxima along the m = 2 components of the density and the angle between them. Figure 2 shows the contours of the m = 2 component of the light distribution for NGC 598. The stars mark the position of the local maxima. They definitely give the impression of being at roughly 90° apart. However, instead of discussing each galaxy individually, which would lead to very small number statistics, I found it more convincing to make a cumulative histogram of the $\Delta\theta$ values between two successive maxima, for all galaxies in the sample. The result is given in figure 3. Although the sample is very small a χ^2- or Kolmogorov-Smirnov test shows that the distribution of the angle differences has a very small probability of being consistent with random. Thus the search for the leading component seems to be on a good track. However it should be noted that the bin containing most of the values is not centered around 90° but around roughly 65°. This shift from 90° could be explained in two different ways. Figure 1c shows the result obtained when adding 90% of the spiral in figure 1a to 10% of its mirror image. Note that the difference in angle between two consecutive maxima is now smaller than 90°. This can be understood by realising that for small ratios of leading to trailing amplitudes the variation of the density along the spiral has to be taken into account when predicting the positions of the maxima. A second alternative is if the leading and trailing spirals do not have the same wavenumber.

The reason one would expect, at first guess, these two wavenumbers to be the same is that the Lin-Shu-Kalnajs dispersion relation (Lin and Shu, 1966; Kalnajs, 1965) as well as the equivalent one for the gas, show a symmetry between leading and trailing waves, i.e. if, for a given ν (r), $k(r)$ is a solution of the dispersion relation then $-k(r)$ is also a solution. However this symmetry can be easily broken by the presence of resonances, dissipation, non linear effects or the presence of an external forcing. Figure 1d shows the result obtained when adding 70% of the spiral in figure 1a to 30% of a spiral with twice the wavenumber and opposite winding. It is thus clear that a substantial difference between the wavenumbers of leading and trailing components could account for a shift of the maximum of the peak of the histogram from 90°.

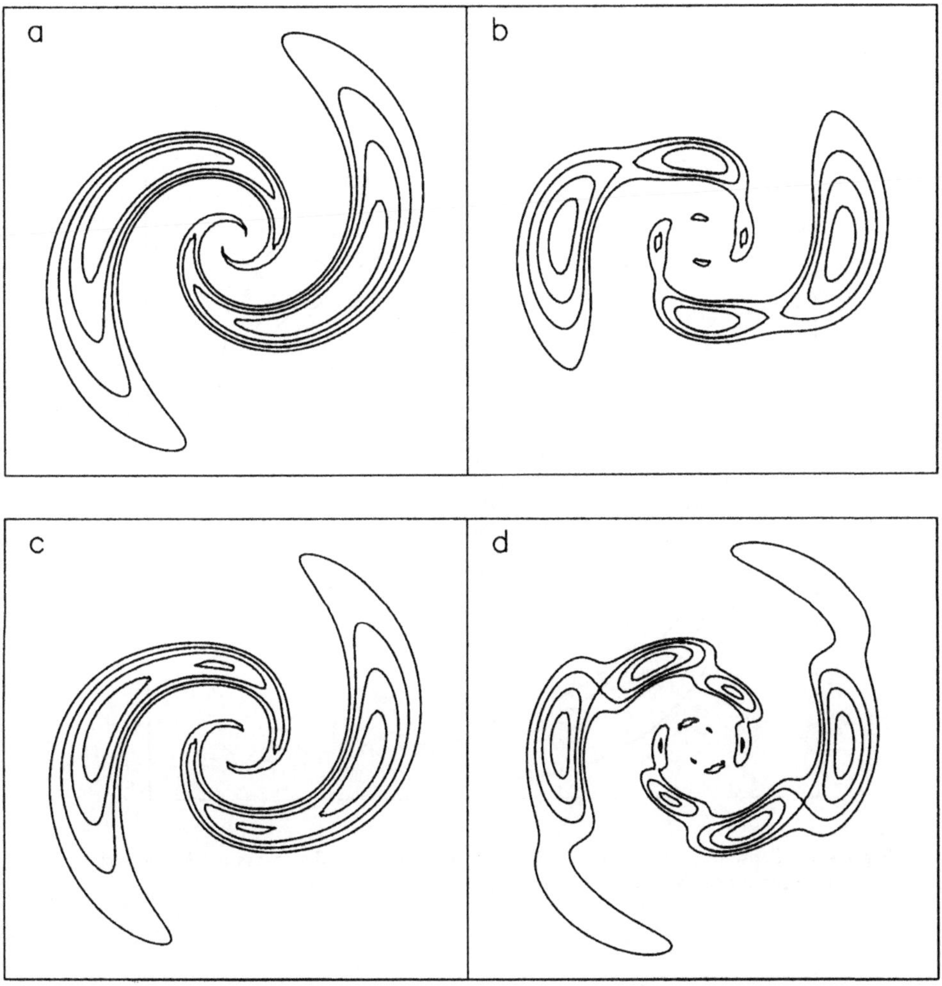

Figure 1. Superposition of trailing and leading spirals (a) 100% trailing (b) 70% trailing and 30% leading (c) 90% trailing and 10% leading (d) 70% trailing and 30% leading with twice the wavenumber.

Such a difference can also be argued from the Fourier spectra of the galaxies given in figure 5 of Considère and Athanassoula (1988), where several peaks can be seen on the leading sides but not necessarily at the positions of the maxima on the trailing side. There are thus at least two possible explanations for the shift of the peak of the distribution from 90°. Which one is dominant may vary from galaxy to galaxy. Furthermore the amplitude of the shift need not be the same for each galaxy, resulting in a spread in the histogram. If the trailing spiral is more tightly wound than the leading one, then the angle difference is larger than 90°, the opposite being true if it is more open. Thus, if the shift in maximum is due to the different wavenumbers of the leading and trailing components, the leading component, in the present sample of galaxies, is on average more tightly wound than the trailing one.

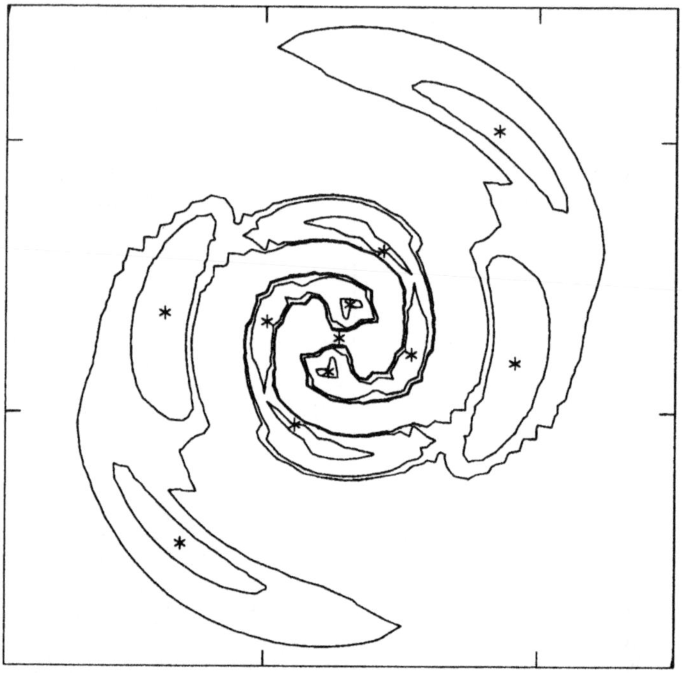

Figure 2. Contours of the m = 2 component of NGC 598. The stars mark the positions of the maxima along the arm and the center of the galaxy.

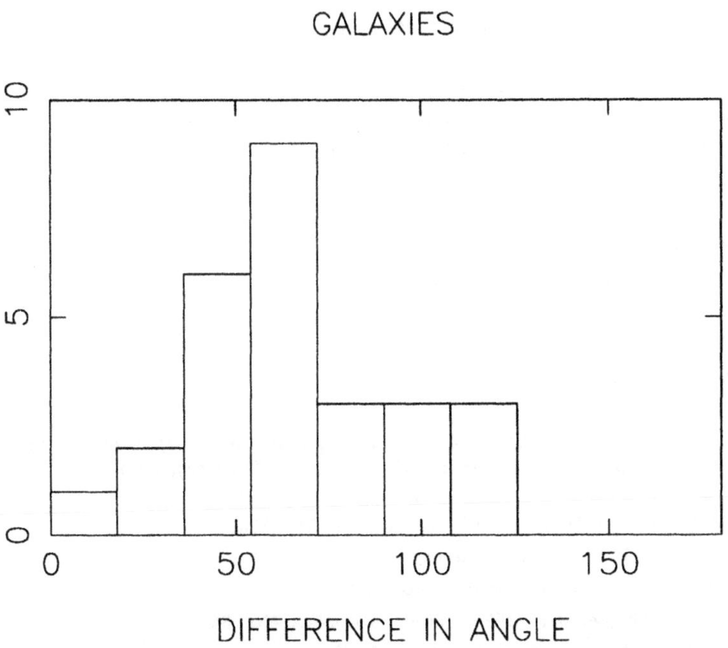

Figure 3. Histogram of the angle differences between two successive maxima in the m = 2 component for all galaxies in the sample discussed here.

Acknowledgements

I would like to thank M. Tagger for interesting discussions and J. Sellwood for sending me his results on groove-modes in advance of publication.

References

Athanassoula, E.: 1978, *Astron. Astrophys.* **69**, 395.

Athanassoula, E.: 1980, *Astron. Astrophys.* **88**, 184.

Athanassoula, E.: 1984, *Physics Reports* **114**, 321.

Athanassoula, E.: 1990, In *Chemical and Dynamical Evolution of Galaxies*, (J. Franco and F. Matteucci, Eds.), World Scientific Publ., in press.

Bardeen, J.M.: 1975, In *Dynamics of Stellar Systems*, (A. Hayli, Ed.), 297, Reidel (Dordrecht)

Bertin, G., Lin, C.C., Lowe, S.A., Thurstans, R.P.: 1989, *Astrophys. J.* **338**, 78 and 104.

Considère, S., Athanassoula, E.: 1988, *Astron. Astrophys. Suppl. Ser.* **76**, 365.

Elmegreen, B.G, Elmegreen, D.M., Seiden, P.E.: 1989, *Astrophys. J.* **343**, 602.

Fridman, A.M.: 1989, In *Dynamics of Astrophysical Discs*, (J.A. Sellwood), 185, Cambridge University Press.

Fridman, A.M., Morozov, A.G., Nezlin, M.V., Pasha, I.I., Polyachenko, V.L., Rylov, A.Y., Snezhkin, E.N., Torgashin, Y.N., Trubnikov, A.S.: 1987, In *Observational Evidence of Activity in Galaxies*, (E.Y. Khachikian et al., Eds.), 147, Reidel (Dordrecht).

Goldreich, P., Lynden-Bell, D.: 1965, *Mon. Not. R. astr. Soc.* **130**, 125.

Julian, W.H., Toomre, A.: 1966, *Astrophys. J.* **146**, 810.

Kalnajs, A.J.: 1965, Ph. D. Thesis, Harvard University.

Kalnajs, A.J.: 1972, *Astrophys. Lett.* **11**, 41.

Lin, C.C., Shu, F.H.: 1964, *Astrophys. J.* **140**, 646.

Lin, C.C., Shu, F.H.: 1966, *Proc. Nat. Acad. Sci.* **55**, 229.

Lindblad, B.: 1958, *Stockholm Obs. Ann.* **20**, No. 6.

Lindblad, B.: 1959, *Handbuch der Physik* **53**, 21.

Lovelace R.V.E., Hohlfeld R.G.: 1978, *Astrophys. J.* **221**, 51.

Mark, J. W.-K.: 1974, *Astrophys. J.* **193**, 539.

Mark, J. W.-K.: 1976a, *Astrophys. J.* **203**, 81.

Mark, J. W.-K.: 1976b, *Astrophys. J.* **205**, 363.

Mark, J. W.-K.: 1976c, *Astrophys. J.* **206**, 418.

Mark, J. W.-K.: 1977, *Astrophys. J.* **212**, 645.

Maxwell, J.C.: 1856, *Adams Prize Essay: On the Stability of the Motion of Saturn's Rings*, reprinted in *Scientific Papers of James Clerk Maxwell, Vol. I*, 288, (W.D.Niven), Cambridge University Press, 1890.

Noguchi, M., Ishibashi, S.: 1986, *Mon. Not. R. astr. Soc.* **219**, 305.

Papaloizou, J.C.B., Lin, D.N.C.: 1989, *Astrophys. J.* **344**, 645.

Quinn, P.J., Goodman J.: 1986, *Astrophys. J.* **309**, 472.

Pellat, R., Tagger, M., Sygnet, J.F.: 1990, *Astron. Astrophys.* **231**, 347.

Roberts, W.W., Shu, F.H.: 1972, *Astrophys. Lett.* **12**, 49.

Sanders, R.H.: 1977, *Astrophys. J.* **217**, 916.

Sanders, R.H., Huntley, J.M.: 1976, *Astrophys. J.* **209**, 53.

Sanders, R.H., Tubbs, A.D.: 1980, *Astrophys. J.* **235**, 803.

Sellwood, J.A.: 1981, *Astron. Astrophys.* **99**, 362.

Sellwood, J.A.: 1985, *Mon. Not. R. astr. Soc.* **217**, 127.

Sellwood, J.A.: 1989, In *Dynamics of Astrophysical Discs*, (J.A. Sellwood), 155, Cambridge University Press.

Sellwood, J.A.: 1990, In *Galactic Models*, (J.R. Buchler, S.T. Gottesman and J.H. Hunter), 158, New York Academy of Sciences

Sellwood, J.A., Sparke, L.S.: 1988, *Mon. Not. R. astr. Soc.* **231**, 25p.

Sellwood, J.A., Lin, D.N.C.: 1989, *Mon. Not. R. astr. Soc.* **240**, 991.

Sellwood, J.A., Kahn, F.D.: 1990, preprint.

Sygnet, J.F., Tagger, M., Athanassoula, E., Pellat, R.: 1988, *Mon. Not. R. astr. Soc.* **232**, 733.

Tagger, M., Sygnet, J.F., Athanassoula, E., Pellat, R.: 1987, *Astrophys. J.* **318**, L43.

Thomasson, M., Donner, K.J., Sundelius, B., Byrd, G.G., Huang, T.Y., Valtonen, M.J.: 1989, *Astron. Astrophys.* **211**, 25.

Toomre, A.: 1969, *Astrophys. J.* **158**, 899.

Toomre, A.: 1977, *Ann. Rev. Astron. Astrophys.* **15**, 437.

Toomre, A.: 1981, In *The Structure and Evolution of Normal Galaxies*, (S.M. Fall and D. Lynden-Bell, Eds.), 111, Cambridge University Press.

Toomre, A.: 1989, In *Dynamics of Astrophysical Discs*, (J.A. Sellwood, Ed.), 153, Cambridge University Press.

Toomre, A.: 1990, In *Dynamics and Interactions of Galaxies*, (R. Wielen, Ed.), 292, Springer Verlag.

remaine, S.: 1981, In *The Structure and Evolution of Normal Galaxies*, (S.M. Fall and D. Lynden-Bell, Eds.), 67, Cambridge University Press.

an Albada, G.D.: 1985, *Astron. Astrophys.* **142**, 491.

an der Kruit, P.C., Searle, L.: 1981, *Astron. Astrophys.* **95**, 105.

ang, T.A.: 1976, Ph. D. Thesis, MIT.

Discussion

. Buta - May I present a photograph and color index map of the southern spiral alaxy NGC 4622, whose morphology seems very relevant to your retrograde encounter mulations. This galaxy has two relatively normal-looking outer arms which seem to rminate in a near circular ring. Inside the ring there appears to be a leading arm which as recently recognized by Byrd, Thomasson, Donner, Sundelius, Huang and Valtonen 1989, *Cel. Mech.* **45**, 31), who also made some related models. Deborah Crocker and imaged the galaxy in B and I and we find that, while the two outer arms and half f the ring are clearly distinguishable in B−I, the leading arm and inner disk show no gnificant color differentiation. Are these the kinds of characteristics you would expect om your models?

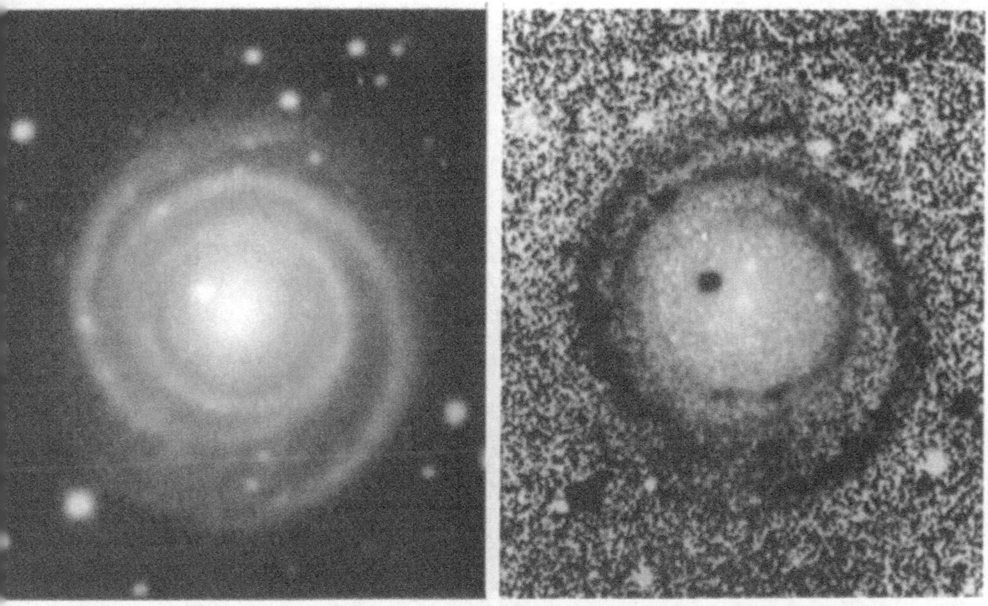

Figure 4. B-band CCD image (left) and B−I color index map (right, where the sense dark to light is blue to red) of NGC 4622, obtained with the CTIO 1.5-m telescope in February 1990.

. Athanassoula - I would expect the m = 1 spiral to be mainly confined to the region urrounding its ILR, and the m = 2 to be dominant further out, where the tidal forcing s greatest. This is, I believe, indeed the case for NGC 4622. I can unfortunately say othing about the colours without doing any detailed modeling.

F. Combes - In the case of driving by a companion, is there a particular relation between the external spiral corotating with the companion, and the inner density-wave (bar)? Is there for instance conjunctions of resonances (ILR and CR) or a better efficiency when that occurs?
E. Athanassoula - In most of the examples I discussed in section 5, there was, at any given time, only one pattern speed present. On the other hand in many other simulations there have been more than one mode present and then resonance overlap is very common. The physics of this has been discussed by Tagger et al. (1987) and Sygnet et al (1988).

I. King - Theoreticians are much enamored of density waves with low m, but we also have flocculent spirals, which probably do not have a density wave. How are we to understand them dynamically?
E. Athanassoula - Indeed flocculent spirals exist and are actually much more numerous than grand design ones. However I would not readily concede that they do not have density waves. The type of local responses discussed in section 6 could well explain the spiral structure in galaxies like NGC 2841 and NGC 5055.

Evangelia Athanassoula during her talk. In the foreground Ron Buta shows a picture of one of his favorite galaxies.

DARK MATTER IN ELLIPTICAL GALAXIES

T. de Zeeuw
Sterrewacht Leiden, The Netherlands

ABSTRACT. Evidence for the existence of dark halos around elliptical galaxies is reviewed, with emphasis on recent observations of stellar kinematics and cold gas. Elliptical galaxies with extended HI disks offer the best prospect of measuring the density profile at large radii. There is little evidence for a variation of the mass-to-light ratio inside two half-light radii. In some elliptical galaxies a modest increase is indicated at larger radii. Radial velocity studies of planetary nebulae and globular clusters will not put strong constraints on mass-to-light ratios, but provide a direct probe of the dynamical structure of the stellar halos out to large radii.

1 Introduction

Studies of the behavior of the mass-to-light ratio $\mathcal{M}/L$ as a function of radius in a galaxy usually focus on one of two issues: a) the existence of dark matter in and around galaxies, and b) the presence of supermassive black holes in the nuclei of galaxies. Here we consider only the former problem and restrict ourselves to a discussion of elliptical galaxies. The evidence for nuclear black holes in ellipticals is reviewed in de Zeeuw and Franx (1991).

The determination of $\mathcal{M}/L$ as function of radius r cannot be separated from the determination of the intrinsic shape and internal dynamics of a galaxy. The luminosity distribution $L(r)$ of a galaxy is generally deduced from the observed surface brightness distribution by assuming an intrinsic shape, and deprojecting. The density distribution $\rho(r)$ is found by measuring the kinematics of a set of tracer objects in the galaxy. If their orbits are known, the observed velocities can be used to find the gravitational potential of the galaxy, from which the density distribution which generates it follows. This again requires assumptions about the shape of the system, and about the velocity distribution of the tracer objects.

It is relatively straightforward to derive an $\mathcal{M}/L$ profile for spiral galaxies. To good accuracy the luminous parts of these systems can be approximated by a flat circular disk so that the inclination of the disk follows from its apparent flattening on the sky and the calculation of $L(r)$ is easy. Cold HI gas occupies the circular orbits in the plane of the disk, and hence provides a nearly ideal kinematic tracer. The rotation curve, and hence the mass distribution, follows immediately from the observed line-of-sight HI radial velocities. The result is well-known: the rotation curves generally stay flat out to many optical radii, and correspond to a density which falls off proportional to $1/r^2$. The light distribution decreases exponentially fast, however, so that $\mathcal{M}/L$ increases outwards,

G. Longo et al. (eds.), Morphological and Physical Classification of Galaxies, 139–152.

indicating the presence of a massive halo of unseen material in spiral galaxies (e.g., van Albada et al., 1985; van Albada, 1988). Deviations of circular symmetry can not reconcile the observed flat rotation curves with constant $\mathcal{M}/L$ models.

Elliptical galaxies are slowly tumbling triaxial systems, so that two viewing angles are needed to define the direction of observation, and two axis ratios need to be specified for the determination of the intrinsic shape. Instead of simply computing an inclination from the apparent flattening, as is done for spirals, a total of four parameters must be obtained to fix the geometry of an individual elliptical galaxy. Furthermore, only a few ellipticals contain detectable amounts of cold gas which extends to large radii. Although these are excellent candidates for $\mathcal{M}/L$ determinations, different morphologies are possible for settled cold gas in a triaxial galaxy (Steiman-Cameron and Durisen, 1982; Merritt and de Zeeuw, 1983), and deriving a mass distribution from the observed gas kinematics is not as simple as for spiral galaxies.

Hot X-ray gas can provide good constraints on $\mathcal{M}/L$ at large radii. However, few temperature profiles have been measured, and as a result the mass determinations are quite uncertain (e.g., Forman, Jones and Tucker, 1985; Trinchieri, Fabbiano and Canizares, 1986; Sarazin, 1987; Fabbiano, 1989). The data that is currently being gathered by ROSAT will help to improve this situation. For this reason we will not discuss the X-ray data here; the reader is referred to Kent (1990) for a concise summary.

In §2 we discuss absorption line measurements of the integrated light, and radial velocity determinations of individual objects in the halos of nearby ellipticals. The kinematics of cold gas is reviewed in §3. §4 summarizes the conclusions.

2 Stellar kinematics

2.1 Integrated light

A widely used method for calculating an $\mathcal{M}/L$ value for an elliptical galaxy is the core-fitting method often referred to as the King method. It is based on the formula (Rood et al., 1972; Richstone and Tremaine, 1986)

$$\frac{M}{L} = \frac{9\sigma^2\eta}{2\pi G I_0 r_c}, \tag{1}$$

where σ is the central line-of-sight velocity dispersion measured for the stars, I_0 is the central surface brightness, r_c is the core radius, where the surface brightness has fallen to half its central value, and G is the gravitational constant. The factor η equals 1 to an accuracy of better than a few percent for a spherical galaxy with constant $\mathcal{M}/L$, a homogeneous core, and an isotropic velocity distribution. These assumptions are not likely to be all valid for elliptical galaxies, so that $\mathcal{M}/L$-values computed with equation (1) should be considered as estimates. Anisotropic velocity distributions and departures from the spherical shape result in errors of up to a factor 3 (Richstone and Tremaine, 1984; Merritt, 1988a).

Bacon, Monnet and Simien (1985) considered oblate galaxy models with a de Vaucouleurs profile and constant $\mathcal{M}/L$, and used the virial theorem to include the effect of flattening in an estimate of $\mathcal{M}/L$. They find

$$\frac{M}{L} = h(q,i)\frac{\langle\sigma^2 + V^2\rangle}{I_e r_e}, \tag{2}$$

where r_e is the half-light radius, and I_e is the surface brightness at r_e. The numerator is the luminosity weighted average $\sigma^2 + V^2$ over the whole galaxy, where V is the observed rotation velocity, and h is a known function of the apparent axis ratio q and the inclination i of the galaxy. This generalizes and improves the earlier result for an isotropic spherical de Vaucouleurs galaxy (Binney, 1982; Tonry, 1983). The quantity $\langle\sigma^2 + V^2\rangle$ is observationally inaccessible, so the authors replace it with the maximum measured value $(\sigma^2 + V^2)_{max}$, and multiply the right hand side of equation (2) with a correction factor which depends not only on q and i but also on the assumed anisotropy, and has to be calculated from models. Using a mean correction factor results in an uncertainty of about 30% in the derived $\mathcal{M}/L$–values.

The core-fitting method and the virial method both produce a "global" value of $\mathcal{M}/L$, which in practice is frequently based on kinematic observations well inside $1r_e$, and therefore should be treated with caution.

A number of studies have sought to use the detailed behavior of the line-of-sight velocity dispersion with radius to detect unseen mass. This idea is based on the fact that the dispersion decreases with increasing radius in an isotropic spherical galaxy with constant $\mathcal{M}/L$ and a de Vaucouleurs profile, and becomes proportional to $r^{-1/2}$ at large radii. If the observed velocity dispersion does not conform to this pattern, and if the velocity distribution can be assumed isotropic, then the density profile of the galaxy must differ from its luminosity profile, and hence $\mathcal{M}/L$ must vary with radius.

In practice, observed velocity dispersion profiles are fit with solutions of the Jeans equations which give the velocity dispersions consistent with an assumed mass model (Binney and Mamon, 1982; Tonry, 1983), or with models based on distribution functions (Richstone and Tremaine, 1984; Bertin, Saglia and Stiavelli, 1988; Saglia, 1990). Not every solution of the Jeans equations is guaranteed to correspond to a physical, i.e., non-negative, distribution function, so the latter method is preferable, even though the former is usually easier to apply. The difficulty with these approaches is that the dynamical models are non-unique, which means that variations in the density profile can be traded off against variations in the velocity anisotropy. It is therefore impossible to derive an $\mathcal{M}/L$ profile from a single velocity dispersion profile, although useful constraints on variation of $\mathcal{M}/L$ with radius can sometimes be found (Katz and Richstone, 1985; Merritt, 1987).

The Jeans approach has received renewed attention recently (Binney, Davies and Illingworth, 1990; van der Marel, Binney and Davies, 1990; van der Marel, 1991). These authors use the observed surface photometry, assume an inclination, and derive the corresponding oblate or prolate luminosity distribution by deprojection. The Jeans equations are then solved subject to an assumption about the velocity anisotropy, after which projection yields the predicted line-of sight velocity dispersion at any point. This is compared with the observed values along different position angles. In this way the two-dimensional velocity dispersion field is used to constrain $\mathcal{M}/L$ and the inclination, while at the same time testing whether the assumed velocity anisotropy is consistent with the data. These studies need to be extended to include a larger variety of velocity anisotropy (Bacon, Simien and Monnet, 1983; Bacon, 1985; Fillmore, 1986; Fillmore and Levison,

1988; van der Marel and Cinzano, 1991), and to include triaxial models, which show a great freedom in the velocity dispersion fields (Statler, 1987, 1991).

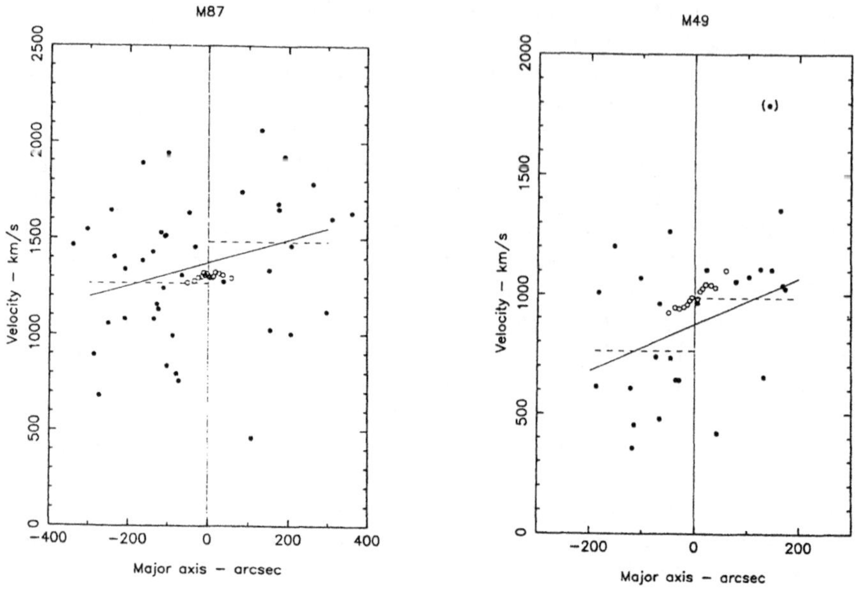

Figure 1. Radial velocities of globular clusters in M 87 and NGC 4472 (M 49) versus projected distance along the photometric major axis (Mould et al., 1990). The straight line is a fit to the data (after exclusion of the largest redshift cluster for NGC 4472). The dashed curve is a similar fit to a flat rotation curve. The rotation curve in integrated light is indicated by open circles (Davies and Birkinshaw, 1988).

Early absorption line measurements usually obtained σ−profiles out to about an r_e, and are consistent with constant $\mathcal{M}/L$ models and at most a mild anisotropy (e.g. Bertin, Saglia and Stiavelli, 1988). Models with massive halos have been fit to data of this kind, but they often have rising σ profiles just outside the last measured point (e.g. Katz and Richstone, 1985), or are based on specific choices for the phase-space distribution function (Saglia, Bertin and Stiavelli, 1991). This does not imply that dark halos should be ruled out, but simply says that the evidence for unseen mass in the luminous parts of ellipticals is not compelling.

In the past few years improvements in detectors have made it possible to measure dispersion profile out to $2r_e$ and beyond. A number of these have been obtained recently in the context of the ESO Key Project on Dark Matter in Elliptical Galaxies (Bertin et al., 1989). The observed dispersion profiles fall gently, and seem to imply no or at most a modest variation of $\mathcal{M}/L$ over the optical body of the galaxy. Detailed modeling of hese data is in progress.

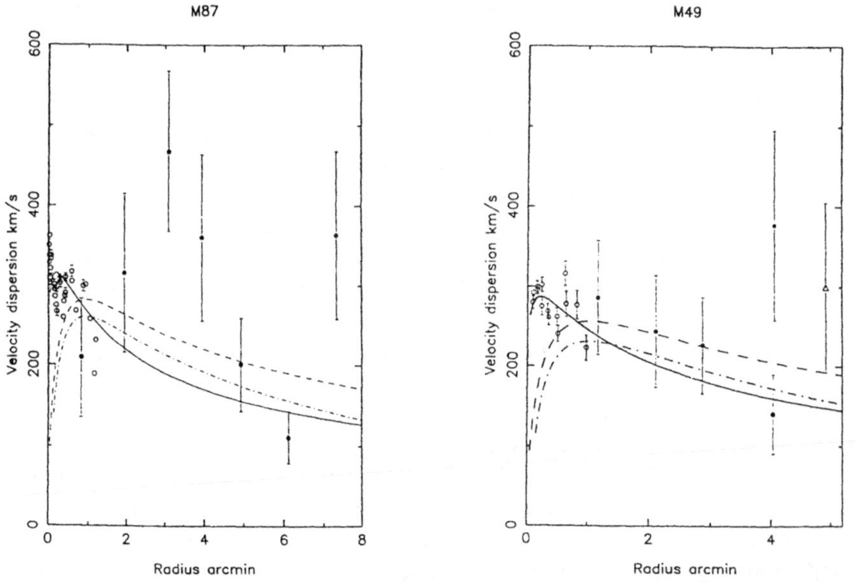

Figure 2. Velocity dispersion as a function of radius for the globular cluster systems of M 87 and M 49=NGC 4472 (Mould et al., 1990). A flat rotation curve has been removed from the cluster data, and the solid symbols represent the resulting velocity dispersions. For NGC 4472 two points are shown at 4′, with and without the largest redshift cluster. The open triangle refers to the four nearest dwarf galaxy companions of this galaxy. The solid line is an isotropic model fitted to the stellar velocity dispersion measurements of Davies and Birkinshaw (1988), represented by open circles. The dashed curves are tangentially anisotropic models.

2.2 Individual object

Recently it has become possible to identify individual globular clusters and planetary nebulae in the halos of nearby galaxies, and to measure their radial velocities. This has been done out to about $4r_e$, but the resulting velocities do not give strong constraints on $\mathcal{M}/L$, partly because of the relatively small number of measurements and partly because of our ignorance of their intrinsic velocity anisotropy. More interestingly, the data indicate that the kinematics of the stellar halo differs from that of the luminous central regions.

Mould et al. (1990) have studied the kinematics of the globular cluster systems of the two Virgo ellipticals M 87 and NGC 4472 (Fig. 1 and 2). The data for M 87 (see also Huchra, 1988) are consistent with a velocity dispersion that is constant with radius, i.e., with a massive halo, and is inconsistent with constant $\mathcal{M}/L$. This agrees with other studies of this galaxy, as summarized e.g., by Kent (1990). A model with $\mathcal{M}/L$ constant out to $2r_e$ cannot be ruled out with great confidence for NGC 4472, however. The globular cluster systems of both galaxies show evidence of rotation, with $v/\sigma \approx 0.3$, which is comparable to the v/σ of the globular cluster system in our own galaxy and in Andromeda, and is consistent with the moderate flattening of the luminous part of the galaxies. These results should be considered preliminary, as the number of observed globular clusters in these two cases is still small, being measured in the few dozens. This is also the likely cause for the apparent difference between the systemic velocity of the globular cluster system and the parent galaxy seen in both cases (Fig. 1).

Radial velocities of about 70 globular clusters in the halo of Centaurus A are known (Hesser et al., 1984; Harris, 1988). In addition, radial velocities have now been measured for over 400 planetary nebulae out to $4r_e$ by Ford and collaborators (Ford et al., 1988; Hui, 1990). The velocity dispersion is reported to decrease outwards, and falls to about 100 km/s at $4r_e$, showing that $\mathcal{M}/L$ is not increasing much with radius. This is consistent with the globular cluster data, and also with the kinematics of the HI gas associated with the dust lane which gives a constant $\mathcal{M}/L_B$ equal to $5M_\odot/L_\odot$ out to $1.2r_e$ (van Gorkom et al., 1986), assuming Centaurus A is at a distance of 3 Mpc (Tonry and Schechter, 1990). The line profiles of the CO associated with the inner dust lane, where the HI has a hole, give the same value for $\mathcal{M}/L$ in the center (Quillen et al., 1991).

The major new result that emerges from the planetary nebulae measurements of Centaurus A is that the whole system of planetaries appears to be rotating with a velocity of 100 km/s at $4r_e$. This may be compared to 40 km/s measured from the stellar absorption lines in the luminous inner region (Wilkinson et al., 1986). Thus, v/σ increases outwards, and approaches 1 at $4r_e$. Furthermore, the rotation axis of the planetary nebula system appears to be strongly misaligned from the photometric axes of the light distribution. It might be possible to explain this as a projection effect due to triaxiality, but it may well imply that the halo of Centaurus A is not in equilibrium. This would not be surprising in view of the recent infall which must have caused the prominent dust lane (Tubbs, 1980; Quillen et al., 1991). The rotational properties of elliptical galaxies are reviewed by Franx elsewhere in this volume.

3 Cold gas

A few elliptical galaxies contain detectable amounts of cold gas (Bertola, 1987; Lees et al., 1991; van Gorkom, 1991). It is thought that in many cases this gas was accreted recently, for example by the capture of a small gas–rich disk galaxy (Athanassoula and Bosma, 1985). Since settling times are long (Steiman–Cameron and Durisen, 1988; de Zeeuw, 1990), it is not always clear whether accreted gas has had time to reach a preferred plane.

3.1 Closed orbit

In many studies of ellipticals with regular gas disks it is assumed that the gas has indeed settled, and occupies the stable simple closed orbits. If the galaxy has a stationary triaxial figure such closed orbits are not circular but elliptic, and exist in two of the three principal planes (Heiligman and Schwarzschild, 1979). Thus, cold gas will be found either perpendicular to the long axis or perpendicular to the short axis of the system. Since in a triaxial galaxy the projected principal axes generally do not coincide with the major and minor axes of the projected surface brightness distribution, one expects a significant misalignment between the kinematic axes of the gas and the photometric axes of the light distribution for most viewing directions. Also, the gas morphology will appear to be twisted and misaligned, due to the variation of the surface density caused by the velocity change along an elliptic orbit. The morphology and two-dimensional velocity field of such a gas disk contain significant information on the intrinsic shape of the host galaxy, and on the variation of $\mathcal{M}/L$ in it (de Zeeuw and Franx, 1989; Lees and de Zeeuw, 1991). If the galaxy has a tumbling triaxial figure, cold gas settles either on elliptic orbits in the equatorial plane or on the so-called anomalous orbits, which do not lay in any of the principal planes, but form a stable configuration which is intrinsically warped (Heisler, Merritt and Schwarzschild, 1982; Magnenat, 1982). The kinematics properties of such gas disks have not been investigated in detail, but it is evident that a wide variety of morphologies and velocity fields can be obtained (van Albada, Kotanyi and Schwarzschild, 1982), and that, conversely, these contain information on $\mathcal{M}/L$-variations and the intrinsic shape. Elliptical galaxies with regular gas disks clearly are prime candidates for studies of $\mathcal{M}/L$.

3.2 Emission line gas

Allowing for the proper geometry is important when deriving density profiles, and hence $\mathcal{M}/L$, from gas kinematics. This was demonstrated for the E3 galaxy NGC 5077, which has a regular disk of ionized gas along its apparent minor axis. Gas velocities have been measured along seven position angles out to $1r_e$ (DeMoulin-Ulrich, Butcher and Boksenberg, 1984; Bertola et al., 1991). A simple model with the gas on circular orbits in a spherical potential fits the observed gas kinematics, but implies that $\mathcal{M}/L$ declines inwards by more than a factor of three inside $0.5r_e$, similar to what has been reported for the E4 galaxy NGC 7097 by Caldwell, Kirshner and Richstone (1986). However, constant $\mathcal{M}/L$ triaxial mass models with non-rotating figures and the gas in either of the two preferred planes can be constructed which are consistent with the morphology and kinematic of both the gas and the stars (Fig. 3). In particular, they reproduce the observed misalignment of 23° between the line of zero velocity of the gas and the apparent major axis of the galaxy as being due to a projection effect. This constrains the intrinsic shape of the galaxy to be either nearly oblate or nearly prolate with in each case the gas on elliptic orbits over the pole. The two different geometries correspond to the two allowed configurations for the gas. A costant $\mathcal{M}/L$ oblate model with the gas in a warped polar ring of the kind discussed by Sparke (1986) can be fitted to the data also. Higher resolution two-dimensional velocity measurements are needed to distinguish between these models. They will also put strong constraints on the intrinsic shape of the galaxy. The current data show already that there is no reason to suspect that $\mathcal{M}/L$ varies over the luminous part of NGC 5077.

Bulges of spiral galaxies may well be triaxial (e.g., Bertola, Rubin and Zeilinger, 1989; Gerhard, Vietri and Kent, 1989), and interpretation of their rotation curves should likewise be done carefully (Gerhard and Vietri, 1986, 1990). It should be noted, however, that the central gas rotation curves may be suspect: the emission-line gas in many bulges sometimes seems to rotate much slower than expected from the observed stellar velocity dispersions (Fillmore, Boroson and Dressler, 1986; Kent, 1988; Schweizer, van Gorkom and Seitzer, 1989). The observed gas velocities in NGC 4594 are smaller even than the mean stellar streaming velocities (Kormendy and Westphal, 1989), which is hard to reconcile with motion on simple closed orbits. These cases show that high resolution velocity fields of emission line gas in bulges and ellipticals are useful, not only for determination of $\mathcal{M}/L$ values, but also for the derivation of intrinsic shapes, but that they require careful interpretation. A study of a sample of ellipticals with regular gas disks is currently being carried out in the context of the ESO Key Project on Dark Matter in Elliptical Galaxies (Bertin et al., 1989).

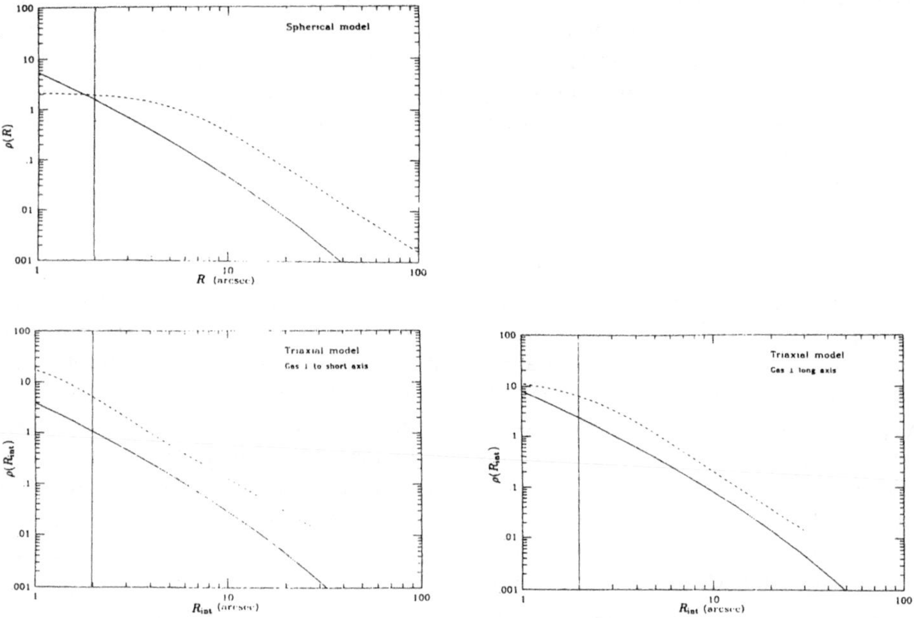

Figure 3. Density profile (dashed curve) of spherical and triaxial mass models which fit the gas velocities measured along seven position angles in NGC 5077 (Bertola et al., 1991), compared with the luminosity profile (solid curve). Three cases are shown: a spherical model, a triaxial model with axial ratios 0.44:0.51:1 and the gas perpendicular to the short axis, and a triaxial model with axial ratio 0.54:0.92:1 and the gas perpendicular to the long axis.

3.3 Atomic hydrogen

Emission line gas velocity fields often can not be measured much beyond $1r_e$, but there are about half a dozen elliptical galaxies with detectable amounts of cold HI gas which extends to many effective radii (van Gorkom, 1991). These are ideal cases for studying the behavior of $\mathcal{M}/L$ at large radii.

Kent (1990) has compared the central $\mathcal{M}/L$–values based on stellar velocities dispersions with those found at large radii from HI for five ellipticals. He concludes that in four of these, NGC 1052 (van Gorkom et al., 1986), NGC 2974 (Kim et al., 1988), NGC 4278 (Raimond et al., 1981), and Centaurus A (van Gorkom et al., 1990), little radial increase in $\mathcal{M}/L$ is evident, despite the fact that HI is observed as far out as $10r_e$ in NGC 1052 and NGC 2974. The exception is IC 2006 (Schweizer, van Gorkom and Seitzer, 1989), an E1 galaxy with a narrow ring of HI at $6.5r_e$, out to which $\mathcal{M}/L$ appears to increase by a factor of three if the ring is assumed to be intrinsically circular.

Traditionally, HI velocity fields have been fitted with planar circular motion or with tilted circular ring models, i.e., just as is customary for spiral galaxies. The results mentioned in §3.2 indicate that this is not necessarily the correct approach for elliptical galaxies. Even though the HI extends to very large radii, deviations from circular motions might still be significant. Indeed, Lees (1991) has fitted both circular and elliptic orbit models to new VLA observations of NGC 4278 which have higher resolution and sensitivity than the earlier Westerbork measurements by Raimond et al. (1981). She finds that a triaxial model with elliptic orbits fits better (as did Raimond et al., 1981), and reproduces the central twist in the velocity field, the morphology of the disk, and the misalignment of the kinematics axes from the photometric axes. The resulting $\mathcal{M}/L$ increases by about a factor of two out to $6r_e$. This is substantially less than in a model with circular orbits.

It is clearly worthwhile to do similar investigations of the other elliptical galaxies with HI at large radii. This includes NGC 5666, which has a twisted velocity field (Lake, Schommer and van Gorkom, 1987), and IC 2006, for which Franx, van Gorkom and de Zeeuw (1991) have obtained a new high resolution HI map. Owing to the modest spatial resolution of the HI data, reported $\mathcal{M}/L$-gradients often rely on a central value that is derived by the core fitting method from stellar velocity dispersion measurements. So, while improving the modeling of the HI kinematics will provide more reliable estimates of $\mathcal{M}/L$ at large radii, it is crucial to obtain more accurate central $\mathcal{M}/L$ values as well, e.g., by detailed modeling of stellar velocity dispersion profiles along a number of position angles (cf §2.1).

4 Conclusions

There is little observational evidence for $\mathcal{M}/L$ variations inside $2r_e$ in normal elliptical galaxies. At larger radii, radial velocity observations of various kinematic tracers are becoming available, but these do not yet provide an unambiguous determination of the $\mathcal{M}/L$ profile; in most cases it is not yet possible to rule out either constant $\mathcal{M}/L$ models or models with massive halos.

The best kinematic tracer of the gravitational potential is cold HI gas, just as for spiral galaxies. However, projection effects and dynamical effect are important for the observed kinematics of cold gas in triaxial galaxies, and result in twisted velocity fields with misaligned kinematic axes. If the gas has had time to settle on the simple closed orbits, then high-resolution observations, combined with detailed modeling, can put significant constraints on the intrinsic shape of the galaxy, and on the variation of $\mathcal{M}/L$ with radius.

The recent work on radial velocity measurements of individual objects in the stellar halos of elliptical galaxies, notably that on the planetary nebulae by Ford, Hui and collaborators, offers the exciting prospect of measuring the rotational state of the stellar halos of elliptical galaxies to radii of up to $5r_e$.

It is a pleasure to acknowledge enlightening discussions with Marijn Franx, Jacqueline van Gorkom, Joanna Lees, and Roeland van der Marel.

References

Athanassoula, E.O., Bosma, A.: 1985, *Ann. Rev. Astron. Astrophys.*, **23**, 147.

Bacon, R.: 1985, *Astron. Astrophys.*, **143**, 84.

Bacon, R., Monnet, G., Simien, F.: 1985, *Astron. Astrophys.*, **152**, 315.

Bacon, R., Simien, F., Monnet, G.: 1983, *Astron. Astrophys.*, **128**, 405.

Bertin, G., Bertola, F., Buson, L.M., Danzinger, I.J., Dejonghe, H., Sadler, E.M., Saglia, R.P., Vietri M., de Zeeuw, P.T., Zeilinger, W.W.: 1989, *The Messenger*, **56**, 19.

Bertin, G., Saglia, R.P., Stiavelli, M.: 1988, *Astrophys. J.*, **330**, 78.

Bertola, F.: 1987, IAU Symposium n. 127, *Structure and Dynamics of Elliptical Galaxies*, ed.: P.T. de Zeeuw, Reidel, p. 135.

Bertola, F., Rubin, V.C., Zeilinger, W.W.: 1989, *Astrophys. J. Letters*, **345**, L29.

Bertola, F., Bettoni, D., Danziger, I.J., Sadler, E.M., Sparke, L.S., de Zeeuw, P.T.: 1991, *Astrophys. J.*, in press.

Binney, J.J.: 1982, *Ann. Rev. Astron. Astrophys.*, **20**, 399.

Binney, J.J., Davies, R.L., Illingworth, G.D.: 1990, *Astrophys. J.*, **361**, 78.

Binney, J.J., Mamon, G.A.: 1982, *Mon. Not. R. Astr. Soc.*, **200**, 361.

Caldwell, N., Kirshner, R.P., Richstone, D.O.: 1986, *Astrophys. J.*, **305**, 136.

Davies, R.L., Birkinshaw, M.: 1988, *Astrophys. J. Supp. Ser.*, **68**, 409.

DeMoulin-Ulrich, M.H., Butcher, H.R., Boksenberg, A.: 1984, *Astrophys. J.*, **285**, 527.

de Zeeuw, P.T.: 1990, *Dynamics and Interactions of Galaxies*, ed. R. Wielen, Springer-Verlag, p. 263.

de Zeeuw, P.T., Franx, M.: 1989, *Astrophys. J.*, **343**, 617.

de Zeeuw, P.T., Franx, M.: 1991, *Ann. Rev. Astron. Astrophys.*, **29**, in press.

Fabbiano, G.: 1989, *Ann. Rev. Astron. Astrophys.*, **27**, 87.

Fillmore, J.A.: 1986, *Astron. J.*, **91**, 1096.

Fillmore, J.A., Boroson, T.A., Dressler, A.: 1986, *Astrophys. J.*, **302**, 208.

Fillmore, J.A., Levison, H.F.: 1989, *Astron. J.*, **97**, 57.

Ford, H.C.,Ciardullo, R., Jacoby, G.H., Hui, X.: 1989, IAU Symposium n. 131, *Planetary nebulae*, ed.: S. Torres-Peimbert, Reidel, p. 335.

Forman, W., Jones, C., Tucker, W.: 1985, *Astrophys. J.*, **293**, 102.

Franx, M.: 1991, this volume.

Franx, M., van Gorkom, J.H., de Zeeuw, P.T.: 1991, in preparation.

Gerhard, O.E., Vietri, M.: 1986, *Mon. Not. R. Astr. Soc.*, **223**, 377.

Gerhard, O.E., Vietri, M., Kent, S.M.: 1989, *Astrophys. J. Letters*, **345**, L33.

Gerhard, O.E., Vietri, M.: 1990, *Dynamics and Interactions of Galaxies*, ed.: R. Wielen, Springer-Verlag, p. 342.

Harris, W.A.: 1988, IAU Symposium n. 126, *Globular Cluster Systems in Galaxies*, eds: J.E. Grindlay, A.G. Davis-Philip, Kluwer, p. 237.

Heiligman, G., Schwarzschild, M.: 1979, *Astrophys. J.*, **233**, 872.

Heisler, J., Merritt, D.R., Schwarzschild, M.: 1982, *Astrophys. J.*, **258**, 490.

Hesser, J.E., Harris, H.C., van der Bergh, S., Harris, G.L.M.: 1984, *Astrophys. J.*, **276**, 491.

Huchra, J.: IAU Symposium n. 126, *Globular Cluster Systems in Galaxies*, eds: J.E. Grindlay, A.G. Davis-Philip, Kluwer, p. 255.

Hui, X.: 1990, private communication.

Katz, N., Richstone, D.O.: 1985, *Astrophys. J.*, **296**, 331.

Kent, S.M.: 1988, *Astron. J.*, **96**, 514.

Kent, S.M.: 1990, *ASP Conference Series Vol 10: Evolution of the Universe of Galaxies*, ed.: R.G. Kron, Brigham Young Univ. Print Services, p. 109.

Kim, D.W. , Guhathakurta, P. , van Gorkom, J.H. , Jura, M., Knapp, G.R.: 1988, *Astrophys. J.*, **330**, 684.

Kormendy, J., Westphal, D.J.: 1989, *Astrophys. J.*, **338**, 752.

Lake, G., Schommer, R.A., van Gorkom, J.H.: 1987, *Astrophys. J.*, **314**, 57.

Lees, J.F.: 1991, *Warped Disks and Inclined Rings around Galaxies*, eds: S. Casertano, P.D. Sackett, F. Briggs, Cambridge Univ. Press, p. 50.

Lees, J.F., Knapp, G.R., Rupen, M.P., Phillips, T.G.: 1991, *Astrophys. J.*, in press.

Lees, J.F., de Zeeuw, P.T.: 1191, in preparation.

Magnenat, P.: 1982, *Astron. Astrophys.*, **108**, 89.

Merritt, D.R.: 1987, *Astrophys. J.*, **313**, 121.

Merritt, D.R.: 1988a, *Astron. J.*, **95**, 496.

Merritt, D.R.: 1988b, *ASP Conference Series Vol. 5: The Minnesota Lectures on Clusters of Galaxies and Large-Scale Structure"*, ed.: J.M. Dickey, Brigham Young Univ. print service, p. 175.

Merritt, D.R., de Zeeuw, P.T.: 1983, *Astrophys. J. Letters*, **267**, L23.

Mould, J.R., Oke, J.B., de Zeeuw, P.T., Nemec, J.M.: 1990, *Astron. J.*, **99**, 1823.

Quillen, A.C., de Zeeuw, P.T., Phillips, T.G., Phinney, E.S.: 1991, *Astrophys. J.*, in press.

Raimond, E., Faber, S.M., Gallagher, J.S., Knapp, G.R.: 1981, *Astrophys. J.*, **246**, 708.

Richstone, D.O., Tremaine, S.D.: 1984, *Astrophys. J.*, **286**, 27.

Richstone, D.O., Tremaine, S.D.: 1986, *Astron. J.*, **92**, 72.

Rood, H.J., Page, T.L., Kintner, E.C., King, I.R.: 1972, *Astrophys. J.*, **175**, 627.

Saglia, R.P.: 1990, *PhD Thesis*, Scuola Normale Superiore, Pisa.

Saglia, R.P., Bertin, G., Stiavelli, M.: 1991, *Astrophys. J.*, submitted.

Sazarin, C.L.: 1987, IAU Symposium n. 127, *Structure and Dynamics of Elliptical Galaxies*, ed.: P.T. de Zeeuw, Reidel, p. 179.

Schwarzschild, M.: 1954, *Astron. J.*, **59**, 273.

Schweizer, F., van Gorkom, J.H., Seitzer, P.: 1989, *Astrophys. J.*, **338**, 770.

Sparke, L.S.: 1986, *Mon. Not. R. Astr. Soc.*, **219**, 657.

Statler, T.S.: 1987, *Astrophys. J.*, **321**, 113.

Statler, T.S.: 1991, *Astron. J.*, submitted.

Steiman-Cameron, T.Y., Durisen, R.H.: 1982, *Astrophys. J. Letters*, **263**, L63.

Steiman-Cameron, T.Y., Durisen, R.H.: 1988, *Astrophys. J.*, **325**, 26.

Tonry, J.: 1983, *Astrophys. J.*, **266**, 56.

Tonry, J., Schechter, P.L.: 1990, *Astron. J.*, **100**, 1794.

Trinchieri, G., Fabbiano, G., Canizares, C.R.: 1986, *Astrophys. J.*, **310**, 637.

Tubbs, A.D.: 1980, *Astrophys. J.*, **241**, 969.

van Albada, T.S.: 1988, *Space Science and Fundamental Physics*, ESA-SP **283**, 165.

van Albada, T.S., Bahcall, J.N., Begeman, K., Sancisi, R.: 1985, *Astrophys. J.*, **295**, 305.

van Albada, T.S., Kotanyi, C.G., Schwarzschild, M.: 1982, *Mon. Not. R. Astr. Soc.*, **198**, 303.

van der Marel, R.P.: 1991, *Mon. Not. R. Astr. Soc.*, submitted.

van der Marel, R.P., Binney, J.J., Davies, R.L.: 1990, *Mon. Not. R. Astr. Soc.*, **245**, 582.

van del Marel, R.P., Cinzano, P.: 1991, this volume.
van Gorkom, J.H.: 1991, this volume.
van Gorkom, J.H., Knapp, G.R., Raimond, E., Faber, S.M., Gallagher, J.S.: 1986, *Astron. J.*, **91**, 791.
van Gorkom, J.H., van der Hulst, J.M., Haschick, A.D., Tubbs, A.D.: 1990, *Astron. J.*, **99**, 1781.
Wilkinson, A., Sharples, R.M., Fosbury, R.A.E., Wallace, P.T.: 1986, *Mon. Not. R. Astr. Soc.*, **218** 297.

Discussion

P. van der Kruit - Your conclusion about the absence of $\mathcal{M}/L$ gradients in the inner parts of elliptical galaxies is of interest for the question of whether spiral bulges are similar to ellipticals. The flat rotation curve of NGC 7814 I showed in my talk indicates a strong $\mathcal{M}/L$ gradient over the optical extend. This measurement is possible in NGC 7814 because the disk mass is very small. So at least one bulge differs in this respect from the ellipticals you described.
T. De Zeeuw - It will be interesting to see what high spatial resolution observations of ellipticals with HI gas disks will show.

R. Buta - With the wonderful new VLA data on the ring of IC 2006, can you comment on how much the geometric and kinematic axes of the ring disagree? This of course may help you to decide on the true shape of the ring, assuming it is not purely expanding or contracting at the same time that it is rotating.
T. De Zeeuw - This is precisely what Franx, van Gorkom and myself are planning to do.

D. Burstein - I am a bit confused. Most galaxies appear to have color or line strength gradients that decrease radially outwards, changing from 2–3 times solar metallicity in the nucleus to 0.5 times solar at $2r_e$. Bill Baum's elegant measurements perhaps show this best. Changes in B$-$V colors also show this, typically changing by 0.2 magnitude over the same radial interval. Thus, if you see a constant $\mathcal{M}/L$ with radius, relative to V light, does this not imply a decreasing stellar $\mathcal{M}/L$ with radius, and hence increasing $\mathcal{M}/L$ of dark to luminous matter?
A. Renzini - Provided the mass function is not a strong function of metallicity, the variation implied by color or line strength gradients is quite small, consistent with the upper limits that Tim has just reported. This is to say that $\mathcal{M}/L$ is not much sensitive to metallicity.
M. Capaccioli - Let me try to add another element to your picture. Cappellaro, Held, and myself succeeded in measuring rotation and velocity dispersion curves for a few bulges of close to edge-on SO's out to $2r_e$. Vietri has examined the data for NGC 3115 using a model similar to Binney, Davies, and Illingworth (1990), showing that they are largely inconsistent with $\mathcal{M}/L$ constant from $0.4r_e$ outwards. We have instead some evidence from the vertical structure of the disk that $\mathcal{M}/L$ is costant inside r_e.

S. Kent - Do you have a velocity dispersion profile in NGC 5077?

T. De Zeeuw - Yes, along a number of position angles. The profiles extend to about $1r_e$, and modeling is in progress.

F. Bertola - Constant mass-to-light ratios may be consistent with dark halos that have the same mass distribution as the luminous matter. What probably rules out dark matter are the rather low values found for $\mathcal{M}/L_B$.
T. De Zeeuw - Agreed. The recent studies of NGC 5077 and Centaurus A that I mentioned both find $\mathcal{M}/L_B$ of 3–5 in solar units out to about $1r_e$.

I. King - We still have a large discrepancy that needs to be faced. In rich clusters of galaxies, the dark matter is more than ten times what would be indicated by the internal $\mathcal{M}/L$ of ellipticals. Dynamical studies, such as that of Simon White, have shown that the dark matter is not in the galaxies, which have an $\mathcal{M}/L \leq 30 - 40$. Since early type galaxies are mainly found in rich clusters, they would seem, even after stripping, to have a lot of dark matter; before stripping the $\mathcal{M}/L$ could have been even larger. I think that your summary has been excellent; thus this discrepancy would appear to require even better observations than we now have of early type galaxies.
J. van Gorkom - Do we know whether dark matter has mainly been stripped from ellipticals or could it also have been stripped mostly from spirals?
T. De Zeeuw - Merritt (1988b) has argued that no more than 15% of the mass in clusters can be attached to individual galaxies. This implies that the total $\mathcal{M}/L$ for a galaxy is 0.15 times the $\mathcal{M}/L$ of the whole cluster, which results in the number mentioned by Ivan, $\mathcal{M}/L \leq 30$. Thus the halos may contain less than three to five times the mass of the luminous parts. This rather low factor may be caused by stripping during the formation of the cluster. Merritt's result shows that in order to answer the question of whether elliptical galaxies have dark halos one has to demonstrate that any dark matter found is in fact related to the galaxy, and has a similar velocity dispersion. This is particularly relevant for one the best studied, and very often quoted, cases, namely that of M 87.

J.–L. Nieto - I am skeptical about the blind use of the King and Minkowski formula to determine $\mathcal{M}/L$ in the core of ellipticals, often as reference value from which the outer value may or may not differ (example IC 2006). Cores are not isothermal (see my review), and as a corollary of this, this formula does not apply and may be erroneous by large amounts.
T. De Zeeuw - Agreed. As I mentioned, the King formula gives a good and useful estimate of $\mathcal{M}/L$, but may easily be off by factors of up to 2 or 3 in individual cases, especially for non-spherical galaxies.

Bertola (bottom left) looks attentively to Tim de Zeeuw who is showing a sample of dark matter.

DARK MATTER IN LATE-TYPE GALAXIES

S.M. Kent
Center for Astrophysics
Cambridge MA, USA

ABSTRACT. The status of our knowledge of the extent and distribution of dark matter in late-type galaxies is reviewed. Recent attempts to alleviate the need for dark matter in late-type galaxies are examined, but none are found to give a satisfactory explanation. The mass distribution and shapes of dark matter halos are still known only poorly. Recent work on halo shapes suggests that they are not as flattened as predicted by cosmological simulations.

1 Introduction

The principal evidence for dark matter in spiral galaxies comes from a comparison of their light profiles and rotation curves. Typically, the light profiles of galaxies are declining exponentially at large radii, whereas the rotation curves are flat, implying that the surface mass density (if the mass were in the disk) is falling only as $1/r$. The local mass/light ratio rises to such high values that one infers the existence of some new, unseen mass component.

It would be impossible to present a complete review of dark matter in spiral galaxies in the space available here. Past reviews of the subject may be found in Faber and Gallagher (1979), Kent (1987b), and IAU Symposium 117. Here, I shall concentrate on just a few selected topics: the evidence for (and against) dark matter in spiral galaxies, correlations between halo and galaxy parameters, and new data on the shapes of galaxy halos.

2 Is dark matter required?

Every few years, someone comes along and challenges the need for dark halos (or a new law of gravity) in spiral galaxies for various reasons. Here I will look at three such challenges.

Do the data demand dark matter? Very little of the early work on dark matter made a quantitative comparison between the luminosity profiles and the mass distribution

G. Longo et al. (eds.), Morphological and Physical Classification of Galaxies, 153–162.

inferred from rotation curves. It came as a surprise, therefore, when Kalnajs (1983) showed that for 4 galaxies with surface photometry and measured rotation curves, the rotation curves could be reproduced by models with a constant M/L ratio and no dark halo, in spite of the fact that the rotation curves appeared to be flat or rising everywhere. In fact, however, three of his four galaxies only had optically determined rotation curves which do not extend as far as HI-determined rotation curves, and it is now recognized that the more extended HI data completely rule out constant M/L ratio models in virtually all cases (i.e., van Albada et al., 1985). Figure 1 shows photometry and rotation curves for two galaxies adapted from Kent (1986, 1987a), one with only optical data and one with HI data, illustrating the problem.

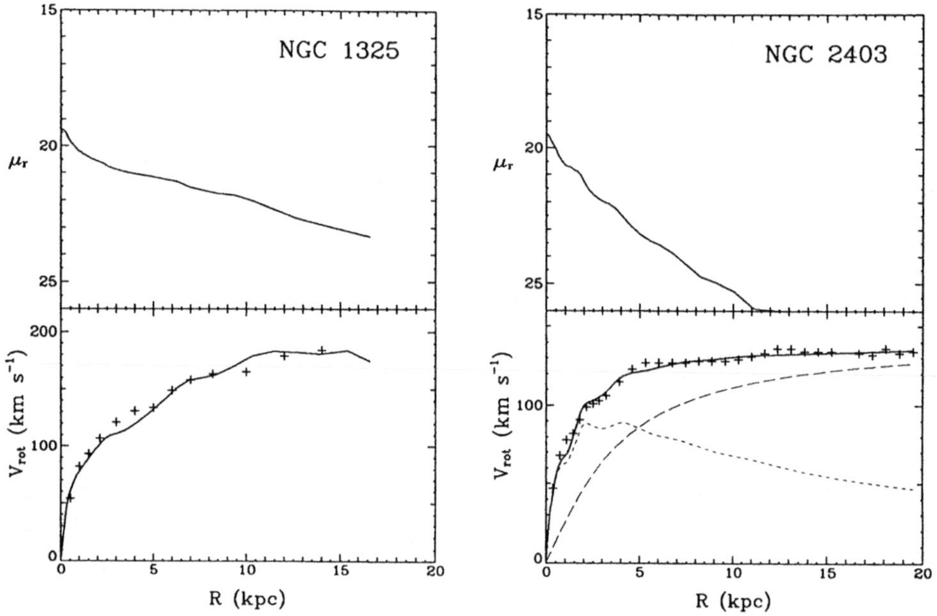

Figure 1. Surface brightness profiles (top panels) and rotation curves (bottom panels) for two representative spiral galaxies. The photometry is from Kent (1986, 1987a); the measured rotation curves (pluses) are from Rubin et al., (1985) and Begeman (1988). The optically-determined rotation curve for NGC 1325 can be fitted by a model with constant M/L ratio and no dark halo; the HI-determined rotation curve of NGC 2403 cannot be so fitted; the separate disk and halo contributions are shown.

Are the data accurate? Binney (1987) noted that the surface brightness of galaxies is typically much lower than that of the night sky, and so if one incorrectly measures the sky background value in regions where there is still light from the galaxy (as could easily be done on narrow-field CCD frames), then one would underestimate the true galaxy surface brightness profile. However, a careful look at how surface photometry is done shows that this conjecture is groundless. For example, figure 2 compares two luminosity profiles measured for the well-known spiral galaxy NGC 3198, one done photographically and the other with a CCD. The CCD was mounted behind a small telescope to ensure hat it had a wide field of view. The two profiles are in good agreement with each other nd both diverge noticeably at large radii from the mass profile that would be needed to eproduce the HI rotation curve.

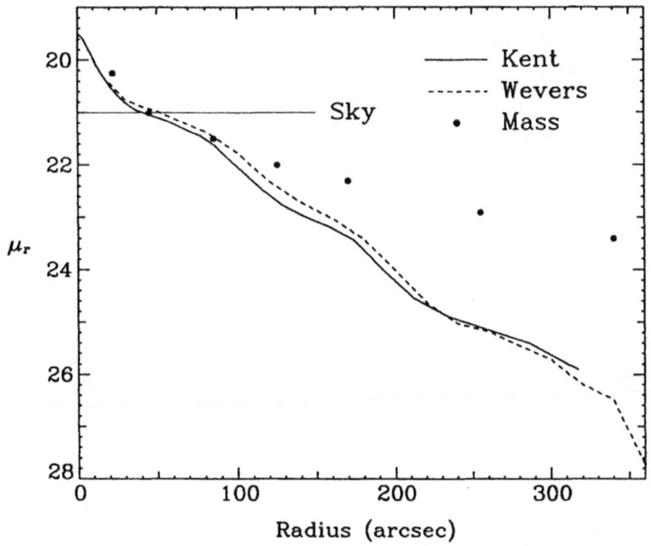

Figure 2. Comparison of the measured surface brightness profile of NGC 3198 with the mass profile required to reproduce the observed HI-determined rotation curve.

Is the dark matter in molecular gas? Recently, Valentijn (1990a,b) has postulated the existence of a population of compact opaque molecular clouds in spiral galaxies with a scale length about 5 times that of the stars which could be sufficiently massive to provide all the dark matter. The basis for this claim comes from a study of the dependence of the surface brightness of spiral galaxies as a function of inclination, from which it appears that spirals are fairly opaque at all radii. Valentijn rules out the possibility that diffuse dust (such as that seen by IRAS) is solely responsible and argues that a second component is needed to provide the necessary extinction. The clouds are presumed to be sufficiently opaque that their extinction cross-section is the same as their geometric cross-section. The opacity as a function of wavelength is more constant than for diffuse dust. The "dark matter" is then in the form of either H_2 or highly shrouded stars buried within the molecular clouds.

The means by which these clouds are inferred to exist are very indirect, and one would like to find more direct observational evidence of their existence. Such evidence, however, is lacking, and there is some evidence to the contrary. The CO emission (which traces molecular gas and much of the dust) often has the same density profile in late-type galaxies as the stars (i.e., Young and Scoville, 1982). Although there is some evidence that the CO/H_2 ratio declines with increasing radius in galaxies, it seems improbable that it could conspire to fall so rapidly that the bulk of H_2 remains invisible. Furthermore, the dust/gas ratio may fall as well, so one might expect the extinction (if not the total H_2 mass) to track the CO unless the clouds have extremely high optical depth. There is as yet no evidence for the existence of large amounts of extinction in a "grey" component. Observations of edge-on galaxies at visible and near-infrared wavelengths show that galaxies become progressively more optically thin with increasing wavelength, and the optically-prominent dust lanes nearly disappear by the time one reaches the infrared K band (i.e., Wainscoat, Freeman, and Hyland, 1989). A comparison of optical and infrared light profiles in galaxies shows that the infrared scale length is actually shorter than that in the optical (Giovanardi and Hunt, 1988), showing that there is no hidden population of stars to provide the missing mass. Figure 3 shows the optical-infrared color gradients in five spiral galaxies, demonstrating that spirals become bluer at large radii.

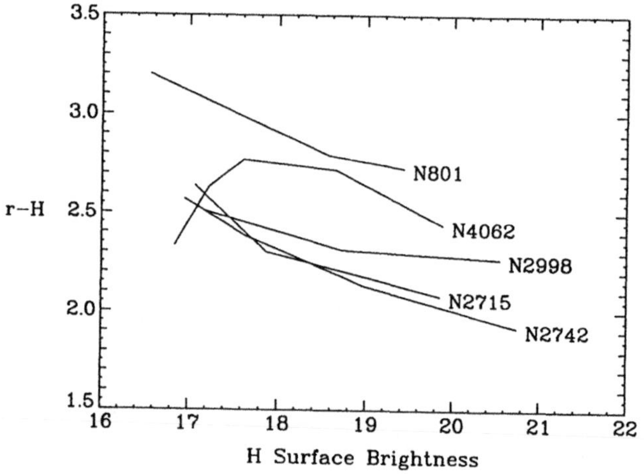

Figure 3. Optical-infrared color gradients in five spiral galaxies. The optical data are from Kent (1986); the infrared data are from Giovanardi and Hunt (1988). H surface brightness rather than radius is used for the abscissa in order to place the galaxies on a comparable scale.

3 How much dark matter is there?

One would like to know what the spatial distribution of dark matter is in spiral galaxies and compare it with other properties of the galaxies. To do so one must resort to some sort of mass modeling procedure. Usually, a galaxy is modeled as the sum of a constant M/L stellar disk (with a light profile determined from observations) plus an isothermal sphere (or equivalent) for the dark halo. In early type spirals a bulge component is needed as well. Free parameters are the M/L ratio of the disk (and bulge if needed) and a core radius and scale density for the halo. These parameters are found by computing the rotation curve contribution from each component and making a least-squares fit to the observed rotation curve. Unfortunately, the parameters are usually highly correlated with one another, so small changes in the observables can lead to large changes in the parameter values. The solution space can be restricted by imposing some additional constraints. Three methods that have been used are

a) *Maximum-Disk solution:* The stellar contribution to the rotation curve is made as large as possible without exceeding the observed velocities (van Albada et al., 1985; Kent, 1986). Usually one can fit the entire rising portion of the rotation curve by this method. Any excess is assigned to the dark halo. Usually one imposes the additional restriction that the halo does not have a hollow center, which reduces the stellar M/L ratio by about 20-35%.

b) *Spiral Structure Constraints:* Athanassoula, Bosma, and Papaioannou (1987) used swing-amplifier theory to determine what range of disk/halo masses are needed to suppress one-armed, two-armed and four-armed spiral structure. Since spirals often show two-armed structure, this sets a narrow range of possible models.

c) *Disk Kinematics:* van der Kruit and Freeman (1984) have attempted to measure the stellar disk mass directly in galaxies from observations of the vertical stellar velocity dispersion and vertical scale height. Unfortunately one cannot measure all the relevant quantities in a single galaxy, so it is necessary to assume some average values for the missing quantities in any given object. Typically, though, they find that the disk masses are of order only 1/3 to 1/2 the maximum-disk values.

Some evidence suggests that the maximum-disk solution with no hollow core in the halo is close to the correct solution. Athanassoula, Bosma, and Papaioannou (1987) find that this solution is often the same as their maximum-disk solution with no one-armed spiral structure. The fact that a maximum-disk solution with no halo can often reproduce a substantial portion of the inner rotation curve (as for NGC 1325 in figure 1) argues that the stellar disk dominates the mass inside about 2 disk scale lengths. Simple-minded analyses also show correlations between some measure of luminosity concentration in galaxies and the steepness of the rise in rotation curves (i.e., Corradi and Capaccioli, 1990). These arguments are not conclusive, but if the low disk masses found by van der Kruit and Freeman are correct, then the dark matter must track the disk mass quite closely in the inner parts of galaxies. These correlations do not extend to the lowest luminosity galaxies that have been studied, however; for example, in the dwarf DDO 154, the halo appears to dominate the mass almost everywhere (Carignan and Freeman, 1988).

To gain further insight into the relation between dark and luminous matter, Athanassoula, Bosma, and Papaioannou (1987) and Kormendy (1990) have searched for correlations between dark halo parameters and other galaxy properties. The halo parameters are a core radius, central density, and characteristic velocity dispersion, although only two are independent. Given the unconstrained flexibility in the models, it is not surprising that one finds a large scatter in most of these correlations. Furthermore, some of the correlations may not be genuine but rather introduced as an artifact of the model fitting process (i.e., the dark matter core radius and central density are tightly coupled in the fitting process). In fact, I have found only a few independent correlations that are convincing. These include halo velocity dispersion versus galaxy luminosity (Athanassoula, Bosma, and Papaioannou, 1987; Kormendy, 1990), halo core radius versus morphological type (Athanassoula, Bosma, and Papaionnou, 1987); and disk/halo mass ratio versus galaxy luminosity (Persic and Salucci, 1990).

The first correlation above is just a variant of the Tully-Fisher law, while the latter two are telling us something new about halo properties. One would like to use these correlations to find out something about, say, the initial perturbation spectrum from which galaxies formed, but models of galaxy formation with dissipational infall of baryons into dark matter halos (Blumenthal et al., 1986; Ryden and Gunn, 1987; Ryden, 1989) show that adiabatic compression of the halos is significant, and so we are likely to learn more about the compression process than the initial conditions from which galaxies form. Furthermore, n-body cosmological simulations do not yet have enough dynamic range to give meaningful predictions about galaxy structure on scales smaller than tens of kiloparsecs.

4 Shapes of dark matter halos

A fairly robust prediction of n-body cosmological simulations is that dark matter halos should be strongly triaxial, i.e., should have three unequal moments of inertia. Frenk et al., (1988) have calculated the distribution of ellipticities for halos that form in cold dark matter simulations. If the halos have principal axes a, b, c with $a > b > c$, then they find that typical ratios are $b/a \approx 0.6$, $c/a \approx 0.5$, albeit with large scatter.

Considerable circumstantial evidence exists that halos are, indeed, not round, but it is difficult to make quantitative statements about how flattened halos actually are. Here, I will review the past evidence and discuss a new program to measure the shapes of spiral disks. In all cases I will quote the axis ratio of the ellipsoidal mass distribution needed to produce a given configuration, even though actual mass distribution, being a combination of a halo and disk, is not ellipsoidal in shape.

Polar rings: Several cases are known of S0 galaxies which possess a ring or disk of gas orbiting in a plane perpendicular to that of the S0 disk. By comparing the orbital velocity of the stars and gas in the two disks at the same radius and assuming that they follow closed orbits in a flattened potential, it is possible to derive the flattening of the gravitational potential and hence that of the underlying mass distribution. For three S0's, Whitmore, McElroy, and Schweizer (1987) find flattenings that correspond to c/a ranging from 0.7 to 1. These values assume that the halo is an oblate spheroid aligned with the S0 disk.

Warps: Edge-on spiral galaxies often show a warp in the faintest optical isophotes and the outer HI. Sparke and Casertano (1988) have shown quantitatively that self-sustained warps can be produced by placing a disk tilted with respect to the symmetry axis in an oblate halo. The amount of warping depends on the halo ellipticity, but also on other parameters that can only be constrained weakly. Consequently, one can place only loose quantitative constraints on the halo flattening, although warps can form only for $c/a > 0.5$.

Shapes of spheroidal systems: Kent, Huchra, and Stauffer (1989) have attempted to use the spatial distribution and kinematics of globular clusters in M31 to deduce the shape of its mass distribution. The globular cluster system has a flattening of 0.3-0.4. If the velocity distribution is isotropic, then the disk and halo have approximately equal masses out to a radius of 30 kpc. The resulting gravitational field can be approximated by that of an oblate spheroid with $c/a = 0.3$. However, if the velocity distribution is not isotropic, then only very loose constraints can be placed on the shape of the mass distribution; at best, Kent et al., could barely rule out a spherical shape. This test also measures the flattening of the mass distribution at much smaller radii, where the disk potential is more dominant, than the previous methods.

Shapes of Galaxy Disks: If a galaxy disk is embedded in triaxial halo with the disk axis aligned, say, with the halo short axis, then stars in the disk will not be orbiting in an axisymmetric potential, and the disk shape should be out of round by an amount that depends on the axis ratio b/a of the halo. For example, a disk embedded in a halo with b/a of 0.6 will have an apparent ellipticity of about 0.25. Richard Del Val at Harvard and I have conducted an imaging survey of 32 galaxies selected to be face-on in order to study the statistics of disk shapes. The galaxies were selected to be face-on by virtue of having a narrow apparent HI line width compared with the expected line width based on the Tully-Fisher relation. We picked objects from the HI surveys of Giovanelli and Haynes (1985 and subsequent papers) and Lewis (1987) and our selection criterion was that the inclination should be less than 26° so that an intrinsically round disk would appear to have an ellipticity $\epsilon < 0.1$. We measured the apparent ellipticity of the disks at an isophotal radius that approximately matches the UGC radius.

Figure 4 shows the histogram of apparent ellipticities for the sample. They are distributed almost uniformly between 0 and 0.25. Table 1 gives the intrinsic ellipticity distribution obtained after unfolding the selection effects in the sample.

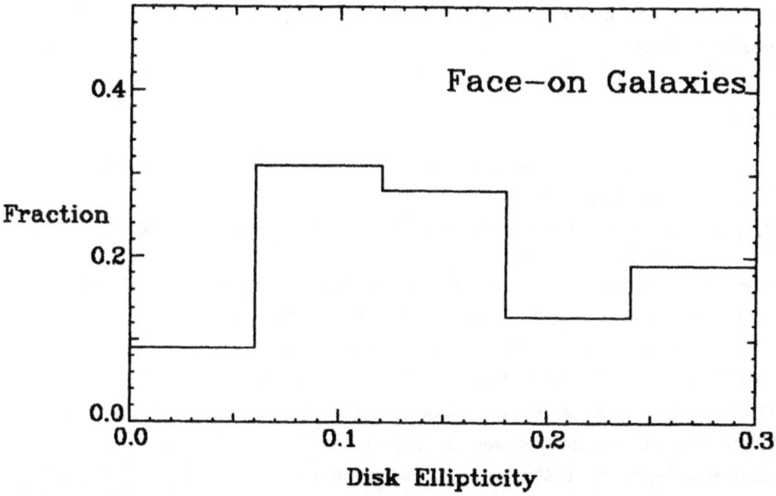

Figure 4. Histogram of the distribution of disk ellipticities for a sample of face-on galaxies.

The mean intrinsic ellipticity is about 0.12. If we interpret the ellipticity as being due to orbital motion in a triaxial halo, then the average halo b/a must be about 0.8, considerably rounder than the halos of Frenk et al. (1988). The ellipticity could equally well be due to nonaxisymmetric structure in the disks (bars or spiral arms) or nonequilibrium motion (tidal distortion by a companion), further reducing the amount attributable to a flattened halo. The conclusion from all this is that if halos are formed as highly flattened systems, then the process of forming disks by the dissipational sinking of baryonic matter must be effective at rounding out the halo, at least close to their centers where we are measuring the disk shapes. It should be fairly straightforward to simulate this process numerically to see if such a rounding can, in fact, be achieved. Conceivably one could even set a lower limit on the disk mass needed.

Table 1 – Intrinsic Disk Ellipticity Distribution

Ellipticity	%
0.00 - 0.10	51
0.10 - 0.20	37
0.20 - 0.30	16

5 Summary

It is somewhat frustrating that, although the need to postulate the existence of dark matter in spiral galaxies is unavoidable, quantifying the shape and distribution of the dark matter is still elusive. One area where significant progress can be made is in using infrared arrays to map the light distribution of galaxies in the near-IR rather than the optical, where variations in the apparent disk M/L ratio due to dust and young stars should be greatly reduced.

References

Athanassoula, E., Bosma, A., and Papaioannou, S.: 1987, *Astron. Astrophys.* **179**, 23.

Begeman, K.: Ph. D. Thesis, University of Groningen.

Binney, J.: 1987, *Material Content of the Universe*, Ed.: J. D. Barrow, P. J. E. Peebles, and D. W. Sciama, Royal Society, London, p. 33.

Blumenthal, G. R., Faber, S. M., Flores, R., and Primack, J. R.: 1986, *Astrophys. J.* **301**, 27.

Carignan, C., and Freeman, K. C.: 1988, *Astrophys. J. Letters* **332**, L33.

Corradi, R. L. M., and Capaccioli, M.: 1990, *Astron. Astrophys.* **237**, 36.

Faber, S. M., and Gallagher, J. S.: 1979, *Ann. Rev. Astr. Ap.*, **17**, 135.

Frenk, C. S., White, S. D. M., Davis, M., and Efstathiou, G.: 1988, *Astrophys. J.* **327**, 507.

Giovanardi, C., and Hunt, L. K.: 1988, *Astron. J.* **95**, 408.

Giovanelli, R., and Haynes, M. P.: 1985, *Astron. J.* **90**, 2445.

Kalnajs, A.: 1983, *Internal Kinematics and Dynamics of Disk Galaxies*, IAU Symposium 100, Ed: E. Athanassoula, Reidel, Dordrecht.

Kent, S. M.: 1986, *Astron. J.* **91**, 1301.

Kent, S. M.: 1987, *Astron. J.* **93**, 816.

Kent, S. M.: 1987, *Nearly Normal Galaxies*, Ed.: S. M. Faber, Springer, New York.

Kent, S. M., Huchra, J. P., and Stauffer, J.: 1989, *Astron. J.* **98**, 2080.

Kormendy, J.: 1990, *Evolution of the Universe of Galaxies*, Ed.: R. Kron, Astron. Soc. Pac., San Francisco.

Lewis, B. M.: 1987, *Astrophys. J. Suppl. Ser.* **63**, 513.

Rubin, V. C., Burstein, D., Ford, W. K., and Thonnard, N.: 1985, *Astrophys. J.* **289**, 81.

Ryden, B. S., and Gunn, J. E.: 1987, *Astrophys. J.* **318**, 15.

Ryden, B. S.: 1989, *Astrophys. J.* **329**, 589.

Sparke, L. S., and Casertano, S.: 1988, *Mon. Not. R. Astr. Soc.* **234**, 873.

Valentijn, E. A.: 1990a, *Nature*, **346**, 153.

Valentijn, E. A.: 1990b, *The Interstellar Disk-Halo Connection in Galaxies*, IAU Symposium 144, Ed.: J. B. G. M. Bloemen, Kluwer, Dordrecht.

van Albada, T. S., Bahcall, J. N., Begeman, K., and Sancisi, R.: 1985, *Astrophys. J.* **295**, 305.

van der Kruit, P. C., and Freeman, K. C.: 1984, *Astrophys. J.* **278**, 81.

Wainscoat, R. J., Freeman, K. C., and Hyland, A. R.: 1989, *Astrophys. J.* **337**, 163.

Wevers, B. M. H.: 1984, Ph. D. Thesis, Groningen University.

Whitmore, B. C., McElroy, D. B., and Schweizer, F.: 1987, *Astrophys. J.* **314**, 439.

Young, J. S., and Scoville, N.: 1982, *Astrophys. J.* **258**, 467.

Discussion

D. Burstein - If one presented you with a set of rotation curves for different galaxies I contend you cannot identify galaxies by either a) Hubble type or b) Bulge-to-disk ratio. I will agree, however, that you can separate galaxies by luminosity.

S. Kent - I would partly agree, but by to point out that rotation curves are not sensitive to spiral arm morphology, which is a major determinant of Hubble type. Also, although luminosity concentration (or bulge/disk ratio) is correlated with Hubble type, the correlation has considerable scatter (as there is with virtually any other physical parameter). Finally, rotation curves in the bulge regions of galaxies can have an anomalous behavior. Hence the fact that one cannot predict the Hubble type with any reliability is hardly surprising.

J. van Gorkom - This is partly in response to the remark of Dave Burstein. Stefano Casertano and I have recently completed a paper on yet another trend for rotation curves. We looked at a sample of very well determined HI rotation curves and plot the slope of curve between about 0.5 Holmberg radius and the last measured point, which is typically 1.5 Holmberg radii. What we find is that there is a strong correlation with circular velocity, or in other words with morphological type. At low circular velocities, less than 100 km/s, the rotation curves are still rising in the outer parts. At circular velocities between 100 and 200 km/s the rotation curves are flat, while at velocities larger than 200 km/s the curves are slightly declining in the outer parts. So going from late to early type spirals the slope of the rotation curves in the very outer parts of the galaxy does change systematically, suggesting that it might be the ratio between the dark and luminous matter that determines the Hubble morphological type.

M. Persic - Indeed, the shape of rotation curves is a key quantity in the phenomenology of spiral galaxies. Even inside the optical radius (equal to approximately 3 disk scalelengths), the shape of rotation curves is strongly correlated with the luminosity, L_B. High-luminosity, high rotation-velocity galaxies have flat curves between approximately 2 and 3 disk lengthscales, while low-luminosity, low rotation-velocity galaxies have steeper curves in the same radial region (Rubin et al., 1985; Persic and Salucci, 1988 and 1991). This very prominent observational result permits to estimate the statistics of dark matter inside the optical radius of spirals by evaluating the disk to halo mass ratio for individual galaxies. We find that low-luminosity galaxies are halo-dominated already in the optical regions, while high-luminosity galaxies are disk-dominated there (Persic and Salucci, 1987 and 1988; see also Salucci, 1986). The scaling low emerging from the analysis of rotation curve profiles is $\mathcal{M}_{disk}/\mathcal{M}_{halo} \propto L_B^{0.7}$ at the optical radius in the luminosity range $-17.5 \gtrsim M_B \gtrsim -23.3$ (Persic and Salucci, 1990a,b).

M. Franx - Is there any convincing evidence that dark matter is dominant in the center of galaxies?
S. Kent - No, although Piet van der Kruit might disagree.

P. van der Kruit - Two remarks: 1) In relation to the discussion on accuracy of faint disk photometry I would emphasize that the comparison of the profiles from your CCD work and our (Wevers et al.) photographic data shows that surface photometry can be done *very* reliably down to 1% of sky.
2) I would emphasize again that the similar rotation curves of NGC 891 and 7814 (see my talk in this meeting) is strong evidence that dark matter dominates within the optical image.
S. Kent - Regarding (2) I should point out that work by myself on Sa galaxies and Kormendy and Westpfal on N4594 shows that in many cases the optical rotation curve in the bulge dominated parts of galaxies is considerably lower than the circular velocity that one predicts based on the observed bulge stellar velocity dispersion and light profile. In those galaxies where the two do agree, there is a good correlation between some measure of luminosity concentration (or bulge/disk ratio) and rotation curve shape. This conclusion is based on the observations of a few dozen galaxies, not just two.

R. van der Marel - I recently used the Jeans equations to study the flattening of the dark halo (DH) of our own Galaxy (van der Marel, submitted). I found that the current observational data on the metal-weak halo (MWH) of our Galaxy implies that the dark halo must have axis ratio $q_{DH} > 0.34$. If one assumes that $q_{DH} = q_{MWH}$, then one must have $q_{DH} > 0.53$.

M. Capaccioli - Could you please comment on the arguments that justify the equivalence "rotation curve = circular velocity". For instance, recent observations of NGC 5252 (Fig. 1) prove that symmetry arguments can be misleading.
S. Kent - Since gas is dissipative, in a steady state it should settle onto closed, periodic orbits in a galaxy. In an axisymmetric galaxy, a symmetry of the rotation curve necessarily follows. As your figure shows, however, symmetry is not sufficient. Although I would not try to interpret your data as measuring circular velocities, there may be other occasions in which we are fooled into interpreting a symmetric rotation as reflecting circular velocities. I have argued elsewhere that such is, in fact the case in the bulges of Sa galaxies. If the gas is in a transient state, we can look for clues such as broadened emission lines. Probably the best evidence that we are measuring circular rotation is by mapping the complete velocity field, not just along a single axis as it is often done.

DARK MATTER IN DWARF GALAXIES

C. Pryor
Department of Physics and Astronomy
Rutgers, the State University of New Jersey, USA

ABSTRACT. This review focuses on dark matter in the low-luminosity dwarf spheroidal galaxies that are satellites of our own Galaxy. Originally controversial, the large velocity dispersions in Draco and Ursa Minor are now incontestable and the dispersions of Carina, Sculptor, and Fornax have increased as the data have improved. Mass-to-light ratios from core fitting range from 5.7 for Fornax to 94 for Draco. These values are larger than those found for globular clusters and for most low-luminosity ellipticals, suggesting that all of the dwarf spheroidal satellites of our Galaxy contain dark matter. The presence of non-stellar dark matter in Draco, Ursa Minor, and Carina seems hard to avoid. The data are consistent with all the galaxies having dark halos with a central density of $\sim$0.1 $M_{\odot}/pc^{-3}$.

1 Introduction

Galaxies become increasingly dominated by dark matter as their luminosity decreases (Tinsley, 1981; Kent, 1987; Persic and Salucci, 1988, 1990; Kormendy, 1988, 1990). Thus dwarf galaxies are excellent locations for studying dark halos. They are also important for probing how the halos vary with galaxy luminosity (Kormendy, 1990). An HI rotation curve and disk surface photometry alone determine both the central density and core radius of the halo in the best cases (Lake and Feinswog, 1989). The lowest luminosity galaxies with well-determined parameters are DDO 154 (Carignan and Beaulieu, 1989), with absolute magnitude $M_V = -14.2$, and DDO 170 (Lake et al., 1990), with $M_V \simeq -15.5$. Both galaxies have dark matter core radii, r_c, of 2 – 3 kpc and central mass densities, ρ_0, of about 0.01 $M_{\odot}$ pc^{-3}. Dwarf irregular galaxies with much lower luminosities tend not to display clear rotation, for example GR8 (Carignan et al., 1990) and M81 dwA (Sargent et al., 1983), both of which have $M_V \simeq -11$. The properties of the dark matter are uncertain in this case, but Carignan et al., (1990) estimate $\rho_0 \simeq$ 0.07 $M_{\odot}$ pc^{-3} for GR8. For even lower luminosity galaxies, one must turn to the dwarf spheroidal satellites of our own galaxy, which range from $M_V = -12.4$ for Fornax to -8.7 for Draco and Ursa Minor (Webbink, 1985). These systems are probably the least luminous members of a class of "diffuse", low surface brightness spheroidal galaxies that are distinct from ordinary elliptical galaxies (Wirth and Gallagher, 1984; Kormendy, 1985). The exponential brightness profiles of these galaxies suggests that they may be dwarf irregular galaxies that have lost their gas or converted it all to stars (Lin and Faber,

G. Longo et al. (eds.), Morphological and Physical Classification of Galaxies, 163–178.

1983; see the discussion and references in Kormendy and Djorgovski, 1989). This review focuses on the evidence for dark matter in the dwarf spheroidals.

2 Dwarf spheroidal satellites of our galaxy: stellar radial velocities

Because these galaxies have very low surface brightnesses, their velocity dispersions can only be determined by measuring velocities for individual stars. Draco, Ursa Minor, Sculptor, Carina, and Fornax, which are between 70 and 150 kpc from the Sun (Webbink, 1985), have all had dispersions measured. At these distances the brightest giants have V = 17–18 and getting more than a few stars requires measuring stars fainter than 17.5. Since the velocity dispersions are about 10 km s^{-1}, resolutions of $\sim 10^4$ are necessary for the individual spectra and this requires large telescopes and long exposures for such faint stars. The very low metal abundances (e.g., Da Costa, 1988) also contribute to the difficulty of obtaining velocities.

2.1 Early velocity measurements

In 1983 Aaronson reported that the velocity dispersion of Draco was ~10 km s^{-1}, implying an M/L of ~100, based on velocities for three carbon stars (C stars, hereafter). The field has remained controversial and exciting ever since. Velocities for C stars in Carina (6 stars), Sculptor (3 stars), and Fornax (5 stars) quickly followed (Cook et al., 1983; Seitzer and Frogel, 1985). The dispersions in these galaxies were all ~6 km s^{-1}, implying a large range of M/L_V's: 0.5 – 10. This early work is review by Aaronson (1986), Kormendy (1987a), Aaronson and Olszewski (1988), and Da Costa (1988). The similarity of the measured dispersions and the range in the M/L's raised the possibility that binary or atmospheric motions were dominating the observed dispersions or that the measurement errors had been underestimated.

Table 1 lists the instruments that have been used to measure radial velocities of dwarf spheroidal stars. The first column gives the line number, referenced in table 2, and the second lists the telescope and spectrograph. These are followed by the detector, the wavelength of the observations, and the expected velocity accuracy. This last was taken to be $1/20^{th}$ of the resolution, a reasonable limiting value for cassegrain spectrographs. Two general approaches are apparent. One is to use high dispersion echelle and coude spectrographs and low signal-to-noise (S/N). This has been made possible by the low dark count of intensified Reticon packages. The other approach is to use lower resolution and a CCD detector. Higher S/N is necessary to improve the accuracy and to get above the CCD read noise. In this case the measurement uncertainty is comparable to the velocity dispersion and must be accurately determined and subtracted in quadrature. This requires repeat measurements of stars or comparisons between velocities obtained with different instruments.

Table 1. Instruments used to measure radial velocities.

	Telescope and Spectrograph	detector	λ (Å)	1/20 resol. (km s^{-1})	
1	MMT, Echelle	intensified Reticon	5180, 5636	0.4	
2	CTIO 4m, RC (600, 1200 l/mm)	GEC CCD	7900	4.5, 2.3	
3	CTIO 4m, RC (790 l/mm)	GEC CCD	8600	4.3	
4	CFHT 3.6m, Coude	RV scanner	4000-4600	0.9	[1]
5	AAT, RGO	IPCS	4732, 5636	7.5	[2]
6	AAT, fiber-fed RGO	IPCS	5200	7.5	
7	LCO 2.5m, Cassegrain	intensified Reticon	4800-5800	3.6	[3]
8	LCO 2.5m, Echelle	2D-Frutti	4600-5600	0.6	

[1] Kormendy (1987a), [2] Godwin and Lynden-Bell (1987), [3] Cook et al., (1983)

By 1986 Aaronson and Olszewski (1987a) had obtained 25 velocities for 11 stars in Draco and 23 velocities for 10 stars in Ursa Minor. Most of these stars were K giants. Some stars with variable velocities were found, but these had not significantly affected the velocity dispersion. These data confirmed the large M/L's and were sufficient to show that measurement error and binary motions were not important (Aaronson and Olszewski, 1987a, Tremaine, 1987). Kormendy (1987a) compared CFHT and MMT velocities for 4 Draco stars and concluded that the agreement was satisfactory. However, Godwin and Lynden-Bell (1987) concluded from comparing their data for four Carina C stars with that from Seitzer and Frogel (1985) and Cook et al., (1983) that the measurement uncertainties had been underestimated by a factor of ~2. These studies had lower velocity resolution than those compared by Kormendy (1987a). The Godwin and Lynden-Bell (1987) and the Seitzer and Frogel (1985) data sets also had few repeat measurements to determine the external uncertainty. The clear lesson is that repeated velocity measurements of the same stars are necessary to produce a believable dispersion.

2.2 Recent velocity measurements

Observations since 1986 have removed many of the problems with the early data. Samples are bigger and there are more repeat measurements. Measurements of the same stars with independent instruments are still rare, but should come within the next few years as fiber-fed spectrographs increase the number of stars that can be observed with a single exposure. Table 2 lists the samples that currently exist. The first 3 columns give the galaxy, the number of stars with measured velocities, N_s, and the total number of velocity measurements, N_v. Next comes the velocity dispersion. These are sometimes for just a central subset of the stars; see the following discussion of the studies. The last two columns are the telescope and instrument used, labeled with the line number from table 1, and the reference for the study.

Table 2. Dwarf spheroidal velocity dispersion studies.

Galaxy	N_s	N_v	σ (km s^{-1})	Instr.	Reference
Draco	23	65	10.5	1	Olszewski (1990)
Ursa Minor	18	41	10.5	1	Olszewski (1990)
Carina	13	27	8.8 ± 2.5	2,3	Seitzer et al., (1990a and pers. comm.)
Sculptor	16	22	6.3 ± 1.2	3	Armandroff and Da Costa (1986)
	7	9	7.2 ± 2.3	1	Aaronson and Olszewski (1987b)
	32	52	7.0 ± 1.2	3	Da Costa and Armandroff (1990)
Fornax	80	80	$9.4^{+1.5}_{-1.4}$	6	Paltoglou and Freeman (1987, 1990)
	30	60	10.5 ± 1.8	2,3	Seitzer et al., (1990b and pers. comm.)
	20	24	10.5	8	Mateo et al., (1990)

Olszewski's (1990) continued measurements have built a very strong case for large velocity dispersions in Draco and Ursa Minor. In Draco, 11 of the 23 stars have 3 or more epochs and only 3 have just one epoch. Two C stars and two K giants appear to have variable velocities. In Ursa Minor, 5 of the 18 stars have 3 or more epochs and 5 stars have just one epoch. The C and CH star in the sample are both velocity variables, as is one K giant. For both galaxies, the stars with velocities far from the mean have been the most heavily observed. The measurement uncertainty is well determined and about 2 km s^{-1}. That the MMT echelle can achieve this accuracy is further supported by recent measurements of faint stars in the globular cluster NGC 6535 (Pryor et al., 1991). These data have a S/N similar to that of the dwarf spheroidal data. Velocities for 20 stars yield a dispersion about the mean of 2.9 km s^{-1}. The cluster dispersion is expected to be ~2 km s^{-1}, so measurement errors are also ~2 km s^{-1}.

An important recent result is the improved velocity dispersion for Carina. Seitzer et al., (1990a and personal communication) have velocities for 7 C stars at three epochs and for 6 K giants at 1 epoch. One C star appears to have a variable velocity. The total sample yields a velocity dispersion, corrected for the measurement uncertainty, of 8.8 ± 2.5 km s^{-1}. Seitzer et al., (1990a) derive an M/L_V of 49.

Da Costa and Armandroff (1990) now have two or more velocities for 13 of their sample of 32 K giants in Sculptor. The dispersion in table 2 is for the 18 stars within the core radius. The 14 stars outside the core have a similar dispersion. They derive a central M/L_V of 8.2 ± 3.7.

Paltoglou and Freeman's (1987, 1990) velocities for 80 K giants in Fornax is by far the largest sample available for a dwarf spheroidal. Obtained with a fiber-fed spectrograph, 40 of the stars are at the center of the galaxy and the other 40 are split between two regions at ±1 core radius (17′) along the major axis. The measurement uncertainties are large, ~10 km s^{-1}, and were determined by comparing velocities obtained with the two halves of the integration. The dispersion in table 2 is for the central field. The two outer fields appear to have a lower dispersion, but the large measurement uncertainties make the value uncertain. The outer fields yield a rotation velocity of 3.5 ± 2.0 km s^{-1}. This limit is sufficient to show that Fornax (which has an ellipticity of 0.30 (Eskridge, 1988c)) is not a rotating disk, but the uncertainty is large enough that the galaxy could be an isotropic oblate rotator (Paltoglou and Freeman, 1990). Paltoglou and Freeman (1990) derive $M/L_V = 3.9 \pm 1$ from core fitting.

Two other groups also have velocity data for the center of Fornax. Seitzer et al., 1990b) have velocities for 25 C stars and 5 globular clusters. They have two epochs for most objects and 3 epochs for six. Mateo et al., (1990) have velocities for 17 stars (mostly K giants) and 3 of the globular clusters. They have repeat measurements for 1 star and the globular clusters, though these were all taken during a single run. These groups find velocity dispersions similar to that of Paltoglou and Freeman (1990). However, their systemic velocities for the galaxy, while agreeing with each other, are ~20 km s^{-1} larger than that of Paltoglou and Freeman (1990). This discrepancy needs to be resolved, though a zero-point error need not affect the measured velocity dispersion.

3 Dwarf spheroidal central densities and mass-to-light ratios

As in spiral and irregular galaxies, the dark matter in dwarf spheroidals is expected to have a more extended spatial distribution than the visible matter. Observations of the visible dwarf spheroidal galaxy can place very few constraints on the amount and distribution of dark matter outside of the galaxy's boundary and thus modeling has concentrated on determining the central mass density, ρ_0, and the central mass-to-light ratio, $(M/L)_0$. Even these quantities have significant uncertainties. These uncertainties arise from both the observational data and the inherent indeterminacy of mass determinations when the shape of the potential is unknown. Better data can eliminate specific models, but even perfectly determined surface brightness and velocity dispersion profiles do not uniquely determine the gravitational potential (Binney and Mamon, 1982; see also Merritt, 1987).

3.1 Uncertainties from the data

Four kinds of observational data are needed to derive the M/L of a dwarf spheroidal: a horizontal branch magnitude to determine the distance, velocities to determine the dispersion, a projected density profile to determine the radial structure of the visible galaxy (or at least a core radius, r_c), and either a total luminosity or a central surface brightness. The increasing amount of velocity data available for the dwarf spheroidal galaxies has meant that the dispersions are no longer the dominant source of uncertainty in the data. However, this is partly a statement about the large uncertainties in some of the radial profiles and, particularly, the luminosities.

Because of the low surface brightness of the dwarf spheroidals, star counts are possible at all radii and are the most accurate method for determining the projected density profile. The first profiles for most of the dwarf spheroidals were determined by Hodge (1961a, 1961b, 1962, 1963, 1964a, 1964b). Demers, Kunkel, and Krautter (1980) have published star counts for Sculptor, as has Eskridge (1988a, 1988c) for Sculptor and Fornax. Carina poses a special problem because of its low galactic latitude. Godwin (1985) produced the first radial profile by using COSMOS stellar photometry in two colors to select members and only counting those. Irwin (personal communication; Lynden-Bell,

1985) has similar APM data for all of the dwarf spheroidals. Regrettably, details of the Carina data and virtually all of the APM data are unpublished.

If r_c is defined as the radius where the surface brightness drops to half of its central value, then the existing profiles typically determine it with an uncertainty of 5%. This definition of r_c is 10% – 20% smaller than the King (1966) model "core radius" for the low-concentration models that match dwarf spheroidal profiles. The exact relation between the two depends on the shape of the entire profile, in particular on the amount of central concentration. The outer parts of the profiles are more uncertain than the inner, because the profiles quickly go below background. The outer profile contributes an uncertainty to the core M/L that is comparable to that from the core shape. The outer profile is also important for the "minimum density" limits discussed in the next section. Large improvements in the outer profiles will only come by increasing the contrast between the galaxy and the background. This can be done either by extending the photometry to the main sequence turnoff (V $\simeq$ 24) or by using color-magnitude data to count only members. Deep star counts using color selection are needed to examine the hints of non-axisymmetric structure in the present data (Eskridge, 1988b, 1988d) and to look for evidence of tidal distortion. This means mapping the dwarf spheroidals using a CCD, which is becoming possible as CCD's increase in size.

Surface photometry of the dwarf spheroidals is difficult because of their large diameters and because even their central surface brightnesses are $\lesssim$10% that of the night sky. Webbink (1985) summarized the existing data and little has changed since then. Fornax (de Vaucouleurs and Ables, 1968; Hodge and Smith, 1974), Sculptor (Hodge, 1966), and Leo II (Hodge, 1982) have photoelectric photometry. The ~0.5 mag disagreement between de Vaucouleurs and Ables (1968) and Hodge and Smith (1974) indicates the difficulty of the measurements. The surface brightnesses of galaxies without photoelectric photometry must be determined by adding up the light from detected stars and then extrapolating to all stars using some luminosity function. Problems with this procedure are the uncertain magnitudes at the faint end of the photographic photometry and uncertainty over which luminosity function to use for the extrapolation. There is excellent agreement between this method and the photoelectric photometry for Sculptor (Hodge, 1966). However, Armandroff and Da Costa (1986) concluded that the central surface brightness for that galaxy is uncertain by ±0.3 mag. The situation is no better for any of the other dwarf spheroidals and could be worse, particularly for Draco and Ursa Minor.

3.2 Uncertainties from the models

If mass follows light and the velocity distribution is isotropic, then it is appropriate to fit the core using King's method (King and Minkowski, 1962; Richstone and Tremaine, 1986), which yields

$$\rho_0 = \frac{9\sigma_{0p}^2}{4\pi G r_c^2} \tag{1}$$

and

$$\left(\frac{M}{L}\right)_0 = \frac{9\sigma_{0p}^2}{2\pi G \Sigma_0 r_c}. \tag{2}$$

Here σ_{0p} is the central projected velocity dispersion, Σ_0 is the central surface brightness, and r_c is the core radius (defined by $\Sigma(r_c) = \Sigma_0/2$). The typical uncertainties of 5% in r_c, 10% in the distance, 16% in σ_{0p}, and 30% in Σ_0 imply an uncertainty of about 40% in ρ_0 and 50% in $(M/L)_0$, when calculated with equations (1) and (2). If mass follows light, then velocity anisotropy can produce an error in the M/L of only about 50% (Merritt, 1988) for spherical galaxies. The relationships between luminosity, velocity dispersion, and radius for elliptical galaxies (Faber and Jackson, 1976; see Kormendy and Djorgovski, 1989 for a recent review) suggest that departures from spherical symmetry contribute less than about 20% uncertainty. However, the large M/L's of Draco, Ursa Minor, and Carina argue that, at least for these galaxies, there is no reason to expect that the mass follows the light. This produces much more freedom in the models.

This freedom has recently been explored by Pryor and Kormendy (1990; PK hereafter) and Lake (1990). The interesting models have the potential dominated by dark matter. If the dark matter core radius is much larger than that of the stars and the stellar velocity distribution is isotropic, then the central density is only reduced to 0.46 of the King's method value (Kormendy, 1987a, PK, Lake, 1990). A lower ρ_0 and $(M/L)_0$ require that the stars be on predominantly radial orbits. These models require that the stars obey $\rho(r) \propto r^{-2}$ in a region outside the core, with the size of the region becoming larger for a lower ρ_0. Thus the surface brightness profile of the galaxy constrains the amount of radial anisotropy and the ρ_0. PK used simple two-component (luminous and dark) King (1966) models to show that the central densities of Draco and Ursa Minor could be no more than ~20× lower than the King's method estimates. Lake (1990) found similar limits. It should be emphasized that these models *require* that the potential be dominated by some mass component that is much more extended than the light. They cannot be used to eliminate the need for dark matter.

Merritt (1987) has shown that the virial theorem implies a model-independent limit on the central density. This "minimum density estimator" is

$$\rho_{min,1} = \frac{9 <\sigma_p^2>}{4\pi G <r^2>}, \tag{3}$$

where $<\sigma_p^2>$ is the mean square projected velocity dispersion of a population of stars used to probe the potential and $<r^2>$ is their mean square actual, not projected, distance from the center. This limit occurs when the matter controlling the gravitational potential has constant density. Since most of the dwarf spheroidals have only a central projected velocity dispersion measured, rather than $<\sigma_p^2>$, PK derived a more extreme limiting density by assuming that the velocity dispersion is constant and isotropic out to a radius r' and zero beyond. In this case the minimum density estimator is

$$\rho_{min,2} = \frac{9 f_c \sigma_{0p}^2}{4\pi G <r^2>}, \tag{4}$$

where f_c is the fraction of the stars projected inside r'. King (1966) models are generally good fits to the profiles of dwarf spheroidals and can be used to evaluate $<r^2>$, $<\sigma^2>$, and f_c. Table 3 gives $\rho_0/\rho_{min,1}$ and $\rho_0/\rho_{min,2}$, where ρ_0 is the central density that would be derived using the King model itself. That latter density is close to what would be derived with eq. (1). The sensitivity of the density ratios to the details of the models is indicated by listing values for both an isotropic King model and a Michie-King model (Michie, 1963, Gunn and Griffin, 1979) with an anisotropy radius equal to the core radius. The ratios are given for concentrations $c \equiv \log(r_t/r_s) = 0.4, 0.5, \ldots, 1.2$, where r_t and r_s are the King model tidal and core radii, respectively. The values for $\rho_{min,2}$ are calculated for $r' = r_s$.

Table 3. Virial theorem minimum central densities.

$\log(r_t/r_s)$		0.4	0.5	0.6	0.7	0.8	0.9	1.0	1.1	1.2
$\rho_0/\rho_{min,1}$	King	4.0	4.5	5.2	6.3	7.7	10	13	18	25
	Michie	4.5	5.4	6.5	8.2	10	13	17	22	30
$\rho_0/\rho_{min,2}$	King	6.7	7.9	9.8	13	17	24	35	52	79
	Michie	7.5	9.2	12	15	19	26	34	45	61

PK applied these results to Draco (c = 0.7) and Ursa Minor (c = 0.4) and found limiting densities similar to those from the two-component models. The virial theorem limits are somewhat weaker than the factor of 20 reduction derived from the two-component models because those limits came from extreme models that were poor fits to the surface brightness data. Since the other dwarf spheroidals have concentrations similar to those of Draco and Ursa Minor, the lower limits on their central densities should also be about $1/20^{th}$ of the King's method values. The exception is Sculptor, for which Eskridge (1988b) finds c=1.0, but notes that the observed profile is above the King model at large radii.

3.3 Discussion of central densities and mass-to-light ratios

Table 4 lists central densities and mass-to-light ratios for the dwarf spheroidals. The first three columns give the name of the galaxy, its absolute magnitude, and its distance. These last two quantities are taken from Webbink (1985). Next come the ellipticity (one minus the axial ratio) and the adopted core radius in arcminutes, taken from the most recent study of the profile. The core radius given is the geometric mean of the values along the major and minor axes, since this is more appropriate to use in eqs. (1) and (2). The next two columns are the V band central surface brightness, in magnitudes per square arcminute, and the central velocity dispersion. These quantities are taken from the studies discussed above with the exception of the central surface brightness for Carina, which is from Seitzer (personal communication) and is based on the Mighell (1990) luminosity function. Then come three calculated quantities: $(M/L_V)_0$, derived with eq. (1); ρ_0, from eq. (2); and $\rho_{0,L}$, the central density of luminous matter if that matter has $M/L_V = 2.0$. The last two columns are the King concentration of the profile, taken from the most recent star counts, and $\rho_{min,2}$, calculated with with eq. (4).

Table 4. Properties of dwarf spheroidal galaxies.

Galaxy	M_V	d (kpc)	ϵ	r_c (′)	μ_0 ($\frac{mag}{\square'}$)	$\sigma_{0,p}$ ($\frac{km}{s}$)	$(M/L_V)_0$ ($\frac{M_\odot}{L_{\odot V}}$)	ρ_0 ($M_\odot/pc^3$)	$\rho_{0,L}$	c	$\rho_{min,2}$ ($\frac{M_\odot}{pc^3}$)
Fornax	–12.4	145	0.30	11.3	14.8	10.	5.7	0.073	0.026	0.8	0.004
Sculptor	–11.4	78	0.34	6.2	15.0	7.0	11	0.41	0.076	1.0	0.01
Carina	–9.2	92	0.31	6.0	16.4	8.8	53	0.50	0.019	0.7	0.03
Draco	–8.9	75	0.29	5.4	16.3	10.5	94	1.3	0.028	0.7	0.09
Ursa Minor	–8.9	69	0.55	6.6	16.4	10.5	83	1.0	0.024	0.4	0.13

The M/L's for Fornax, Sculptor, and Carina in table 4 are all larger than those given in Kormendy (1987a). The principal reason is that, with more velocities, the velocity dispersions have increased. Carina has joined Draco and Ursa Minor in having both a large M/L and a high central density.

The presence of some kind of dark matter in at least some of the dwarf spheroidals seems inescapable. Take Draco as an example. There is only a 5% probability that the dispersion is low enough to reduce the M/L to 0.53 times the value in table 4, assuming a χ^2 distribution with 19 degrees of freedom. Assuming further reductions of 30% for the uncertainty in the distance, 10% for that in r_c, a factor of 1.8 for that in μ_0, and a factor of two for uncertainties in the modeling, yields $M/L_V = 8$. These are very extreme limits on the uncertainties, all acting in the same direction. A larger reduction in the M/L is possible with the anisotropic models discussed by PK and Lake (1990), but these models *presuppose* an extended, dominant, dark component.

Modified Newtonian dynamics, MOND, (Milgrom, 1983a; Bekenstein and Milgrom, 1984) has been proposed as an explanation for the large M/L's of dwarf spheroidals (Milgrom, 1983b). However, Gerhard and Spergel (1990) have recently shown that the range of dwarf spheroidal M/L's is impossible to accommodate within MOND. A less exotic way to avoid dark matter is to assume that the large M/L's are due to a lack of virial equilibrium, possibly due to resonant orbital coupling (Kuhn and Miller, 1989) or overwhelming tidal forces. The problem with these mechanisms is that, even if the stellar M/L_V is 6, the observed velocity dispersions exceed those expected in virial equilibrium by a factor of ~3 and the galaxies must be unbound and freely expanding. Then Draco, Ursa Minor, and Carina will disappear in less than 10^8 years and it is surprising that we see so many galaxies caught in the act of destruction (Aaronson and Olszewski, 1987b). There are several possible ways to test for tidal disruption. The most direct is to search the outer envelopes of the galaxies for non-axisymmetric distortions. Another is to look for streaming motions. If a galaxy is no longer bound because it has been pulled apart by tidal forces, then it seems likely that all of its stars are moving radially outward. In that case the observed flattenings are due to more rapid expansion along one axis and simple models suggest that this streaming motion would cause the galaxy to appear to rotate with a velocity comparable to the dispersion, unless the axis is nearly perpendicular to the line of sight.

Could the dark matter in dwarf spheroidals be the population of stars we see? Armandroff and Da Costa (1986) calculate the M/L_V for an old stellar population that formed in a single burst with a power-law IMF, $dN \propto m^{-(1+x)}dm$, extending between 50 and 0.2 $M_\odot$. They find $M/L_V \simeq 6$ for x $\simeq$ –0.5 or 2.8. Extending these calculations to a lower mass limit of 0.025 $M_\odot$ yields $M/L_V = 6$ for x = 1.35 and $M/L_V = 50$ for x = 2.3 (see also Aaronson and Olszewski, 1987b). Chernoff and Weinberg (1990) show that stellar mass loss will disrupt stellar systems with x $\leq$ 0.5. They were studying globular clusters, but their results should apply equally well to dwarf spheroidals without dark halos. Thus the dark matter theoretically could be low-mass stars but not stellar remnants. Observationally, however, there is little evidence for stellar populations with such large M/L's.

Kormendy (1987b) finds an upper limit of $M/L_V \simeq 10$ for the cores of bulges and ellipticals. This limit occurs for the most luminous, and thus most metal rich, galaxies. Since M/L_V is expected to decrease with decreasing metallicity (Tinsley, 1978), the fairer comparison is with low luminosity ellipticals. Galaxies with $M_B > -20$ have $(M/L_V)_0 \lesssim 6$. The least luminous elliptical in Kormendy's (1987b) sample is M32 (M_B = –15.2), which has $M/L_V = 2.5 \pm 0.1$ (Richstone, 1987). Even this galaxy has [Fe/H] = –0.2, based on $Mg_2 = 0.185$ (Davies et al., 1987) and the relation between Mg_2 and [Fe/H] in Terlevich et al., (1981). Globular clusters provide the most extensive set of M/L's for old, metal poor stellar populations. They typically have $(M/L_V)_0 = 1 - 2$ (Illingworth, 1976; Pryor et al., 1988). However, mass segregation could be concentrating giants in cores of globular clusters, reducing $(M/L)_0$ (though if there are enough heavy remnants present, mass segregation increases $(M/L)_0$). The global M/L's of globular clusters are more uncertain than is widely appreciated. Large numbers of low mass stars can be added to the models without affecting the fits (Gunn and Griffin, 1979) and very large global M/L's are usually possible. The exceptions are low concentration clusters, where mass segregation can have little effect. Pryor et al., (1990) have shown that $M/L_V \leq 3$ for several such clusters. Globular clusters may have lost their low-mass stars through evaporation (Chandrasekhar, 1942) or cluster expansion (Chernoff and Weinberg, 1990). Because of its long relaxation time and large mass, ω Cen should not have been much affected by either of these processes, nor should it have much mass segregation. It has $(M/L_V)_0 \simeq 2.5$ (Meylan and Mayor, 1986). Though velocity dispersions are needed for more clusters with long relaxation times, the weight of the evidence is that globular clusters have a global $M/L_V \simeq 2$. The above observations strongly suggest, but do not prove, that all of the dwarf spheroidals in table 4 contain non-stellar dark matter.

Interestingly, NGC 205, a close companion of M31 and the most luminous dwarf spheroidal in the Local Group, provides a possible example of a large stellar M/L in a moderately metal poor population. Carter and Sadler (1990) and Held et al., (1990) have recently measured the velocity dispersion outside of the semi-stellar nucleus. They derive $M/L_R = 9.4$ and 7, respectively. These are uncomfortably large for a galaxy with $M_V = -16.5$ and $<[Fe/H]> \simeq -0.85$ (Mould, Kristian, and Da Costa, 1984). The density implied, again outside of the nucleus, is $\sim 30\ M_\odot\ pc^{-3}$, which is much higher than the dark halo densities of both more and less luminous galaxies. This argues that the large M/L is not due to dark matter. Internal extinction (e.g., Hodge, 1973) may be at least partly responsible for the high M/L, but more studies of this and others of the more luminous dwarf spheroidals would be interesting.

Assuming that exotic dark matter is present, then the simplest picture which is consistent with the data in table 4 is one in which all five galaxies are embedded in dark halos whose central densities increase from 0.07 $M_{\odot}$ pc^{-3} for Fornax to 1 $M_{\odot}$ pc^{-3} for Draco and Ursa Minor. These central densities should be reduced by at least a factor of two if the dark matter is much more spatially extended than the stars (Kormendy, 1987a, PK, Lake, 1990), but cannot be much lower than 0.1 $M_{\odot}$ pc^{-3} because of the minimum central densities for Draco and Ursa Minor (the final column of table 4; PK; Lake, 1990) and the need to keep the center of Sculptor (which has $\rho_{0,L} = 0.076$ $M_{\odot}$ pc^{-3} from table 4) dominated by dark matter to avoid having to produce $M/L_V = 11$ with stars. Another picture that is consistent with the data gives all five galaxies dark halos with $\rho_0 \simeq 0.1$ $M_{\odot}$ pc^{-3}, with perhaps a slight increase from Fornax to Draco. This would require that the velocity dispersion tensor of the stars become increasingly radial as the luminosity of the galaxy decreased. Such models, with the assumption of a large dark matter core radius, predict a specific velocity dispersion profile.

Large velocity samples will be needed to test these predictions of the anistropic models (perhaps 250 stars; PK), so the tests used should be as efficient as possible. Taking velocity moments and binning throws away information. The best approach is probably to compare the data with a model N(R,V), the distribution of stars in projected radius and velocity. How large an improvement in efficiency can result from this is currently unknown. Another possibly interesting test is to use the stars with the largest velocities to place constraints on the potential in the same way that high velocity stars in our own Galaxy are used to constrain the local escape velocity. Measuring the proper motions associated with the internal velocities would be a decisive test (Leonard and Merritt, 1989; PK), but these are extremely small: ~0.003 arcsec century^{-1}.

Phase space constraints (Tremaine and Gunn, 1979) on the dark matter cannot be directly applied to dwarf spheroidal galaxies because the core radii of the halos are unknown. However, Gerhard and Spergel (1990) have shown that if the dark matter is neutrinos with a rest mass of 30 eV and if the halos have $\rho_0 = 0.1$ $M_{\odot}$ pc^{-3}, then the core radii of the halos must be at least 10 kpc and the total masses of the dwarf spheroidal galaxies are so large that dynamical friction causes them to spiral into the center of our Galaxy in less than 10^{10} years. As emphasized by Kormendy (1987a), dark halos with $\rho_0 \simeq 0.1$ $M_{\odot}$ pc^{-3} are larger than those seen in almost all giant galaxies. That density is consistent with the scaling laws for dark matter found by Kormendy (1990) for more luminous galaxies, though these remain very uncertain. More generally, the large densities and M/L's of the dwarf spheroidals support the observation that smaller galaxies are increasingly dominated by dark matter (Tinsley, 1981; Kent, 1987; Persic and Salucci, 1988, 1990; Kormendy, 1988, 1990).

In summary, the more we study dwarf spheroidal galaxies, the more interesting they seem to become. They appear to have much to teach us about galaxy formation and evolution and the properties of the dark matter. A rich reward of increased understanding is likely to result from obtaining better velocity dispersions, surface brightnesses, and profiles. Happily, improving technology should make this data available within the next few years.

Acknowledgements

I am deeply grateful to Ed Olszewski, George Paltoglou, Ken Freeman, Pat Seitzer, Taf Armandroff, Gary Da Costa, and Mario Mateo for letting me quote their data before publication. Without their generosity this would have been a much less interesting review. David Merritt, Bob Schommer, and John Kormendy offered comments on a draft of this review. I thank everyone above for many useful conversations.

References

Aaronson, M.: 1983, *Astrophys. J. Letters* **266**, L11.

Aaronson, M.: 1986, In Stellar Populations (C. A. Norman, A. Renzini, and M. Tosi, Eds.), 45, Cambridge.

Aaronson, M. and Olszewski, E.: 1987a, In *Dark Matter in the Universe*, IAU Symposium No. 117 (J. Kormendy and G. R. Knapp, Eds.), 153, Reidel.

Aaronson, M. and Olszewski, E. W.: 1987b, *Astron. J.* **94**, 657.

Aaronson, M. and Olszewski, E.: 1988, In *Evolution of Large Scale Structure in the Universe*, IAU Symposium No. 130 (J. Adouze, M.-C. Pelletan, and A. Szalay, Eds.), 409, Kluwer.

Armandroff, T. E. and Da Costa, G. S.: 1986, *Astron J.* **92**, 777.

Bekenstein, J. and Milgrom, M.: 1984, *Astrophys. J.* **286**, 7.

Binney, J. and Mamon, G. A.: 1982, *Mon. Not. R. Astr. Soc.* **200**, 361.

Carignan, C. and Beaulieu, S.: 1989, *Astrophys. J.* **347**, 760.

Carignan, C., Beaulieu, S. and Freeman, K. C.: 1990, *Astron. J* **99**, 178.

Carter, D. and Sadler, E. M.: 1990, *Mon. Not. R. Astr. Soc.* **245**, 12P.

Chandrasekhar, S.: 1942, *Principles of Stellar Dynamics*, Univ. Chicago Press, Chicago, ch. V.

Chernoff, D. F. and Weinberg, M. D.: 1990, *Astrophys. J.* **351**, 121.

Cook, K, Schechter, P. and Aaronson, M.: 1983, Bull. Am. Astron. Soc. **15**, 907.

Da Costa, G.: 1988, In *Globular Cluster Systems in Galaxies*, IAU Symposium No. 126 (J. Grindlay and A. G. D. Philip, Eds.), 661, Reidel.

Da Costa, G. S. and Armandroff, T. E.: 1990, in preparation.

Davies, R. L., Burstein, D., Dressler, A., Faber, S. M., Lynden-Bell, D., Terlevich, R. J. and Wegner, G.: 1987, *Astrophys. J. Suppl. Ser.* **64**, 581.

Demers, S., Kunkel, W. E. and Krautter, A.: 1980, *Astron. J.* **85**, 1587.

de Vaucouleurs, G. and Ables, H. D.: 1968, *Astrophys. J.* **151**, 105.

Eskridge, P. B.: 1988a, *Astron. J.* **95**, 1706.

Eskridge, P. B.: 1988b, *Astron. J.* **96**, 1336.

Eskridge, P. B.: 1988c, *Astron. J.* **96**, 1353.

Eskridge, P. B.: 1988d, *Astron. J.* **96**, 1614.

Faber, S. M. and Jackson, R. E.: 1976, *Astrophys. J.* **204**, 668.

Gerhard, O. E. and Spergel, D. N. (1990), submitted to *Astrophys. J. Letters.*

Godwin, P. J.: 1985, In *Dynamics of Star Clusters*, IAU Symposium No. 113 (J. Goodman and P. Hut, Eds.), 77, Reidel.

Godwin, P. J. and Lynden-Bell, D.: 1987, Mon. Not. R. Astr. Soc. **229**, 7P.

Gunn, J. E. and Griffin, R. F.: 1979, *Astron. J.* **84**, 752.

Held, E. V., Mould, J. R. and de Zeeuw, P. T.: 1990, *Astron. J.* **100**, 415.

Hodge, P. W.: 1961a, *Astron. J.* **66**, 249.

Hodge, P. W.: 1961b, *Astron. J.* **66**, 384.

Hodge, P. W.: 1962, *Astron. J.* **67**, 125.

Hodge, P. W.: 1963, *Astron. J.* **68**, 470.

Hodge, P. W.: 1964a, *Astron. J.* **69**, 438.

Hodge, P. W.: 1964b, *Astron. J.* **69**, 853.

Hodge, P. W.: 1966, *Astron. J.* **71**, 204.

Hodge, P. W.: 1973, *Astrophys. J.* **182**, 671.

Hodge, P. W.: 1982, *Astron. J.* **87**, 1668.

Hodge, P. W. and Smith, D. W.: 1974, *Astrophys. J.* **188**, 19.

Illingworth, G.: 1976, *Astrophys. J.* **204**, 73.

Kent, S. M.: 1987, *Astron. J.* **93**, 816.

King, I. R.: 1966, *Astron. J.* **71**, 64.

King, I. R. and Minkowski, R.: 1972, In *External Galaxies and Quasi-Stellar Objects*, IAU Symposium No. 44 (D. S. Evans, Ed.), 87, Reidel.

Kormendy, J.: 1985, *Astrophys. J.* **295**, 73.

Kormendy, J.: 1987a, In *Dark Matter in the Universe*, IAU Symposium No. 117 (J. Kormendy and G. R. Knapp, Eds.), 139, Reidel.

Kormendy, J.: 1987b, In *Structure and Dynamics of Elliptical Galaxies*, IAU Symposium 127 (T. de Zeeuw, Ed.), 17, Reidel.

Kormendy, J.: 1988, In *Origin, Structure, and Evolution of Galaxies*, (Fang Li Zhi, Ed.), 252, World Scientific.

Kormendy, J.: 1990, In *The Evolution of the Universe of Galaxies*, (R. G. Kron, Ed.), Astron. Soc. Pacific, in press.

Kormendy, J. and Djorgovski, S.: 1989, *Ann. Rev. Astron. Astrophys.* **27**, 235.

Kuhn, J. R. and Miller, R. H.: 1989, *Astrophys. J. Letters* **341**, L41.

Lake, G.: 1990, *Mon. Not. R. Astr. Soc.* **244**, 70.

Lake, G. and Feinswog, L.: 1989, *Astron. J.* **98**, 166.

Lake, G., Schommer, R. A. and van Gorkom, J. H.: 1990, *Astron. J.* **99**, 547.

Leonard, P. J. T. and Merritt, D.: 1989, *Astrophys. J.* **339**, 195.

Lin, D. N. C. and Faber, S. M.: 1983, *Astrophys. J. Letters* **266**, L21.

Lynden-Bell, D.: 1985, In *The Milky Way Galaxy*, IAU Symposium 106 (H. van Woerden, R. J. Allen and W. B. Burton, Eds.), 461, Reidel.

Mateo, M., Olszewski, E., Welch, D. and Fischer, P.: 1990, in preparation.

Merritt, D.: 1987, *Astrophys. J.* **313**, 121.

Merritt, D.: 1988, *Astron. J.* **95**, 496.

Meylan, G. and Mayor, M.: 1986, *Astron. Astrophys.* **166**, 122.

Michie, R. W.: 1963, *Mon. Not. R. Astr. Soc.* **125**, 127.

Mighell, K. J.: 1990, *Astron. Astrophys. Suppl. Ser.* **82**, 207.

Milgrom, M.: 1983a, *Astrophys. J.* **270**, 365.

Milgrom, M.: 1983b, *Astrophys. J.* **270**, 371.

Mould, J., Kristian, J. and Da Costa, G. S.: 1984, *Astrophys. J.* **278**, 575.

Olszewski, E.: 1990, in preparation.

Paltoglou, G. and Freeman, K.: 1987, In *Structure and Dynamics of Elliptical Galaxies*, IAU Symposium 127 (T. de Zeeuw, Ed.), 447, Reidel.

Paltoglou, G. and Freeman, K.: 1990, *Astron. J.* submitted.

Persic, M. and Salucci, P.: 1988, *Mon. Not. R. Astr. Soc.* **234**, 131.

Persic, M. and Salucci, P.: 1990, *Mon. Not. R. Astr. Soc.* **245**, 577.

Pryor, C. and Kormendy, J.: 1990, *Astron. J.* **100**, 127.

Pryor, C., McClure, R. D., Fletcher, J. M. and Hesser, J. E.: 1988, In *Globular Cluster Systems in Galaxies*, IAU Symposium No. 126 (J. Grindlay and A. G. D. Philip, Eds.), 661, Reidel.

Pryor, C., McClure, R. D., Fletcher, J. M. and Hesser, J. E.: 1990, *Astron. J.* submitted.

Pryor, C., McClure, R. D., Hesser, J. E. and Fletcher, J. M.: 1991, in preparation.

Richstone, D. O.: 1987, In *Structure and Dynamics of Elliptical Galaxies*, IAU Symposium No. 127 (T. De Zeeuw, Ed.), 261, Reidel.

Richstone, D. O. and Tremaine, S.: 1986, *Astron. J.* **92**, 72.

Sargent, W. L. W., Sancisi, R. and Lo, K.-Y.: 1983, *Astrophys. J* **265**, 711.

Seitzer, P., Da Costa, G., Armandroff, T. E. and Light, R. M.: 1990a, in preparation.

Seitzer, P., Da Costa, G., Armandroff, T. E. and Light, R. M.: 1990b, in preparation.

Seitzer, P. and Frogel, J. A.: 1985, *Astron. J.* **90**, 1976.

Terlevich, R., Davies, R. L., Faber, S. M. and Burstein, D.: 1981, *Mon. Not. R. Astr. Soc.* **196**, 381.

Tinsley, B. M.: 1978, *Astrophys. J.* **222**, 14.

Tinsley, B. M.: 1981, *Mon. Not. R. Astr. Soc.* **194**, 63.

Tremaine, S.: 1987, In *Nearly Normal Galaxies* (S. M. Faber, Ed.), 76, Springer-Verlag.

Tremaine, S. D. and Gunn, J. E.: 1979, *Phys. Rev. Letters*, **42**, 467.

Webbink, R. F.: 1985, In *Dynamics of Star Clusters*, IAU Symposium No. 113 (J. Goodman and P. Hut, Eds.), 541, Reidel.

Wirth, A. and Gallagher, J. S.: 1984, *Astrophys. J.* **282**, 85.

Discussion

E. Athanassoula - One should keep in mind that the M/L values you have quoted rely on the assumption that the dwarf spheroidals are in equilibrium. Numerical simulations that S. Morin, K. Prendergast and I have been doing show that this may not necessarily true.
C. Pryor - Yes, I agree that alternate reasons for the large observed dispersions must be kept in mind. I believe that the short lifetimes of out-of-equilibrium systems argue against this explanation. However, we need direct observational tests of these theories. For example, better observations of the shapes of these galaxies might provide evidence that they are coming apart.

I. King - Can you use the tidal force of the Milky Way to constrain the extent of dark halos in these systems? The maximum size should depend on the mean density of total matter (and of course on what you believe about the total mass of the Milky Way).
C. Pryor - Until we know the radial distribution of the dark matter, we cannot use tidal force arguments to limit the total amount of dark matter. Tides limit the mean density of a system, as long as the central densities are enough for stability, the core radius (and thus the mass) can be anything.

J. Melnick - Many years ago Roberto Terlevich and I showed that the trend in M/L with metallicity observed in dwarf spheroidals can be explained if the MF slope depends on metallicity and if dwarf galaxies have lost a large fraction of their mass, presumably in the form of galactic winds. This is in any event needed to explain the mass-metallicity relation (ref. ESO Workshop *Dynamics and chemical evolution of the outer halo of the Galaxy*, 1987)

C. Pryor - This is one of the wide range of possible explanations for the MIL'S of the one galaxy's dwarf spheroidals.

G. Djorgovski - The local dSph's obey an excellent correlation between luminosity and metallicity. It is hard to understand how such a correlation could survive if there was a substantial stripping of the luminous material. On the other hand, such relation is predicted in galactic wind models, e.g. by Arimoto, Yoshii, and others.

S. Kent - If you plot N 205 on your figure, then the correlation of M/L with L falls apart.

Gustav Tammann and the Melnick's at a wine-break.

Participants waiting for the bus to Pompei.

ELLIPTICAL GALAXIES IN X-RAYS

L. Ciotti[1], A. D'Ercole[2], S. Pellegrini[1], and A. Renzini[1]
[1]*Dipartimento di Astronomia, Università di Bologna, Italy*
[2]*Osservatorio Astronomico di Bologna, Italy*

ABSTRACT. We present a new class of 1D hydrodynamical evolutionary sequences for the gas flows in elliptical galaxies. The model galaxies are described by either King's or de Vaucouleurs' virialized density distributions, and are surrounded by variable amounts of dark matter. Two source terms operate: mass loss from evolving red giants, and a secularly declining heating by type I supernovae. The resulting model galaxies evolve through up to three consecutive evolutionary stages, that we call wind, outflow, and inflow phase. Agreement with the observed trends of the X-ray luminosity with the optical characteristics of the galaxies is achieved for models in which the dark to visible mass ratios is around ten, when the standard SNI rate is assumed.

1 Introduction

As first revealed by *Einstein* observations, normal (non-cD) elliptical galaxies, can be powerful X-ray sources, with 0.5 – 4.5 KeV luminosities L_X ranging from $\sim 10^{39}$ to $\sim 10^{42}$ erg s^{-1} (Forman, Jones and Tucker, 1985; Canizares, Fabbiano and Trinchieri, 1987, hereafter CFT; Fabbiano, 1989).

The X-ray emission strongly correlates with the optical (e.g. blue) luminosity of the observed galaxies, and CFT find $L_X \propto L_B{}^{1.7\pm0.3}$, but there is substantial scatter at any luminosity, as shown in figure 1 which reproduces CFT data. Since the scatter is substantially larger than expected from measuring errors, the actual variety of astrophysical conditions prevailing in elliptical galaxies could hardly be reduced to the average correlation. Moreover, for $L_B \lesssim 3 \times 10^{10} L_\odot$ the observed L_X is consistent with an emission due entirely to discrete sources (see also CFT), while only for higher luminosities the evidence for an additional contribution from a hot interstellar medium (ISM) is actually compelling. There is little doubt that evolving stars shed a considerable amount of mass, and therefore one important astrophysical question is how and where elliptical galaxies manage to dispose this gas. Early studies have shown that with standard SN rates (and no Dark matter!) all elliptical galaxies would support a supersonic wind (Mathews and Baker, 1971), and therefore galaxies would contain very little hot gas, and would not be powerful X-ray emitters. One way to prevent the development of galactic winds is to deepen the gravitational potential well, e.g. by assuming the presence of massive dark

G. Longo et al. (eds.), Morphological and Physical Classification of Galaxies, 179–190.

halos, and/or reducing the supernova heating (e.g. Nulsen, Stewart, and Fabian, 1984 Sarazin and White, 1987).

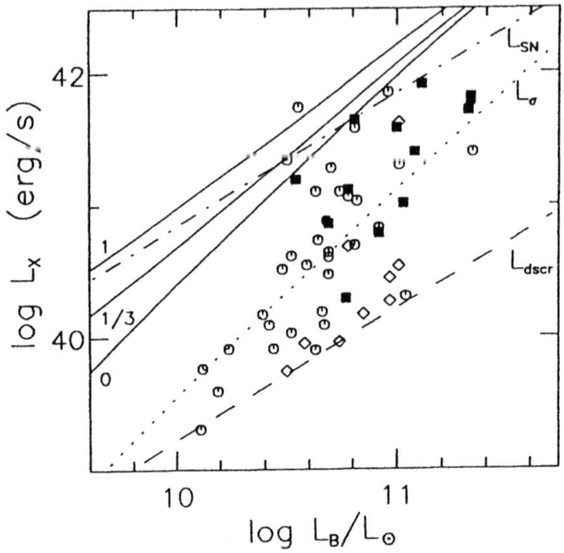

Figure 1. The L_X vs. L_B plot using CFT data. Filled squares and open diamonds refer to galaxies with boxy and disky isophotal shapes, respectively. Open circles refer to galaxies for which this morphological detail is unknown. The dot-dashed line shows L_{SN} at the standard rate ($\vartheta_{SN} = 1$). L_σ is shown by the dotted line, while L_{dscr} represents the expected contribution from discrete stellar X-ray sources (dashed line). Finally the solid lines represent the expected L_X of steady state cooling flow models, for various assumed SN I rates, each labelled by the factor ϑ_{SN}.

The resulting steady state cooling flow models are however far from giving a satisfactory account for the X-ray properties of elliptical galaxies, as the bulk of them are much fainter than the models would predict (see figure 1).

Evolutionary, hydrodynamical calculations which include the time dependence of the source terms have been first constructed by Loewenstein and Mathews (1987b, hereafter LM). LM assumed a nearly constant type I supernova rate, and the mass of the dark matter halo to exceed by a factor of 9 that of visible (stellar) matter; their models develop an inflow from the very beginning, but their ultimate fate is to revert the inflow into an outflow (which would later develop into a supersonic wind), as the stellar mass loss rate secularly declines while the supernova heating remains nearly the same. Apparently, these models are less affected by the X-ray overluminosity syndrome affecting steady-state cooling flow solutions, but this comes from a reduction of both the type I supernova rate and the stellar mass-loss rate by a factor ~ 4 over standard values.

Besides the puzzling large dispersion of L_X for given L_B, other major embarrassments affect inflow models. First, over a Hubble time an amount of material comparable to the mass of stars in the galactic core flows into the centers of these galaxies, while the corresponding distortions of the central optical surface brightness and velocity dispersion are not observed. Second, inflow models exhibit very high central X-ray surface brightness, with the X-ray profile being much more centrally concentrated than the optical profile, in contrast with the rather similar surface brightness distributions exhibited by the three well resolved galaxies (Trinchieri, Fabbiano, and Canizares, 1986; Sarazin and White, 1988).

However, by allowing the SN I rate to decline sufficiently fast with time, massive early inflows can be replaced by winds and outflows, thus preventing the accumulation of large amounts of unseen mass at the center, producing smoother X-ray surface brightness distributions, and allowing large variations in X-ray luminosity at each optical luminosity (Renzini, 1989; D'Ercole et al., 1989, hereafter Paper I; Ciotti et al., 1991, hereafter Paper II). Under this very simple assumption model elliptical galaxies can experience up to three subsequent evolutionary phases, in which an initial supersonic wind is followed by a subsonic outflow until a central cooling catastrophe leads eventually to the onset of an inflow. The duration of the transient outflow phase turns out to be extremely sensitive to virtually every model structural parameter, thus offering a natural explanation for the large scatter in X-ray luminosities.

2 Model galaxies and source terms

Like in previous evolutionary- or stationary-flow studies we assume the distribution of stellar and dark matter to be time-independent, therefore neglecting the effects of stellar mass loss and gas flows on the shape and depth of the gravitational potential well. Similarly neglected are any interactions with surrounding galaxies which might affect the distribution of matter. Star formation is supposed to have occurred all in a single burst at $t = 0$, and then the *passive* evolution of the stellar population is followed. We then consider two distinct classes of models, one in which the distribution of the bright matter density is represented by a King model (King, 1972), and another in which it is represented by the $r^{\frac{1}{4}}$ law (de Vaucouleurs, 1948).

For a first group of models that we call *King Galaxies* (KGs) we assume the stellar distribution to be described by the same density profile used by LM: ρ_{*o} is the central stellar density, r_{c*} the core radius, and the distribution is truncated at the tidal radius $r_t = \delta r_{c*}$; then ρ_{*o}, r_{c*} and the central stellar velocity dispersion σ_* are related by the virial condition $\rho_{*o} r_{c*}^2 = 9\sigma_*^2/4\pi G$. For a second family of models, that we call *de Vaucouleurs' Galaxies* (dVGs), we use the spatial density distribution given in Mellier and Mathez (1987), truncated at $r_t = \delta r_e$, r_e being the effective radius. As for KG models, also for dVG models we use the virial theorem to connect the total stellar mass to the central velocity dispersion.

The density distribution of the dark matter halo is assumed to follow a quasi-isothermal law: as in LM, ρ_{ho} is the central dark matter density, r_{ch} is the core radius and the tidal radius r_t is assumed to be the same as for the stellar distribution. The ratio of the total stellar mass and the total dark matter mass is given by $R = M_h/M_*$, $\gamma = \rho_{ho}/\rho_{*o}$ for the KGs and $\beta = r_{ch}/r_e$ for the dVGs.

In order to both reduce the number of model parameters, and explore the fraction of the parameter space actually occupied by real galaxies, we have constrained our models to lie on the Fundamental Plane defined by:

$$L_B = 3.6 \times 10^3 r_e \sigma_*^{1.3}/h \quad (L_\odot), \tag{1}$$

where L_B is the total blue luminosity in solar units, r_e is in pc, σ_* in km s^{-1} and $h = H_\circ/100$ (Djorgovski and de Carvalho, 1989). For all the KG models we have assumed $r_e = 12 r_{c*}$. In conclusion, we identify each KG model by the set of independent parameters $(L_B, \sigma_*, \delta, R, \gamma)$ and each dVG model by $(L_B, \sigma_*, \delta, R, \beta,)$. Note that for KG models $\beta = r_{ch}/r_{c*}$ is uniquely related to R, γ, and δ.

The only secularly evolving ingredients are in our models the source terms: stellar mass loss and supernova heating. Stellar mass loss provides the raw material for our galactic flows, and we study the behavior of these flows under the influence of a time-varying SN heating. For quick calculation, the mass return to the ISM in a single-age stellar population is given by:

$$\dot{M}_*(t) \simeq -1.5 \times 10^{-11} L_B t_{15}^{-1.3} \qquad (M_\odot \mathrm{yr}^{-1}), \tag{2}$$

where t_{15} is the age in unit of 15 Gyr. This expression is valid in the age range from ~ 0.5 to over 15 Gyr (see Paper II).

It is well known that only type I supernovae (actually type Ia) have so far been observed in elliptical and S0 galaxies. A secularly declining SN I rate follows quite plausibly from current views about SN I progenitors (see Paper II for an extensive discussion) and we correspondingly assume:

$$R_{SN}(t) = 2.2 \times 10^{-13} \vartheta_{SN} L_B t_{15}^{-s} \qquad (\mathrm{yr}^{-1}) \tag{3}$$

where ϑ_{SN} is a parameter, such that for $\vartheta_{SN} = 1$ and $t_{15} = 1$ the standard SN I rate in ellipticals is recovered, i.e 0.22 SNU (1 SNU = 1 SN per century per $10^{10} L_\odot$; Tammann, 1982). Moreover we assume s values such that R_{SN} decreases slightly faster than $\alpha_* = \dot{M}_*/M_*$. In practice, this implies $s \gtrsim 1.3$ when the stellar IMF follows the Salpeter distribution. We further assume that each SN injects a total kinetic energy $E_k = 10^{51}$ erg into the ISM of galaxies, and that this energy is completely turned into heat, an excellent approximation for the hot diluted gas we are dealing with (Mathews, 1989). Therefore, the global SN heating (hereafter SN luminosity) is given by:

$$L_{SN}(t) = E_k R_{SN}(t) \simeq 3.24 \times 10^{43} R_{SN}(t) = 7.1 \times 10^{30} \vartheta_{SN} L_B t_{15}^{-s} \qquad (\mathrm{erg\,s^{-1}}). \tag{4}$$

Heat to the ISM is also provided by collisions between stellar winds and/or between winds and the ambient gas, that thermalize relative to the galaxy the kinetic energy carried by stellar mass loss. This process provides heat at a global power:

$$L_\sigma = \frac{3}{2}\alpha(t) \int_0^{r_t} 4\pi r^2 \rho_*(r) \sigma_*^2(r) dr, \tag{5}$$

where $\sigma_*(r)$ is the local, one-dimensional stellar velocity dispersion, $\alpha(t) = \alpha_*(t) + \alpha_{SN}(t)$, and $\alpha_{SN}(t) = 1.4 R_{SN}(t)/M_*$, having assumed that each SN I ejects $1.4 M_\odot$ of material.

We further define L_{grav}^- as the power required to steadily extract from the galaxy the gas shed by stars, and L_{grav}^+ as the power generated by the steady inflow of this gas through the galactic potential well, down to the center of the galaxy. Thus we have:

$$L_{grav}^- = -\alpha(t) \int_0^{r_t} 4\pi r^2 \rho_*(r) \phi(r) dr \tag{6}$$

and $L^+_{\rm grav} = -L^-_{\rm grav} - \dot{M}_*\phi(0)$. In a realistic situation stationary conditions are only seldom verified, or even approached. Nevertheless, these quantities can be very useful to predict several global features of the flows and their evolution (see also CFT; Renzini, 1989). The integrals appearing in these definitions can be expressed as a product of two terms: a dimensional factor, and a dimensionless function of R, β, and δ:

$$L^{\pm}_{\rm grav} = \dot{M}_*(t)\sigma_*^2 F^{\pm}(R, \beta, \delta). \tag{7}$$

In equation (7) the dimensionless *form factors* $F^{\pm}$ are different functions of (R, β, δ) respectively for the KG models (with $\beta = r_{\rm ch}/r_{\rm c*}$ and $\delta = r_t/r_{\rm c*}$), and for the dVG models (with $\beta = r_{\rm ch}/r_{\rm e}$ and $\delta = r_t/r_{\rm e}$). Note that, as expected given our assumptions, the time dependence of $L^{\pm}_{\rm grav}$ is totally controlled by $\alpha(t)$, while the other terms on the r.h.s. of equation (7) describe the properties of the potential well.

We also introduce the function $\chi(t)$, as the ratio of the power required to steadily extract the gas from the galaxy, to the power actually made available by kinetic and supernova heating; so for both KG and dVG models we have:

$$\chi(t) = \frac{L^-_{\rm grav}(t)}{L_\sigma(t) + L_{\rm SN}(t)} \simeq \frac{t_{15}^{s-1.3}}{\vartheta_{\rm SN}} \times \left(\frac{\sigma_*}{300}\right)^2 \times \tilde{F}^-(R, \beta, \delta), \tag{8}$$

where we have neglected $\alpha_{\rm SN}$ in front of α_*, and L_σ in front of $L_{\rm SN}$, and $\tilde{F}^-$ differs from F^- only by a numerical factor.

When the gas lost by stars is allowed to flow freely down to the center of the galaxy (such as in cooling flow models) the global rate at which energy has to be radiated away is given by:

$$L_{\rm inflow} = L_\sigma + L^+_{\rm grav} + L_{\rm SN}, \tag{9}$$

and an X-ray luminosity $L_{\rm X} \simeq L_{\rm inflow} + L_{\rm dscr}$ is expected (see CFT). In figure 1 $L_{\rm inflow}$ is displayed for three values of $\vartheta_{\rm SN}$. To connect σ_* to $L_{\rm B}$ we have used the Faber-Jackson relation. One can clearly appreciate that complete inflow models are far too bright in X-rays than indicated by the observations, a well known embarrassment for cooling flow models. Figure 1 shows that even with no supernovae at all ($\vartheta_{\rm SN} = 0$) $L_{\rm inflow}$ is still too large, basically because $L^+_{\rm grav}$ is too large: if all the gas falls in then too much gravitational energy is liberated, and galaxies become too bright in X-rays.

In the case of outflowing galaxies the global energy balance will indeed imply

$$L_{\rm X} \simeq L_{\sigma_*} + L_{\rm SN} - L^-_{\rm grav} + L_{\rm dscr} - L_{\rm out}, \tag{10}$$

where $L_{\rm out}$ is the enthalpy flow at the edge of the galaxy, and where now the gravitational well acts as a sink of energy, contrary to the inflow case described by equation (9). A much wider variety of situations can therefore be expected in the case of outflowing galaxies, and even more if partial winds are allowed (Loewenstein and Mathews, 1987a; Renzini, 1989).

When evaluated at the present epoch, equation (8) provides useful indications about the present character of the flow for every set of model parameters. We note that for $\vartheta_{\rm SN} = 1$ and low values of R one has $\chi \ll 1$ for every reasonable combination of the parameters, i.e. with the standard SN rate and without dark matter all galaxies would sustain a supersonic wind and the classical Mathews and Baker (1971) result is recovered. We also note that for low values of σ_* the present χ is less than unity with the standard SN rate, i.e. there is ample energy input to sustain a supersonic wind in these fainter galaxies, even with e.g. 10 times more dark than visible matter, and low X-ray luminosities are predicted. On the contrary, at the opposite extreme ($\sigma_* \gtrsim 300\,\mathrm{km\,s^{-1}}$) i.e. for optically bright galaxies, one has $\chi \gtrsim 1$ even for modest amounts of dark matter SN heating may be insufficient to prevent inflows, and large X-ray luminosities are then expected. For intermediate luminosities, a transition between the two regimes has necessarily to take place, and a large sensitivity of L_X to the galactic model parameters is therefore anticipated.

3 The evolutionary sequences

To describe the evolution of the hot gas in the model elliptical galaxies we solve the time-dependent, Eulerian equations of gasdynamics with galactic source terms (see Paper II). We assume the model galaxies to be initially devoid of gas.

So many parameters are required to fully determine a model galaxy that testing the influence of each possible combination of them would have been a rather cumbersome matter. We therefore focused on few such combinations, trying to get some general insight from their effects on our models. Our hydrodynamical models reproduce the pattern in figure 1 provided that the parameter combinations imply present χ values about unity for medium-luminosity galaxies ($L_B \simeq 5 \times 10^{10} L_\odot$); we further restrict to only one value of the other supernova parameter s, adopting $s = 1.5$ for all our simulations (see Paper II). Indeed, for χ values much different from unity most galaxies would now be in either a wind or an inflow phase, and so they would be either too faint or too bright X-ray emitters. This requirement restricts the possible combinations of ϑ_{SN} and R: so, for example for $\vartheta_{SN} = 1$, a dark to visible mass ratio R$\approx$ 10 is required, a value which drops to ≈ 3.3 if one assumes $\vartheta_{SN} = 0.5$. These figures are somewhat sensitive also to the relative distributions of the dark and visible mass, i.e. to the parameter β.

The X-ray evolution of all the $L_B = 5 \times 10^{10} L_\odot$ KG models is shown in figure 2. Models differing only for their central velocity dispersion show that the larger σ_*, the sooner the model galaxies experience their cooling catastrophe and the associated X-ray brightening. So, while at $t = 15$ Gyr some models have already experienced their transition to the inflow phase, others are still in their outflow or wind phase. For this particular set of parameters we therefore expect a very high sensitivity of L_X to variations of σ_* in the range $\sim 240 - 265$ km s^{-1}, with L_X increasing from $\sim 10^{38}$ to over 10^{41} erg s^{-1}. We emphasize that this extreme sensitivity of L_X on σ_* is the key characteristic of our model galaxies, and can therefore account for the very steep correlation of L_X with L_B, see figure 1. As already mentioned, the larger σ_* the deeper the potential well (i.e. the larger the binding energy per unit mass), the harder for the gas to leave the galaxy, and the sooner in the evolution the transition to the X-ray bright evolutionary phases. In turn, the Faber-Jackson relation ensures that – on average – the brighter the optical luminosity of the galaxy, the larger σ_*, and therefore the larger also the X-ray luminosity. Note, however, that models in the critical range of σ_*'s will also be very sensitive to changes in every other parameter; see Paper II for a thorough discussion of he action of each parameter. There is a close similarity to the case of early galactic wind models for the chemical evolution of elliptical galaxies (e.g. Larson, 1974; Arimoto and Yoshii, 1987), according to which in the more massive (and now brighter) galaxies the deeper potential well ensures star formation and metal enrichment to proceed further before being discontinued by the development of a wind, thus establishing the observed metallicity-luminosity correlation.

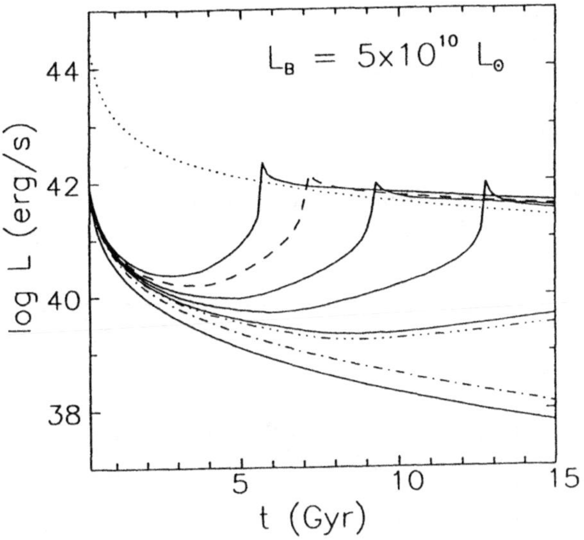

Figure 2. The X-ray evolution of various models with $L_B = 5 \times 10^{10} L_\odot$. Solid lines represent sequences which differ only for their value of the central velocity dispersion, from 220 to 280 km s^{-1}: the larger σ_*, the earlier the transition to the inflow phase marked by the sharp peak of L_X. Models with lower X-ray luminosity are at present still in their wind or outflow phase at present.

In essence, we propose that *both the $L_X - L_B$ correlation, and the metallicity-luminosity correlation, are two manifestations of the same underlying cause, i.e. of the fact that optically brighter galaxies are also gravitationally more tightly bound, as dictated by the Faber-Jackson relation.* The presence of metallicity gradients in ellipticals add further support to this scenario (see Franx and Illingworth, 1990).

In figure 3, just analogous to figure 1, we plot the X-ray luminosity at $t = 15$ Gyr for all the models constructed for this display, and characterized by $0.8 \leq \chi \leq 1.4$ and various optical luminosities. A comparison between these two figures makes clear the degree of success of the present class of evolutionary sequences in accounting for the observed X-ray luminosities of elliptical galaxies. Such a comparison indicates a remarkable prevalence of outflows, with the X-ray faint galaxies (those with $L_X \simeq L_{dscr}$ in figure 1) being still in their wind phase, and perhaps only a handful among the brightest galaxies having already experienced the transition to the inflow phase. We conclude that this class of models is able to successfully reproduce the richness of X-ray properties exhibited by real galaxies, just where cooling flow models had failed.

4 Conclusions

The main results of our study can be summarized as follows.

• The gas flows in model galaxies experience up to three consecutive evolutionary phases: the wind, outflow, and inflow phases. The first phase is characterized by supersonic winds, which tend to decelerate until a subsonic outflow is established. This outflow phase can be maintained for many Gyr until a central cooling catastrophe triggers the transition to the inflow phase.

• The epoch of the cooling catastrophe is very sensitive to the structural parameters describing the models, and to the adopted SN rate, and so is the present X-ray luminosity which reflects the amount of hot gas in the galaxies. The large dispersion in L_X at given values of L_B is naturally accounted for in terms of modest galaxy-to-galaxy differences in any of the structural parameters, including small differences in the amount of dark matter and in its distribution. With a standard SN rate ($\vartheta_{SN} = 1$) the mass of the dark matter surrounding elliptical galaxies has to be some 10 times larger than that of the visible matter.

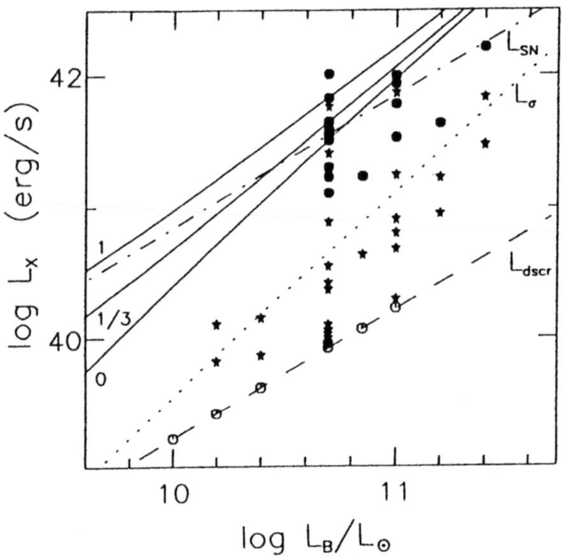

Figure 3. The L_X vs. L_B plot for our whole set of model galaxies at t=15 Gyr, in the same format as figure 1. L_X includes the expected contribution from discrete sources (L_{dscr}). Filled circles represent model galaxies which are in their inflow phase, stars and open circles represent galaxies respectively in their outflow and wind phase.

• The correlation of X-ray and optical luminosities, and the $L_X - \sigma_*$ correlation are seen as an ongoing manifestation of the fact that brighter galaxies are characterized by a larger binding energy per unit mass, i.e. as an implication of the Faber-Jackson $L_B - \sigma_*$ relation, not unlike the L_B-metallicity correlation which is established at the formation epoch of elliptical galaxies. In this scenario, type II SNe are responsible for the early establishment of galactic winds, and type I SNe for their maintainance long after any star formation has ceased, with most galaxies supporting outflows up to the present epoch and beyond.

• The main difference between the behavior of the flows in KGs and dVGs is the presence of a small *mini-inflow* only ~ 300 pc in size, near the center of the dVG models. This is due to the very centrally peaked concentration of matter near the center, which favors an early, localized cooling catastrophe. Therefore, for most of the time dVG models present small mini-inflows coupled to outer winds or outflows, much similar to those typical of KG sequences. As in the case of KG sequences, a global, much more extended cooling catastrophe takes place later, and we can still speak of a wind, an outflow, and an inflow phase just as for KG sequences.

In Paper II we also address a number of other relevant aspects. These include the evolution of the X-ray surface brightness profile of the model galaxies, the interaction of galactic flows with the intracluster medium, the effect of interactions with other galaxies and merging events, the problem of the fate of the cooling gas during the inflow phase, and finally the constraint on the model parameters ϑ_{SN} and s which in principle can be set by the iron abundance in the galactic flows and in the intracluster medium.

References

Arimoto, N., Yoshii, Y.: 1987, *Astron. Astrophys.* **173**, 23.

Canizares, C., Fabbiano, G., Trinchieri, G.: 1987, *Astrophys. J.* **312**, 503 (CFT).

Ciotti, L., D'Ercole, A., Pellegrini, S., Renzini, A.: 1991, in preparation, Paper II.

D'Ercole, A., Renzini, A., Ciotti, L., and Pellegrini, S.: 1989, *Astrophys. J. Letters* **341**, L9 (Paper I).

de Vaucouleurs, G.: 1948, *Ann. Ap.* **11**, 247.

Djorgovski, S., and de Carvalho, R.: 1989, preprint.

Fabbiano, G.: 1989, *Ann. Rev. Astron. Astrophys.* **27**, 87.

Forman, W., Jones, C., and Tucker, W.: 1985, *Astrophys. J.* **293**, 102.

Franx, M., and Illingworth, G.: 1990, *Astrophys. J. Letters* **359**, L41.

King, I.R.: 1972, *Astrophys. J. Letters* **174**, L123.

Larson, R.B.: 1974, *Mon. Not. R. Astr. Soc.* **166**, 585.

Loewenstein, M., and Mathews, W.G.: 1987a, in *Nearly Normal Galaxies*, ed. S.M. Faber (New York: Springer-Verlag), p. 96.

Loewenstein, M., and Mathews, W.G.: 1987b, *Astrophys. J.* **319**, 614 (LM).

Mathews, W.G.: 1989, *Astron. J.* **97**, 42.

Mathews, W.G., and Baker, J.: 1971, *Astrophys. J.* **170**, 241.

Mellier, Y., and Mathez, G.: 1987, *Astron. Astrophys.* **175**, 1.

Nulsen, P.E.J., Stewart, G.C., and Fabian, A.C.: 1984, *Mon. Not. R. Astr. Soc.* **208**, 185.

Renzini, A.: 1989, in *Evolutionary Phenomena in Galaxies*, ed. J.E. Beckman and B.E.J. Pagel (Cambridge Univ. Press), p. 422.

Sarazin, C.L., and White, R.E.III: 1987, *Astrophys. J.* **320**, 32.

Sarazin, C.L., and White, R.E.III: 1988, *Astrophys. J.* **331**, 102.

Tammann, G.: 1982, in *Supernovae: A Survey of Current Research*, ed. M. Rees and R. Stoneham (Dordrecht: Reidel), p. 371.

Trinchieri, G., Fabbiano, G., and Canizares, C.R.: 1986, *Astrophys. J.* **310**, 637.

Discussion

R. Bender - It is interesting that very luminous S0 objects, like Sombrero, have comparably low X-ray luminosity when compared to E's of the same luminosity. The same is true for disky ellipticals which have lower X-ray luminosity than boxy ellipticals in the range where the luminosity function of these two types overlap. So, there may be a dependence of X-ray luminosity on other parameters like M_*/L_B (see Bender et al. 1990) or shape. All these further parameters are difficult to test from observations, but may possibly be tested with your models or modifications of them.
A. Renzini - Indeed. In figure 1 boxy and disky galaxies are represented with different symbols, and there is some indication in the sense that you mention, although the statistics is certainly poor. Supposing that the X-ray–boxyness correlation exists, we have envisaged two possible, albeit still speculative explanations. One possibility is that merging events (supposed to cause the boxy isophotal distortion) could trigger an earlier transition to the inflow phase, with the associated X-ray brightening. The other possibility, is that the early-type galaxies which are in part rotationally supported (i.e. the diskies) may be characterized by a lower binding energy per unit luminosity (actually a lower $L^-_{\rm grav}/L_B$ ratio) compared to boxies. If so, their χ would be systematically smaller, and therefore they would be more likely to support winds rather than inflows.

P. Pişmiş - Have any of the galaxy you considered active nuclei, at least in a mild manner in the form of jets, lobes etc.? If yes is there a correlation between activity and their X-ray behavior?
A. Renzini - Our models neglect any nuclear activity which could contribute to the heating of the ISM. However, our models show a great variety of inflow rates, from no inflow at all during wind/outflow phases, to quite small rates in the case of nuclear mini-inflows, to quite large rates during the global inflow phase. This may help explaining the great variety of nuclear activity behaviors exhibited by ellipticals and bulges. Observationally, the optical, X-ray, and radio luminosity all correlate to each other to some degree.

M. Franx - How well is the contribution from discrete sources known? It appears that the analysis is quite dependent on this.
A. Renzini - The discrete source contribution adopted in figure 1 is just the lower envelope of the data points. This is somewhat fainter than the contribution adopted by CFT. None of our conclusion is particularly dependent on this assumption.

M. Franx - Are the points near the line of the discrete source contribution upper limits?
A. Renzini - Yes, some of them are upper limits. This further indicates that in these galaxies the contribution from a hot ISM is very small, much smaller than predicted by inflow models.

J. Melnick - The catastrophic cooling phase should be associated with optical emission. Therefore, if as you propose, the most X-ray luminous galaxies are in the inflow phase one would expect a substantial fraction of them to harbor active nuclei, or have optical filaments in the central regions. Is that the case?

A. Renzini - Yes, to some extent. As I mentioned a moment ago an X-ray–optical–radio correlation appears to exist. Our models are not designed to follow the fate of the cooling material, in particular whether it is used to form stars or jupiters, or it is accreted by a central black hole. However, we argue that, if there is a time in the evolution of these galaxies when an intense burst of star formation may take place in the nucleus, quite naturally this is the moment of the central cooling catastrophe, when up to $\sim 10^9 M_\odot$ of material rapidly cool off. We speculate that this may coincide with a major optical display, quite in the mood of the AGN model that you and Terlevich have proposed.

J. Melnick - The X-ray surface brightness distribution of ellipticals follows quite closely that of the optical light. Can you reproduce this feature in your models?
A. Renzini - Yes, for the three galaxies for which there exist sufficient *Einstein* data. As shown in the figure below, the X-ray surface brightness distribution is subject to major changes in the course of the evolution, through the three stages. In particular, during the outflow phase it follows quite closely the optical distribution (represented by the solid line), while during the inflow phase it is much more centrally peaked, as well known also for the case of cooling flow models. We conclude that the three galaxies are most likely still in their outflow phase.

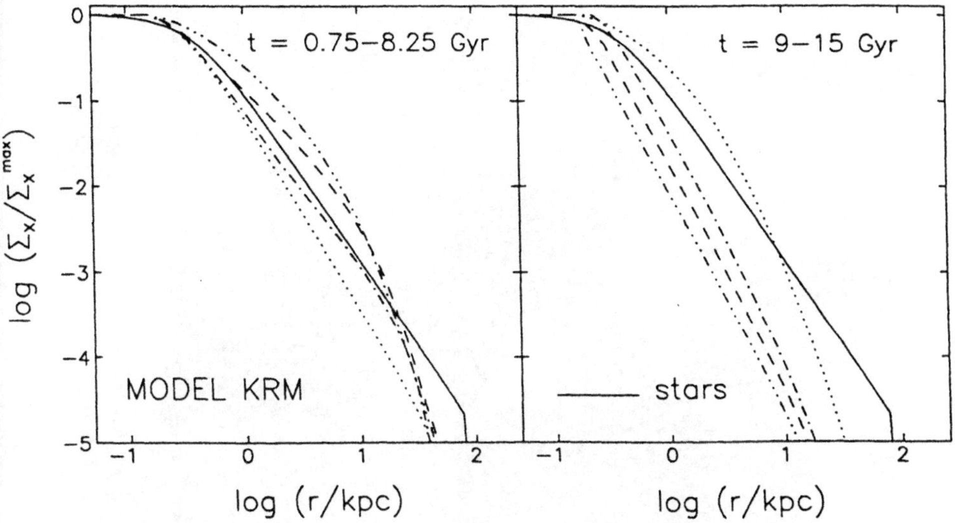

Figure 4. The evolution of the X-rays surface brightness profile for a selection of representative times during the various evolutionary phases of the "KRM" model defined by ($L_B, \sigma_*, \delta, R, \gamma, \vartheta_{SN}, s$)=($5 \times 10^{10}$ergs^{-1}, 270 km s^{-1}, 180, 9, 0.0125, 2/3, 1.5). In the left panel models in the wind phase (dotted line) and outflow phase (other lines) are displayed. In the right panel a model towards of the outflow phase is shown (dotted line) together with a few models in their inflow phase (other lines). The optical surface brightness profile is also shown for reference (solid line).

I. King - I would like to clarify your distinction between King models and de Vaucouleurs models. At a central concentration of 2.2-2.3 they are nearly identical. Thus your

distinction should really be central concentration.
A. Renzini - Indeed, the difference is in the center only. Basically, our $r^{\frac{1}{4}}$-law models have much higher central stellar densities compared to *King* models.

I. King - Ah, I see. You have fitted each kind of model to the same density curve. In that case it is just the difference in central cusp that matters.
A. Renzini - Yes, for your second comment. But we have used two different density curves for the two kinds of models (i.e. we have not approximated a *de Vaucouleurs* density law by adjusting the parameters of a *King's* density law).

Alvio Renzini answering Bender's question.

THE WARM COMPONENT OF THE ISM OF ELLIPTICAL GALAXIES

F. Macchetto[1], W.B. Sparks
Space Telescope Science Institute
Baltimore, USA

ABSTRACT. The characteristics of the optical emission-line gaseous component of elliptical galaxies are reviewed. We focus on three questions — the origin and fate of the gas and its excitation. The relative merits of accretion or infall events and cooling-flows are compared with regard to these questions. The cooling-flow model has the largest number of hurdles to overcome.

1 Overview

Elliptical galaxies often contain an extended, usually low excitation, component of ionized gas in their ISM. In recent years it has become apparent that this warm, $T \approx 10^4$K, gas may provide the key to understanding many diverse physical processes and observations related to elliptical galaxies, their interstellar material, environment and activity. Investigation of the problems of the *origin* of such gas, its ultimate *fate* and the *excitation mechanisms* responsible for the emission cover a suprisingly large variety of topics in extragalactic research including the frequency and relevance of mergers in galaxy evolution, the theory of 'cooling-flows', the onset of nuclear activity and the beaming of radiation from active nuclei together with more prosaic subjects such as the survival of dust in a hostile environment and the stellar IMF in unusual environments.

The determination of inter-relationships between different gas phases is crucial. On the one hand accretion or infall events have been cited as an attractive explanation for many of the observed phenomena, while on the other hand it has been proposed that line emitting filaments are thermally unstable gas condensing from a hot corona pervading the galaxy, a cooling-flow.

These two pictures, currently the two most popular scenarios, are incompatible. One demands a quiescently evolving cluster in which gas, over the age of the Universe, cools and condenses, while the other implies a dynamic interplay between galaxies and their subcomponents with episodic heating, mergers, starbursts and so forth. The basic astrophysical processes of relevance are quite different, and it is because of this that the topic assumes an importance beyond simply understanding the particular issue of ionized gas in ellipticals.

[1]On assignment from the Astrophysics Division, Space Science Department European Space Agency

G. Longo et al. (eds.), Morphological and Physical Classification of Galaxies, 191–206.

This review attempts to summarise the observational 'facts' which provide the foun dations for understanding, and to assess the compatability of those observations wit these two very different viewpoints. The cooling-flow model has the greatest numbe of problems to overcome, while the infall picture can work well particularly if electro conduction of heat from the coronal phase to the cooler filaments is effective. Alternativ excitation processes are considered, Section 4.

2 Observational material

Historically, the existence of an ionized gaseous phase in elliptical galaxies has beer known for some time from spectroscopic work (Mayall 1936, Humason, Mayall anc Sandage 1956, Minkowski and Osterbrock 1959, Osterbrock 1960). Humason, Mayal and Sandage (1956) found that approximately 18% of the 82 elliptical galaxies they observed showed optical line emission. Minkowski and Osterbrock (1959) carried out a detailed study of NGC 1052 and considered photoionization by horizontal branch stars and tapping of turbulent motions as possible energy sources. Caldwell (1984) observed the $[OII]$3727Å region for about 100 galaxies and detected $\approx 40\%$. Phillips et al. (1986) obtained spectra of over 200 E and S0 galaxies and estimate 55—60% of the galaxies have $[NII]$6584Å with an equivalent width > 0.5Å.

Imaging of elliptical galaxies through narrow-band emission-line isolating filters has more recently lead to a better appreciation of the phenomenon. Arp (1967) discovered extended $H\alpha$ emitting filaments in M87 using photographic techniques with interference filters, however it is really only with the advent of more sensitive detectors, CCDs in particular, that emission-line imaging of usefully large numbers of galaxies has become feasible. With the work of Ford and Butcher (1979), Demoulin-Ulrich et al. (1984), Sparks, Macchetto and Golombek (1989), Trinchieri and di Serego Alighieri (1989) and Kim (1989) together with comparable studies directed principally at clusters of galaxies by Heckman (1981), Cowie et al. (1983), Hu, Cowie and Wang (1985), Heckman et al. (1989), and radio galaxies, Baum et al. (1988), McCarthy (1988), it has become obvious that in order to correctly determine the total extent and luminosity of warm gas in elliptical galaxies, full two-dimensional coverage is essential, although as recently as 1986 Schweizer could state that "virtually all our knowledge about the warm component...stems from optical spectroscopy". Long slit spectroscopy provides vital information on kinematics, line ratios and simply whether or not the ionized gas is extended, but it is only with imaging that the full extent of the phenomenon can be properly appreciated and quantified.

Arp (1967) first observed M87 using interference filters and discovered filaments located opposite to the well-known synchrotron jet. Ford and Butcher (1979) presented a thorough spectroscopic and imaging analysis of that same system, and discussed many of the potential excitation and origin options. They also found overabundances of nitrogen with respect to hydrogen. Demoulin-Ulrich et al. (1984) obtained both imaging and spectroscopy of 12 elliptical galaxies, detecting emission in six. Baum et al. (1988) surveyed a large number of low-power radio galaxies. Heckman (1981), Heckman et al. (1989) surveyed cooling-flow galaxies and clusters, and Cowie et al. (1983) and Hu, Cowie and Wang (1985) surveyed clusters with and without cooling-flows. Kim (1989) presented the results of narrow-band CCD imaging of 26 early-type galaxies selected from IRAS detections, and Trinchieri and di Serego Alighieri (1989) observed 13 elliptical galaxies with known X-ray properties for optical line emission, detecting extended line emission in 10 of them.

A deep survey for line emission in early-type galaxies is being conducted as part of the ESO Key Program "Towards a Physical Classification of Galaxies", Bender et al., using the ESO 3.6-m. and NTT, see Fraix-Burnet, Macchetto and Sparks (1991), Sparks et al. (1991), this conference, for progress reports. Without the sensitivity resulting from use of a large telescope, a significant fraction of the emission is easily lost, e.g. compare NGC 1052 in Kim (1989) to Forbes, Sparks and Macchetto (1990) reproduced in figure 3.

Knowledge of the distribution of dust is vital in this study, and there have been a number of surveys for dust in elliptical galaxies, e.g. Hawarden et al. (1981), Sparks et al. (1985) and Bertola (1986) for a review. Kim (1989) based his study of optical line emission on IRAS detected galaxies and found a high fraction of detections, possibly indicative of an association between dust and emission-line gas.

Figures 1 through 4 illustrate a number of archetypal examples of the phenomenon, where we show the gas and dust in NGC 4696, NGC 1052 and IC 1459. These illustrate a number of important general points:

- The quantity of ionized gas (where filament systems have been imaged) lies in the range 10^2 to $10^7 M_\odot$, e.g. Kim (1989), Heckman et al. (1989), with the highest values occurring in the classical 'cooling-flow' clusters such as Perseus (NGC 1275).
- Luminosities range from 10^{38} to 10^{42}erg s^{-1} in $H\alpha + [NII]$ emission alone.
- Our observations and those of Kim (1989) show that spiral structure within the emission-line gas is common, particularly one-armed spirals. See figure 1 for examples, and also Fraix-Burnet et al. (1991). Even the traditionally chaotic systems such as NCG4696 and M87 conform to this rough morphological description.
- Emission systems are frequently (always?) associated with dust. All galaxies shown have dust associated with their emission filament systems.
- Other indicators of merged subunits are common. IC 1459 has a counter-rotating stellar core (Franx and Illingworth 1988, Balcells & Quinn 1990), NGC 4696 has a discrete stellar subsystem while NGC 1052 has a large radius HI ring around the minor axis of the galaxy.
- The angular momentum vector of gas is usually completely misaligned with respect to the stellar angular momentum vector, and of significantly higher specific angular momentum.
- Malin-Carter shells are found in a number of emission-line ellipticals, again usually indicative of a merger.

In summary, despite early thoughts that elliptical galaxies are to all intents and purposes devoid of gas, we know that a substantial fraction of order 50% contain detectable ionized gas often associated with dust. For later type galaxies the ISM appears to be inextricably involved in the morphological class of the galaxy and we would like to know if this is also true for elliptical galaxies. There are indications that line emission is associated with signs of accretion or merging.

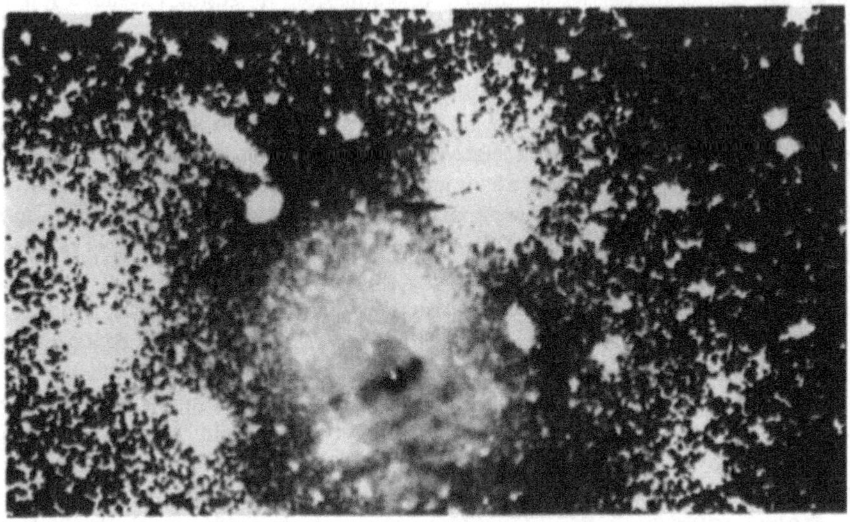

Figure 1. Direct images showing (a) a CCD direct frame of NGC 4696, (b) the same after subtraction of a model galaxy, (c) an image of the $H\alpha + [NII]$ emission and (d) a map of the optical depth at V, from Sparks et al. (1989).

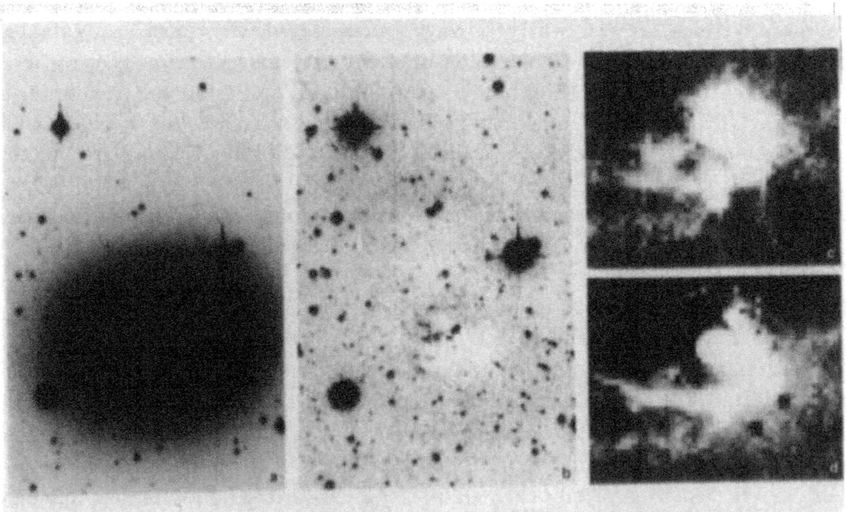

Figure 2. Image of the dust, shown as dark regions, in NGC 1052. The frame is ≈ 3.5 arcmin in height.

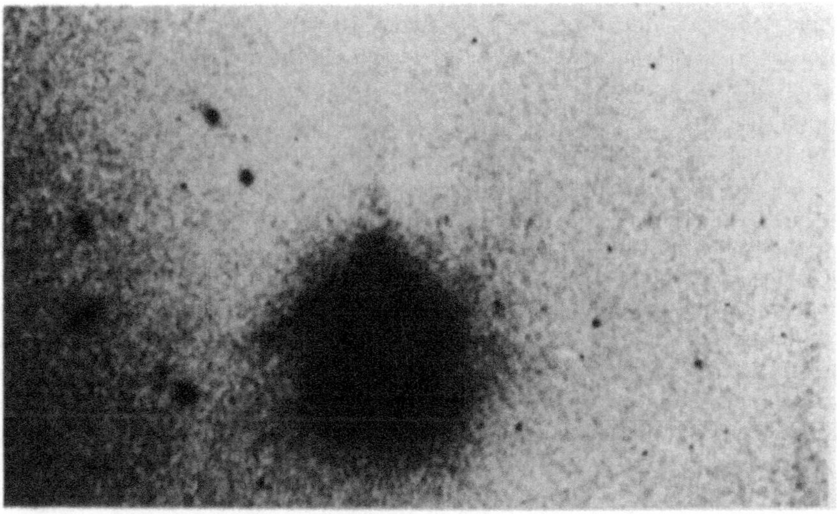

Figure 3. Line emission, $H\alpha + [NII]$ in NGC 1052, scale as for figure 2.

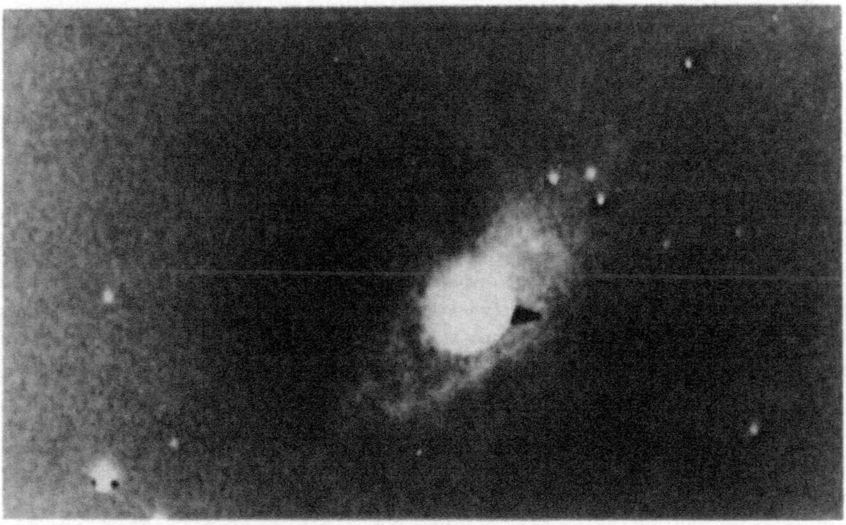

Figure 4. $H\alpha + [NII]$ emission in IC 1459, scale as for figure 2.

3 Statistical results

3.1 Kinematical and morphological properties

Consistent with the theme of the meeting, Baum, Heckman and van Breugel (1989) an Tadhunter et al. (1989) have attempted to categorize the emission-line components o the basis of their morphology and kinematics. Systems may be identified which ar (i) rotators, e.g. M84, 3C33, (ii) calm non-rotators and (iii) chaotic non-rotators, e.g M87. The former are well-ordered gas discs and appear to occur within the most powerful active ellipticals, external to rich clusters. The excitation can be high and rotatio velocities fast. The chaotic non-rotators are irregular filamentary systems whose velocit can be dominated by turbulence or at least random motions. These occur in elliptica galaxies at the centres of rich clusters and typically have a LINER-type spectrum, wit $[NII] > H\alpha$. The interesting question is whether in this categorization we are witnessing the same phenomenon in different environments, or a different phenomenon in th different types of galaxy. The concept of a purely descriptive classification at this stag is a sensible approach — it would be premature to impose physics on the classification The ultimate goal is to understand why these types occur of course, but the first step i to identify the types.

The gaseous kinematics bear no discernable relationship in general with the kinematics of the underlying stellar population, e.g. Davies (1986), Davies and Illingwort (1986), Bertola and Bettoni (1988), Franx and Illingworth (1988), Franx, Illingwort and Heckman (1989). Gas angular momentum can be high, unlike the stars, and wit a direction often orthogonal or even counter to the stellar rotation. Morphologically, a described above, our own and other data have shown that spiral structure within the emission-line gas is common, even in supposedly 'chaotic' systems.

3.2 Correlation with X-ray properties

Numerous authors have found evidence for a relationship between the X-ray and optical properties. These are thoroughly reviewed by Heckman et al. (1989). The consensus appears to be that although there is a large intrinsic scatter in the relationship, there is a highly significant correlation between the presence of optical line radiation and X-ray excess. Heckman et al. (1989) find a significance level of 0.06% (confidence level 99.94%) for a correlation between $\dot{M}$ and $L_{H\alpha}$. To summarize Heckman et al.'s analysis: (i) only clusters with high values of $\dot{M}$ and L_X(excess) have emission nebulae. (ii) the cooling-flow nebulae are significantly more luminous than those in typical ellipticals. (iii) brightest cluster galaxies in cooling-flows have more line emission than radio galaxies with the same radio power, implying a 'special' quality for the cooling-flow objects. Balanced against this are a huge dispersion in line luminosity, which is also generally much higher than if atoms recombine only once (the H_{rec} problem) and the extent of the line emission is much less than the supposed cooling region.

Phillips et al. (1986) note that despite low numbers (only eight X-ray detected galaxies in their sample) all but one showed line emission, a result significant at the 5% level. Hu, Cowie and Wang (1985) and Hu (1988), figure 5, show convincingly that mission lines only occur in clusters with short X–ray cooling times.

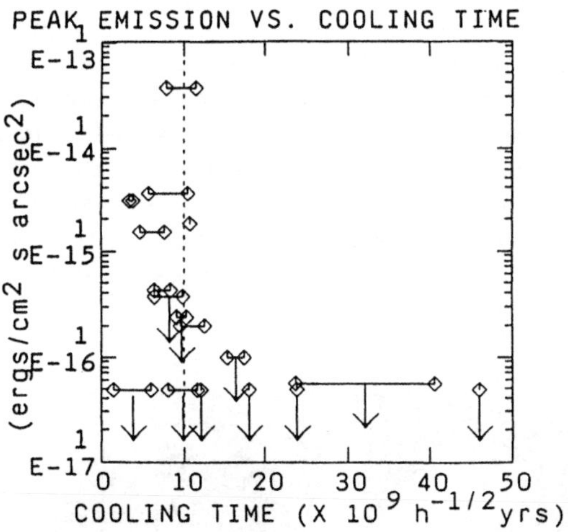

Figure 5. Peak emission line surface brightness vs. X–ray cooling time, from Hu (1988).

Trinchieri and di Serego Alighieri (1989) find essentially no correlation between the presence of $H\alpha$ emission and other global parameters of the elliptical galaxies they observed (interpreted by them as evidence for an independent origin for the gas) with the exception of X–ray emission. There, indeed they find "a clear relationship", although again numbers are rather small (13 galaxies), figure 6.

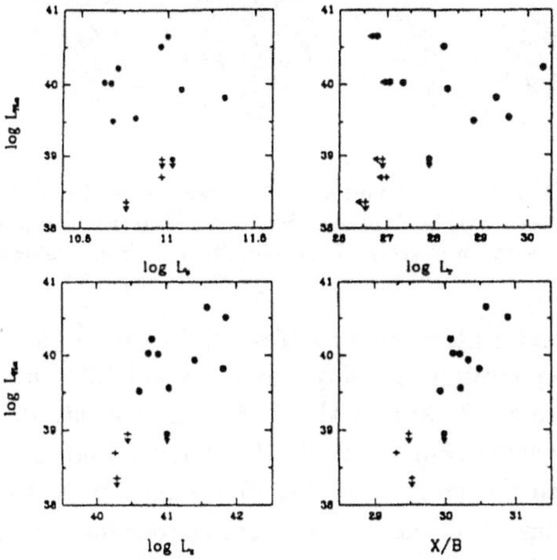

Figure 6. Correlation of optical emission line flux against optical, radio and X–ray luminosity, and X–ray-to-optical luminosity ratio, from Trinchieri and di Serego Alighieri (1989).

3.3 Optical peculiarities and evidence for infall

Schweizer and Seitzer (1989), Schweizer et al. (1990) have examined a large number of early-type galaxies for optical peculiarities conventionally interpreted as evidence for a merger or accretion event. They find that some 50% of the galaxies show such evidence. It is often the case that the galaxies having optical emission filaments fall into this category. More work is needed to establish how generally applicable this statement is, but for example, we note the following cases: NGC 1316 is probably the archetypal merging galaxy (Schweizer, 1980), showing minor axis chaotic dust, Malin-Carter shells and other large scale loops. Schweizer also showed that there is extensive $[OII]$ throughout the inner regions, and $H\alpha$ imaging and dust-imaging was presented by Sparks and Macchetto (1989), see figure 7.

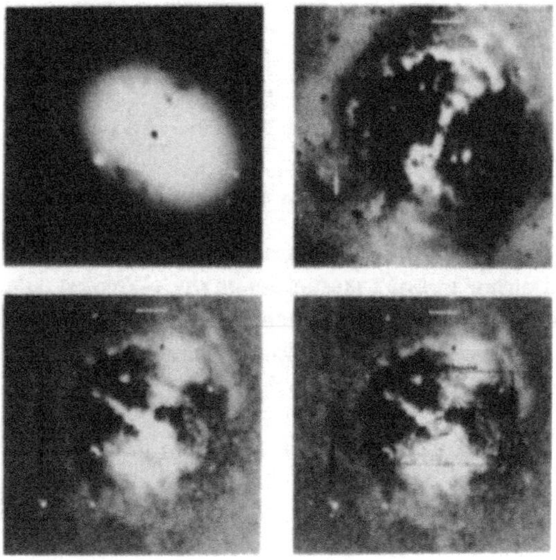

Figure 7. NGC 1316 (Fornax A) showing top left, direct B-band image, top right, dust absorption (shown as white areas), bottom left, $H\alpha + [NII]$ after continuum subtraction, and bottom right, the same with an overlay of the radio source from Geldzahler and Fomalont (1984).

There is indeed much chaotic filamentary emission, having a close correspondence with the dust distribution. Very similar in many ways is NGC 1052, figures 2 and 3, with not only irregular minor axis dust (Knapp et al. 1978, Sparks et al. 1985, Forbes et al. 1990) but also extensive line emission (Forbes et al. 1990) and a high angular momentum large radius HI disc or ring, van Gorkom et al. (1986). There are also suggestions of a tidal distortion in the HI. Forbes et al. note other strong reasons for supposing an accretion event has occurred in the two other flat-spectrum radio galaxies they studied: not only do all contain emission filaments and dust — one, IC 1459, has a counter-rotating stellar core, see above, while the other, NGC 6958, has clear Malin-Carter shells. Thomas et al. (1986) had proposed that the galaxies NGC 1316, NGC 1052, IC 1459 and other nearby ellipticals host small-scale cooling-flows.

The variety of line-strength gradients in elliptical galaxies found by Gorgas et al. (1990) lead them also to favour mergers of subunits as the formation process for elliptical galaxies.

Finally, a direct observation is provided by van Gorkom et al. (1989) — they show that for *all* eight radio galaxies in which HI absorption is detected, there is a redshifted and therefore infalling component.

3.4 Association with dust

Another important feature of emission systems is that they are frequently (always?) associated with dust. Kim (1989) from his observations of 26 galaxies states that "...ionized gas is detected in all dust-lane galaxies, and a dust-lane is detected in all galaxies with emission lines." This has implications for the origin of the gas, the cooling energy budget and as a side issue may be used to constrain geometrical properties (Sparks, Macchetto and Golombek 1989). With the realisation from IRAS that many elliptical galaxies contain dust (although of an unknown spatial distribution), the levels of emission are such that the dust *is* an important coolant, typically equalling or exceeding the X–ray flux and optical line emission.

NGC 4696 is a classical 'cooling-flow' galaxy located at the centre of the Centaurus cluster which was found by Fabian et al. (1982) to contain extensive emission filaments. The galaxy also contains dust, and it was shown by Sparks et al. (1989) that the dust and emission-line gas are co-spatial to the limits of the observations, see figure 1. They also demonstrated that not only are the emission filaments closely associated with dust but that the dust extinction properties are normal. Their observations lead them to favour an external origin for the gas and dust. In other nearby galaxies such as NGC 1316 and NGC 1052 discussed above, the dust and line-emitting gas always occur together. In extreme cooling-flows such as NGC 1275, dust is clearly important. This strongly suggests that the origins of the warm gas and the dust are related.

3.5 Summary

To sumarise, there are two clear correlations, one between emission-line gas and X-ray excess and the other with the presence of dust. There also appear to be correlations between the kinematic and spectroscopic character of the emission, and galactic environment, Baum et al. (1989) and there are indications that the gaseous component occurs along with other optical peculiarities that are normally taken to be signs of a merger.

4 Physical processes and discussion

4.1 Origin of the gas

The common occurrence of dust within line-emitting filaments is easily understood if the gas has an external origin such as stripping of the outer parts of a passing spiral or the accretion of a gas-rich dwarf. This too would naturally account for the normal extinction properties of the dust and the wide variety of morphology exhibited by dust-lanes in early-type galaxies, Hawarden et al. (1981), Bertola (1986). The frequent simultaneous occurence of Malin-Carter shells and other peculiarities, Schweizer et al. (1990), high angular momentum, e.g. Tadhunter et al. (1989), presence of large radius HI features van Gorkom et al. (1986), and lack of correlation between cold gas and most global properties have all been taken as evidence that infall events are common.

On the other hand, the short cooling times inferred from X–ray data together with the correlation between optical line emission and X–ray excess has been taken as evidence of a cooling-flow origin for the filaments. Although such a model may explain the X-ray properties, including, for example, the presence of a distribution in temperature in the coronal gas, several difficulties arise when confronted with optical and infrared observations. Dust is a a problem in the cooling-flow model. Theoretical dust sputtering times are short in the coronal X-ray gas, and so thermally unstable condensations in the simple picture should be dust-free, which is evidently not the case. It would be necessary to demonstrate that dust can form within the cooling filaments, perhaps as stellar ejecta from star formation within the cooler gas, or perhaps condensing behind shocks within the filaments. In either event, the energetic consequences of the additional coolant may need to be considered. No obvious explanation has been raised to account for the angular momentum and morphological properties alluded to above, and the energy balance is not good, with optical lines being typically 10 to 100 times more luminous than inferred from the X–ray cooling rate. Equally, the emission line surface brightness distribution does not correspond to the X–ray inferred cooling radius (returned to below, Section 4.2).

It has been suggested that the dust and gas are the relic of inefficient heating of stellar ejecta to coronal temperatures during the process of normal stellar evolution within the galaxies. It seems to us that this is implausible given the coherent nature of the filament systems over many kiloparsecs, even tens of kiloparsecs, and also the high angular momentum of the gas which is quite unlike that of the stars.

4.2 Excitation mechanisms

Amongst the competing theories are internal shocks, photoionization by an active nucleus, photoionization by young stars within the filaments, photoionization by hot old stars, photoionization by X–ray photons of the coronal phase and ionization either directly or indirectly by hot electrons of the coronal gas. Ionization directly by the relativistic plasma associated with radio emission appears unlikely given a lack of spatial coincidence between optical and radio emission.

The optical spectrum is very similar to Galactic shock excited filaments, which has lead a number of people to favour shock excitation for the line-emission. In addition, there is ample energy available in gaseous bulk motions, and the problem of requiring multiple recombinations in the cooling-flow model is reasonably well-explained if gas undergoes repressurizing shocks, although on a quantitative level, it remains somewhat ad hoc and predictions and tests are hard to come by.

Sparks et al. (1989) proposed excitation from a conduction front at the interface of cool gas and hot coronal gas. This works well energetically, and can successfully predict the optical surface brightness of line emission with radius — traditionally a problem for cooling-flow and shock models, Sparks and Macchetto (1990). The filaments are much more centrally peaked than expected if they arise from cooling throughout the 'cooling region'. It is unclear whether or not the correct optical spectrum would be produced by this mechanism since realistic modelling has not been carried out, and the viability of conduction depends on the topology of the interstellar magnetic fields. Faraday rotation measurements suggest ordered fields on kiloparsec scales which would favour effective conduction and various authors have calculated only modest suppression factors for tangled fields, Rosner and Tucker (1989), Tribble (1989). An attractive feature of the model is that infall events may be retained as the primary origin for line emission material but the thermal interaction between the cooler filaments and the hot coronal phase may lead to correlations between optical and X–ray properties that had previously been interpreted as due to a cooling flow.

Heckman et al. (1989) have proposed that photoionization by X–ray photons associated with the coronal gas may be important in ionizing the filaments. This is based on the inference that X–ray and optical line-emission surface brightnesses are comparable. In calculating optical surface brightness, however, Heckman et al. assumed a unity covering factor for optical line emission within the emission radius. Inspection of typical images, e.g. Ford and Butcher (1979) for M87, shows that the actual covering factor is only a few percent. Coupled with a rather low assumed value for total line emission to $H\alpha$ line emission (5–10% in $H\alpha + [NII]$ rather than more typically 1%) we see in fact that the actual optical surface brightness is some two orders of magnitude higher than the X–ray which argues strongly against such a model.

Although photoionization by an active nucleus is popular, e.g. Robinson et al. (1987), Heckman et al. (1989) "conclusively rule out photoionization by a dilute nuclear photoionization source" on the grounds of the radial variations of line ratios. Nevertheless particularly in the rarer high excitation extended emission regions, nuclear photoionization could be playing a rôle. Spatial anisotropies or temporal variations in the nuclear radiation field may further influence Heckman et al.'s conclusion.

Photoionization of cool gas by UV photons from horizontal branch stars has been proposed, Minkowski and Osterbrock (1959), Trinchieri and di Serego Alighieri (1989). The energy budget may be adequate, although this has been disputed, e.g. Demoulin-Ulrich et al. (1984), Kim (1989). In any event, there remain the problems that this model does not in any obvious way lead to the correlation between line emission and X–ray cooling time, nor does it give the observed LINER spectrum.

In summary, internal shocks are feasible as an excitation mechanism, with the possibility of thermal conduction either directly or indirectly energizing the filaments as an interesting alternative that is somewhat more predictive.

4.3 The ultimate fate of the warm gas

Two possible sinks for the gas at least need to be considered — star formation and accretion onto the nucleus. Taken to its extreme, if the derived mass inflow rates in the cooling-flow model are correct and result in star formation, then whole galaxies would be built over the age of the Universe. On the other hand, if star formation *is* the sink for cooling gas at the rate inferred from X–ray observations, there are serious difficulties with hiding the newly formed stars. Johnstone et al. (1987) found evidence from absorption line strengths in a number of cooling-flow galaxies for ongoing star formation. It has been suggested that the IMF is biased to low mass stars to avoid large disturbances to galaxy colours. Gorgas, Efstathiou and Aragón (1990) however found no difference between magnesium line strengths of normal and cooling-flow galaxies. Their limit is jus consistent with the results of Johnstone et al. (1987), and implies, as others includin Johnstone et al. have found, that only a small fraction of the current star formation if proceeding at the rate given by the cooling-flow models, could be coming from star with a normal IMF. There would need to be an upper mass cutoff around $0.5M_{\odot}$. In NGC 1275 there is at least one 'conventional' HII region, Shields and Filippenko (1990) but estimates by these authors and by Shields, Filippenko and Basri (1990) show that the IMF is *not* devoid of massive stars and that if due to a cooling-flow, the mass accretion rate must be substantially lower than derived from the X–ray emission. They favour an aging starburst as the origin on star formation rather than a cooling-flow model.

If the emission-line gas has as its origin infall, then the amounts of matter involved are substantially less than those of the cooling-flow model. Typically some 10^6—$10^7M_{\odot}$ of gas may be involved. Indeed, in the presence of support at 10^4K by conduction, the Jean's mass within the warm gas is characteristic of the mass of a globular cluster, then star formation may occur but only where masses exceed the Jean's mass and therefore give rise to the formation of globular clusters preferentially in giant elliptical galaxies Sparks et al. (1989).

It is fairly well-established that optical line emission enhances the probability that a galaxy will be active, e.g. Phillips et al. (1986), Baum and Heckman (1989). Morphologically, in central dominant galaxies in clusters at least, the one-armed spiral distribution argues for a one-way trip down onto the active nucleus, while much more directly, van Gorkom et al. (1989) find HI absorption which is redshifted (indicating infall) in all eight active elliptical galaxies in which it is detected. The rather small quantities of gas involved in the infall picture can probably be accomodated with a mixture of accretion onto an active nucleus together with some star formation.

Alternative fates for the gas include expulsion by the active nucleus, or evaporation if strong thermal interaction between the coronal and cooler gas occurs.

5 Summary and key questions

Of the theories outlined, the cooling-flow model has the largest number of hurdles to overcome, requiring ad hoc re-pressurizing shocks to overcome excess line luminosity compared to X–ray emission, a yet-to-be determined mechanism for giving the emission-line surface brightness with radius, a convincing sink for the cooling material, an explanation for the presence of dust, high angular momentum gas and discrete stellar subsystems and the presence of colder gas such as HI and CO. It does successfully account for X–ray emission peaked to the centres of clusters and a good coincidence is obtained between cooling rates derived from X–ray lines and global considerations.

The infall model can simply account for the presence of dust and other kinematic peculiarities such as high angular momentum, counter-rotation in both stellar cores and gas discs and the presence of Malin-Carter shells and cold gas. The lack of correlation with global characteristics, properties of dust and morphology of dust lanes all favour an external origin. On the other hand, it fails to account for correlation between X–ray emission and line emission, nor does it explain the lower temperature X–ray peak at the cores of clusters in its simplest form.

The addition of thermal conduction as a physical process in the infall picture can satisfy essentially all of these requirements. Energetically it works well, and in detail can successfully reproduce emission-line surface brightness profiles without special pleading. The cooler X–ray emitting gas in that picture is simply induced by energy losses through radiation in the vicinity of the infalling material, which is itself excited into emission by the energy transfer. It remains to be seen if the characteristic LINER spectrum can arise from a conduction front. Other tests of that model are straightforward — a detailed spatial relationship between soft X–ray emission and optical line emission is required, and coronal lines of intermediate temperature should be found close to the filaments rather than diffusely throughout the 'cooling radius'.

A number of key questions remain that should be answered in the next year or two:

- How good is the dust/line-emission correlation?
- What *is* the excitation mechanism of the ionized gas?
- What are the abundances in the filaments?
- Is it true that dust is destroyed when immersed in the coronal phase at the rate predicted by theory?
- How well-founded is the relationship between X–ray and optical emission?
- What is the detailed spatial relationship between all of the X–ray, optical and infra-red radiation?
- Can the elusive intermediate temperature coronal lines be detected and if so, what is their spatial distribution?
- What are the magnetic field strengths and topology within the ISM of elliptical galaxies?
- What is the spatial distribution of dust and is there a diffuse component either associated with the hot phase or evaporating from an infalling cloud?

In conclusion, we have shown that a large fraction of ellipticals possess a warm component to their ISM. Correlations have been described between such a component and both X–ray excess and the presence of dust. Mergers appear the most likely origin of material, and viable excitation mechanisms include internal shocks or electron conduction. The fate of the gas is unclear, but observations suggest modest star formation together with infall to an active nucleus.

References

Arp, H.: 1967, *Astrophys. Letters* **1,** 1.

Balcells, M. and Quinn, P.: 1990, *Astrophys. J.* **361,** 381.

Baum, S.A., Heckman, T. and van Breugel, W.: 1989, *ESO Workshop on 'Extranuclear Activity in Galaxies'.*, Eds. E.J.A. Meurs and R.A.E. Fosbury.

Baum, S., Heckman, T., Bridle, A., van Breugel, W. and Miley, G.: 1988, *Astrophys. J. Suppl. Ser.* **68,** 643.

Baum, S. and Heckman, T.: 1989, *Astrophys. J.* **336,** 702.

Bertola, F.: 1986, *in 'Structure and Dynamics of Elliptical Galaxies'.*, IAU# 127, Ed. T. de Zeeuw, p. 135.

Bertola, F. and Bettoni, D.: 1988, *Astrophys. J.* **329,** 102.

Caldwell, N.: 1984, *Publ. Astron. Soc. Pac.* **96,** 287.

Cowie, L.L., Hu., E.M., Jenkins, E.B. and York, D.G.: 1983, *Astrophys. J.* **272**, 29.

Davies, R..: 1986, *in 'Structure and Dynamics of Elliptical Galaxies'.*, IAU# 127, Ed. T. de Zeeuw, p 63.

Davies, R. and Illingworth, G.D.: 1986, *Astrophys. J.* **302**, 234.

Demoulin-Ulrich, M.-H., Butcher, H.R. and Boksenberg, A.: 1984, *Astrophys. J.* **285**, 527.

Fabian, A.C, Atherton, P.D., Taylor, K. and Nulsen, P.: 1982, *Mon. Not. R. Astr. Soc.* **201**, 17P.

Forbes, D.A., Sparks, W.B. and Macchetto, F.: 1990, *IAU # 124 'Paired and Interacting Galaxies'.* NASA CP 3098, Eds. J.W. Sulentic, W.C. Keel, C.M. Telesco, p. 431.

Ford, H.C. and Butcher, H.R.: 1979, *Astrophys. J. Suppl. Ser.* **41**, **147**.

Fraix-Burnet, D., Macchetto, F., Sparks, W.B.: 1991, *Morphological and Physical Classification of Galaxies.*

Franx, M. and Illingworth, G.: 1988, *Astrophys. J.* **327**, L55.

Franx, M., Illingworth, G. and Heckman, T.: 1989, *Astrophys. J.* **344**, 613.

Geldzahler, B.J. and Fomalont, E.B.: 1984, *Astron. J.* **89**, 1650.

Gorgas, J., Efstathiou, G. and Aragón, A.: 1990, *Mon. Not. R. Astr. Soc.* **245**, 217.

Hawarden, T.G., Longmore, A.J., Tritton, S.B. and Corwin, H.G.: 1981, *Mon. Not. R. Astr. Soc.* **196**, 747.

Heckman, T.M.: 1981, *Astrophys. J.* **250**, L59.

Heckman, T.M., Baum, S., van Breugel, W.J.M. and McCarthy, P.: 1989, *Astrophys. J.* **338**, 48.

Hu, E., Cowie, L.L. and Wang, Z.: 1985, *Astrophys. J. Suppl. Ser.* **59**, **447**.

Hu, E.: 1988, *Cooling Flows in Clusters and Galaxies, ed. A.C. Fabian.*

Humason, M.L., Mayall, N.U. and Sandage, A.: 1956, *Astron. J.* **61**, 97.

Johnstone, R.M., Fabian, A.C. and Nulsen, P.E.J.: 1987, *Mon. Not. R. Astr. Soc.* **224**, 75.

Kim, D.: 1989, *Astrophys. J.* **346**, 653.

Knapp, G.R., Kerr, F.J. and Williams, B.A.: 1978, *Astrophys. J.* **222**, 800.

Mayall, N.U.: 1936, *Publ. Astron. Soc. Pac.* **48**, 14.

McCarthy: 1988, *Thesis, University of California.*

Minkowski, R. and Osterbrock, D.: 1959, *Astrophys. J.* **129**, 583.

Osterbrock, D.E.: 1960, *Astrophys. J.* **132**, 325.

Phillips, M.M., Jenkins, C.R., Dopita, M., Sadler, E. and Binette, L.: 1986, *Astron. J.* **91**, 1062.

Robinson, A., Binette, L., Fosbury, R.A.E. and Tadhunter, C.N: 1987, *Mon. Not. R. Astr. Soc.* **227**, 97.

Rosner, R. and Tucker, W.H.: 1989, *Astrophys. J.* **338**, 761.

Schweizer, F.: 1980, *Astrophys. J.* **237**, 303.

Schweizer, F.: 1986, *in 'Structure and Dynamics of Elliptical Galaxies'.*, IAU# 127, Ed. T. de Zeeuw, p. 109.

Schweizer, F. and Seitzer, P.: 1989, *in 'Dynamics and Interactions of Galaxies'.*, Heidelberg, Eds. A. Toomre and R. Wielen.

Schweizer, F., Seitzer, P., Faber, S.M., Burstein, D., Dalle Ore, C. and Gonzalez, J.: 1990, *preprint.*

Sparks, W.B., Wall, J.V., Thorne, D.J., Jorden, P.R., van Breda, I.G., Rudd, P.J. and Jorgensen, H.E.: 1985, *Mon. Not. R. Astr. Soc.* **217,** 87.

Sparks, W.B., Macchetto, F., Golombek, D.: 1989, *Astrophys. J.* **345,** 153.

Sparks, W.B. and Macchetto, F.: 1989, *in 'Dynamics and Interactions of Galaxies'.*, Heidelberg, Eds. A. Toomre and R. Wielen.

Sparks, W.B. and Macchetto, F.: 1990, *IAU # 124 'Paired and Interacting Galaxies'.*, NASA CP 3098, Eds. J.W. Sulentic, W.C. Keel, C.M. Telesco, p. 403.

Sparks, W.B., Macchetto, F., Fraix-Burnet, D., Bender, R., Capaccioli, M. and Nieto, J.-L.: 1991, *Morphological and Physical Classification of Galaxies.*

Shields, J.C. and Filippenko, A.V.: 1990, *Astrophys. J.* **353,** L7.

Shields, J.C., Filippenko, A.V. and Basri, G.: 1990, *Astron. J.*, in press.

Tadhunter, C.N., Fosbury, R.A.E. and Quinn, P.J.: 1989, *Mon. Not. R. Astr. Soc.* **240,** 225.

Thomas, P.A., Fabian, A.C., Arnaud, K.A., Forman, W. and Jones, C.: 1986, *Mon. Not. R. Astr. Soc.* **222,** 655.

Tribble, P.C.: 1989, *Mon. Not. R. Astr. Soc.* **238,** 1247.

Trinchieri, G. and di Serego Alighieri, S.: 1989, *in Extranuclear Activity in Galaxies.*, Eds. E.J.A. Meurs and R.A.E. Fosbury.

van Gorkom, J.H., Knapp, G.R., Raimond, E., Faber, S.M. and Gallagher, J.S.: 1986, *Astron. J.* **91,** 791.

van Gorkom, J.H., Knapp, G.R., Ekers, R.D., Ekers, D.D., Laing, R.A. and Polk, K.S.: 1989, *Astron. J.* **97,** 708.

Discussion

J. van Gorkom - I think it might not be too easy to divide galaxies in chaotic non rotators and rotators. We have very nearby an example of a radio galaxy that is both a rotator, the Hα emission from the disk, and a chaotic non rotator, the Hα emission near the radio lobes.

J. Melnick - I'd like to remind you that in M 87 we directly see a massive star cluster in the central region. The total mass is $\sim 10^9\ M_\odot$. Therefore I think that in many galaxies photoionization by stars and supernovae must be an important mechanism for the under gas.

D. Macchetto - Photoionization by hot stars along can produce the necessary UV photons, and the energy balance is adequate. However it does not reproduce the LINER spectrum nor does it explain the X-ray/emission line correlations.

A. Renzini - (During the discussion in Macchetto paper) Is not it a problem in suggesting that the Hα emission is due to stellar mass loss, that the observed specific angular momentum in the Hα gas is always much larger than the specific angular momentum of the stars?

D. Burstein - I think it is important to keep in mind that most of evidence for gaseous disks and/or merger effects produce little light compared to the parent galaxies. The

absorption line-strength variations studied in the Schweizer et al. paper are small perturbations on the spectra of these galaxies. In addition, although the Mg2 index does vary intrinsically relative to central velocity dispersion in the *7 Samurai* sample, I find no evidence that these variations sistematically affect peculiar motions.
D. Macchetto - I agree with your comment.

I. King - These small amounts of gas are interesting, and significant in what they reveal. Nevertheless we should not be sight of the fact that a much larger amount of gas was lost from stellar evolution in the early life of the galaxy, in spite of the quite high escape velocity. That has always been puzzling. But perhaps the scenario presented by Renzini will take care of this problem.
D. Macchetto - It is true that the amount of gas associated with the emission lines is relatively small ($10^5\ M_\odot - 10^9\ M_\odot$). Nevertheless it is important in the life of the elliptical today. Its presence accounts for the X-ray/optical line emission correlation. Furthermore a large number of these infall events would obviously leave a major impact on the evolution of the elliptical galaxy.

L. Buson - Obviously kinematical information could give essential clues to disentangle among possible origin of the ionized gas. At which extent do you plan to include kinematical data in the analysis of your present sample?
D. Macchetto - Our sample is currently built around Hα (and OIII) narrow-band emission line observations, as well as broad V and R measurements. In the future it will be very important to follow up this work with detailed kinematical studies.

D. Merrit - Is the quoted mass a nuclear mass, a dynamical or a luminous mass?
D. Macchetto - The mass of the core of NGC 7457 within 0.1″ is $\approx 30000\ M_\odot$ based on its luminosity.

Duccio Macchetto ready to defend his points of view.

THE BAR COMPONENT IN DISK GALAXIES

A. Bosma
Observatoire de Marseille
Marseille, France

ABSTRACT. I will first review the recent work on the kinematics of barred disk galaxies, as traced by the gas (HI, Hα, and CO) and by the stars. Then I will consider the properties of the bar, as can be inferred from studies of sub-components like rings, dust lanes, spiral arms, etc., and discuss the problem of lenses. Recent work points to the possibility of complex patterns involving the coupling of several modes of different radial extent. Finally, I will touch briefly on the question of bars as drivers of secular evolution in disk galaxies.

1 Introduction

The bar component in disk galaxies is relatively ubiquitous. About one-third of the spirals are known to be strongly barred, while another third are classified as intermediate between ordinary and barred spirals, which means that they have weak bars or oval distortions. Even some of the ordinary spirals could be barred, as witnessed by NGC 1068, classified as RSA(rs)b in the Second Reference Catalogue (De Vaucouleurs et al., 1976), but for which recent observations in the near-infrared clearly show the presence of a strong bar in the central parts (e.g. Thronson et al., 1989).

Although it is sometimes thought that the multitudes of forms frequently associated with bars, such as inner rings, outer rings, dust lanes, spiral arms, plumes, etc., are just different sorts of "icing on the cake", it becomes increasingly clear that bars play an important role in the dynamical evolution of disk galaxies, and that the "icings" are closely connected with the exact state of the cake itself. Understanding these forms may thus help us to assess the state of evolution of disk galaxies in general.

2 Kinematics of barred disk galaxies

In the last few years a lot of new information on the kinematical properties of barred spirals and SB0's has become available. Due to the nature of bars, however, the information given by each particular tracer, be it neutral hydrogen, ionized hydrogen, the CO molecule, or the stars, is only very partial, and difficult to treat in a quantitative framework without a large effort in modelling. The qualitative signature of streaming motions associated with bars or oval distortions is well understood, however, and is readily recognizable in particular in two-dimensional velocity fields.

G. Longo et al. (eds.), Morphological and Physical Classification of Galaxies, 207–220.

2.1 21-cm line studies

The advent of the VLA and its spectral line capability has made it possible to overcome the circumstance that few northern barred spirals were suitable for detailed study in the 21-cm line, and the pioneering study of NGC 5383 with the Westerbork telescope (Sancisi et al., 1979) has now been followed up by several others. M83, NGC 1365, and NGC 1097 were studied by Ondrechen and Van der Hulst (1985, 1989a,b), and NGC 3992, NGC 4731, NGC 3359, NGC 1300, and NGC 1073 by the Florida group (Gottesman et al., 1984, Ball, 1986, Hunter et al., 1988; England, 1989, 1990; England et al., 1990, see also Contopoulos et al., 1989).

HI studies of oval distortions (term generally used in case no bar is visible, but a clear cos 2θ distortion is present), started by the work on NGC 4151 and NGC 4736 (Bosma et al., 1977a,b), drew the attention to the presence of outer rings and pseudo- rings, and the recent work by Van Driel et al. (1989) on the SB0/a galaxies NGC 1291 and NGC 5101 clearly show the presence of a large amount of HI in the outer rings around these galaxies.

The general results of all these studies can be summarized as follows : 1) the velocity field shows general rotation on which characteristic deviations from circular motions are superimposed. This characteristic shows itself as a skewing of the isovelocity contours around the systemic velocity towards the bar, and a corresponding S-shape distortion in the major axis. Contrary to kinematical warps, there is general misalignment between the kinematical and photometric axes (Bosma, 1978, 1981). 2) The rotation curves remain flat in the outer parts, similar to those of ordinary spiral galaxies, so that in the usual interpretation dark haloes are necessary. 3) The HI is strongly concentrated in the arms or rings, and for later types like NGC 4731 or NGC 1073 also in the bar itself.

2.2 Hα-emission

The early work with long slit spectra was hampered by the fact that the Hα-emission is very inhomogeneously distributed. In fact, most of the bars in earlier type spirals do not show any emission in the bar. Only in later types it is possible to obtain coverage of the bar region. In this respect, the work on NGC 5383 (Peterson et al., 1978) was most helpful for further development of studies of the gas flow in barred spirals, and several detailed models have been produced for this galaxy (e.g. Huntley, 1978; Sanders and Tubbs, 1980; Duval and Athanassoula, 1983).

The full benefit of Fabry-Perot interferometry in producing two-dimensional velocity fields has been used by Teuben et al. (1986) to study NGC 1365. Data for the nuclear region at high resolution (5") and data for most of the galaxy at lower resolution (necessary to get a better signal-to-noise ratio) clearly indicate that an inner Lindblad resonance is present in this galaxy, thus helping to settle an important discrepancy between the earlier gas flow studies, and newer results using a higher order scheme with lower viscosity (see below). Other galaxies studied in this manner include NGC 253 (Pence, 1981), NGC 4027 (Pence et al., 1988), and NGC 5728 (Schommer et al., 1988).

2.3 CO studies

Currently the larger mm-telescopes are capable of high resolution and sensitivity in the main molecular lines due to CO, and the present generation of mm-interferometers have been used to map the central regions of already quite a few galaxies. Two results stand out from these studies. First, the detection of CO in the dust lanes of M83 (Handa et al., 1990) and NGC 1530 (Downes et al., 1990). It can be expected that sensitive mm-interferometer measurements can give us considerable detail on the gas flow across the dust lanes in these galaxies. Second, a number of ordinary spiral galaxies show in their central parts the typical velocity deviations expected for a bar, and in some of the interferometer maps the CO itself has a barlike distribution. The presence of these bars is difficult to discern on optical photographs, however. Good examples are NGC 6946 (e.g. Ball et al., 1985; Sofue et al., 1988, see also the Hα-study by Bonnarel et al., 1988), IC 342 (e.g. Ishizuki et al., 1990), and Maffei II (Weliachew et al., 1988).

2.4 Stellar velocity data

Absorption line spectroscopy of the stellar component of barred galaxies is difficult, and up till now only the higher surface brightness galaxies with little gas and dust, i.e. the SB0's, have been studied. The best studied case is NGC 936 (Kormendy, 1982a, 1983; Kent, 1987; Kent and Glaudell, 1989), but recent work by Bettoni, Galletta and their collaborators have permitted to extend the conclusions about this galaxy to several other cases (e.g. Bettoni, 1989 and references therein). The data are consistent with streaming motions along elliptical orbits elongated parallel to the bar. The velocity dispersions are relatively high in the bar, but in general in reasonable agreement with what is expected from some N-body models. The ratio of bulge velocity dispersion to bulge rotation is lower in bulges of barred galaxies than in bulges of normal galaxies (Kormendy, 1982b), suggesting that perhaps disk material has been used in the build-up of bulges in barred galaxies (see also below). Slight asymmetries exist in the rotation data : the symmetry center of the position-velocity curves along the bar in several galaxies does not coincide exactly with the bulge center.

In several SB0's also emission lines due to gas have been studied, and in two cases, NGC 4546 (Galletta, 1987) and NGC 2217 (Bettoni et al., 1990), the gas appears to be counterrotating with respect to the stars. This has been taken to argue that perhaps this gas originated from a captured dwarf galaxy.

The only study of velocity dispersions in late type barred spirals is the one by Bottema (1990) of NGC 3059. For this face-on SBc galaxy he finds a nearly constant velocity dispersion in the bar, of order 50 km s^{-1}. This value is significantly lower than the values of order 140 km s^{-1} found in the inner parts of the bars of several face-on SB0's by Jarvis et al. (1988).

3 Properties of the bar

The most fruitful way to model the bar of a barred galaxy is to consider it as a density wave, rotating with a certain pattern speed Ω_p, in an otherwise axisymmetric disk. Such a configuration arises naturally as end result of unstable disks in N-body simulations. A lot of structure can occur in such a configuration, and resonances can be set up naturally.

3.1 Rings and Resonances

Statistics of size ratios of rings have led to the recognition that they are associated with resonances : the outer ring with the outer Lindblad resonance (OLR), and the inner ring with the ultraharmonic resonance just inside corotation (CR) (Athanassoula et al. 1982). Already Kormendy (1979) had found a typical ratio of outer ring size to bar size of 2.2, and the statistics of the ratio of outer ring size to inner ring size shows again the same preferred ratio. Data on new samples of galaxies by Schwarz (1984c) and in particular Buta (1986a) showed further that the statistics of the angles between the bar and the major axis of the inner ring favours the idea that inner rings are elongated along the bar, with axial ratios between 0.6 and 1.0, and that there are two types of outer rings, one family orientated perpendicular to the bar, and one family parallel to the bar. Nuclear rings are found as well, which in general can be associated with the inner Lindblad resonance (ILR).

Theoretical underpinning of the statistics on rings comes especially from the work of Schwarz (1979, 1981, 1984ab, 1985ab), who used a sticky particle code to simulate the behaviour of gas in a disk with an oval distortion. In general the gas distribution quickly evolves from a spiral form to a distribution in rings, which are located at the principal resonances. In a way this evolution is too fast, since there are barred galaxies without rings. Presumably this is related to the way the gas is treated (Athanassoula, 1983; Van Albada, 1985).

Further case studied of barred ringed galaxies have been carried out by Buta in a series of papers and summarized in Buta (1990 and references therein). In a first series of papers he concentrated on rings in barred spirals, and found them to be sites of star formation. For NGC 1433 the inner ring shows an azimuthal colour variation. Most of these galaxies show evidence for non-circular motions in the bar region. In a second series of papers he concentrated on weakly barred ring galaxies, the properties of which are perhaps less readily explained by current theories. For NGC 3081 the inner ring is very strong and elongated, but the bar seems very weak. Peculiarities exist also for NGC 7187 and NGC 7020, the latter one having a ring with a hexagonal shape.

3.2 Orbits

The study of orbits in barred spiral potentials can tell us about the main families of periodic orbits which exist in them, and around which a lot of matter can be trapped. Much work on this has been done by Contopoulos and his collaborators (e.g. Contopoulos and Papayannopoulos, 1980), also Van Albada and Sanders (1982), Athanassoula et al. (1983), Teuben and Sanders (1985), Pfenniger (1984), etc. The main families of periodic orbits are the x_1-family, which supports the bar and extends roughly out to corotation, an x_2-family which exists inside the ILR (if the pattern speed is low enough for it to exist), the x'_1-family outside corotation, which can be unstable thereby resulting in the depopulation of the orbits between corotation and the OLR, and several others of lesser importance. Since the x_1-family is the only one capable of supporting the bar, its presence is a necessary requisite in any realistic self-consistent bar potential. E.g. for increasing axial ratio a/b of certain prolate shapes the importance of quasi-periodic orbits trapped around the x_1-family diminishes in favour of stochastic orbits; this puts an upper limit to the axial ratio in real bars (Athanassoula et al., 1983)

3.3 Dust lanes

The association of dust lanes with shocks in the gas flow was made in the early sixties by Prendergast, but only in the late seventies the first attempts could be made to produce models realistic enough to compare with real observations. Even the beam scheme models from Sanders and Tubbs (1980), designed to fit the data on NGC 5383, suffered from some problems : though it was conjectured that an ILR was necessary to form shocks which are offset in the leading direction with respect to the bar, a subsequent orbit analysis by Van Albada and Sanders (1982) did not find the x_2-family of orbits in the corresponding potential. Later work using a new second-order scheme with low numerical viscosity hinted that the presence of an ILR might be necessary to get nice looking dust lanes (Van Albada, 1985; Athanassoula, 1985).

An extensive study of the conditions under which realistic shocks identifiable with dust lanes are formed has been made by Athanassoula (1985, 1989, 1990b). The main results of her study are that the presence of an x_2-family is not only necessary, but that to get the offset of the correct amplitude in the case of straight dust lanes, the extent of the family should be about the size of the minor axis of the bar. Another result is that the Lagrangian radius (corotation radius) has to be roughly 1.2 times the size of the bar major axis in order to get the most realistic shapes for the shock region. Variation of the axial ratio of the bar allows curved dust lanes in the case of more oval shapes.

Observational evidence for the association of dust lanes with shock regions comes from three different tracers : 1) long slit spectroscopy along directions perpendicular to the dust lanes show large velocity jumps when the dust lane is crossed. This effect has been seen in data of NGC 6221 (Pence and Blackman, 1982) and NGC 1365 (Lindblad and Jörsäter, 1988). In fact, these high velocities may cause the colliding clouds to disperse rather than collapse, thus explaining the very inhomogeneous distribution of Hα-emission (Tubbs, 1982). 2) Radio continuum emission is seen to be clearly enhanced for the dust lanes in M83 and NGC 1097 (Ondrechen and Van der Hulst, 1984; Ondrechen, 1985). 3) There is now direct evidence for the concentration of molecular material (CO) in dust lanes (section 1.3).

3.4 Mass of the bar

The mass of the bar can, in principle, be obtained from surface photometry, provided reasonable assumptions can be made for the behaviour of the axisymmetric components and for the mass-to-light ratios. Several studies have been made in order to do this, but the results are not very satisfactory. It is generally found that the bar presents roughly 10 - 30% of the disk mass, and a smaller amount for later types (e.g. Ann and Lee, 1988).

Elmegreen and Elmegreen (1985) made a study of 34 barred spirals, and concluded that the form of the radial distribution of light along the bar differs for early types vs.

late types : in early types the light profile does not decrease very rapidly along the bar while in later types (where the bar size is usually smaller with respect to the disk size) the light profile follows usually an exponential drop-off similar to or even steeper compared to the disk profile.

3.5 Lenses

A lens is a component in disk galaxies which has a shallow surface brightness gradient and a steep edge. The role of lenses is not very well understood, and at least five theories can be found in the literature.

1. Kormendy (1979) notes that in 17 out of 20 cases, the bar fills a lens in one dimension and suggests that bars may evolve into lenses within a Hubble time.
2. Athanassoula (1983) argues on the basis on N-body simulations that lenses may result from instabilities of galactic disks. The initial hotter parts of the disks would evolve to a lens, just as the bar results from the colder parts of an unstable disk.
3. Teuben and Sanders (1985) suggest that the lens part of a barred spiral consists of stars on stochastic orbits, while the bar part is formed by stars belonging to the x_1-family.
4. Buta (1990) suggests that lenses and inner rings are the same phenomena, but of different amplitude. I might add here that since most inner rings are sharp features with signs of recent star formation, they may evolve into a more diffuse component due to orbital diffusion. A similar evolution has been envisaged for outer rings (cf. discussion after Athanassoula, 1983). Note that Kormendy (1979) already found that lenses are more frequent in galaxies of type SB0-a and inner rings more frequent in types SBa-bc.
5. Bosma (1983) notes that lenses also exist in non-barred galaxies, and that their edges are like the sharp edges found in disks by Van der Kruit and Searle (1982). So lenses could be a primary component, with the outer disk added later.

Clearly further work is possible to sort out these various ideas (some of which are not mutually exclusive).

3.6 Shapes

Most of the theoretical work described above has been done assuming simple analytical forms for the potential of the bar, either of the form of a 2θ-perturbation, or of a prolate ellipsoid. However, it has been noted often that the geometric shapes of some bars are rather rectangular, more than ellipsoidal. This has been recently quantified by Athanassoula et al. (1990), in which they generalize an ellipse to obey the equation

$$\left(\frac{|x|}{a}\right)^c + \left(\frac{|y|}{b}\right)^c = 1$$

where c is a shape parameter. For $c = 2$ the shape is ellipsoidal, for $c < 2$ the shape is "disky", and for $c > 2$ the shape is "boxy". This method allows isophote shapes to be fitted exactly, if they have the form of a generalized ellipse (contrary to what is done in determining the a_4-parameter in elliptical galaxy work). For a sample of a dozen SB0-galaxies Athanassoula et al. find that the parameter c varies between 2.5 - 5.5 in the region of the bar. Ohta et al. (1990) for a sample of six later type barred spirals try to determine the bar isophotes by subtracting the axisymmetric (bulge and disk) components, and also conclude that bars are rectangular.

3.7 Potentials

Instead of assuming a convenient analytical function for the potential of a bar, the modern trend is now to take the observations as a starting point for its derivation. Kent and Glaudell (1989) and Kent (1990) studied the bars in NGC 936 and NGC 4596, resp. However, they decompose the galaxy into a bulge part, a disk part, and a bar part, and fit functions to each of them, which are then used to determine the potential. Since the decomposition process is difficult, and rather arbitrary, it is not clear anymore whether the thus determined potentials are realistic.

Athanassoula and Wozniak (1990) use another method, by doing a double Fourier transform on the rectified image of a galaxy (assuming a constant mass-to-light ratio) and solving Poisson's equation in Fourier space. The resulting potentials for a sample of SB0-galaxies may not resemble in detail the potential of the models which until recently were used for studies of orbits and of the gas flow. Orbit calculations for NGC 936 show a wealth of detail (Wozniak and Athanassoula, 1990). This method seems promising and might shed some more light on the question of the precise manner in which the onset of stochasticity occurs (cf. discussion in Athanassoula, 1990a).

3.8 Pattern speeds and multiple patterns

Bar pattern speeds can be determined from the identification of resonances : an outer ring can be assumed to be at the OLR, an inner ring just inside CR, and a nuclear ring at the outer ILR. Knowledge of the conditions of shock formation might help to identify the position of the outer ILR from the change in curvature of the dust lanes, or the position of CR from the end of the bar. Orbit studies tend to converge on the statement that bars end roughly at corotation.

An elegant method independent of the identification of the resonance positions has been developed by Tremaine and Weinberg (1986). If it can be assumed that the disk is flat, and that the bar has a well defined pattern speed, and if the surface brightness of the tracer obeys the continuity equation, the bar pattern speed can be derived from photometry and kinematical data. A good tracer are e.g. the stars in an SB0-galaxy. Kent (1987) tried this for the galaxy NGC 936, but found a rather large error margin. The method seems to work well for N-body results.

Multiple patterns are possible, and are seen in some of the N-body simulations. Tagger et al. (1987) and Sellwood and Sparke (1988) show that in Sellwoods simulations of the bar-instability in a cold disc the outer spiral can have a lower pattern speed than the bar. In fact, there is a mode coupling going on, in which the corotation of the bar overlaps with the ILR of the spiral. A further analytical development of this has been presented by Sygnet et al. (1988). In this spirit, it is thus not surprising if one assumes that the bar ends roughly at corotation and yet finds that the dust lanes are still located on the concave side of the outer arms.

More complicated galaxies can be understood in this sense as well. NGC 1068 is a good example (Baldwin et al., 1987). The main body of the galaxy is an oval surrounded by an outer ring roughly twice as large. In the inner parts the recent 1.6 and 2.2 μm data by Thronson et al. (1989) show a hidden bar, which extends only to about 16" to either side of the nucleus. It is tempting to identify this extent as the corotation radius of the bar, and as the radius of the outer ILR of the oval. NGC 4736 and NGC 1566 could be explained in a similar manner. Finally, there are several SB0's with bars within bars, NGC 1291 being the best known example (de Vaucouleurs, 1975, see also Jarvis et al., 1988).

Late type barred spirals are often seen to be asymmetric (e.g. de Vaucouleurs and Freeman, 1972), with the bar displaced from the center of rotation of the disk. This displacement is much larger than the displacements seen in the stellar rotation data of SB0-galaxies reported by Bettoni (1989). A prototypical case is NGC 4027, which is well studied photometrically and kinematically (Pence et al., 1988). A recent attempt to interpret such galaxies has been made by Tagger and Athanassoula (1990), in terms of a coupling between m = 1 and m = 2 spirals.

4 Bar as driver of secular evolution

The possibility of secular evolution in disk galaxies was strongly advocated by Kormendy at various occasions (e.g. Kormendy, 1979, 1982), and he identified the bar as one of its main agents. Various arguments can be marshaled in favour of this : the bar instability in disk galaxies in general, the stimulation of bar formation during interactions, the gas transport towards the central regions seen in gas flow studies, the possibility of building up a bulge from disk material suggested by some studies, and the frequent association of bars with starbursts and active galactic nuclei. I will only briefly touch upon the main points concerning this topic.

4.1 Bar instability

The bar instability in disk galaxies has been widely studied by N-body techniques, and is now well understood. The respective role of the stabilizing nature of the dark halo (originally proposed by Ostriker and Peebles, 1973), and the velocity dispersions in the disk have been sorted out by Athanassoula and Sellwood (1986). Once the instability is fully developed, the end product is a realistic bar, as shown by the dissection of an N-body bar by Sparke and Sellwood (1987). Even the squarish character of the bars seen in SB0's is well reproduced (Athanassoula et al., 1990). Usually there is a decrease in pattern speed after the bar has developed, but otherwise stellar N-body bars are very robust, and do not disappear even after many rotation periods (Carnevali, 1983).

The bar instability can be triggered and/or amplified by an interaction. This was first shown by Noguchi (1987), and extensively studied by Gerin et al. (1990). In fact, further work by Athanassoula (1990c) seems to allow for the identification of definite criteria to predict when the result of an interaction is a spiral, and when it is a bar.

4.2 Buildup of the central parts

The gas flow models clearly indicate that gas can be transported to the central parts : in both particle codes and fluid codes this occurs, and in fact some "tricks" are build n to avoid too much build-up, by assuming that star formation occurs at high density peaks (thus removing gas), and that stars loose mass (thus producing gas more or less uniformly over the galaxy). This possibility has stimulated a lot of research into linking statistically the occurrence of a bar with the presence of nuclear activity. Already Sersic and Pastoriza (1965) noted that hot spot nuclei occur more frequently in SAB or SB galaxies. Simkin et al. (1981) clearly linked the appearance of some Seyfert galaxies to the ring models produced by Schwarz (1979). Recently, Arsenault (1989) again tries to show that a sample of starburst galaxies are more frequently associated with bar and rings than a control sample of normal galaxies.

Further support for this comes from studies of the central parts of some galaxies. A good case is NGC 1097, where the hot spots are distributed in a nuclear ring, with strong non-thermal emission (e.g. Hummel et al., 1989) and CO-emission (Gerin et al., 1989). Statistical support of this idea comes from analysis of IRAS data (Hawarden et al., 1986) and subsequent radio-continuum work (Puxley et al., 1987), where again it was argued that bars enhance the star formation, in particular in the central parts, and are somehow important in understanding starbursts. However, a recent study by Pompea and Rieke (1990), specifically designed to find in the 2.2μ K-band the possible bars hidden at optical wavelengths, shows that bars cannot be the only cause of enhanced far infrared emission.

An early 3D N-body simulation by Combes and Sanders (1981) showed quite clearly the formation of 'box' or 'peanut' shaped bulges as an end product of a bar instability. Theoretical work by Pfenniger (1984) indicates that the vertical resonances are important in 3D-bars, and can help deflect stars in planar orbits up into orbits away from the plane. Recent simulations by Combes et al. (1990) seem to confirm this : it seems quite natural that the bar instability first forms in the plane of the galaxy, and after a while the vertical instabilities cause an evolution of the bar shape towards a fat boxy or peanut shaped object, which seen face-on would be called a bulge. They argue that the current statistics of the frequency of box or peanut bulges does not contradict their results.

The time lag seen between the development of the bar instability to observable levels and the development of the bulge is important to circumvent the contradiction with the thought that bars are thin. This latter belief is based on arguments concerning the structure of the edge-on galaxies NGC 4762 (Wakamatsu and Hamabe, 1984) and NGC 4452 (Hamabe and Wakamatsu, 1985), where from the stepwise radial luminosity profile and its behaviour in the vertical direction it is inferred that these galaxies have a bulge, a bar, a lens, and an outer disk. The bar is very thin indeed : for NGC 4762 a thickness of 155 pc is estimated, and an equivalent velocity dispersion in the z-direction of only 19 km/sec; for the lens a thickness of 500 pc and a z-velocity dispersion of 42 km/sec is proposed. It should be possible to check the proposed thinness of the bar by measuring velocity dispersions in face-on galaxies. Note that the hidden bars in some galaxies, which show up only in the near infrared pictures (e.g. NGC 1068), cannot be too thick or of peanut shape, since otherwise we should have seen them in the optical images already as a sort of triaxial bulge.

The link between the effect of bar dissipation and bulge formation and subsequent activation of an active galactic nucleus has been pushed further in recent studies of Hassan and Norman (1990) and Pfenniger and Norman (1990). In the first study, the effect of introducing a more and more massive black hole in the center on the behaviour of the main families of periodic orbits has been found to destroy gradually the importance of the x_1-family sustaining the bar. In the second study, weakly dissipative orbits, representing the behaviour of gas clouds, are computed, in a barred galaxy model with a central mass concentration such as a black hole or a secondary inner bar. Rapid accretion across the ILR is found, and large excursions outside the plane are possible there, due to the presence of broad horizontal and vertical resonances. Thus the growth of a central mass concentration in a barred galaxy might induce resonance regions which could enhance the fueling rate of AGN's and starbursts. Due to the complexity of the problem, neither study is yet self-consistent. It may well be that there is a regulation mechanism which limits the growth of the bulge due to the bar action, in such a way as to assure the correlation between bar length and bulge size already discussed by Athanassoula and Martinet (1980).

To drive home the possibility of secular evolution once more, it is of interest to consider the fate of M51. The recent data on the interaction (Rots et al., 1990), showing a huge HI tail, are now opening up the possibility that the spiral structure in NGC 5194 was stimulated during the passage. Yet in the infrared the main disk of NGC 5194 is already distinctly barred (McLean and Wright, 1988), which could also be due to the interaction. If the companion escapes or disintegrates before it has time to merge with NGC 5194, the process of bulge building advocated by Combes et al. (1990) and Pfenniger and Norman (1990) might do its job, and turn this Sc-galaxy into an Sb-galaxy. Alternatively, if the companion is captured, its central parts might be dense enough to sink all the way into the center, and, if the disk survives, the end result will be similar.

5 Concluding remarks

The study of bars in disk galaxies has already shown that a lot of morphological details traditionally translated in some of the classification schemes can be understood in terms of components, both primary like the bar (and may be the lens), and secondary such as rings, (maybe lenses), dust lanes, spiral arms, plumes, etc. Most of the understanding is necessarily qualitative, but the tools are now sharpened to get quantitative fits to the observational data. The intriguing effects of secular evolution due to a bar will be a subject of fruitful study in the coming years, in particular since simulations with three-dimensional codes will become readily feasible. Further work in this area can be expected, and will hopefully lead to predictions which can be tested observationally.

Acknowledgements

Thanks are due to Lia Athanassoula, not only for ten years of lively discussions which certainly increased my knowledge of bars, but also for allowing me to talk about her work in advance of publication.

References

Ann, H.B, Lee, S.W.: 1987, *Journal of Korean Astron. Soc.* **20**, 49.

Arsenault, R.: 1989, *Astron. Astrophys.* **217**, 66.

Athanassoula, E.: 1983, In *Internal Kinematics and Dynamics of Galaxies, I.A.U. Symp. 100* (E. Athanassoula, Ed.), 145, Reidel (Dordrecht)

Athanassoula, E.: 1985, In *New Aspects of Galaxy Photometry* (J.L. Nieto, Ed.), 155, Springer Verlag, Berlin.

Athanassoula, E.: 1989, In *Proceedings of the joint Varenna-Abastumani International School and Workshop on Plasma Astrophysics*, **1**, 341.

Athanassoula, E.: 1990a, In *Galactic Models, Annals of the New York Academy of Sciences*, **596**, (J.R. Buchler, S.T. Gottesman, and J.H. Hunter, Eds.), 181.

Athanassoula, E.: 1990b (in preparation)

Athanassoula, E.: 1990c (in preparation)

Athanassoula, E., Bosma, A., Crézé, M., Schwarz, M.P.: 1982, *Astron. Astrophys.* **107**, 101.

Athanassoula, E., Bienaymé, O., Martinet, L., Pfenniger, D.: 1983, *Astron. Astrophys.* **127**, 349.

Athanassoula, E., Martinet, L.: 1980, *Astron. Astrophys.* **87**, L10.

Athanassoula, E., Morin, S., Wozniak, H., Puy, D., Pierce, M.J, Lombard, J., Bosma, A.: 1990, *Mon. Not. R. Astr. Soc.* **245**, 130.

Athanassoula, E., Sellwood, J.A.: 1986, *Mon. Not. R. Astr. Soc.* **221**, 213.

Athanassoula, E., Wozniak, H.: 1990, (in preparation).

Baldwin, J.A, Wilson, A.S., Whittle, M.: 1987, *Astrophys. J.* **319**, 84.

Ball, R.J.: 1986, *Astrophys. J.* **307**, 453.

Ball, R., Sargent, A.I., Scoville, N.Z., Lo, K.Y., Scoot, S.L.: 1985, *Astrophys. J.* **298**, L21.

Bettoni, D.: 1989, *Astron. J.* **97**, 79.

Bettoni, D., Fasano, G., Galletta, G.: 1990, *Astron. J.* **99**, 1789.

Bonnarel, F., Boulesteix, J., Georgelin, Y.P., Le Coarer, E., Marcelin, M., Bacon, R., Monnet, G.: 1988, *Astron. Astrophys.* **189**, 59.

Bosma, A.: 1978, *Ph.D. Thesis, University of Groningen.*

Bosma, A.: 1981, *Astron. J.* **86**, 1825.

Bosma, A.: 1983, In *Internal Kinematics and Dynamics of Galaxies, I.A.U. Symposium No. 100*, (E. Athanassoula, Ed.), 253, Reidel, Dordrecht.

Bosma, A., Ekers, R.D., Lequeux, J.: 1977a, *Astron. Astrophys.* **57**, 97.

Bosma, A., Van der Hulst, J.M., Sullivan, W.T.: 1977b, *Astron. Astrophys.* **57**, 373.

Bottema, R.: 1990, *Astron. Astrophys.* **233**, 372.

Buta, R.J.: 1986a, *Astrophys. J. Supp. Ser.* **61**, 609.

Buta, R.J.: 1990d, In *Galactic Models, Annals of the New York Academy of Sciences*, **596**, (J.R. Buchler, S.T. Gottesman, and J.H. Hunter, Eds.), 58.

Carnevali, P.: 1983, *Astrophys. J.* **265**, 701.

Combes, F., Debbasch, F., Friedli, D, Pfenniger, D.: 1990, *Astron. Astrophys.* **233**, 82.

Combes, F., Sanders, R.H.: 1981, *Astron. Astrophys.* **96**, 164.

Contopoulos, G., Papayannopoulos, Th.: 1980, *Astron. Astrophys.* **92**, 33.

Contopoulos, G., Gottesman, S.T., Hunter, J.H., England, M.N.: 1989, *Astrophys. J.* **343**, 608.

de Vaucouleurs, G., Freeman, K.C.: 1972, *Vistas in Astronomy* **14**, 163.

de Vaucouleurs, G.: 1975, *Astrophys. J. Supp. Ser.* **29**: 193.

de Vaucouleurs, G., de Vaucouleurs, A., Corwin, H.G.: 1976, *Second Reference Catalogue of Bright Galaxies*, Univ. of Texas Press (Austin).

Downes, D., Solomon, P.M., Radford, S.: 1990, (in preparation).

Duval, M.F, Athanassoula, E.: 1983, *Astron. Astrophys.* **121**, 297.

Elmegreen, B.G., Elmegreen, D.M.: 1985, *Astrophys. J.* **288**, 438.

England, M.N.: 1989, *Astrophys. J.* **337**: 191.

England, M.N.: 1990, *Astrophys. J.* **344**, 669.

England, M.N., Gottesman, S.T., Hunter, J.H.: 1990, *Astrophys. J.* **348**, 456.

Galletta, G.: 1987, *Astrophys. J.* **318**, 531.

Gerin, M., Combes, F., Athanassoula, E.: 1990, *Astron. Astrophys.* **230**, 37.

Gerin, M., Nakai, N., Combes, F.: 1988, *Astron. Astrophys.* **203**, 44.

Gottesman, S.T., Ball, R., Hunter, J.H., Huntley, J.M.: 1984, *Astrophys. J.* **286**, 471.

Hamabe, M., Wakamatsu, K.: 1985, *Astrophys. J.* **339**, 783.

Handa, T., Nakai, N., Sofue, Y., Hayashi, M., Fujimoto, M.: 1990, *Publ. Astr. Soc. Japan* **42**, 1.

Hassan, H., Norman, C.: 1990, *Astrophys. J.* **361**, 69.

Hawarden, T.G, Mountain, C.M., Leggett, S.K., Puxley, P.J.: 1986, *Mon. Not. R. Astr. Soc.* **221**, 41P.

Hummel, E., van der Hulst, J.M., Keel, W.C.: 1987, *Astron. Astrophys.* **172**, 32.

Hunter, J.H., Ball, R., Huntley, J.M., England, M.N., Gottesman, S.T., 1988, *Astrophys. J.* **324**, 721.

Huntley, J.M.: 1978, *Astrophys. J.* **225**, L101.

Ishizuki, S., Kawabe, R., Ishiguro, M., Okamura, S.K., Morita, K.I., Chikada, Y., Kasuga, T.: 1990, *Nature* **344**, 224.

Jarvis, B., Dubath, P., Martinet, L., Bacon, R.: 1989, *Astron. Astrophys. Suppl. Ser.* **79**, 513.

Kent, S.M.: 1987, *Astron. J.* **93**, 1062.

Kent, S.M.: 1990, *Astron. J.* **100**, 377.

Kent, S.M., Glaudell, G.: 1989, *Astron. J.* **98**, 1588.

Kormendy, J.: 1979, *Astrophys. J.* **227**, 714.

Kormendy, J.: 1982a, *Astrophys. J.* **275**, 529.

Kormendy, J.: 1982b, *Astrophys. J.* **275**, 57.

Kormendy, J.: 1982c, In *Morphology and Dynamics of Galaxies*, (L. Martinet and M. Mayor, eds.), 115, Geneva Observatory.

Kormendy, J.: 1984, *Astrophys. J.* **286**, 132.

Lindblad, P.O., Jörsäter, S.: 1988, In *Evolution of Galaxies, Proc. of a meeting at the 10th European regional I.A.U.*, (J. Palous, Ed.), 289.

Noguchi, M.: 1987, *Mon. Not. R. Astr. Soc.* **228**, 635.

Ohta, K., Hamabe, M., Wakamatsu, K.: 1990, *Astrophys. J.* **357**, 91.

Ondrechen, M.P.: 1985, Ph.D. Thesis, Univ. of Minnesota.

Ondrechen, M.P.: 1985, *Astrophys. J.* **90**, 1474.

Ondrechen, M.P., Van der Hulst, J.M.: 1983, *Astrophys. J.* **269**, L47.

Ondrechen, M.P., Van der Hulst, J.M.: 1989, *Astrophys. J.* **324**, 29.

Ondrechen, M.P., Van der Hulst, J.M., Hummel, E., *Astrophys. J.* **342**, 39.

Ostriker, J.P, Peebles, P.J.E.: 1973, *Astrophys. J.* **186**, 467.

Pence, W.D.: 1981, *Astrophys. J.* **247**, 473.

Pence, W.D., Blackman, C.P.: 1984, *Mon. Not. R. Astr. Soc.* **207**, 9.

Pence, W.D., Taylor, K., Freeman, K.C., de Vaucouleurs, G., Atherton, P., 1988, *Astrophys. J.* **326**, 564.

Pence, W.D., Taylor, K., Atherton, P.D., *Astrophys. J.* **357**, 415.

Peterson, C.J., Rubin, V.C., Ford, W.K., Thonnard, N.: 1978, *Astrophys. J.* **219**, 31.

Pfenniger, D.: 1984, *Astron. Astrophys.* **134**, 373.

Pfenniger, D., Norman, C.: 1990, *Astrophys. J.* **363**, 391.

Pompea, S.M., Rieke, G.: 1990, *Astrophys. J.* **356**, 416.

Puxley, P.J., Hawarden, T.G, Mountain, C.M.: 1987, *Mon. Not. R. Astr. Soc.* **231**, 465.

Rots, A.H., Bosma, A., Van der Hulst, J.M., Athanassoula, E., Crane, P.C., 1990, *Astron. J.* **100**, 387.

Sancisi, R., Allen, R.J., Sullivan, W.T.: 1979, *Astron. Astrophys.* **78**, 217.

Sanders, R.H., Tubbs, A.D.: 1980, *Astrophys. J.* **235**, 803.

Schommer, R.A., Caldwell, N., Wilson, A.S., Baldwin, J.A., Phillips, M.M., Williams, T.B., Turtle, A.J.: 1988, *Astrophys. J.* **324**, 154.

Schwarz, M.P., *Ph.D. Thesis, Australian National University.*

Schwarz, M.P.: 1981, *Astrophys. J.* **247**, 77.

Schwarz, M.P.: 1984a, *Mon. Not. R. Astr. Soc.* **209**, 93.

Schwarz, M.P.: 1984b, *Proceed. Austr. Soc. Astr.*, **5**, 464.

Schwarz, M.P.: 1984c, *Astron. Astrophys.* **133**, 222.

Schwarz, M.P.: 1985a, *Mon. Not. R. Astr. Soc.* **212**, 677.

Schwarz, M.P.: 1985b, **Proc. Austr. Soc. Astr.**, **6**, 202.

Sellwood, J.A., Sparke, L.S.: 1988, *Mon. Not. R. Astr. Soc.* **231**, 25P.

Sersic, J.L., *Publ. Astron. Soc. Pac.* **85**, 103.

Sersic, J.L., Pastoriza, M.: 1965, *Publ. Astron. Soc. Pac.* **77**, 287.

Sersic, J.L., Pastoriza, M.: 1967, *Publ. Astron. Soc. Pac.* **79**, 152.

Simkin, S.M., Su, H.J., Schwarz, M.P.: 1981, *Astrophys. J.* **237**, 404.

Sofue, Y., Doi, M., Ishizuki, S., Nakai, N., Handa, T.: 1988, *Publ. Astr. Soc. Japan*, **40**, 511.

Sparke, L.S., Sellwood, J.A.: 1987, *Mon. Not. R. Astr. Soc.* **225**, 653.

Sygnet, J.F., Tagger, M., Athanassoula, E., Pellat, R.: 1988, *Mon. Not. R. Astr. Soc.* **232**, 733.

Tagger, M., Athanassoula, E.: 1990, In *Dynamics of Galaxies and Molecular Clouds Distribution*, (F. Casoli et al., ed.) (in press).

Tagger, M., Sygnet, J.F., Athanassoula, E., Pellat, R.: 1987, *Astrophys. J.* **318**, L43.

Teuben, P.J., Sanders, R.H: 1985, *Mon. Not. R. Astr. Soc.* **212**, 257.

Teuben, P.J., Sanders, R.H., Atherton, P.D., Van Albada, G.D.: 1986, *Mon. Not. R. Astr. Soc.* **221**, 1.

Thronson, H.A., Hereld, M., Majevski, S., Greenhouse, M., Johnson, P., Spillar, E., Woodward, C.E., Harper, D.A., Rauscher, B.J.: 1989, *Astrophys. J.* **343**, 158.

Tremaine, S., Weinberg, M.D.: 1984, *Astrophys. J.* **282**, L5.

Tubbs, A.D.: 1982, *Astrophys. J.* **255**, 458.

Van Albada, G.D.: 1985, *Astron. Astrophys.* **142**, 491.

Van Albada, T.S., Sanders, R.H.: 1982, *Mon. Not. R. Astr. Soc.* **201**, 303.

Van der Kruit, P.C.: 1976, *Astron. Astrophys.* **52**, 85.

Van der Kruit, P.C., Searle, L.T.: 1982, *Astron. Astrophys.* **110**, 61.

Van Driel, W., Rots, A.H., Van Woerden, H.: 1989, *Astron. Astrophys.* **231**, 465.

Weliachew, L., Casoli, F., Combes, F., *Astron. Astrophys.* **199**, 29.

Wakamatsu, K., Hamabe, M.: 1984, *Astrophys. J.* **56**, 283.

Wright, G.S., McLean, I.S: 1988, *Sky and Telescope* **76**, 346.

Discussion

R. Buta - May I, by way of comment, summarize a few other points that could help us understand barred galaxies. The first point is that while statistics of apparent bar/ring orientations suggest that inner rings are aligned parallel to bars on average (e.g. NGC 1433), there exist probable cases of intrinsic misalignment (e.g. NGC 4319), where the angle between the bar and ring could be between 0 and 90 degrees. Such cases may reflect one of the possible effects of interactions. The second point is that in some barred galaxies, an outer ring or pseudo-ring can be weakly dimpled near the bar axis; this is a possible signature of the OLR. The final point is there exist cases where a bar enveloped by an inner ring does not fill the ring. Such cases may be important for identifying the possible resonances which limit bar extent.

S. Kent - Many of the features you showed in your galaxy sketches look remarkably like the shapes of individual orbits within barred potentials, suggesting that somehow material has piled up onto these orbits.

D. Burstein - Regarding bars and lenses in SB0 galaxies, I have two comments. First I now think that the component I called a 'thick disk' in S0's, in my thesis, is what Kormendy later called an 'outer lens'. Thus, if correct, it indicates that the vertical structure of the lens is constant with radius in S0's. Second, many years ago, Sandy Faber and I measured the Mg_2 gradient in NGC 4762, the famous edge-on 'thin' S0 The Mg_2 gradient appears to follow the "kinks" and "ledges" in the luminosity profile indicating separate stellar populations in the separate components. As we were (and are) uncertain about this result, it remains unpublished. However, it indicates that absorption line strength gradient information on S0's and SB0's would be of interest in the studies of bars and lenses.

A. Bosma - Most people think that bars and lenses have the same colour, and thus consist of similar stellar populations. It would indeed be important to follow up on your old result, which contradicts this.

Albert Bosma lecturing.

PECULIAR KINEMATICS

M. Balcells
Kapteyn Lab., Groningen, and
Observatorio del Roque de los Muchachos
Islas Canarias, Spain

ABSTRACT. Kinematically peculiar cores in ellipticals are described, and the evidence for mergers is discussed. Results from simulations show that counterrotating cores can form when an elliptical accretes a lower luminosity elliptical or bulge. The models reproduce well the observed properties of peculiar core ellipticals, including the asymmetries observed in spectral line broadening functions. These asymmetries arise from the absorption of orbital angular momentum by the primary material.

1 Introduction

I will address peculiar kinematics at the cores of ellipticals. Other kinematic peculiarities such as misaligned rotation and photometric axes have been reviewed by Marijn Franx. Cores with rotation axes not aligned with the rotation in the rest of the galaxy are understood as one of the strongest evidences for mergers in ellipticals. Thanks to high-quality observations and numerical modelling, we are beginning to understand what physics is involved in this important aspect of elliptical evolution. We will review current ideas on (a) what are these objects, where do they fit in the global classification of galaxies, and (b) what can we learn about their formation mechanism and in general about the formation of ellipticals.

Peculiar cores have been addressed recently in reviews by Kormendy and Djorgovski (1989) and Bender (1990a; hereafter, B90). The present paper will emphasize the comparison of N-body calculations to observations. In §2 I review observational information. Section 3 addresses formation mechanisms and the results of simulations. Section 4 deals with classification issues and §5 contains a summary.

2 Observations

The importance of core peculiar kinematics in ellipticals was revealed by a series of 1988 papers: Franx and Illingworth (1988, hereafter FI88), Jedrzejevski and Schechter (1988) and Bender (1988). (See also Efstathiou, Ellis and Carter, 1982; Wagner, Bender and Möllenhoff, 1988; Davies and Birkinshaw, 1988; Franx, Illingworth and Heckmann,

G. Longo et al. (eds.), Morphological and Physical Classification of Galaxies, 221–232.

1989.) Rotation curves show a variety of shapes. In some cases, the core rotates more rapidly than the main body of the galaxy. In others, the rotation curves actually reverse at the core. The latter indicate a significant misalignment between the spins of core and main body. Constraints on the relative orientations of the two spins are obtained from rotation curves for two or more slit positions. Cases are known of ellipticals with core spin parallel, antiparallel or orthogonal to the rotation axis of the galaxy (B90).

Kinematically peculiar cores in ellipticals are probably unrelated to the rapidly rotating stellar *nuclei* (Kormendy and Djorgovski, 1989) found in some nearby bulges. Comparison may prove useful for understanding both types of systems. The latter are much smaller, they are embedded in the photometric core, and none is known to counterrotate.

Fifteen ellipticals are known today to show core peculiar kinematics. Observational biases are important due to the small angular sizes of peculiar cores and the paucity of nearby ellipticals. Nevertheless, a consistent set of global properties can be identified for these galaxies (B90). All of them have absolute magnitudes $M_B < -20.5$, belonging to the domain of the giant ellipticals. They are slow rotators, with $(V/\sigma)^* < 0.2$ in 80% of the cases. The typical radius of the peculiar core, which I will call the kinematic radius r_k, is several times larger than the core radius r_c, and smaller than the effective radius r_e (with one exception: NGC 4278, $r_k \approx r_e$, Davies and Birkinshaw, 1989). The velocity dispersion profiles show diverse forms, including peaks (IC 1459) and dips: NGC 5322 (Bender, 1988), NGC 5813 (Efstathiou, Ellis, and Carter, 1982). The case of NGC 5813 is interesting. Recent high spatial resolution data from the WHT in La Palma (Carter, 1991) confirm the existence of a dip in the velocity dispersion profile inward of $r = 7''$, and reveal a sharp inner peak inside $r = 2''$. This peak cannot be explained as the effect of a steep rotation curve alone, and suggests that further substructure exists at scales not accessible to present earth based observations.

Surface brightness humps have been sought as signatures of an added component. In practice interpretation is not straightforward because of uncertainties as to what the underlying brightness distribution might be. Kormendy (1984) fits a superposition of King (1966) models to the surface brightness profile of NGC 5813. Surma and Bender (1990) study deviations from an $r^{1/4}$–law in the region of the peculiar kinematics; excesses are common when the $r^{1/4}$–law is fitted to the region external to the peculiar core. Additional information on superposition of components may be found by isophotal analysis. The ellipticity profiles show kinks or inflections at the kinematic radius (Bender, 1988). Isophotal twists throughout the face of the galaxy are common. These two features could be tracers of the progenitors' isophotal parameters (Kormendy, 1984), or they could be created by the merger (Balcells and Quinn, 1990a; §3.4). Color differences, which would indicate a different age or metallicity of the core population, are difficult to measure due to the effects of seeing (Franx, Illingworth and Heckman, 1989; Peletier et al., 1990) and the presence of dust. As far as know, the color profiles of ellipticals with peculiar cores are not different from those normally found in ellipticals (which do not show a very homogeneous behaviour, but, on the mean, imply negative color gradients [bluer outward]: Bender and Möllenhoff, 1987; Franx, Illingworth and Heckman, 1989; Peletier et al., 1990). Much work needs to be done to find out the true colors of peculiar cores. Preliminary results from CFHT in a poster by Surma et al. (this conference) show a significant inward reddening in the kinematic core of NGC 5322. The effects of seeing have not been modelled yet, but the kinematic radius of this galaxy is large ($r_k \approx 10''$, Bender, 1988), thus the result may be real. The main bodies of galaxies with kinematically peculiar cores have a strong tendency towards boxy or irregular isophotes (see details in B90). The cores themselves have pointed isophotes in most cases, as shown by Nieto (this volume) using high spatial resolution data from CFHT.

Peculiar kinematics are primarily observed along the photometric major axes. If peculiar cores are disks, their orientation may provide some information on the true shapes of the inner parts of the parent galaxies (Illingworth and Franx, 1989), in a similar way as dust lanes in ellipticals contain shape information (Bertola and Galletta, 1978; de Zeeuw, 1990a, 1990b). In practice many spectra are needed to constrain the shape of the core and the exact orientation of its spin.

Important clues are provided by the discovery of galaxies in which the broadening function of the absorption lines is not symmetric (IC 1459, FI88; NGC 5322, B90). The effect is also seen in ellipticals with normal rotation curves (Bender, 1990b) and could in fact be quite common. Spectral lines to the right and to the left of the galaxy center are mirror images of each other, the asymmetry having the same sign as the mean velocity. When the two are superimposed, the wings of the lines coincide quite well, only the peaks differ (see figure 4 in B90). The asymmetries can be modelled with the superposition of a broad Gaussian at zero velocity plus a narrow Gaussian at high velocity, thus they are consistent with an inner disk-like component embedded in the elliptical. When such cores counterrotate the most natural origin for the disk material is external, in an accretion or merger event. This "disk/bulge" interpretation is not unique. An alternative is suggested by numerical models, in which line asymmetries result from absorption of orbital angular momentum during the merger (§3.4).

Have we found suitable "smoking guns" of ongoing mergers with kinematically peculiar cores? There are a few candidates. Schweizer (1990) proposes NGC 7252, a prototype of advanced spiral-spiral merger, which contains an inner gas disk spinning at nearby right angle to the main body of the galaxy (Schweizer, 1982).

NGC 520 (Arp 157), an ongoing, young merger of two spirals, has a nuclear CO disk which counterrotates with respect to one of the bodies of the system (Sanders et al., 1988); counterrotation may also be present in the stellar velocity field (Stanford and Balcells, 1990). Finally, NGC3656 (Arp 155), a galaxy with shells, loops and a dust band, which is in the process of accreting an irregular satellite, contains a core spinning about an axis perpendicular to the rotation axis of the dust band and to the rotation axis of the main body of the galaxy (Balcells and Stanford, 1990).

The gas motions are not always related in a simple way to the stellar motions. In IC 1459, the ionized gas at the core does non counterrotate with respect to the main body, thus gas and stars counter-stream with respect to each other at the core. The stars and gas must have different origin (FI88). In NGC 520, the emission line spectra of Stockton and Bertola (1980) show that the ionized gas does not counterrotate in the region of counterrotating CO disk. In NGC 3656 the arrangement of rotation axes suggests independent origin for gas and stellar core. Since the gas is supplied by the ongoing merger, it follows that the peculiar core is unrelated to the ongoing merger. These orientation effects, paradoxical if we expect the stellar cores to have grown out of infallen gas, may reflect accidents in the individual histories of the galaxies, or they could indicate that ionized gas and dust are unrelated to the formation of kinematically peculiar cores.

Knowing what fraction of ellipticals have core peculiar kinematics is important for estimates of how frequently galaxies merge. This number is uncertain because, as FI88 point out, most studies do not have adequate spatial or velocity resolution. Jedrzejewski and Schechter (1988) and Bender (1988) estimate the fraction of ellipticals with core peculiar kinematics at about 1/3.

The data collected so far addresses only a handful of objects. We should try to expand the sample to allow for statistical studies. Correlations with gas, dust, radio and X-ray luminosity, color gradients, and with merger signatures such as shells, will likely provide important clues to the structure and the formation of kinematically peculiar cores. Also, the known cases need to be studied in more detail. Slits at several position angles, as well as offsets from the center constrain the global velocity field and the shape of the region with peculiar kinematics.

3 Formation of kinematically peculiar cores

Core peculiar kinematics, in particular counterrotation, strongly suggest a merger. It is hard to imagine an origin in a homogeneous collapse. If peculiar cores formed during initial collapse, such collapse had to be very clumpy, essentially a multi-merger of sub-galactic lumps into the protogalaxy at the galaxy formation epoch. Consequently, most ideas (not all) on the formation of peculiar core kinematics propose some sort of merger. The question are, what type of merger, and, can we distinguish between cores formed via different mechanisms, if more than one actually take place in nature.

3.1 Proposed mechanisms

Because several disk signatures are seen in some peculiar cores (§2), mergers involving gas dynamics have been proposed. These could involve either the accretion of a small spiral or a gas rich irregular companion (sE merger) or a spiral-spiral *major merger* (SS merger). The SS merger origin is advocated by Schweizer (1990) based on the observation of two kinematic subsystems in the merger remnant NGC 7252. In the sE merger, the gas would settle into one of the principal plane of the galaxy, dissipate to the center and turn into stars, maybe in a starbust-like event (Illingworth and Franx, 1989). This idea is in line with the conventional wisdom that gas dissipates to the center and *feeds the monster* in active galaxies. The relation of this process to ellipticals with dust-lanes (e.g. Bertola, 1987) is uncertain, as the latter are preferentially minor-axis structures. Predictions include disky isophotes and asymmetric line profiles. Numerical studies of the evolution of gas clouds in the potential of the elliptical are addressed in §3.3. Gas-rich accretions are often said to create redder, high metallicity cores. In fact, predictions on metallicity gradients are not clear, as the new stars would form out of the dwarf gas, which is metal poor. The outcome probably depends on the violence of the star formation event and on the depth of the potential well (Franx and Illingworth, 1990).

Among those origins not involving gas dynamics, Kormendy (1984) proposes the merger between a high- and a low-luminosity elliptical galaxy (eE merger), which creates a core-within-a-core system. This mechanism is analyzed by Balcells and Quinn (1990a) with numerical merger models (see §3.4). Another possibility is a merger in which the satellite disrupts before reaching the center, its debris settling into a disk (another "disk/bulge" description; Illingworth and Franx, 1989). A few numerical experiments with disrupting satellites have not managed to reproduce the effect (Balcells and Quinn, 1989; §3.4), but, by fine-tuning the disruption event, the mechanism could work. Weak counterrotation is seen for some orbit configurations in bars (Bettoni, 1989). This mechanism does not require a merger.

Tidal torquing during galaxy formation ("primordial misalignment") as seen in cosmological simulations (Barnes and Efstathiou, 1987) may misalign the spin of the envelope, but is not likely to misalign that of the core, where the tidal field is weakest. It seems an unlikely mechanism for the formation of peculiar cores.

3.2 Evidence from simulations

Simulations of some of these mechanisms have been carried out. They have already provided important clues on the formation of peculiar cores. Most interesting are the studies of the evolution of the gas phases in mergers, addressed in §3.3, and the results on formation of counterrotating cores in eE mergers, described in §3.4. The latter give the first detailed account of the physics and the observational consequences of the formation process.

3.3 Mergers with gas

Modelling of the evolution of gas in a merger is done with smooth particle hydrodynamics (SPH; Habe and Ikeuchi, 1985; Hernquist and Katz, 1989). SPH describes the evolution of the hot and warm ($T > 10^4$K) gas phases, which gives some information on possible sites of star formation. Barnes (1990) shows that, after an SS merger, large amounts of gas can reach the center of the remnant. For some initial configurations (antiparallel initial spins) a counterrotating disk of gas appears (Barnes, this conference).

Numerical studies of the formation of dust lanes in ellipticals provide results on the evolution of gas accreted on the galaxy via an sE merger. Habe and Ikeuchi (1985, 1988) study the evolution of an oblique ring of gas in the potential of a triaxial galaxy using SPH. The gas ends up in a ring of elliptic closed orbits in one of the preferred planes of the galaxy; for it to reach the center, the potential needs to be sufficiently elongated. Cloud orbits become elongated if the galaxy has a finite core (de Zeeuw, 1990b), and this, too, may supply gas to the center. These results are usually discussed in the framework of feeding active galactic nuclei, but instability, turbulence and shocks in gas systems are probably related as well to the onset of star formation. The form of this relation and the consequences for the formation of stellar peculiar cores is largely unknown.

3.4 Mergers of ellipsoidal components

When two elliptical stellar systems of different luminosities merge, the secondary core, being denser (e.g. Kormendy, 1987), may never disrupt in the tidal field of the primary. Then the relevant mechanism is the tidal friction of the secondary core against the primary stars, which sends the self–gravitating core to the remnant center. The resulting system of two embedded galaxies may well look like a galaxy with core peculiar kinematics (Kormendy, 1984). The formation of counterrotating cores via this mechanism was studied by Balcells and Quinn (1990a, hereafter BQ90; see also Balcells and Quinn, 1989, 1990b; Balcells, 1991). The goal was to reproduce the formation of counterrotating cores as well as to compare the merger remnants to ellipticals with peculiar cores in a wide range of observable properties. A series of eE mergers between oblate rotator model galaxies were run varying the mass ratio of the galaxies and the orbital parameters. The evolution was computed self–consistently using the *treecode* of Hernquist (1987, 1988) with $N = 50,000$ particles. The models successfully reproduce the properties of ellipticals with peculiar cores, including line profile asymmetries. They explain counterrotating cores as a transformation of the primary during the merger rather than the result of adding either a spheroid or a disk of secondary particles onto the elliptical.

Two conditions seem necessary for counterrotation; first, the merger orbit has to be retrograde with respect to the primary spin, because the secondary spin does not survive the merger (it ends aligned with the orbital angular momentum J_{orb}), and second, J_{orb} must be smaller than the total spin content of the primary, otherwise the whole primary reverses its spin. This sets a lower limit on the mass ratio at about 10. Thus, counterrotation does not follow from initially opposite spins in the progenitors; the secondary spin is irrelevant. Because the merger orbit is retrograde, the remnant's rotation is slower than that of the progenitor.

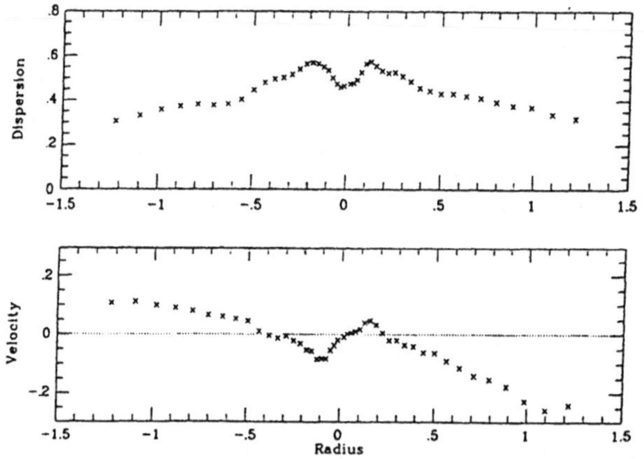

Figure 1. Rotation curve and velocity dispersion profile after a merger of two rotating ellipticals with mass ratio 10. The unit of length is $1.7r_e$, and the unit of velocity is $463(M/5\ 10^{11} M\odot)^{1/2}(r_e/6\ kpc)^{-1/2}(\mathrm{km\ s^{-1}})$. The curves extend to more than $2r_e$. Counterrotation occurs out to $r = 0.5r_e$. The dip in the velocity dispersion profile occurs out to $r \simeq 0.15r_e$.

A typical rotation curve from a retrograde merger between two galaxies with mass ratio of 10 is shown in figure 1. Counterrotation is seen inward of $r \simeq 0.5r_e$, with an amplitude of about 40 km/s. Note the dip in the velocity dispersion profile. This is seen in all the eE merger remnants. It corresponds to the mass being dominated by a less massive ystem. The secondary dominates the surface density out to less than $r = 0.15r_e$, and a mall hump is visible in the surface density profile.

Counterrotation does not involve counterstreaming of primary and secondary particles at the core; *both* primary and secondary particles counterrotate. This is emphasized by the fact that the region of peculiar kinematics is typically larger than the region where he secondary dominates the surface density.

Because lower luminosity ellipticals are bluer (e.g. Visvanathan and Sandage, 1977), he eE merger could introduce a color difference between the secondary–dominated core and the main body of the remnant. This does not apply to the entire kinematic core, typically larger than the secondary–dominated region. In practice, the color–magnitude elations of ellipticals have too much intrinsic scatter to imply systematic color gradients, even in (u-V).

The eE merger modifies the isophotal structure of the precursor galaxies. Not only he isophotes of the secondary but also those of the primary change due to the absorption of orbital energy and angular momentum. From most viewing angles the central region out to about $1r_e$ ends up elongated in the direction of the merger orbit. This means hat no sharp isophote twist is to be expected at the kinematic radius. Instead, gradual sophote twists appear over the face of the galaxy; the latter result from the deposition of orbital angular momentum at progressively smaller radii.

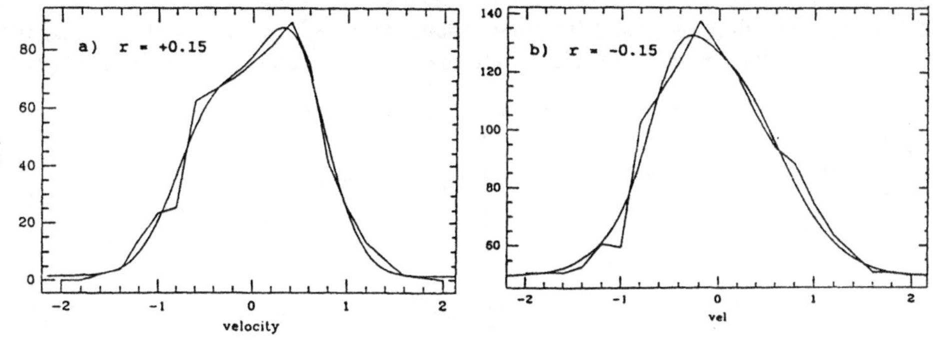

Figure 2. Line-of-sight velocity distributions for a merger remnant which contains a counterrotating core. Ordinates are in particles per bin, abscissae in reduced velocity units. Velocity distributions at two positions $r = \pm 0.15 = \pm 0.25 r_e$ right and left of the center are clearly asymmetric. The smooth curves are two-Gaussian fits to the velocity distributions.

The line-of-sight velocity distributions at the cores of eE merger remnants are asymmetric much in the same sense as is observed in real peculiar cores (see details in Balcells, 1991). Two such velocity distributions are shown in figure 2. For a line of sight where the mean velocity is positive, the velocity distribution has an excess at positive velocities (Fig. 2a). Where the mean velocity is negative, the velocity distribution has an excess at negative velocities (Fig. 2b). As one moves from left to right across the center of the galaxy, the asymmetry first grows, then disappears at the galaxy center, and reverses sign at the other side. The situation is identical to that seen in IC 1459 (see figure 2 in FI88) or in NGC 5322 (Fig. 4 in B90; §2). As was done with these galaxies, I have modelled the velocity distributions in figure 2 with a superposition of two gaussians, one broad at zero velocity, plus one narrow at high velocity which changes sign as one moves across the center of the galaxy. Figure 2 shows the two-Gaussian least-squares fits (smooth curves) superimposed on the velocity distributions. The fits are satisfactory. This emphasizes the similarity between the merger models and the observations.

We now study the contributions of primary and secondary particles to the total observed velocity distribution. Figure 3 shows the velocity distributions of primary (Fig. 3a) and secondary particles (Fig. 3b) separately, corresponding to the line-of-sight velocity distribution portrayed in figure 2a. The velocity distribution of primary particles is asymmetric, while the secondary particles are mapped onto line center. Thus, we cannot identify the narrow, rapidly rotating Gaussian with the secondary material. There is no (visible) disk of secondary particles. *The asymmetry in the total velocity distribution is not due to a rapidly rotating component of material added in the merger (secondary material); it is intrinsic to the primary particles.* The primary velocity distribution becomes asymmetric because of the absorption by the primary particles of (most of) the orbital angular momentum J_{orb} during orbit decay. This process increases their component of angular momentum parallel to J_{orb}.

With the asymmetry of the velocity distributions in mind, I calculate the maximum counterrotation velocity that may be obtained from eE mergers. The peak of the narrow components of the line profiles shows in figure 2 occurs at $v \approx 0.4 = 185(M/5\cdot 10^{11}M_{\odot})^{1/2}(r_e/6\ \mathrm{kpc})^{-1/2}$ (km/s), which, scaled for IC 1459 ($\sigma = 308\ \mathrm{km\cdot s^{-1}}$, $r_e =$ 6.24 kpc (Franx, Illingworth and Heckman, 1989), $M = 9\sigma^2 \times r_e/G = 1.24\cdot 10^{12}\ M_{\odot}10^{12}$), gives $v_{narrow} = 285\ \mathrm{km\cdot s^{-1}}$. This is close to the $v_{narrow} = 300\ \mathrm{km\cdot s^{-1}}$ of IC 1459 (FI88).

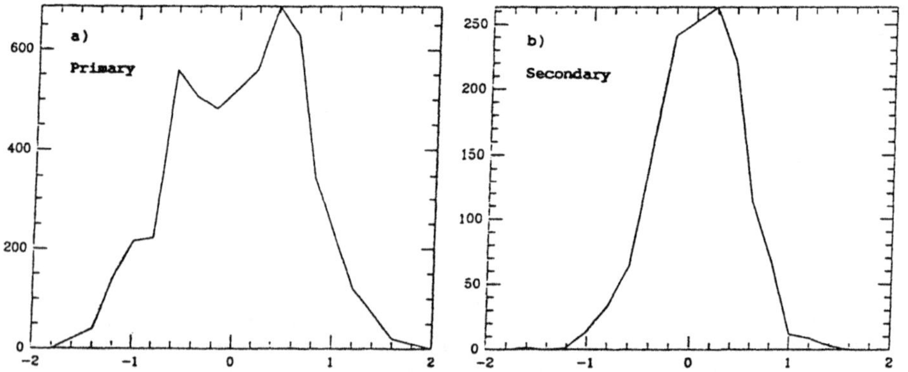

Figure 3. The contribution of (*a*) primary and (*b*) secondary particles to the asymmetric line-of-sight velocity distribution shown in figure 2a. Ordinates are in (particles per bin)×10, abscissae in reduced velocity units. The velocity distribution of primary particles is asymmetric, while that of secondary particles is narrow, projected onto line center, and cannot account for the asymmetry of the total velocity distribution.

There is an uncertainty of ~ 2 in these calculations; nevertheless, they show that extreme systems such as IC 1459 could result from an eE merger, without the need to invoke satellite disruption or gas dynamics.

In summary, retrograde eE mergers create counterrotating cores, and the resulting galaxies have similar properties to ellipticals with peculiar cores: they rotate slowly, may have weak surface brightness excesses at small radii, do not show systematic color differences core-envelope, show isophotal twists, and have asymmetric line-of-sight velocity distributions. We conclude that ellipticals with peculiar core kinematics could very well be remnants of the merger between a high- and low-luminosity elliptical galaxy. Minor disagreements with observations remain. The models always show central velocity dispersion dips, while some galaxies show dips and other peaks. The boxiness of peculiar core ellipticals and the ellipticity profiles of Bender (1988) has not been reproduced (yet: work underway). Finally, it is worth emphasizing that this study does not rule out formation via other mechanisms.

Physically, counterrotation is a result of the absorption of the orbital angular momentum J_{orb} by the primary particles. As the secondary core approaches the center, less and less primary material is available to absorb J_{orb}, and, if the orbit is retrograde, the rotation curve gets finally reversed.

Survival appears essential, otherwise J_{orb} does not get deposited in the core. Homologous scaling ($mass \propto r^2$) may cause disruption (for mass ratio 10). A counterrotating secondary disk forms, but it is too faint to be observable; and the primary material counterrotate. The BQ90 models used $mass \propto r^{1.25}$ as observed in Virgo ellipticals (Binggeli, Sandage and Tarenghi, 1984).

4 Classification issues

Ellipticals with core peculiar kinematics emphasize the usefulness of Bender's boxy/disky classification criterion since the vast majority of ellipticals with peculiar cores fit in one Bender's class: the boxy, slowly-rotating ellipticals (§2). This class has well defined morphological, as well as physical, attributes. If these attributes are due to common formation processes, the merger origin of core peculiar kinematics gives an indication that boxy ellipticals as a class are merger remnants.

Both the disk/bulge and the core-within-a-core description of counterrotating core are applications of the "independent component" approach to galaxy classification (Kormendy, 1977). This approach, useful for understanding galaxies, breaks down for counterrotating cores formed via eE mergers, as counterrotating cores cannot be understood as the superposition of largely unperturbed systems. But peculiar cores are useful for component analysis; they give evidence that non-trivial velocity fields are normal attributes of (some) spheroidal components of galaxies.

5 Conclusions

The origin of peculiar cores is most likely a merger. The role of gas and post-merger star formation is uncertain at present. But, interestingly, gas dynamics need not be invoked to explain observations, as these are accounted for by the accretion of a compact elliptical or bulge onto an elliptical. In the eE merger, counterrotation results from the absorption of orbital angular momentum by the primary particles rather than the superposition of the secondary at the core of the primary. This distinction makes the link between peculiar cores and mergers more subtle than originally thought. Asymmetric line profiles, which could be interpreted as evidence for a disk of accreted material, are not due to a disk in the models but, again, to the absorption of J_{orb} by the primary. Probably, then, pointy isophotes in peculiar cores (§2) are not evidence that they are cold, disky dynamical systems. Open questions remaining are the systematics of the a_4 isophotal coefficient and the relative orientations of rotation and photometric axes. Understanding the role of gas is the next frontier in this field.

Acknowledgements

I would like to thank Ralf Bender, Peter Quinn, Marijn Franx, Tjeerd van Albada and Reynier Peletier for many interesting discussions on core peculiar kinematics. I acknowledge a grant of CRAY time from the San Diego Supercomputer Center to run the merger simulations.

References

Balcells, M., Quinn, P.J.: 1989, *Astrophys. Space Sci.* **156**, 133.

Balcells, M., Quinn, P.J.: 1990a, *Astrophys. J.* **361**, 381 (BQ90).

Balcells, M., Quinn, P.J.: 1990b, in *Dynamics and Interaction of Galaxies*, (R. Wielen, ed.), 210, Springer-Verlag.

Balcells, M., Stanford, S.A.: 1990, *Astrophys. J.* **362**, 443.

Balcells, M.: 1991, *Astron. Astrophys.* , submitted.

Barnes, J.: 1990, in *Dynamics and Interaction of Galaxies*, (R. Wielen, ed.), 186, Springer-Verlag.

Bender, R., Möllenhoff, C.: 1987, *Astron. Astrophys.* **177**, 71.

Bender, R.: 1988, *Astron. Astrophys.* **202**, L5.

Bender, R.: 1990a, in *Dynamics and Interaction of Galaxies*, (R. Wilen, ed.), 232, Springer-Verlag.

Bender, R.: 1990b, *Astron. Astrophys.* **229**, 441.

Bertola, F., Galletta, G.: 1978, *Astrophys. J. Letters* **226**, L115.

Bertola, F.: 1987, in *Structure and Dynamics of Elliptical Galaxies*, IAU Symp. n. 127 (T. de Zeeuw, ed.), 135, Reidel Publishing Co.

Binggeli, B., Sandage, A., Tarenghi, M.: 1984, *Astron. J.* **89**, 64.

Carter, D.: 1991, in preparation.

Davies, R.L., Birkinshaw, M.: 1989, *Astrophys. J. Suppl. Ser.* **68**, 409.

de Zeeuw, T.: 1990a, *Ann. Rev. Astron. Astrophys.* **28**, 000.

de Zeeuw, T.: 1990b, in *Dynamics and Interaction of Galaxies*, (R. Wielen, ed.), 263, Springer-Verlag.

Efstathiou, G, Ellis, R.S., Carter, D.: 1982, *Mon. Not. R. Astr. Soc.* **201**, 975.

Franx, M., Illingworth, G.D.: 1988, *Astrophys. J. Letters* **327**, L55.

Franx, M., Illingworth, G.D., Heckman, T.: 1989, *Astron. J.* **98**, 538.

Franx, M., Illingworth, G.D.: 1990, in *Dynamics and Interaction of Galaxies*, (R. Wielen, ed.), 253, Springer-Verlag.

Habe, A., Ikeuchi, S.: 1985, *Astrophys. J.* **289**, 540.

Habe, A., Ikeuchi, S.: 1988, *Astrophys. J.* **326**, 84.

Hernquist, L.: 1987, *Astrophys. J. Suppl. Ser.* **64**, 715.

Hernquist, L.: 1988 *Computer Phys. Comm.* **48**, 107.

Illingworth, G.D., Franx, M.: 1989, in *Dynamics of Dense stellar Systems*, (D. Merrit, ed.), 13, Cambridge, University Press.

Jedrzejewski, R., Schechter, P.L.: 1988, *Astrophys. J. Letters* **330**, L87.

King, I.: 1966, *Astron. J.* **71**, 64.

Kormendy, J.: 1977, *Astrophys. J.* **214**, 359.

Kormendy, J.: 1984, *Astrophys. J.* **287**, 577.

Kormendy, J.: 1987, in *Structure of Elliptical Galaxies*, IAU Symp. n. 127, (T. de Zeeuw, ed.), 17, Reidel Publishing Co.

Kormendy, J., Djorgovski, S.: 1989, *Ann. Rev. Astron. Astrophys.* **27**, 235.

Peletier, R.F., Davies, R.L., Illingworth, G.D., Davies, L.E., Cawson, M.: 1990, *Astron. J.* **100**, 1091.

Sanders, D.B., Scoville, N.Z., Sargent, A.I., Soifer, B.T.: 1988, *Astrophys. J. Letters* , **324**, L55.

Schweizer, F.: 1982, *Astrophys. J.* **252**, 455.

Schweizer, F.: 1990, in *Dynamics and Interaction of Galaxies*, (R. Wielen, ed.), 60, Springer-Verlag.

Stanford, S.A., Balcells, M.: 1990, *Astrophys. J.* **355**, 59.

Stockon, A., Bertola, F.: 1980, *Astrophys. J.* **235**, 37.

Surma, P., Bender, R.: 1990, in *Dynamics and Interaction of Galaxies*, (R. Wielen, ed.), 277, Springer-Verlag.

Varnas, S.R., Bertola, F., Galletta, G., Freeman, K.C., Carter, D.: 1987, *Astrophys. J.* **313**, 69.

Wagner, S.J., Bender, R., Möllenhoff, C.: 1988, *Astron. Astrophys.* **195**, L5.

Discussion

M. Franx - I think we cannot say yet that there is a typical scale for kinematically distinct systems. I think there is a serious observational bias because we do not have (much) data that goes beyond r_e. I have several questions regarding your simulations: (1) Along which axes do you "measure" counterrotation? The figure you showed has counterrotation along an axis near to the minor axis, which is rarely observed. (2) Do your simulation produce systems like NGC 4406, which shows minor axis rotation in the outer part and major axis rotation in the inner parts? (3) Is the material that produces the asymmetry in the line profiles in a disk? You stated that it was not, but, did you check your numerical results? You might, for example, look in the energy-angular momentum plane for a signature. Before you do that, you cannot conclude it is *not* in a disk.

M. Balcells - I agree that our ability to detect peculiar cores is limited by seeing on the small side, and by signal-to-noise on the big side. However, if peculiar cores are much larger than r_e we would probably call them "peculiar envelopes". The model suggest that peculiar cores from eE mergers cannot be much larger than r_e; for mass ratios of 5, enough J_{orb} is deposited in the envelope that the rotation axes of core and envelope differ very little (Fig. 8 in BQ90). Regarding (1), there is some tendency in the BQ90 models for minor axis counterrotation. This is in part due to the satellite being too massive, which reorganizes the kinematics of most of the primary. Less massive secondaries have less effects. Precession of the merger orbit toward the galaxy principal plane might provide a mechanism to preferentially generate major axis counterrotators. Precession is observed, but I don't know if it alone can account for the statistics on major/minor axis counterrotation. (2) The models show a wide variety of orientations between rotation and photometric axes. (And most show isophote twists.) The only general rule is that the peculiar core spins along its major axis. In the other parts ther is much freedom; so, yes, NGC 4406 can be reproduced. (3) No, I don't think there is disk. The satellite torque increase J_z but does not decrease the z-velocity dispersion o the primary particles. They remain a hot system. If we manually select those particle which make the line profile asymmetric, then from the two-Gaussian fits already know the answer: $V/\sigma \approx 2$ (similar to IC 1459). But this seems artificial to me. My poin was that there is no disk *of secondary particles,* i.e., the asymmetry does not trace new material added to the galaxy; it traces the *effects* of the accretion on the primary.

J.L. Nieto - (1) You are certainly extremely right to warn us that pointed isophotes do not necessarily mean that there is a disk. In the sample that R. Bender, P. Surma and I have analyzed (see my review), the best case of photometrically decoupled "core" is certainly that of NGC 3115, where Kormendy and Richstone have subtracted the bulge and shown a clear disk, as in NGC 4594. It would be surprising if these features looking like disks are not disks. (2) It turns out, also, that galaxies that appear to be merger remnants (including those having kinematically decoupled cores), have flatter central regions and larger apparent "core radii", at variance with the predictions of the numerical simulations of Farouki et al. (1983). It would be extremely informative if you could check these effects on your simulations.
M. Balcells - (1) Strong departures from ellipses are clear disks. Small positive a_4 values are less clear. By definition, any galaxy with positive a_4 will, upon subtraction of an elliptical model, result in a disk. That's what you showed in NGC 3115. My point was, and still is, we don't know that the underlying isophotes are pure ellipses. (2) Numerical models cannot explain core radii, which are unresolved in today's numerical simulations. BQ90 avoid this problem by measuring the evolution of the galaxy concentration at r_e, measuring $\langle SB \rangle_e$ à la Binggeli, Sandage and Tarenghi (1984). The eE mergers cause an increase of r_e and a decrease $\langle SB \rangle_e$ in agreement with observations.

D. Merrit - Have you thought about how your mechanism scales to other systems? For instance, the central dominant galaxies in clusters are thought to form from massive galaxies that spiral into the cluster core. Would you predict that galaxy clusters have counterrotating cores?
M. Balcells - The initial conditions need to be clumpy, e.g., a compact group of galaxies falling into a large cluster. The group needs to be denser than the cluster to survive, and massive, for dynamical friction to send the group to the cluster center. These conditions might be realistic. Let's suppose there is no dominant galaxy there. The group will form a low-velocity dispersion system at the cluster core. I think this is a plausible scenario for the formation of a low- to moderate-mass cD since the group would easily merge. If the infall orbit was retrograde, the central group, or the resulting dominant cluster galaxy, would counterrotate. If counterrotating dominant cluster galaxies are not observed, this might indicate that clusters form by mergers of more-or-less equal size subclusters; a low-velocity dispersion central system forms as well in this case with the most tightly bound objects of each subcluster.

COLD GAS IN ELLIPTICAL GALAXIES

J.H. van Gorkom
Department of Astronomy
Columbia University, USA

ABSTRACT. A significant fraction (at least 10%) of ellipticals contains detectable amounts of HI and CO. For ellipticals neither the content nor the kinematics of the gas are related to the stellar component, quite contrary to the situation for spiral galaxies. The total amounts of gas are small, typically between 10^8 and 10^9 $M_\odot$. These gaseous disks do resemble gaseous disks in spirals, the HI is very extended, while the CO is centrally peaked and very regular rotation occurs.

1 Introduction

In this talk I shall concentrate on the properties of cold gas in ellipticals, rather than on the more general topic of cold gas in early type galaxies, as has been suggested by the organizing committee. While ellipticals and S0s are morphologically very different (S0s do have a cold stellar component (the disk), ellipticals are diskless), the gaseous properties of ellipticals and S0s are in fact quite similar and they are very different indeed from those of Sa galaxies. A major difficulty in assessing the difference in gaseous properties of Sa's, S0s and ellipticals in the lack of a reliable optical catalog. Whether a galaxy gets classified as an E or S0 still seems to depend somewhat on who does the classifying. Although some people would argue with this, it seems to me that the current classification depends to some extent on the cold interstellar medium. Thus any difference found in gaseous properties of Es and S0s should be looked at with scepticism, it might just be a result of the classification. Although this is an important point I shall ignore it for the time being and try to restrict myself as much as possible to pure ellipticals.

Two excellent reviews, one on gas and dust in ellipticals by Schweizer (1987) and one specifically on the cold ISM in ellipticals by Knapp (1987) appeared in IAU Symposium 127. The main difference between now and then is, that while Knapp concluded by saying that little work had been done on molecular gas, this has now changed dramatically and the detection rate of Es in CO might be as 50%. Since these results are so important and so recent, I shall describe them in more detail.

In this talk I shall first review the global properties of the cold gas, then I shall discuss what we have learned from detailed imaging of a few selected systems and finally I shall say a few words about the relation between cold gas and radio activity in ellipticals.

G. Longo et al. (eds.), Morphological and Physical Classification of Galaxies, 233–238.

2 Global properties of cold gas

Ellipticals were long thought to be devoid of cold gas, but finally in the late seventies small amounts of HI (typically a few times $10^8 M_\odot$) were found in a few nearby ellipticals. Among the first galaxies to be detected were NGC 4278 (Bottinelli and Gouguenheim, 1977; Gallagher et al., 1977) and NGC 1052 (Knapp et al., 1978; Fosbury et al., 1978). Encouraged by these results people embarked on more sensitive searches for HI in large samples of ellipticals. Currently, despite much efforts and improved receivers, it is still true that most ellipticals (90%) remain undetected in HI. The HI masses, that have been found, range from a few times 10^4 $M_\odot$ to several times 10^9 $M_\odot$. The most recent statistical analysis of these data (Knapp et al., 1985) shows, that although the data are not inconsistent with a bimodal distribution, in which some ellipticals are gas rich while most are gas poor (Sanders, 1980), they do not suggest it. By properly taking into account detections and upper limits, Knapp et al. can derive the form of the distribution function for the relative HI content, M_{HI}/L_B for ellipticals. The data are best described by a shallow power law $N(M_{HI}/L_B) \sim (M_{HI}/L_B)^{-1.5}$. This distribution is very different from that of spirals, where $N(M_{HI}/L_B)$ has a well defined mean value and a small dispersion for a given morphological type. This result suggests that the gas and star contents of ellipticals are decoupled, a result which will be confirmed again and again by looking at other components of the cold gas and by comparing the kinematics of the gas and stars.

Dust has also been found in ellipticals. About 40% have dust absorption (Sadler and Gerhardt, 1985; Bertola, 1987). A similar fraction of bright ellipticals is detected at long IRAS wavelength (Knapp et al., 1989).

The real big new is that ellipticals are now being detected in CO as well. The most extensive survey to date are the observations made by Lees et al. (1991) with the Caltech Submillimeter Observatory (CSO). Since in spiral galaxies the cold gas mass is closely related to the global FIR emission, these authors chose the observe a sample of FIR bright ($S_{100\mu} > 1\ Jy$) ellipticals. This approach turned out to be very successful, 8 out of 20 ellipticals were detected. The H_2 masses are typically $10^7 - 10^8$ $M_\odot$. Lees et al. present a statistical analysis of all the currently available data on CO observations of ellipticals, similar to the analysis of the HI data by Knapp et al. (1985). The sample consists of 17 ellipticals, which have been detected in CO and 27 for which upper limits exist. So we are still talking about small number statistics, but these turn out to be quite interesting. The relative CO content of ellipticals CO/L_B exhibits an even more shallow power law than for HI in ellipticals, $N(M_{H_2}/L_B) \sim (M_{H_2}/L_B)^{-1/3}$. Just like for HI in spirals the relative CO content in disk galaxies has a well defined mean and a gaussian distribution (Verter, 1987). So the global properties of molecular and atomic gas appear to be similar in ellipticals as well as in spirals. Young and Knezek (1989) found that M_{H_2}/M_{HI} decreases by a factor 7 going from Sa to Sdm galaxies. In contrast no such effect is found by Lees et al. (1991) for E, S0, S0a and Sa galaxies. They find instead that the mean ratio is constant for these different types and similar to the value found by Young and Knozek (1989) for Sb and Sc galaxies. These conflicting results might well be due to small number statistics, possibly the ratio is constant from Es to Scs and then decreases for the very late type galaxies.

The detection rate in CO of the bluer low luminosity ellipticals is twice as high as of the brighter ellipticals. This could be a selection effect, more luminous ellipticals will on average have larger linewidths, while most current CO receivers have a relatively small velocity baseline (e.g. 660 km s^{-1} at the CSO). CO detected ellipticals are stronger at FIR wavelengths than the undetected ellipticals. Thus FIR strength might be a good predictor of the presence of detectable amounts of molecular gas, but this also implies that the high detection rate found by Lees et al. (1991) for the FIR bright elliptical sample, will not be found for just any sample of ellipticals.

3 Distribution and kinematics of the cold gas

While the global content of HI and H_2 are quite similar, it is in their detailed distribution that they differ. HI is usually distributed in a very extended disk of ring, the CO tends to peak in the center (Lees et al., 1991; Quilled et al., 1991). For several ellipticals the detailed distribution and kinematics of the cold gas, including HI, HII and CO, is now known (e.g. NGC 1052, van Gorkom et al., 1986; Davies and Illingworth, 1986, NGC 5128; van Gorkom et al., 1987; Quillen et al., 1991, NGC 4278; Raimondi et al., 1981; Demoulin-Ulrich et al., 1984; Lees, 1991 and IC 2006, Schweizer et al., 1989). The gaseous disks are often highly inclined to the stellar galaxy, the specific angular momentum of the gas is usually much larger than that of the stars, ruling out for once and for all stellar mass loss as the origin of the gas. In one galaxy, IC 2006, the stars and gas rotate in opposite directions. What is surprising is that, despite the often rather chaotic distribution of the gas (see e.g. NGC 1052), the kinematics behaves in a regular way, with the ionized gas, the HI and CO all perfectly aligned in velocity, e.g. in NGC 1052 the rotation axis of the gas remains constant over a factor 10 in radius (Davies and Illingworth 1986). This implies that although it is tempting to suggest that the cold gas has been captured recently, it must have had time to settle into this regular rotation pattern. In the outer parts the rotation periods are quite large, typically 10^9 years or so. So then the suggestion by Schweizer et al., 1989 that this gas might be left over from the merger that formed the elliptical in the first place becomes almost plausible.

The rotation curves as derived from the gas kinematics are flat. For a discussion of the derived M/L ratios see de Zeeuw (this conference). Using very high quality data (both in angular and velocity resolution) one can actually begin to derive the intrinsic shape of the elliptical galaxy. This has been done for NGC 5128 using CO by Quillen et al. (1991) and for NGC 4278 using VLA HI observations by Lees (1991). For NGC 5128 a warped disk in a prolate potential provides the best fit to the data, the profile shapes are inconsistent with planar circular and non circular motions. The profiles of NGC 4278 are well fit by a trivial model, although a centrally warped disk cannot be ruled out.

4 Cold gas and the presence of a radio active nucleus

A puzzle that has been with us for quite some time now is, why it is that some ellipticals become powerful radio galaxies, while others of similar optical luminosity and even in the same environment are radio quiet down to very low levels. A possible scenario is that all ellipticals contain a central engine, some of which are quiet now, because of lack of fuel. The ones that are active now must have a fresh supply of cold gas as fuel. Observational support for this hypothesis keeps growing. Dust lanes tend to be perpendicular to radio jets (Kotanyi and Ekers, 1979), indicating that the rotation axis of the gas and radio jets are aligned. Radio active ellipticals are more likely to be detected in HI (Knapp and Wardle, in preparation). Radio active ellipticals are relatively strong FIR emitters (Golombek et al., 1988, Knapp et al., 1990) and this is due to thermal emission from dust and is not caused by non thermal emission from the nucleus, as was convincingly shown by the millimeter and submillimeter observations of Knapp and Patten (1991).

So the fuel might indeed be there, what is more we might even see it falling in toward the center. HI absorption surveys toward the nuclei of radio active ellipticals find systematically absorption components at velocities which are redshifted with respect to the systemic velocity of the galaxies, blueshifted components have not yet been found making it likely that we do see infall (van Gorkom et al., 1989).

5 Conclusions

Finally all the components of the cold ISM as we know it from spiral galaxies have been found in ellipticals as well. The properties of these gas disks are very similar to those of spirals, except that the total gas masses involved are much smaller. The most important change in the last five years is that now also molecular gas gets detected in ellipticals. CO has mostly been found in the centers of ellipticals, filling up the holes in the HI disks, but in fact hardly any data exist on the total extent of the molecular gas. The kinematics of the disks are surprisingly regular.

The cold gas and hot stellar component appear to be totally uncorrelated. This is perhaps not surprising, for spirals it is the stellar disk component that is closely related to the gas, in ellipticals we don't have such a component to compare the gas to. The origin of the gas in ellipticals is still not very clear. While stellar mass loss can be ruled out, the regular velocity fields suggest that if this gas was accreted after the elliptical was formed, it probably was not that long after.

Acknowledgements

I am very grateful to the organizers of the meeting for inviting me to a most enjoyable meeting. I thank Jill Knapp, Tim de Zeeuw and Francois Schweizer for many interesting discussions and I am especially grateful to Jill Knapp and Joanna Lees for allowing me to use their data in advance of publication. This work was supported by NSF grant AST-89-17744 to Columbia University.

References

Bland, J. Taylor, K., Atherton, P.D.: 1987, *Mon. Not. R. Astr. Soc.* , **228**, 595.

Bertola, F.: 1987, in IAU Symp. 127, *"Structure and Dynamics of Elliptical Galaxies"*, ed. P.T. de Zeeuw, p. 135.

Bottinelli, L., Gouguenheim, L.: 1977, *Astron. Astrophys.* , **54**, 641.

Davies, R.L., Illingworth, G.: 1986, *Astrophys. J.* , **302**, 234.

Demoulin-Ulrich, M.H., Butcher, H.R., Boksenberg, A.: 1984, *Astrophys. J.* , **285**, 527.

Fosbury, R.A.E., Mebold, U., Goss, W.M., Dopita, M.A.: 1978, *Mon. Not. R. Astr. Soc.* , **183**, 549.

Gallagher, J.S., Knapp, G.R., Faber, S.M., Balick, B.: 1977 *Astrophys. J.* , **215**, 463.

Golombek, D., Miley, G.K., Neugebauer, G.: 1988, *Astron. J.* , **95**, 26.

Knapp, G.R.: 1987, in IAU Symp. 127, *"Structure and Dynamics of Elliptical Galaxies"*, ed. P.T. de Zeeuw, p.145.

Knapp, G.R., Gallagher, J.S., Faber, S.M.: 1978, *Astron. J.* , **83**, 189.

Knapp, G.R., Turner, E.L., Cuniffe, P.E.: 1985, *Astron. J.* , **90**, 454.

Knapp, G.R., Guhathakurta, P., Kim, D.W., Jura, M.: 1989, *Astron. Astrophys. Suppl. Ser.* , **70**, 257.

Knapp, G.R., Patten, B.M.: 1990, preprint, POP-384.

Knapp, G.R., Bies, W.E., van Gorkom, J.H.: 1990, *Astron. J.* , **99**, 476.

Kotanyi, C., Ekers, R.D.: 1979, *Astron. Astrophys. Letters* , **73**, L1.

Lees, J.F., Knapp, G.R., Rupen, M.P., Phillips, T.G.: 1991, preprint, POP-372.

Lees, J.F.: 1991, PhD thesis Univ. of Princeton.

Quillen, A.C., de Zeeuw, P.T., Phinney, E.S., Phillips, T.G.: 1991, preprint.

Raimond, E., Faber, S.M., Gallagher, J.S., Knapp, G.R.: 1981, *Astrophys. J.* **246**, 108.

Sadler, E.M., Gerhardt, O.: 1985, *Mon. Not. R. Astr. Soc.* , **214**, 177.

Sanders, R.H.: 1980, *Astrophys. J.* , **242**, 931.

Schweizer, F.: 1987, in IAU Symp. 127, *"Structure and Dynamics of Elliptical Galaxies"*, ed. P.T. de Zeeuw, p. 109.

Schweizer, F., van Gorkom, J.H., Seitzer, P.: 1989, *Astrophys. J.* , **338**, 770.

van Gorkom, J.H., Knapp, G.R., Ekers, R.D., Ekers, D.D., Laing, R.A., Polk, K.: 1989, *Astron. J.* , **97**, 708.

van Gorkom, J.H., Knapp, G.R., Raimond, E., Faber, S.M., Gallagher, J.S.: 1986, *Astron. J.* , **91**, 791.

van Gorkom, J.H., van der Hulst, J.M., Haschick, A.D., Tubbs, A.D.: 1990, *Astron. J.* , **99**, 1781.

Verter, F.: 1987, *Astrophys. J. Suppl. Ser.* , **65**, 555.

Young, J.S., Knezek, P.M.: 1989, *Astrophys. J. Letters* , **347**, L55.

Discussion

D. Burstein - As this is a meeting on classification of galaxies, I would reinforce your comment about a galaxy being 'truly' an elliptical. In the original Knapp and Wardle papers on H in ellipticals the galaxies are really a mixture of RC2 types, EE, S0, Sa and Sb, I would urge you in your printed version of this paper, to use RC2 types for the galaxies.

R. Buta - Concerning the accuracy of the classification of early type galaxies brought up by Dave Burstein, I think it would be prudent for observers to know what material was used for the typing. The problem with classifying early type galaxies is not so much the classification systems as it is the inadequacy of the image material. The moral is, that we should know the sources of the types we use!

J.H. van Gorkom - I agree.

Jaqueline van Gorkom testing microphones.

COLD GAS IN LATE-TYPE GALAXIES

R. Sancisi
Kapteyn Astronomical Institute
University of Groningen, The Netherlands

1 Introduction

In the last two decades a large number of spiral galaxies have been studied in the 21-cm line. These studies have revealed the density distribution and kinematics of HI and have provided the measurements of the total mass and the mass distribution in those systems. This has led to the classic picture for HI disks: spiral arm structure, warping in the outer parts and flat rotation curves. Naturally, in the early studies attention was drawn to the large-scale properties and regularities. In recent studies, a more detailed picture has been emerging. Furthermore, the connection between the effects of environment and peculiarities in the HI distribution are being investigated. Here, the "classic" picture will be briefly described: the main emphasis will be on some of the more recent, less well-known findings.

2 Large-scale structure and kinematics

The radial distribution of HI often shows a deficiency in the central regions and a large extent in the outer parts, far beyond the optical image. The most sensitive observations (Briggs and Wolfe, 1989; Van Gorkom et al., 1991) indicate that in most galaxies the HII layer does not extend beyond 2 Holmberg radii (to the detection limit of $\sim 10^{19}$ cm^{-2}). The HI maps of the nearby spirals M 101, NGC 628 and NGC 6946 clearly show that the outer gas layer has spiral arm structure extending continuously from the bright inner parts. Several systems show pronounced, large-scale asymmetries. The vertical distribution of HII is characterised by a bending and thickening of the disk in the outer parts. In the best studied edge-on galaxy, NGC 891, which is not significantly warped, HI emission is found up to about 5 kpc on either side of the equatorial plane (Broeils et al., 1991).

The kinematical picture is also well-known. The velocity fields derived from 21-cm line observations show predominantly differential rotation and the rotation curves are approximately flat out to the last measured points. Deviations from circular motion are

G. Longo et al. (eds.), Morphological and Physical Classification of Galaxies, 239–244.

present in a large number of systems. They show up as local distortions or wiggles in the velocity fields along the spiral arms or as larger-scale effects indicating, for instance oval distortions or asymmetries.

3 Small-scale structure. HI holes

The maps of HI column densities of the well-studied nearby spiral galaxies such as M 31 M 101 (Fig. 1) and NGC 6946 show a large number of holes (van der Hulst and Kamphuis 1991).

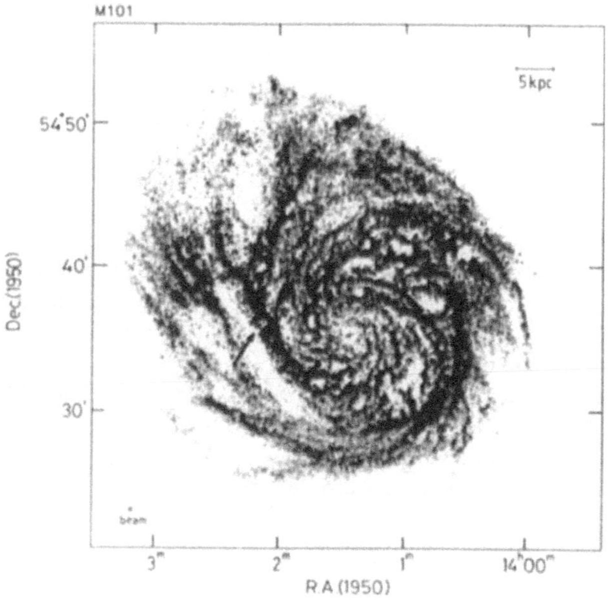

Figure 1. Map of the HI column density distribution in M 101 (Kamphuis et al., 1991). The angular resolution is 13″ × 16″(∼ 500 pc). The arrow indicates the position of the superbubble.

The presence of such holes had already been noticed in the early Westerbork HI maps of M 101 (Allen and Goss, 1979). The more sensitive and higher resolution recent data confirm the earlier indications and allow a more detailed and quantitative study. Some of these holes are normal features —gas-poor interarm or intercloud regions— of the irregular gas distribution in spiral disks. Others appear to have been created recently by violent processes, such as supernova explosions or collisions with gas clouds, as suggested by the presence of shell-like structures and peculiar velocities. One of the most spectacular cases discovered so far is indicated by the arrow in figure 1 (Kamphuis et al., 1991). A complete, expanding shell of 1.5 kpc diameter, $3 \cdot 10^7\ M_\odot$ HI mass and a few 10^{53} erg kinetic energy is found associated with the hole (Fig. 2). The properties of holes, their number, their distribution in the disks, and the relation with regions of star formation are being investigated at present. It is clear, however, from their number and the kinetic energies involved that these holes and associated shells form an important component of the gaseous disks of spiral galaxies and play a non-negligible part in the energy balance of the interstellar medium. Perhaps they are the key for understanding the effects of star formation in the disks of spiral galaxies and the disk-halo connection.

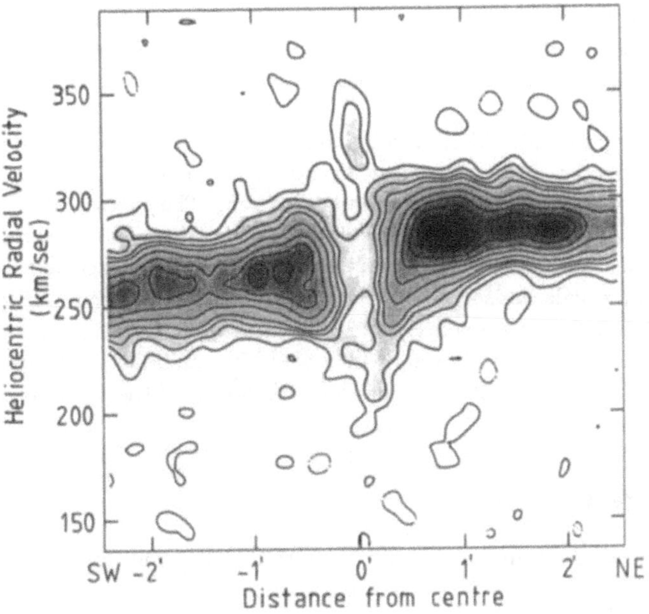

Figure 2. Position-velocity map of HI brightness temperature in M 101, at position angle 30° and centered on the superbubble indicated in figure 1 (Kamphuis et al., 1991). The map shows the 1.5 kpc hole and the 50 km/s expanding shell.

4 Peculiar structures

A large number of multiple systems have been observed in the 21-cm line and their properties are as well-known as those of isolated galaxies. Examples are our own galaxy and the Magellanic Clouds, the Leo Triplet, the M81, M82, NGC3077 and NGC 4631, 4656, 4627 groups of galaxies. In these systems one finds, in addition to the gas contained by the individual members, intergalactic HI complexes with bridge-or tail-like morphologies. These probably result from gravitational interaction.

Similar features —HI tails, peculiar velocities and structures— are also found, surprisingly, in a number of apparently isolated systems (Sancisi, 1990). In these cases there is no bright nearby companion and no clear evidence of a tidal interaction. They do seem to suggest, however, an environmental effect, perhaps collisions with intergalactic gas clouds or capture of gas-rich, dwarf companions. Some of these cases are well-known and new ones are being discovered in increasingly large number in recent, higher sensitivity observations. To the objects already listed, such as M 101, Mkn 348, NGC 1023, 3067,

1961, 5635, 3359, one should add NGC 628 (Kamphuis and Briggs, 1991) and NGC 4565 (Sancisi, 1991). Also the elliptical and SO galaxies with peculiar HI (NGC 1052, 5128 4472, 4694) are examples. Perhaps, in this connection, one should not forget the puzzling spindle-type galaxies, or the many lopsided systems like NGC 891, or even the common warped disks. Although these all look more regular and seem less obvious cases, they may be the result of the same mechanism of gas accretion and simply be in a more advanced stage of the capture and merger process. After all, the more anomalous and peculiar configurations of the few objects mentioned above —those being caught now in the act— would be smeared out and almost disappear within one or two orbital periods (10^{8-9} yr).

These cases may provide the clue to the process of gas acquisition and disk formation in galaxies. The process of gas accretion would not be a continuous drizzle of gas but likely consists of episodic infall of large lumps of gas. The masses of gas involved are typically a few 10^8 $M_\odot$, very small as compared to the HI or total masses of the main galaxy and unlikely to cause any serious damage to its disk. To account for the present gas contents of spiral galaxies such infall events must have been occurring for a long time and were probably more frequent in the past. It is possible that a population of objects around galaxies at high redshifts exists which could be candidates for such infall. Could the heavy-element or perhaps the Lyman-alpha absorbers of the QSO spectra be such a population?

Acknowledgements

I would like to thank J. Kamphuis, B. McNamara and R. Rand for helpful comments.

References

Allen, R.J., Goss, W.M.: 1979, *Astron. Astrophys Suppl.* **36**, 135.

Briggs, F., Wolfe, A.: 1989, private communication.

Broeils, A., Sancisi, R., van Albada, T.S.: 1991, preprint.

van Gorkom, J., van Albada, T.S., Cornwell, T., Sancisi, R.: 1991, in preparation.

van der Hulst, T., Kamphuis, J.: 1991, in *The Interstellar Disk-halo Connection in Galaxies*, IAU Symp. n. 144, Leiden, 18–22 june 1990.

Kamphuis, J., Briggs, F.: 1991, preprint.

Kamphuis, J., Sancisi, R., van der Hulst, T.: 1991, *Astron. Astrophys.* **244**, L29.

Sancisi, R.: 1990, in *Windows on Galaxies*, eds. Fabbiano, G. Gallagher, J.S. Renzini, Kluwer: Dordrecth, p. 199.

Sancisi, R.: 1991, in preparation.

Discussion

I. King - A question and a comment: (1) How do your HI holes in galaxies compare with the bubbles observed by Heiles? (2) I find it hard to believe that the rather low-mass cloud at 150 km/s could have made a hole through the disk of M 101. It has too little mass compared with what has been removed.

R. Sancisi - The HI holes and shells observed in nearby galaxies are probably somewhat larger, but similar to the bubble structures described by Heiles. The total mass of the high-velocity HI gas observed in M 101 is about $2 \times 10^8\ M_\odot$ (van der Hulst and Sancisi, 1988). This corresponds roughly to the amount of "missing" gas in the hole. Of course we don't known anything about the effects on the other disk components.

P. Pişmiş - The isophotal map of NGC 4565 which you showed has an interesting morphological peculiarity. The peaks at the outer isophotes are systematically inclined in the direction perpendicular to the disk. Such a phenomenon is also shown in optical isophotes. Could you comment on this peculiarity?
R. Sancisi - The effect you have noticed is due to the warping of the disk which in this case is also present in the optical image.

D. Burstein - The picture you present of the HI hole in M 101 is very similar to the current picture we have of the superbubble in which we live. The string of HII regions is like the string of HII regions including Orion and Per OB2. The tails of HI high velocity at the galactic poles are like the high latitude gas at the outer parts of the M 101 hole. The squeezing of HI at the edge of the hole is seen as "walls" surrounding one edge of our bubble. How much this analogy could be pursued quantitatively would be very interesting.
van Woerden - (Comment to Burstein) The high-velocity clouds with anomalous velocities of 100 km/s and more cannot be due to a local superbubble. If they were, they should a) be metal-rich, b) be within a few hundred pc. *Some* HVCs are metal-rich, others are poor. But *none* are found in absorption against stars within 1 kpc distance. In fact, no reliable case of CaII or NaI absorption by an HVC against a stellar probe is available and the problem of HVC distances remains unsolved.

Alvio Renzini and Renzo Sancisi.

Shopping in Pompei during happy hours.

ABUNDANCE GRADIENTS AND RELATED PROPERTIES IN GALAXIES

F. Matteucci
Max Planck Institut für Astrophysik,
Istituto di Astrofisica Spaziale,
C.N.R., Frascati, Italy

ABSTRACT. Abundance gradients and related properties in galaxies are reviewed. In particular, radial abundance gradients in spirals and ellipticals are discussed together with possible interpretations.

1 The galaxy

1.1 Observations

Radio and optical observations of HII regions in the disc of our Galaxy (Shaver et al., 1983), in a galactocentric distance range 3.5–13.5 kpc with the Sun at 10 kpc, suggest the existence of Oxigen and Nitrogen abundance gradients. In particular, the Oxigen abundance gradient is $\simeq -0.07 \pm 0.015$ dex kpc^{-1} and that of Nitrogen is not much steeper, $\simeq -0.09 \pm 0.015$ dex kpc^{-1}. Rubin et al. (1988) derived N/O abundance ratio in galactic HII regions from far infrared observations and found a much greater value for this ratio than Shaver et al. (1983). In particular, although the trend for N/O is the same (e.g. almost constant with galactocentric distance), its value is 0.2–0.3 rather than 0.05–0.1, suggesting the possibility that the previous optical abundance determinations were systematically underestimated.

On the other hand, Lester et al. (1987) found, from far-infrared measurements of N/O in galactic HII regions, that this ratio varies significantly with galactocentric distance. The observed HII regions inside 7 kpc have N/O, on average, a factor of 3 higher than those observed at larger radii. These data would argue in favour of a purely "secondary" nature of N, as a result of quiescent CNO processing in stellar envelopes starting from C and O originally present in the star.

Measurements of CNO isotopes generally come from radio and millimetric spectroscopy of interstellar molecules made of CNO atoms. The two molecules which have received the widest attention are carbon monoxide and formaldehyde. Generally, although there is no general consensus on the value of $^{12}C/^{13}C$ in the ISM (from 43 to 75, see Matteucci, 1990a for references), there is evidence for a relative enrichment in time of ^{13}C with respect to ^{12}C. The same is true for the enrichment of ^{14}N with respect to ^{15}N

G. Longo et al. (eds.), Morphological and Physical Classification of Galaxies, 245–260.

and for ^{18}O with respect to ^{17}O and ^{16}O. This is usually interpreted as a nucleosynthesi effect, due to the fact that some elements such as ^{13}C, ^{14}N and ^{18}O have a "secondary" nature. The better established gradient is that of $^{12}C/^{13}C$. Recently, Langer and Penzias (1990) have analyzed data from optically thin CO emissions and found a systemati gradient in the $^{12}C/^{13}C$ isotopic ratio across the Galaxy, ranging from about 30 in the inner part at 5 kpc from the center to about 70 at 12 kpc, with a galactic center value of 24, in agreement with previous studies (Tosi, 1982; Henkel et al., 1982).

Sulphur appears to have a significantly flatter gradient, $\simeq -0.01 \pm 0.02$ dex kpc^{-1} but the data are relative only to the outermost regions of the galactic disc. Recen measurements of S gradients in external galaxies (Evans, 1986; Vilchez et al., 1988) seem also to indicate a similar trend (but see Garnett, 1989).

The gradient of Ne seems to be also relatively flat, at least in our Galaxy, but again data exist only for the external regions of the disc. The gradient of Ar, on the other hand, seems to be very similar to that of O (Shaver et al., 1983).

No information is available about the gradients of elements such as Mg and Si.

Finally, gradients in the Fe abundance are measured from stellar photometry. In particular, from photometry of open clusters Panagia and Tosi (1981) obtained -0.1 ± 0.03 dex kpc^{-1} in a galactocentric distance range 6.5–10.2 kpc, with the solar distance assumed to be 8.5 kpc. With similar spatial resolution Luck (1982) found a similar gradient from spectroscopic observations of 50 late type supergiants.

For a recent critical discussion of abundance gradients derived from stars see Grenon (1987).

1.2 Theory

From the theoretical point of view, many models of chemical evolution have been computed with the aim of reproducing the observed abundance gradients across the galactic disc and many different hypotheses have been proposed:

i) Variation of the IMF across the disc. Unfortunately, there are no clear indications for that.

ii) Variation in the the gas fraction (which is the original idea of the "simple model"). However, the predicted gradients, in this case, are too small when compared to the observed ones (Pagel, 1989).

iii) Variation in the star formation rate to infall rate ratio (Tosi, 1988; Matteucci and François, 1989, hereafter MF).

iv) Inward gas flows caused by infall of low angular momentum material (Lacey and Fall, 1985). In this case, however, radial inflow velocities should not exceed 2 km s^{-1}.

v) Gas flows caused by viscous tranfer of angular momentum across the disc coupled with a treshold in the star formation process (Clarke, 1989).

MF, Matteucci et al. (1989) and Pagel (1989) are the most recent studies concerning abundance gradients of single elements. In particular, MF and Matteucci et al. (1989) have presented theoretical predictions of abundance gradients of O, N, Ne, S and Fe. Their conclusions can be summarized in the following way.

Abundance gradients are obtained by means of a ratio between the star formation rate and the collapse rate increasing with decreasing galactocentric distance, when the initial mass function (IMF) is assumed constant in space and time. The collapse rate corresponds to the rate of mass accretion and is assumed to be a function of galactocentric distance. In particular, the time scale of gas accretion is increasing with radius. This assumption is justified by the results of dynamical models of Larson (1976) and it represents, as we will see better later, a fundamental assumption in order to reproduce abundance gradients as well as the majority of evolutionary differences between the outermost and the innermost galactic regions.

This formulation of the problem leads to conclusions similar to Clarke (1989), who studied the chemical evolution of viscously evolving discs and showed that a metallicity gradient can be obtained with a model where radial flows play an important role only if star formation is suppressed at large radii. In this case the gradient results from inflow of low metallicity material from beyond the cut-off at late times. The net effect of this is the same as the effect of assuming longer time scales of disc formation at larger radii, since in this case the outer disc is still experiencing accretion of primordial/low metallicity material at the present time.

MF's model well reproduces the gradients of O, N and Fe on the galactocentric distance range 4–14 kpc and with an assumed solar distance of 10 kpc. The model provides also an explanation for the differences between abundance gradients of different elements in terms of nucleosynthesis and lifetimes of stellar progenitors of each element. These differences, in fact, cannot be due to the assumed star formation and/or infall rate, since they are the same for each element at a given radius. Therefore, elements mainly produced in massive stars such as O, Ne and S have a flatter gradient than elements mainly produced in single low and intermediate mass stars, such as C and N, or in type I SNe (C–deflagration in white dwarfs in binary systems), such as Fe. This is due to the delay with which these latter are restored to the ISM, which enhances the differences between abundances in the inner and in the outer regions of the disc. Moreover, "secondary" (or even partly "secondary") elements have generally steeper gradients than purely "primary" ones, so that species such as ^{14}N and ^{13}C, which are partly or entirely "secondary" as well as mainly produced in long living stars, have the steepest gradients.

MF found also differences (although small) between gradients of elements produced in massive stars. MF's best model predicts in the range 4–14 kpc S/O and Ne/O gradients of $\simeq +0.015$ dex kpc^{-1} and $\simeq +0.017$ dex kpc^{-1}, respectively. The predicted S/O gradient is smaller, although of the same sign, than those observed in M 101 (Evans, 1986) and M 33 (Vilchez et al., 1988). Concerning our Galaxy, it is difficult to make a comparison because, as already said, the data cover only a restricted galactocentric distance range. On the other hand, Garnett (1989) found negligible differences in the S/O ratio among dwarf irregular galaxies.

MF interpreted these differences as due to the fact that Oxigen comes from a wider range of stellar masses than sulphur and neon, as indicated by current nucleosynthesis calculations (Woosley and Weaver, 1986). This is reflected in a wider range of stellar lifetimes which has the effect of a small delay in the element production, so steepening the Oxigen gradient with respect to those of Ne and S.

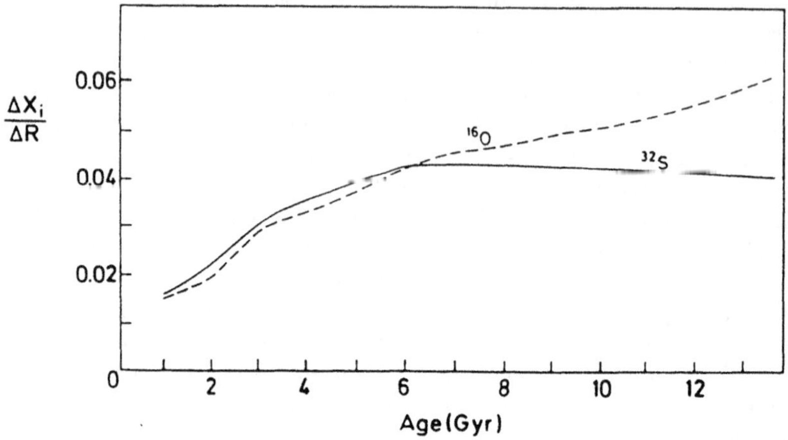

Figure 1. Predicted age-abundance gradient (modulus) relation in the disc of our Galaxy.

In figure 1 is plotted the evolution in time of the gradients of O and S, always in the range 4–14 kpc. The figure shows how gradients are predicted to increase in time but at different rates, due to their different nucleosynthesis and stellar progenitors. As a consequence of this, the observed abundance gradients can help to put constraints on nucleosynthesis theories.

The hypothesis of a variable IMF in order to explain the possible gradients in S/O has been explored by Garnett (1989) by means of a simple model of chemical evolution. His model investigated two hypotheses concerning the variability of the massive star IMF:

1) that the upper mass limit for star formation decreases with increasing metallicity,

2) that the slope of the IMF becomes steeper with increasing metallicity.

The results of his models show a positive gradient of S/O with O/H, contrary to what seems to be observed.

Evans (1986) suggested that the positive sign of the S/O gradient with decreasing radius along the disc of M 101 could be explained if there was some mechanism which enhances the S abundance relative to the O abundance in the outer regions of the disc. In particular, he implied that there must be proportionately more high mass stars than intermediate mass ones at larger radii in the disc of M 101, and that S should be mostly produced in low and intermediate mass stars. Unfortunately, this is at variance with generally accepted results of nucleosynthesis.

In our opinion, every assumption about a variable IMF should be considered quite "ad hoc" in the light of the current knowledge, and it seems more reasonable and natural to explain the positive S/O gradient, if real, as due to differences in nucleosynthesis and stellar progenitors of these two elements.

In table 1 are listed the gradients of ^{4}He, ^{12}C, ^{13}C, ^{16}O, ^{14}N, ^{20}Ne, ^{24}Mg, ^{28}Si, ^{32}S, and ^{56}Fe, obtained with a least–square fit, as predicted by MF's best model with a disc age of 13 Gyr (Sandage, 1988).

This model assumes a star formation law which is a function of both surface gas

Table 1. Predicted and Observed Gradients in the Galaxy (4-14 kpc, ⊙ at 10 kpc)

	Model	Observations
$\frac{\Delta \log(He/H)}{\Delta R}$ [dex kpc^{-1}]	-0.0085	-0.01 ± 0.008[1)]
$\frac{\Delta \log(N/H)}{\Delta R}$ [dex kpc^{-1}]	-0.085	-0.09 ± 0.015[1)]
$\frac{\Delta \log(^{12}C/H)}{\Delta R}$ [dex kpc^{-1}]	-0.066	—
$\frac{\Delta \log(^{13}C/H)}{\Delta R}$ [dex kpc^{-1}]	-0.075	—
$\frac{\Delta(^{12}C/^{13}C)}{\Delta R}$ [dex kpc^{-1}]	+1.5	+4.7[4)];+5.6[5)]
$\frac{\Delta \log(O/H)}{\Delta R}$ [dex kpc^{-1}]	-0.065	-0.07 ± 0.015[1)]
$\frac{\Delta \log(Ne/H)}{\Delta R}$ [dex kpc^{-1}]	-0.048	
$\frac{\Delta \log(Mg/H)}{\Delta R}$ [dex kpc^{-1}]	-0.05	—
$\frac{\Delta \log(Si/H)}{\Delta R}$ [dex kpc^{-1}]	-0.058	
$\frac{\Delta \log(S/H)}{\Delta R}$ [dex kpc^{-1}]	-0.05	-0.01 ± 0.02[1)]
$\frac{\Delta \log(Fe/H)}{\Delta R}$ [dex kpc^{-1}]	-0.07	-0.05[2)];-0.01[3)]

1) Shaver et al. (1983)
2) Janes (1979)
3) Panagia and Tosi (1981)
4) Tosi (1982)
5) Henkel et al. (1982)

density and total surface mass density. The exponent of the surface gas density is $k = 1.$ and it is fixed by reproducing the majority of observational constraints both in the sola neighbourhood and in the whole disc (see MF). In this context, it is worth noting that th present radial distribution of gas (molecular+ neutral hydrogen) across the galactic dis is one of the most uncertain constraints and also one of the most difficult to reproduce What is clear is that star formation laws assuming $k > 1.1$ produce gas distribution which grossly differ from the observations.

On the other hand, models with $k = 1.0$ give an acceptable gas distribution but pre dict too shallow abundance gradients, also when the timescale of disc formation increase with distance from the galactic center.

From analysis of table 1 one can notice that the steepest gradients are those of N and ^{13}C and this is due to their assumed mostly secondary nature, coupled with the fac that they are mostly produced in low and intermediate mass stars.

In the adopted nucleosynthesis prescriptions for low and intermediate mass star (Renzini and Voli, 1981, their case $\alpha = 1.5$, $\eta = 0.33$) part of ^{13}C and ^{14}N has also a primary origin, as a result of the third dredge-up acting in conjunction with the so-callec hot bottom burning in intermediate mass stars during the asymptotic giant branch phase (AGB). Stars which experience such conditions lie in a narrow mass range. The uppe mass limit of this range, M_{up}, e.g. the maximum mass of stars which ignite carbon in a degenerate core and therefore experience the AGB phase, can vary from 5 to 8 $M_\odot$ The lower limit depends on the assumed value for the mixing length in the envelope (α in Renzini and Voli). Both the lower and upper limit are also sensitive to the treatment of convection in the stellar core and to the assumed initial stellar composition. As a consequence, theoretical predictions concerning C and N are quite uncertain.

The predicted $^{12}C/^{13}C$ gradient in the galactocentric distance range 4–14 kpc looks too shallow in comparison with the observed one (Langer and Penzias, 1990). The same problem was found and discussed by Diaz and Tosi (1986). A possible solution would be to assume that the primary fraction of ^{13}C is smaller than assumed or zero. On the other hand, the predicted gradient of ^{14}N is in quite good agreement with that determined by Shaver et al. (1983), and this makes the situation confusing because ^{13}C and ^{14}N are both produced in the CNO cycle in stellar envelopes. Alternatively, ^{13}C could be substantially produced in novae (Starrfield et al., 1978). Therefore, no firm conclusions can be drawn before more observational data and new theoretical calculation will appear.

The predicted gradient of O is in very good agreement with the observed one (Shaver et al., 1983).

The predicted gradient of Iron is similar to that of Oxigen, although slightly steeper, and agrees with observational estimates.

The gradients of Ne, Mg and S are all shallower than the gradient of O although the differences do not exceed $\simeq 0.017$ dex kpc^{-1}. The gradient of Si is steeper than the others and this can be attributed to the fact that type I SNe produce relatively more Si than Ne, Mg and S. The same effect is evident in the predicted |Si/Fe| vs. |Fe/H| relation in the solar neighbourhood (MF, Matteucci, 1990b), where the |Si/Fe| ratio for

|Fe/H| < −1.0 is lower than the ratios of Ne, Mg and S with respect to Iron.

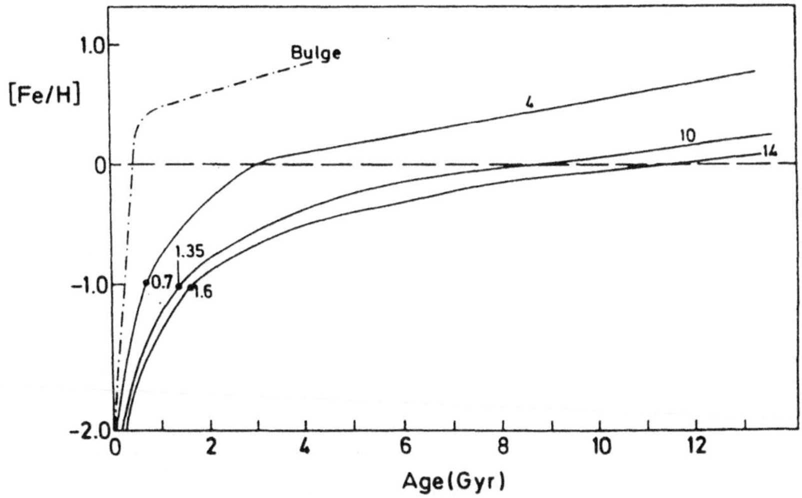

Figure 2. Predicted age-metallicity relations at different galactocentric radii (4, 10 and 14 kpc) in the disc, and for the galactic bulge. The bulge prediction is from the model of Matteucci and Brocato (1990). The timescales at which |Fe/H| = −1.0 is reached are marked on the curves referring to the disc. This time scale is much shorter for the bulge.

4.3 Others gradients across the disc

In figure 2 we show the predicted age-metallicity (metallicity is measured by |Fe/H|) relations at various radii. The solar neighbourhood is assumed to be at 10 kpc from the galactic center and the predicted age-metallicity relation at this radius agrees well with the observed one (Twarog, 1980; Nissen et al., 1985), although the latter is known with large uncertainties only.

It is clear from figure 2 that the age-metallicity relation becomes steeper with decreasing galactocentric distance, from 14 to 4 kpc, and the differences among the various relations are more evident for |Fe/H| > −1.0, which corresponds to the disc phase. However, the timescale at which |Fe/H | = −1.0 is reached is different at different radii, as due to the different star formation history and to the different timescales of mass accretion. The time when the Fe abundance is a factor of ten lower than in the Sun varies from ≃ 0.7 Gyr at 4 kpc to ≃ 1.6 Gyr at 14 kpc, and represents an indication of the duration of the halo phase at those galactocentric distances.

This trend seems to be confirmed by observations of open clusters outside the solar circle which indicate that the chemical enrichment history of the galactic anticenter looks more like that of the Magellanic Clouds than that of the solar neighbourhood (Geisler, 1987).

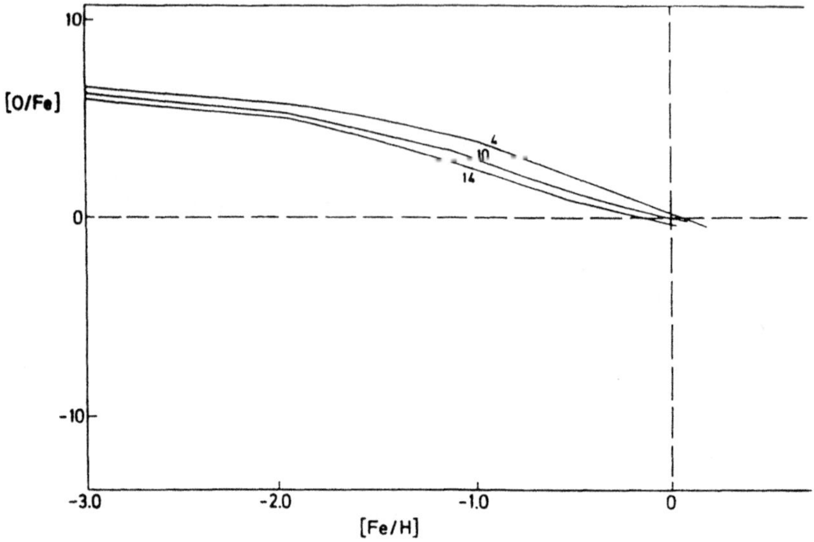

Figure 3. Predicted |O/Fe| vs. |Fe/H| relations at different galactocentric distances in the disc.

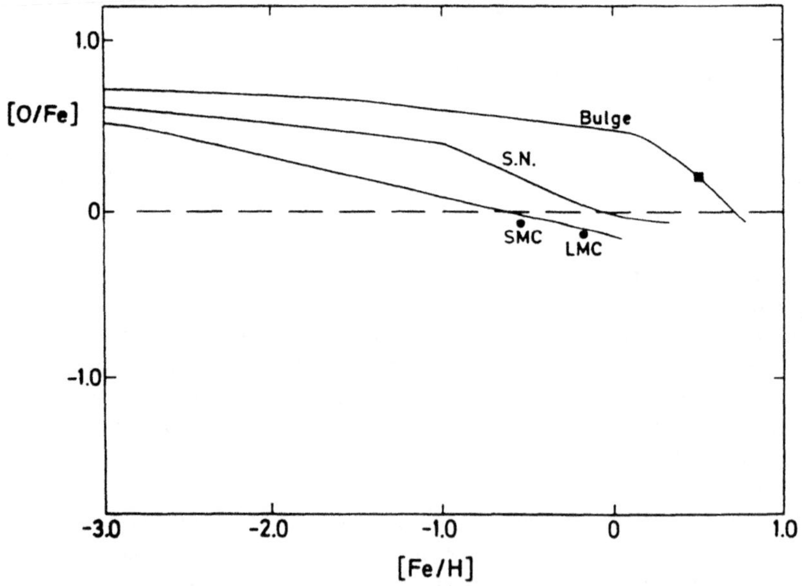

Figure 4. Predicted |O/Fe| vs. |Fe/H| relations for the bulge, the solar neighbourhood (at 10 kpc) and the Magellanic Clouds. The square represents observations from Barbuy and Grenon (1990), whereas the dots represent observations from Russel et al. (1988) and Spite and Spite (1990).

In figure 3 we show the predicted |O/Fe| vs. |Fe/H| relations at various radii. In this case the difference among the various relations is less evident. In particular, very little or no spread is predicted for halo stars. The point at which the slope changes drastically is slightly different at different radii as due to the different age-metallicity relations.

It is worth noting, on the other hand, that a successful model for the galactic bulge (Matteucci and Brocato, 1990) requires an age-metallicity and a |O/Fe| vs. |Fe/H| relation substantially different from those in the disc. This model predicts, in fact, that bulge stars, a large fraction of which has metallicities greater than solar, should have an overabundance of O with respect to Fe as halo stars, which are much more metal poor. This is the result of a substantially different evolutionary history that the bulge must have had, with respect to the disc (see figure 3), in order to account for the observed metallicity distribution of bulge K giants (Rich, 1988, 1990). In figure 4 we show the predicted |O/Fe| vs. |Fe/H| relations in the bulge, in the solar neighbourhood and in the Magellanic Clouds (MCs). The chemical evolution of the MCs corresponds to the evolution of the shell at 20 kpc from the galactic centre in MF's model.

Both the predictions for the bulge and the MCs are in very good agreement with observations, as shown in figure 4. The MCs show, contrarily to the bulge, an underabundance of Oxigen with respect to Iron relative to the solar neighbourhood of the Galaxy. This trend comes out naturally from our models by assuming a much slower evolution for the Clouds with respect to the solar vicinity and to the bulge.

2 External galaxies

2.1 Spirals

The discs of spiral galaxies show radial gradients in the most abundant elements (O, N, S etc.). Oxigen gradients seem to be steeper in the inner zones of the disc ($R/R_{eff} \leq 1.5$–1.75) and are very flat in the outer parts. The gradient seems to be related to the effective radius of the disc, that is the radius containing half of the light, in the sense that galaxies with larger effective radii show a flatter Oxigen gradient and viceversa. Nitrogen gradients are, in general, somewhat steeper than those of Oxigen and there is a well defined trend of S/O vs. O/H with the first ratio increasing with decreasing Oxigen abundance at least from $12 + \log(O/H) \leq 8.5$ (Diaz, 1989).

The interpretation of the abundance gradients given for the Galaxy holds for other spirals. Therefore, gradients can be interpreted as due to an increasing ratio between the rate at which stars form and the rate at which material accumulates to form the disc with decreasing galactocentric distance. On this line, the fact that different spirals show different gradients for the same chemical element can be explained as due to different initial conditions from galaxy to galaxy, leading to different star formation rates and infall rates, as long as the IMF is universal. On the other hand, differences between gradients of different elements in the same galaxy should depend only on stellar nucleosynthesis and the IMF.

2.2 Ellipticals

Radial variations in the metal content of stars between the central and outer parts of ellipticals and S0 galaxies are evident in both line-strength and color gradients. However clear gradients are seen only in luminous ellipticals ($M_B \leq -18$). Moreover, some ellipticals show no or very small gradients.

Generally, colors redden toward the centre although some ellipticals show bluer nuclei (the nuclear colors, in this case, are probably due to recent star formation and/or to non-thermal sources), and line-strengths increase toward the centre.

In a recent work, where UBR photometry of 39 ellipticals and line-strength gradients for 9 giant ellipticals are presented, Peletier (1988) draws the following conclusions:

The gradients in different optical colors correlate well with each other, as do visual-infrared with optical color gradients. This means that the gradients in all colors are well established and are caused by the same physical process (see also Kormendy and Djorgovski, 1989, for a review on the subject).

The color-color relations at different positions within one galaxy are the same as the color-color relations between the nuclei of different galaxies. Therefore, it is likely that the same process that causes one galaxy to be redder than another is responsible for the radial color gradient as well. The color gradients in almost every galaxy can be explained by metallicity decreasing outwards by an average $\Delta \log Z/\Delta \log r = -0.13$. No clear correlation is found in this study between color gradients and galactic luminosity. On the other hand, Vader et al. (1988) find a correlation between color gradients and luminosity and rotational velocity.

Most galaxies show a decrease in Mg_2, <Fe>, NaD and other lines when going outward in radius.

Although gradients in <Fe> are difficult to measure and up to now their quality has not been good enough to derive astrophysically interesting information, Peletier (1988) notices that the Mg_2 gradient and the <Fe> gradient do not correlate well with each other. He suggests that this may be because Mg and Fe do not vary in lock-step, as it has been indeed predicted for elliptical galaxies by Matteucci and Tornambè (1987) and Matteucci and Brocato (1990). They predict that most stars in elliptical galaxies should show overabundances of O and α-elements with respect to Iron. However, no conclusions can safely be drawn on abundance trends on the base of line-strength, because we do not know the precise relation between these metallicity indicators and the real abundance of the element they refer to. The only way to have an estimate of the metallicity is to rely on empirical and theoretical calibrations (Mould, 1978; Burstein, 1979), these latter being especially uncertain. Therefore, differences between gradients of different metallicity indicators should not necessarily reflect the same differences in the corresponding abundances. One common calibration is the one of Burstein (1979):

$$|\mathrm{Fe/H}| = 3.9\mathrm{Mg}_2 - 0.9.$$

This relation is derived from data on globular clusters in the low metallicity range and from theoretical models of Mould (1978), which use stellar tracks with solar abundance ratios, in the high metallicity range. This relation is therefore very uncertain, especially if stars in ellipticals have non-solar ratios. Unfortunately, no information is available on a possible relation between Mg_2 and Mg abundance, $|\mathrm{Mg/H}|$.

From the theoretical point of view we expect the Fe gradient to be steeper than the Mg one, as for the Galaxy, since differences in gradients of different elements depend mostly on stellar nucleosynthesis, when the IMF is assumed constant in a given galaxy (Matteucci and Tornambè, 1987). This assumption seems reasonable because there is no clear evidence insofar that the IMF should vary with the galactocentric distance in ellipticals.

Peletier (1988) finds that line-strength gradients are independent on morphology including deviations from ellipticity, dynamics, global metallicity, etc.., but that the Mg_2 gradient increases with increasing luminosity, although a strong intrinsic scatter is present in this relation. On the other hand, Davies and Sadler (1987) seem to find the contrary. If the metallicity gradient increases with galaxian luminosity, this would argue in favour of dissipational collapse as the mechanism responsible for the creation of metallicity gradients in ellipticals. This is the conclusion of Peletier (1988), who also suggests that the observed spread could be explained by occasional mergers.

Finally, Danziger et al. (1990) are presently calculating line strength gradients in 28 ellipticals. Preliminary results suggest that the Mg_2 gradient strongly varies from galaxy to galaxy and that gradients along different axes in the same galaxy are different.

2.2.1 Theories to explain gradients in ellipticals

a) Dissipative model

In this model elliptical galaxies are thought to be formed as isolated gas clouds in which stars formed very rapidly. The gas is allowed to dissipate energy by falling into the centre before making new stars. These new stars, when formed, will be enriched in heavy elements with respect to the previous generation and so on, thus originating a metallicity gradient. This process lasts until the remaining gas cannot cool and is ejected from the galaxy (Larson, 1975; Carlbergh, 1984).

b) Merger model

Ellipticals are thought to be the result of merging of disc systems (Toomre and Toomre, 1972) without dissipation. The merger model offers less clear predictions about gradients. White (1980) showed that population gradients are only moderately affected by mergers since violent relaxation preserves the ordering of stars in binding energy. Discs of spirals have usually large metallicity gradients so the merger model can probably explain the observed gradients in ellipticals. However, mergers cause initial gradients to decrease, thus one would expect that, in contrast to what is predicted by the dissipative model, gradients diminish with increasing luminosity, since the number of mergers that a galaxy undergoes is expected to increase with luminosity.

c) Cooling flow model

This model predicts that at least a significative fraction of stars in galaxies form from cooling flows (Fabian et al., 1982; Sarazin and O' Connel, 1983; Valentijn, 1983). In this model star formation is still going on, but the bulk of stars must have masses smaller than 0.7 $M_{\odot}$, since otherwise the observed colors would be too blue.

In this picture, no metallicity gradients but age gradients or different mass distributions of stars are expected. Valentijn and Morwood (1985) predict an increase of the upper cut-off of the mass distribution with radius, which then causes the observed color and line-strength gradient.

A strong argument against this model and the assumption that the gradients reflect differences in age and not in metallicity is that the model predicts that recent star formation should take place in the outer regions, and there is no observational evidence for that.

2.3 Irregular galaxies

Abundance gradients in these systems are small or absent. There is at most only a weak gradient in the LMC (equivalent to ≤ -0.04 dex kpc^{-1} at a distance of 42 kpc). There is no evidence for any gradient in SMC (Edmunds, 1989).

3 Summary and conclusions

Abundance gradient observations across the disc of the Galaxy have been reviewed.

A model of chemical evolution for the galactic disc has been presented. The model assumes a star formation rate proportional to both surface gas density and total surface mass density, and that the time scale of disc formation increases with galactocentric distance. The IMF is constant in space and time. Radial gas flows are neglected.

Detailed nucleosynthesis from type I and II SNe is taken into account and the evolution of several elements is computed in space and time. This model reproduces the majority of observational constraints both in the solar vicinity and across the disc. In particular it predicts:

a) negative gradients of O, ^{14}N and Fe in very good agreement with observations, in the galactocentric distance range 4–14 kpc.

Negative gradients of Ne, Mg, Si and S.

Positive, but small, S/O , Ne/O and Mg/O gradients, of the order of +0.015 dex kpc^{-1} +0.017 dex kpc^{-1}, and +0.015 dex kpc^{-1}, over the same galactocentric distance range.

Under the assumptions made, the existence of abundance gradients is due to the fact that the star formation rate/infall rate ratio increases with decreasing radii.

The existence of a positive gradient in the S/O ratio is, on the other hand, mainly due to the different lifetimes of the stellar progenitors of these two elements and to their nucleosynthetic processes, as well as for any gradient in elemental ratios.

The predicted ^{13}C/^{12}C gradient is smaller than observed. This can be attributed to the poor knowledge of the nucleosynthesis of C in low and intermediate mass stars, the main point being if ^{13}C is purely "secondary" or if part of it is produced also as a "primary" element. New calculations of nucleosynthesis in low and intermediate mass stars are necessary before drawing firm conclusions.

All the predicted gradients increase in time, but at a different rate depending on their progenitors and nucleosynthesis. This prediction agrees with observational estimates which indicate a steeper gradient of Fe in young clusters than in older ones (Panagia and Tosi, 1981).

b) Different age-metallicity ([Fe/H]) relations are predicted at different radii. In particular, the growth of metallicity in time is predicted to be shallower at larger galactocentric distances, as a consequence of the slower evolutionary rate of these regions with respect o the inner ones and the bulge. There is observational evidence going in this direction Geisler, 1987). The greatest difference in the various age-metallicity relations occurs for Fe/H| > −1.0, and is of the order of ≃ 0.5 dex, which is greater than typical observaional errors (≃ 0.2 dex).

Therefore, if the solar neighbourhood disc stars were a mixture of stars born at lifferent radii (Grenon, 1990), the observational age-metallicity relation should show an ntrinsic spread of ≃ 0.5 dex.

) Different |O/Fe| vs. |Fe/H| relations are also predicted at different radii, although the pread is not more than 0.14 dex, which is well inside the observational errors.

l) The comparison between model results and data strongly suggests a picture for Galaxy ormation very similar to that proposed by Larson (1976), where the spheroidal componeents of the Galaxy (like halo and bulge) and the innermost regions of the disc formed irst by means of higher collapse and star formation rates, whereas the outermost regions ormed more slowly and may be they are still forming now.

Differences in the time scales of halo formation at different radii are also found, although they are negligible with respect to the differences in the timescales of formation of different regions of the disc, and they are well inside the uncertainties in stellar age determinations.

Colour and line strength gradients in elliptical galaxies have also briefly been discussed. The main conclusions can be summarized as follows:

) The most simple and straightforward interpretation of such gradients is that they are produced by abundance variations. Several scenarios have been proposed to explain gradients in ellipticals but the most convincing one is the dissipational collapse model. The observational evidence does not exclude that some ellipticals may be the results of merging of spirals, although it seems difficult to believe that mergers represent the fundamental process in elliptical formation.

) Abundance gradients in ellipticals, as in spirals, can be a good test of stellar nucleosynthesis and SN progenitor theories. Unfortunately, the relation between abundance indicators and real abundances as well as between abundances and colors are still very uncertain, so no firm conclusions can be drawn about differences in gradients of different elements.

References

Barbuy, B., Grenon, M.: 1990, in *Bulges of Galaxies*, ESO–CTIO Workshop, ed. B.J. Jarvis and D.M. Terndrup, (E.S.O. Press).

Burstein, D.: 1979, *Astrophys. J.* **232**, 74.

Carlbergh,R.G.: 1984, *Astrophys. J.* **286**, 403.

Clarke, C.J.: 1989, *Mon. Not. R. Astr. Soc.* **238**, 283.

Danziger, I.J., Brocato, E., Buson, L., Matteucci, F.: 1990, in preparation.

Davies, R.L., Sadler, E.M.: 1987, in *Structure and Dynamics of Elliptical Galaxies*, IAU Symp. n. 127, ed. T. De Zeeuw, p. 441 (Dordrecht: Reidel).

Diaz, A.I., Tosi, M.: 1986, *Astron. Astrophys.* **158**, 60.

Diaz, A.I.: 1989, in *Evolutionary Phenomena in Galaxies*, ed. J.E. Beckmam and B.E.J. Pagel, p. 37 (Cambridge University Press).

Edmunds, M.G.: 1989, *Rev. Mod. Astron.* **2**.

Evans, I.N.: 1986, *Astrophys. J.* **309**, 544.

Fabian A.C., Nulsen, P.E.J., Canizares, C.R.: 1982, *Mon. Not. R. Astr. Soc.* **201**, 933.

Garnett, D.R.: 1989, *Astrophys. J.* **345**, 282.

Geisler,D.: 1987, *Astron. J.* **94**, 84.

Grenon, M.: 1087, *J. Astrophys. Astr.* **8**, 123.

Grenon, M. : 1990, in *Bulges of Galaxies*, ESO–CTIO Workshop, ed. B.J. Jarvis and D.M. Terndrup, p. 143, (E.S.O. press).

Henkel, C., Wilson, T.L., Beiging, J.: 1982, *Astron. Astrophys.* **109**, 344.

Janes, K.A.: 1979, *Astrophys. J. Suppl.* **39**, 135.

Kormendy, J., Djorgovski, S.: 1989, *Ann. Rev. Astron. Astrophys.* **27**, 235.

Langer, W.D., Penzias, A. : 1990, *Astrophys. J.*, in press.

Lacey, C.G., Fall, S.M.: 1985, *Astrophys. J.* **290**, 154.

Larson, R.: 1975, *Mon. Not. R. Astr. Soc.* **173**, 671.

Larson, R.: 1976, *Mon. Not. R. Astr. Soc.* **176**, 31.

Lester, D.F., Dinerstein, H.L., Werner, M.W., Watson, D.M., Genzel, R., Storey, J.W.V.: 1987, *Astrophys. J.* **320**, 573.

Luck, R.E.: 1982, *Astrophys. J.* **256**, 177.

Matteucci, F.: 1990a, in *Chemistry in Space*, ed. M. Greenberg, (Kluwer Academic Publishers: Dordrecht), in press.

Matteucci, F.: 1990b, in *Frontiers of Stellar Evolution*, ed. D.L. Lambert, (P.A.S.P.), in press.

Matteucci, F., Brocato, E.: 1990, *Astrophys. J.*, in press.

Matteucci, F., François, P.: 1989, *Mon. Not. R. Astr. Soc.* **239**, 885.

Matteucci, F., Franco, J., François, P., Treyer, M.A.: 1989, *Rev. Mex. Astron. Astrofis.* **18**, 145.

Matteucci, F., Tornambè, A.: 1987, *Astron. Astrophys.* **185**, 51.

Mould, J.R.: 1978, *Astrophys. J.* **220**, 434.

Nissen, P.E., Edvardsson, B., Gustafsson, B.: 1985, in *Production and Distribution of CNO Elements*, ed. I.J. Danziger, F. Matteucci and K. Kjär, (E.S.O.: Munich), p. 31.

Pagel, B.E.J.: 1989, *Rev. Mex. Astron. Astrofis.* **18**, 162.

Panagia, N., Tosi, M.: 1981, *Astron. Astrophys.* **96**, 306.

Peletier, R.F.: 1988, Ph. D. Thesis, University of Groningen, The Netherlands.

Renzini, A., Voli, M.: 1981, *Astron. Astrophys.* **94**, 175.

Rich, R.M.: 1988, *Astron. J.* **95**, 828.

Rich, R.M.: 1990, in *Chemical and Dynamical Evolution of Galaxies*, ed. F. Ferrini et al., (Giardini: Pisa), in press.

Rubin, R.H., Simpson, J.P., Erickson, E.F., Haas, M.R.: 1988, *Astrophys. J.* **327**, 377.

Russel, S.C., Bessel,M.S., Dopita, M.A.: 1988, in *The Impact of Very High S/N Spectroscopy on Stellar Physics*, ed. G. Cayrel de Strobel and M. Spite, p. 545, (Dordrecht: Reidel).

Sandage, A.: 1988, in *The Calibration of Stellar Ages*, ed. A.G. Davis Philip, p. 43, (L. Davis Press).

Sarazin, C.L., O'Connell, R.W.: 1983, *Astrophys. J.* **268**, 552.

Shaver, P.A., McGee, R.X., Newton, L.M., Danks, A.C., Pottasch, R.S.: 1983, *Mon. Not. R. Astr. Soc.* **204**, 53.

Spite, M., Spite, F.: 1990 *Astron. Astrophys.* **234**, 67.

Starrfield, S., Truran, J.W., Sparks, W.M., Arnould, M.: 1978, *Astrophys. J.* **222**, 600.

Toomre, A., Toomre, J.: 1972, *Astrophys. J.* **178**, 623.

Tosi, M.: 1982, *Astrophys. J.* **254**, 699.

Tosi, M.: 1988, *Astron. Astrophys.* **197**, 33.

Twarog, B.A.: 1980, *Astrophys. J.* **242**, 242.

Vader, J.P., Vigroux, L., Lachieze–Rey, M., Souviron, J.: 1988, *Astron. Astrophys.* **203**, 217.

Valentijn, E.A.: 1983, *Astron. Astrophys.* **118**, 123.

Valentijn, E.A., Moorwood, A.F.M.: 1985, *Astron. Astrophys.* **143**, 46.

Vilchez, J.M., Pagel, B.E.J., Diaz, A.I., Terlevich, R., Edmunds, M.G.: 1988, *Mon. Not. R. Astr. Soc.* **231**, 257.

White, S.D.M.: 1980, *Mon. Not. R. Astr. Soc.* **191**, 1p.

Woosley, S.E., Weaver, T.A.: 1986, in *IAU Coll. n. 89*, ed. D. Mihalas and K.H. Winkler, p. 91, (Dordrecht: Reidel).

Discussion

P. van der Kruit - You skipped rather quickly over the possibility of a variable IMF, but there is the suggestion for bimodal star formation by Larson and Sandage, where the later also discussed observational evidence. Locally we still have the dichotomy of B- and T-associations which suggest different sites of large- and low-mass stars. Can you rule out bimodal star formation?
F. Matteucci - I can certainly rule out, on the basis of chemical evolution models, all the parametrizations proposed insofar for bimodal star formation, since none of them can reproduce the majority of properties in the galactic disc. On the other hand, I cannot rule out the possibility that star formation in the real world is indeed bimodal.

D. Burstein - Since there are a number of people now trying/have to measure/measured absorption line gradients in galaxies, I would like to emphasize two points.
1) The Fe/H–Mg_2 correlation that I developed in 1979 must be done better. It would be a minor miracle if the first "guess" was correct. 2) In a review I wrote in 1985/P.A.S.P., I point out that Mg_2 strength in an integrated stellar population is much more strongly related to the giant branch temperature, as opposed to the Fe/H dependence of Mg_2 itself.

S.G. Djorgovski - You implied that some ellipticals show different metallicity gradients along different axes. Could you tell then whether the isometallicity contours have the same shape as the isophotes or not? This is predicted in many dissipative formation models, but to my knowledge has never been confirmed.
F. Matteucci - Unfortunately, at the moment we do not have enough informations to confirm this important point.

P. van der Kruit - How many free parameters does your model for the chemical evolution of the Galaxy have?
F. Matteucci - The time scale for the formation of the disc and the star formation rate. Any way, the number of observational constraints that the model reproduces are much more than two.

Francesca Matteucci lecturing.

STRUCTURE AND METALLICITY OF LOW-LUMINOSITY EARLY-TYPE GALAXIES

Ph. Prugniel[1], E. Bica[2], D. Alloin[3]
[1]*European Southern Observatory, Garching, Germany*
[2]*University of Porto Alegre, Brazil*
[3]*Observatoire de Meudon, France*

ABSTRACT. We collected photometric and spectroscopic data for a sample of early-type galaxies ($M_B < -19$, $H_0 = 100$) distributed over the region between giant, compact and dwarf elliptical galaxies in a mean surface brightness vs. luminosity diagram. They appear to join the sequences of giants and dwarfs in the *central* surface brightness vs. luminosity diagram. We also give evidence for a segregation of the metallicity with surface brightness at a given luminosity: a lower surface brightness is associated with a weaker metallicity.

1 Motivations

The motivation for studying low-luminosity early-type galaxies (LLE) is twofold. First, there is an old controversy concerning the physical relation between the different types of elliptical galaxies - giants, compacts and dwarfs - (e.g. Wirth and Gallagher, 1984). Second, elliptical galaxies present a wider range of values for their fundamental parameters - surface brightness and velocity dispersion - as one deals with lower luminosities. Thus, the LLE are well-suited for analyzing the parameters that determine their characteristics.

At low luminosities, elliptical galaxies split in two classes: the compact ellipticals (hereafter CE, as in the case of M32) and the dwarfs (hereafter dE, like NGC 205), forming a sequence of decreasing surface brightness which ends with the dwarf spheroidal galaxies (such as Fornax). We call LLE all the elliptical galaxies fainter than $M_B = -19$ (H_0=100) including dE and CE.

We initiated a program for studying the structure and stellar content of LLE, with the particular aim of bridging, if possible, the gap between dE and CE. We discuss here the structure of 16 LLE in the Virgo cluster with $-16 < M_B < -17$, by means of CCD photometry, and the metallicities of 55 field LLE with $-17 < M_B < -19$, from CCD spectrophotometry. The two samples are shown in the mean surface brightness vs. luminosity diagram (Fig. 1) where M_B values have been computed using group velocities and a Great-Attractor model. It is clear from this plot that the Virgo LLE fill the surface brightness gap between dE and CE.

G. Longo et al. (eds.), Morphological and Physical Classification of Galaxies, 261–264.

2 The structure of LLE

We measured the radial velocities of 20 CE candidates from the list of Binggeli et a (1985), using the CARELEC spectrograph at the 1.93m telescope in Observatoire Haute-Provence. Out of these, 16 appeared to be LLE, for which we subsequently o tained CCD images from the 3.6m CFH and 2m Pic du Midi telescopes. We have used figure 1 the mean surface brightness within the effective aperture, SB_e, and the asym totic magnitude derived from this new surface photometry. We have determined as w the central surface brightness and core radius corrected for seeing effects. Our observ tions span a range in seeing of 0.7 arcsec $<$ FWHM $<$ 2.0 arcsec, and the r_c/σ^* paramet (Kormendy, 1985) is in the range $1.8 < r_c/\sigma^* < 5$. We applied a seeing correction aft Kormendy (1985), to both the CFH and Pic du Midi observations. These correctio are large on average, from 0.5 to 2 magnitudes. The resulting central surface brightne and core radius are displayed in figure 2, as a function of the asymptotic magnitude, f the present sample and for the galaxies observed by Kormendy (1985). *The LLE form sequence joining the E to the dE.*

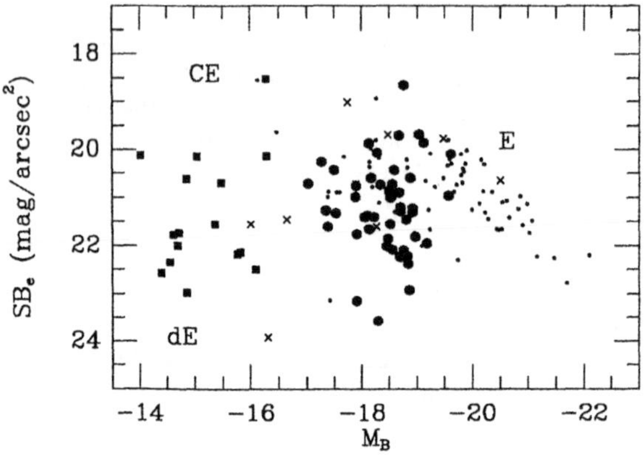

Figure 1. *Effective surface brightness vs. absolute magnitude.* Small dots correspond to the old sample studied by Bica and Alloin. Large dots are the present sample of LLE from the field, crosses are the galaxies with a young stellar population. Squares are the LLE in Virgo.

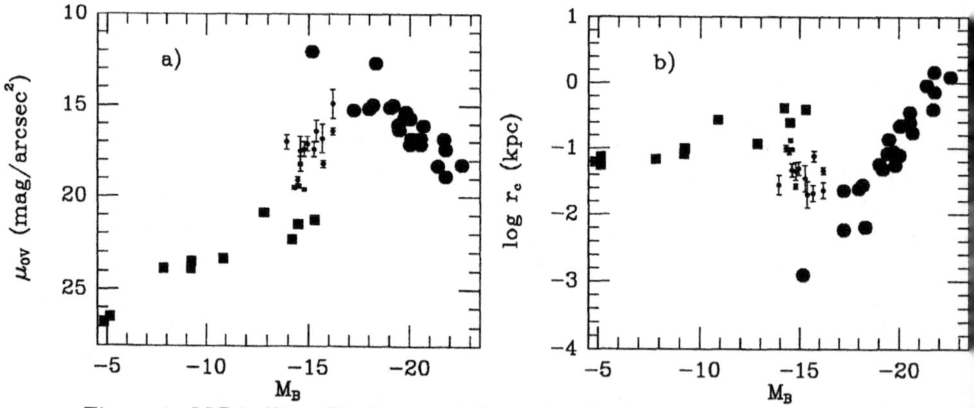

Figure 2. *LLE in Virgo.* The large symbols are the galaxies from Kormendy (1985). The Virgo galaxies are the small dots with error bars corresponding to the uncertainty on the seeing correction. a) Central surface brightness and b) logarithm of the core radius (both corrected for seeing effect) vs. absolute B-magnitude.

3 The metallicity of LLE

Ve selected LLE in the ESO-LV catalogue complemented by other lists for $-37 < \delta < 0$. Ve obtained long-slit CCD spectra for 56 of these galaxies using the ESO 1.52m telescope. he spectral coverage is 3500 Å to 10000 Å with a resolution of 15 Å. Equivalent widths W) have been measured as in Bica and Alloin (1986, 1987a). Detailed population yntheses will be presented elsewhere. We have detected 7 galaxies with significant ractions of young/intermediate age populations as well as one new Seyfert 2. We have plit the remaining LLE in two metallicity groups according to $W(Mg + MgH)$. The ow metallicity group corresponds to 7.0 Å $< W(Mg + MgH) <$ 8.5 Å(18 galaxies) and he high metallicity one to 8.5 Å $< W(Mg + MgH) <$ 10.5 Å (30 galaxies).

Figure 3 consists of an SB_e vs. M_B diagram in which the two groups have been epresented with different symbols. They are clearly separated in the sense that the netal-rich one shows a higher surface brightness. The average SB_e for the metal rich nd the metal poor groups are respectively: $SB_e = 20.9 \pm 0.7$ and 21.8 ± 0.8.

This is evidence for a *dependence of the metallicity on surface brightness at a given uminosity*, at least for the luminosity interval defined earlier. We emphasize that this rend was found thanks to a surface brightness range larger in our sample than in usual amples of giant ellipticals where SB_e is correlated with M_B (see figure 1).

Thus, in addition to the well-known correlation between metallicity and luminosity e.g. Bica and Alloin, 1987b), the present result suggests a two-parameter relation nvolving also the surface brightness. This corresponds to the intuitive idea of a *simple lependence of the metal enrichment on the central escape velocity* V_{esc} (V_{esc} can be xpressed as a combination of SB_e and M_B).

The existence of an SB_e vs. metallicity relation implies that there is no strong tructural evolution in the LLE after their bulk of stellar formation epoch.

4 Bridging E, CE and dE

E and CE galaxies exhibit similar (high) metallicities, in contrast with the systematically ow metal content of dE (Faber, 1973; Bica et al., 1990). E and CE galaxies also present imilar ($r^{1/4}$) profile shapes, whereas those of dE galaxies are exponential (Wirth and Gallagher, 1984). Results by Kormendy (1985) support a continuity between E and CE n the SB_0 vs. M_B diagram, and a discontinuity between E and dE. Yet, there is also strong evidence for an E to dE continuity, for example, in the SB_e vs. M_B diagram (Binggeli et al., 1984), where the CE lay apart from the general E to dE sequence. These conflicting pieces of evidence led to the controversy about the physical relations between the 3 classes of galaxies.

Let us point out that this discussion is substantially misled by selection effects. On one hand, there is a strong bias against the inclusion of intermediate and high surface brightness LLE in many studies, since these galaxies can only be distinguished from background giant ellipticals through redshift measurements. Indeed, LLE become apparent only in carefully selected complete samples like that for Virgo (Binggeli et al., 1985). On the other hand, the E to CE continuity is artificially enhanced by the presence of bulges of spiral galaxies on the classical diagrams by Kormendy (1985).

From the present results, we find that none of the three groups, E, CE, dE, appear to be isolated, either in metallicity, surface brightness or luminosity. Considering the existence of LLE bridging the E, dE, CE gaps as well as the smooth E to dE metallicity transition, we conclude that the most plausible picture is that of a global continuity between E, dE and CE.

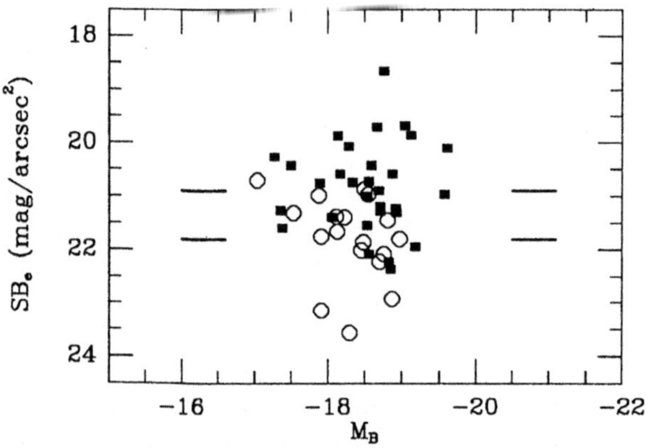

Figure 3. *Dependence of metallicity on SB_e.* The squares are the high metallicity group and the circles the low-metallicity. The horizontal bars note the average SB_e for each group.

References

Bica, E., Alloin, D.: 1986, *Astron. Astrophys.* **162**, 21.
Bica, E., Alloin, D.: 1987a, *Astron. Astrophys. Suppl. Ser.* **70**, 281.
Bica, E., Alloin, D.: 1987b, *Astron. Astrophys.* **181**, 270.
Bica, E., Alloin, D., Schmidt, A.A.: 1990, *Astron. Astrophys.* **228**, 23.
Binggeli, B., Sandage, A., Tarenghi, M.: 1984, *Astron. J.* **89**, 64.
Binggeli, B., Sandage, A., Tammann, G.A.: 1985, *Astron. J.* **90**, 1681.
Faber, S.M.: 1973, *Astrophys. J.* **179**, 731.
Kormendy, J.: 1985, *Astrophys. J.* **295**, 73.
Wirth, A., Gallagher, J.S.: 1984, *Astrophys. J.* **282**, 85.

INTERACTIONS: WEAK

F. Combes
Department de Radioastronomie
Observatoire de Meudon, Paris, France

1 Introduction

This short review is restricted to some recent work on the domain of galaxy interactions. First, I will make a brief point on dynamical friction at large distance, and show that it is difficult to predict the sinking rates of ellipticals and spirals. The Chandrasekhar approach gives resonable orders of magnitude, but does not include all the essential physics: it sometimes over- and sometimes under-estimates the relative decay rates of two galaxies. I will then describe the direct evidence of tides between spheroidal systems, and show that these tides are relatively well understood. The phenomenon of shells in ellipticals will not be discussed here (see the review by Prieur, 1990). Then I will address some of the complex problems relative to spiral galaxies; the complexity comes from the gas dynamics and the multi-components: stability, spiral waves, resonances, rings. I will then consider tidal interactions with dwarfs and the evidence of gas infall on galaxies. Groups and clusters will not be mentioned here, recent reviews can be found by White (1990) and Tremaine (1990).

2 Dynamical friction

Any massive object traveling through a background of collisionless particles suffers a braking force, due to the wake of particles strongly deviated by its passage and trailing behind him. For a mass M moving with uniform velocity v through a homogeneous background of density ρ, the braking force and the sinking rate can be obtained from the Chandrasekhar formula (Chandrasekhar 1943, hereafter C43): the deceleration γ is proportional to $\rho(< v)\ M/v^2\ ln(\Lambda)$, where Λ is the ratio of max to min impact parameter b_{max}/b_{min}. However, this approach is much too simple when confronted to the interaction between two galaxies, where the background density is far from being costant. Even the dominant physical phenomena in dynamical friction are not well known and have been questioned recently:

G. Longo et al. (eds.), Morphological and Physical Classification of Galaxies, 265–276.

a) self-gravity

First Lin and Tremaine (1983) surprisingly confirmed the validity of the Chandrasekhar formula by multiple three-body technique (neglecting self-gravity). But in order to eliminate relaxation effects, they have fixed the center of the target-galaxy in an inertial frame. When this condition is released, simulations show that the braking is then much less, whether one let the background particles interact with themselves or only with a mean-field. After some controversy (White 1983; Bontekoe and van Albada 1987; Zaritsky and White 1988), it appears that self-gravity is indeed negligible for mass ratio of the order of 0.1 between the satellite and the target. This means that collective effects in the target galaxy negligibly enhance the gravitational braking of a satellite (Fig. 1). Weinberg (1986, 1989) using linear perturbation theory, and Hernquist and Weinberg (1989) with simulations, confirmed these results, although they claim that the introduction of self-gravity slows the orbital decay of a satellite. But this is only a question of point of view. In the logic of analytical perturbations development, they also fix the target in an inertial frame, taking the center of mass shift as an aspect of self-gravity. However, it is more interesting to dissociate the two effects, which are not of same nature: first, the displacement of the barycenter of the target (which gives rise to an inertial force); second the true deformation of the primary, which takes the orbital energy and is the only decelerating agent.

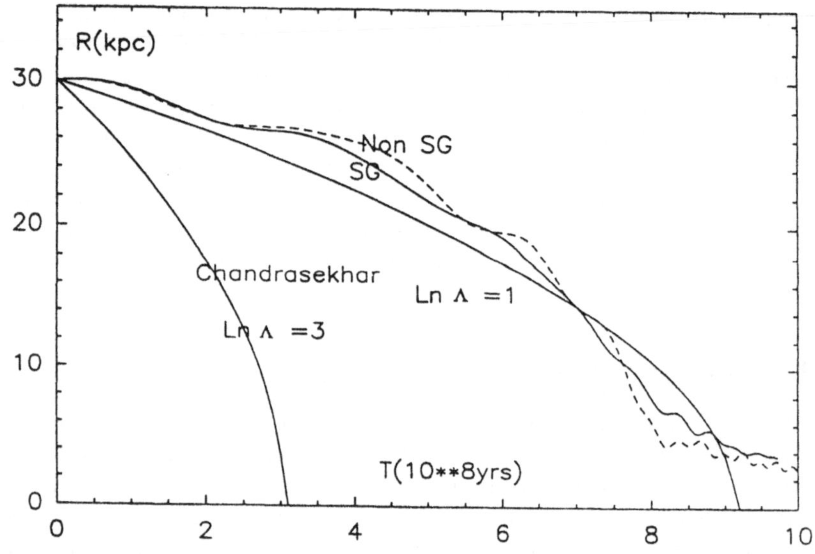

Figure 1. Rate of decay of a small rigid companion (mass ratio 0.1) on an elliptical galaxy of mass $10^{12} M_o$, initially in a circular orbit. The N-body simulation (SG) and the test-particle approach (non-SG) give the same results. The decay rate deduced from the Chandrasekhar approach is also shown.

In the application of the C43 formula, the target is also fixed. This comes from the approximation of the infinite medium, with infinite mass, that would not recoil on the satellite passage. Therefore the essential term in the deformation that brakes the companion is then of dipolar form, which corresponds mainly to a shift of the barycenter, and not to a real deformation. But a mere shift cannot decelerate the relative orbital motion. The true deformation must be computed in the center-of-mass frame, where the dipolar term is always zero. The essential term in the deformation is now quadrupolar, and the deceleration of smaller magnitude. In that respect, the C43 formula *overestimates the friction*. Simulations can still be compared with the C43 formula, where the decelerating force is known within a free proportionally factor $ln\,\Lambda$. The latter has been varied by a factor 10 or so, and even varied during the sinking of the satellites (Bontekoe and van Albada 1987).

b) The global-local character

The Chandrasekhar approach is only local, in the sense that the braking force depends only on the local density ρ. This is not realistic of course, and especially for the friction at distance, where the galaxies are not overlapping. There the "dynamical friction" is due to the tidal deformations of the target galaxy, which are of global nature. The approach is quite different from the local one, where there is no tides. Now the Chandrasekhar formula *under-estimates the friction!*

Figure 2 shows the kind of deformation expected for the target, when developed in the fixed unperturbed center (Weinberg 1989): the $l = 1$ dipole term (corresponding mainly to the displacement of the barycenter of the target), the $l = 2$ quadrupole term, a braking term because of its phase lag with the satellite, than the higher order terms forming the local wake behind the satellite. The relative importance of these two contributions depends strongly on the relative distance, and on the satellite extension (Prugniel and Combes, in prep.).

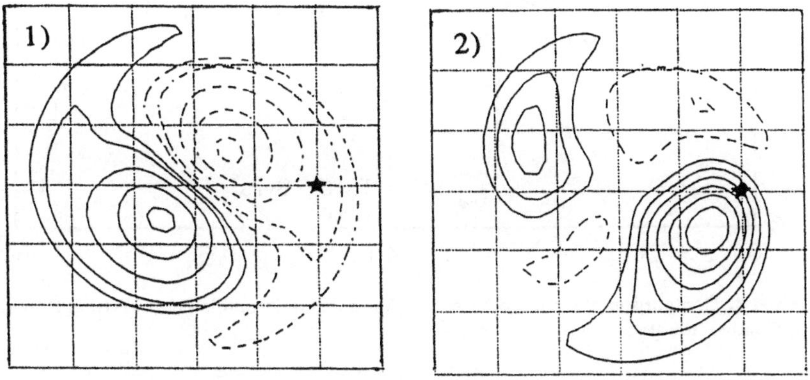

Figure 2. Representations of the deformation of the target: 1) dipole (l=1) deformation due to the center of development non coinciding with the barycenter 2) quadrupole (l=2) and higher-order terms, when l=1 is subtracted (from Weinberg 1989). The companion is presented by a star.

c) Rigid companion

The application of the C43 formula for the braking force between two galaxies consider the companion as a point mass. But of course, it is extended. This implies first an over estimate of the friction, because the impact parameter of background particles are in fac limited, and also a dynamical stripping: the companion center is more braked than th outer parts (Mulder 1983). But these are small effects. The dominant one, which ha not yet been fully investigated (recent work on the decay rates of companions represen the latter as a rigid Plummer potential), is the tidal deformation of the companion. Thi deformation is much more important than that of the target, because of the mass ratio We have simulated this phenomenon and found a tail-like structure for the companio (Prugniel and Combes 1990); this deformation takes away the orbital energy and angula momentum of the relative motion (mainly in escaped particles). The decay time-scale can be reduced by factors 2 or 3 (Fig. 3). This has also been found by Balcells and Quinn (1990), who notice that the sinking rates of two comparable-mass galaxies car be multiplied by 2. They also see significantly different sinking rates for retrograde and prograde encounters, when one of the elliptical galaxy is rotating.

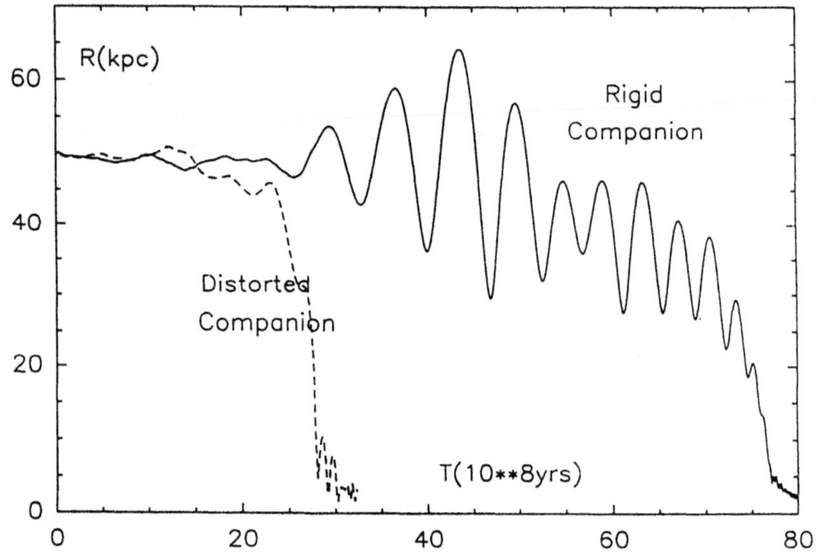

Figure 3. Comparison of the decay rates for a rigid and deformable satellite (from Prugniel and Combes 1990).

d) Disk-galaxy interactions

Dynamical friction is then more complex, since resonance effects can be dominant (Palmer and Papaloizou 1982, 1985). These resonance effects could transfer energy and angular momentum to the satellite, which would then move away from the target instead of being braked. Simulations however show that the friction is dominant, and that results are well represented by the impulsive approximation for parabolic or hyperbolic orbits (Byrd et al., 1986; Valtaoja, 1990), but self-gravity reduces the decay time by a factor 2 for bound orbits (Quinn and Goodman, 1986).

In summary, only orders of magnitudes can be given for the decay rates of interacting galaxies; the decay time-scale is of the order of their orbital periods.

3 Tides in elliptical galaxies

3.1 Stripping and luminosity profiles

The outer luminosity profiles of elliptical galaxies could be largely due to tides (cf. the tidal classes of Kormendy, 1977). Contrary to the theoretical expectations (tidal truncation, King, 1962), the outer radii of ellipticals can be puffed up by tidal actions (Miller, 1986; McGlynn, 1990). Also, the central density of the victims are significantly decreased: this means that energy is added to the cores of the systems (they would be more and more sensitive to subsequent encounters). Galaxy interactions, at least between stellar systems, are not the mechanism to produce compact galaxies, although compact ellipticals appear gathered around brighter ones (Nieto and Prugniel, 1987; Prugniel et al., 1989). Dissipative phenomena involving gas would be necessary.

Of course, the initial concentration of the perturbed system determines the intensity of the produced damage. Spherical halos around galaxies are very fragile, and can be quickly stripped by the cluster tides (Merritt, 1984), while dwarf-ellipticals will only be stripped in their outer parts. Miller (1986) for moderately concentrated spheroids, find that tidal disruption is a sudden phenomenon: the galaxy either suffers little mass loss or is totally disrupted. The orbit plays also a fundamental role: a steady tidal field, such as experimented during a circular orbit around a target, is much less catastrophic than when the system is on a non-circular orbit (McGlynn, 1990). The remnant density profile is found to vary in r^{-4}. In any case, predictions are difficult, since no simple overall scaling of mass loss and energy change in function of collision strength have been found: Aguilar and White (1985) derive cross-sections depending very little on the internal velocity structure of galaxies, but strongly on surface density profiles in the outer parts.

3.2 Characteristic deformations

Tidal effects are directly observed in physical binaries, where have been recently emphasized symmetrical distortions of isophotes (Davoust and Prugniel, 1988; Borne and Hoessel 1988): one side of each galaxy is compressed, while the opposite side is distended. These distortions are most likely due to tides, as seen in the simulations (Borne et al., 1988; Balcells et al., 1988). Indeed for non-rotating spherical systems of comparable mass, the distortions are expected to be symmetric, due to the action-reaction principle. There is no privileged direction, except that defined by the orbit. This is very different from the distortions observed in disk-galaxies, where complete asymmetry is found between tidal from the pair members. Kinematically, these systems are also perturbed, their rotation curve reveals a U-shape. This could be interpreted as tidal friction in action (Borne and Hoessel, 1988).

The question can be raised of the life-time of similar binaries, but supergiant and in the center of clusters: the dumb-bell galaxies. According to numerical simulations, rapid merging should occur in one or two orbital periods (White, 1978). However the observed abundance of multiple systems in clusters leads to an "empirical" life-time larger than 4×10^9 yrs according to Rood and Leir (1979). Rix and White (1989) recently proposed that equilibrium models of dumb-bell galaxies could be built to match the observations. However, these models would concern only symmetric and un-distorted systems.

4 Disk galaxies and spiral structure

4.1 Encounters of self-consistent disk-halo galaxies

For a long time, interactions between disk-galaxies have been better modeled by test-particles, à la Toomre and Toomre (1972). This comes from the difficulty to stabilize a self-consistent cold disk. Indeed, to avoid spontaneous instabilities of the $m = 2$ type (bars or spirals) or even higher m, simulated disks must be heated (large-velocity dispersion) or be surrounded by a heavy spherical halo. Both remedies have the consequence to decrease the sensitivity of the system to perturbers; the response then reduces more to that of spheroidal systems: wide and fat tails, tidal bulges, and not to the thin filaments characteristics of tides in spiral galaxies. The solution resides in the consideration of several components: not only hot spheroidal haloes, but several populations of stars including the young "cold" population recently formed out of the gas component, which forms the thin tails and filaments. But their proper treatment requires a lot more particles than were available at the beginning, the more so as instabilities can be easily triggered by the $\sqrt{N}$ fluctuations (Clutton-Brock, 1972; Farouki and Shapiro, 1981; Gerhard, 1981). Barnes (1988) began to address these questions first by purely stellar dynamics. He built bulge-disk-halo fully self-consistent 3D systems with enough particles to follow interactions and merging of spiral galaxies. The important point to stress here is that he simulated the haloes self-consistently, in contrast with previous N-body simulations with even more particles (a few 10^5, with FFT-codes), which considered haloes as rigid potentials. Indeed he showed that haloes are efficient recipient of binding energy and angular momentum, which considerably helps the merging of the luminous components, by letting them keep cool. His successful simulation of the Antennae reveals long and thin tails in the cold-disk stellar component. Dynamical friction is so strong that the orbital decay is less than one initial orbital period: this demonstrates how efficient can be the tidal friction (especially with the massive and extended haloes used). Once the orbital energy and angular momentum have been absorbed by the haloes, the visible remnant can then be much more concentrated, and less rotating than previously thought. This dissipates one of the main objections against elliptical galaxies bring the result of merging between spirals (Kormendy, 1989).

4.2 Density waves

An aspect in which a self-consistent approach is absolutely needed is the excitation of spiral patterns in the inner parts of the galaxies. Although tides were suspected to trigger the grand design in M51 for example, it was only in 1981 that Toomre demonstrated how the tidal material arms in the outer parts could propagate by the swing amplification towards the interior. The simulations of Zang were beautiful and convincing: the time-scale of the amplification in the inner parts is so short (proportional to the rotation period at this distance) that the spiral wave is even stronger than in the outer parts, and culminates before the companion has time to go aways. An attempt to really simulate and fit the M51 tidal interaction and consequent morphology has revealed harder than expected, however (Hernquist, 1990). The parameters to be explored are numerous; the mass distribution is essential for the disk stability and the rate of swing amplification. A special tuning is needed to fit the right wavelength of the spiral. A similar attempt to fit M81 (Fig. 4) reveals that only significant tidal perturbations can trigger the observed spiral pattern (Brouillet et al., 1990).

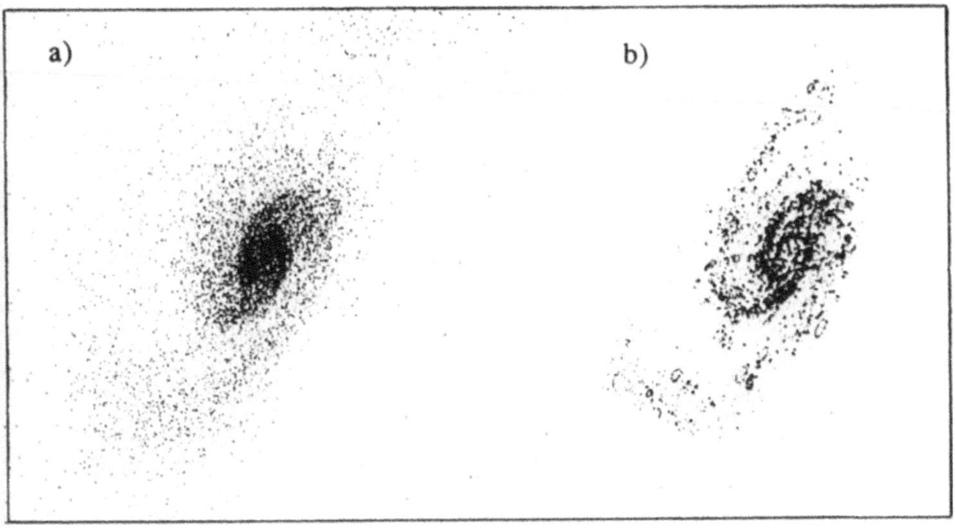

Figure 4. Self-consistent simulations of the interaction M81-NGC3077: a) stars; b) gas (from Brouillet et al. 1990).

A density wave very easy to trigger by tidal interactions is the bar (Noguchi, 1987, 1988; Gerin et al., 1990). Confusion is however brought by the spontaneous bar instabilities, easily amplified from the noise in N-body simulation. Contrary to the spiral waves, bars are robust and do not vanish away. Only a steady tidal forcing can decrease their strength (Combes et al., 1990). The angular velocity of the bar, and the induced outer two-arms spiral have an angular velocity characteristic of the mass-distribution of the galaxy. When a bar instability is driven by a companion, the frequency of the bar is the same, in particular it does not depend on the companion velocity, or orbit. When a spiral pattern is induced by a bound companion in the outer parts, its angular velocity is different and lower than that of the central bar. Although the bar instability sets in differently with or without a companion (Fig. 5), the result in the inner parts is exactly the same. The excitation of the bar by the companion could be attributed to non-linear mode coupling, as discussed by Tagger et al. (1987). The outer spiral structure, corotating with the companion has its inner Lindblad resonance in the vicinity of the bar corotation.

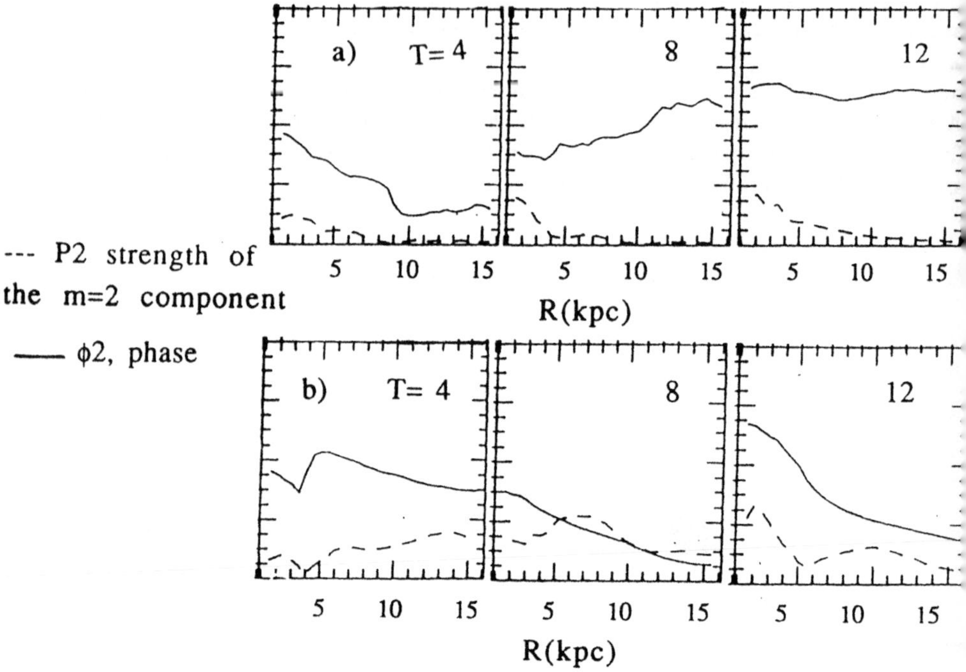

Figure 5. Strength of the m=2 Fourier component in function of radius in a) an isolated galaxy; b) the same with a companion. The bar m=2 wave is excited from outside for the second case. Time is in units of 12 Myr.

4.3 Gas dynamics

Self-consistent simulations including gas have only recently been developed, although schematical attempts had previously been reported (Miller et al., 1970; Negroponte and White, 1983). Barnes (1990) presents an experiment based on the SPH hydrodynamics (Gingold and Monaghan, 1982; Hernquist and Katz, 1989). We have chosen to represent the interstellar gas by a cloudy medium, which corresponds well to the molecular component and part of the atomic one. It is not a continuous fluid, but rather ballistic particles for which energy is dissipated in collisions. The energy reinjected by star-formation compensates of this dissipation, such that an axisymmetric gaseous distribution (without gravitational torques) remains stable, without infalling towards the center. In the presence of non-axisymmetric potentials (either bars or companions), gravitational torques can drive the gas radially, and the latter inside the corotation region accumulates in nuclear rings (Combes et al., 1990). After a merger, a large fraction of the total gas mass is concentrated to the nucleus, and could be responsible of the starbursts conspicuous in ultra-luminous IRAS objects.

The bar and ring formations in gas dynamics during a tidal interaction is supported by the growing evidence of such features in the molecular gas in the center of galaxies (M82, IC342, NGC 1068, NGC1097, etc... e.g. Combes, 1990a). The fact that the gas is becoming more self-gravitating in the center (where it represents up to 20% of the dynamical mass) could explain the high frequency of molecular bars.

4.4 Ring-galaxies

Head-on collisions are now well-known to be responsible for ring morphologies in disk galaxies (Lynds and Toomre, 1976). The first test-particles simulations have revealed the propagation of a density wave, from the center to the outskirts. A self-consistent approach has been reported by Appleton and James (1990); they show that the introduction of haloes allows the formation of a ring with an impact parameter as high as 40% of the disk radius, and an inclination of 60° on the disk normal. The role of dynamics on star-formation processes can be studied in details in these systems: in particular, a strong compression on the leading edge of the expanding ring is expected, and quasi-periodic starbursts are predicted to follow the formation of subsequent rings (Appleton and Struck-Marcell, 1987). The number and morphology of the various rings could provide information on the gravitational potential (Struck-Marcell and Lotan, 1990).

5 Interactions with dwarfs

It has been advocated that interactions of disk galaxies with gas-rich dwarf irregular galaxies could provide gas infall on spirals (Sancisi et al., 1990). Infall of fresh gas is indeed needed to explain the existence of a majority of spirals in the sky. Several arguments converge to this point: 1) the star-formation rate is now such that gas will be consumed and locked up in low-mass stars in 10^9 yrs or so (Larson, 1988); 2) successive mergers or accretions could have formed bulges of spirals, but the disk would not have survived without heating and thickening, unless fresh gas has infallen afterwards; 3) the maintenance of a spiral structure, even in sheared pieces, requires a certain infall of fresh and cold gas (Toomre, 1990); 4) metallicity problems (Pagel and Edmunds, 1981); 5) polar rings and warps (Binney, 1990), in particular warps have a wind-up time that allows an estimation of the accretion rate, of the order of $1M_\odot$ per yr. Until now, no evidence have been found of continuous infall from external primordial gas; intergalactic HI clouds remain undetected, except in small interacting groups, where one expects HI tidal debris. Thus gas infall might be constituted by the sole dwarfs accretion.

Such kind of infall might be in action for the Milky Way, in the form of HI high velocity clouds (HVC), that appears to come from the Magellanic Clouds (Mirabel, 1982). The Magellanic Stream could be a polar ring in formation. To support the hypothesis of gas infall, it would be interesting to find evidence for similar HVCs from external galaxies. So far, negative results have been reported (Wakker et al., 1987). Recently, van der Hulst and Sancisi (1988) claimed to have detected some infalling clouds towards the M101 plane. However, in the case observations are well explained by a tidal perturbation from a small dwarf companion NGC 5477. The high-velocity gas is the neutral atomic gas from the plane of M101 itself, perturbed by the crossing of the plane by the companion

(Combes, 1990b). This produces a circular wave similar to that of head-on collisions i ring-galaxies. Some gas is expelled from the plane, and a hole can be seen in the M10 HI plane, surrounded by the high-velocity clouds (Fig. 6). What is to be emphasize here is that close interactions with dwarfs are not directly efficient to provide gas infall Indeed, the dwarf itself is not largely perturbed by the plane crossing: this is due to th higher degree of mass concentration of the dwarf. The latter experiences a gravitationa interaction with an impact parameter much larger than its typical tidal radius, whil from the point of view of the target spiral, it is a nearly head-on collision. Simple three body simulations confirm that the dwarf is only slightly stripped (Combes et al., 1988) If the two galaxies are bound, the central spiral must await the dynamical decay followe by the accretion of the dwarf to get its gas!

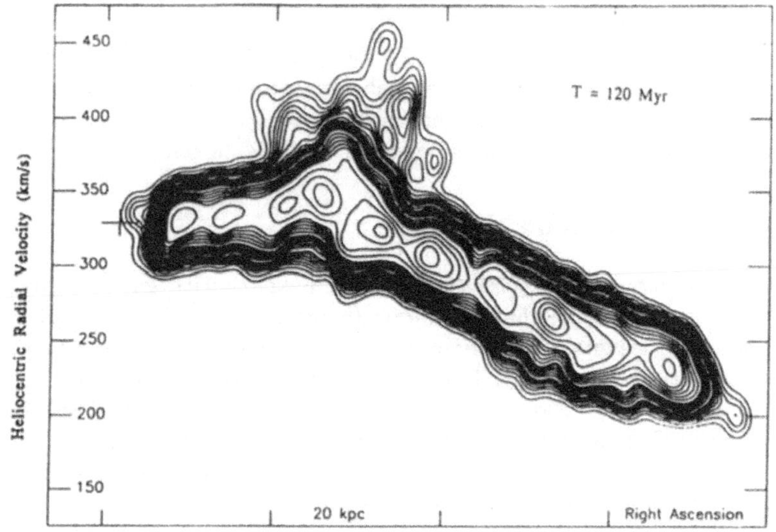

Figure 6. Simulation of the M101 plane crossing by its dwarf companion NGC 5477: evidence for high-velocity gas from the M101 plane.

References

Aguilar, L., White, S.D.M.: 1985, *Astrophys. J.* **295**, 374.

Appleton, P.N., James, R.A.: 1990, in *"Dynamics and Interactions of Galaxies"* ed. R. Wielen (Springer-Verlag) Heidelberg, p. 200.

Appleton, P.N., Struck-Marcell, C.: 1987, *Astrophys. J.* **318**, 103.

Balcells, M., Quinn, P.J.: 1990, preprint.

Balcells, M., Borne, K.D., Hoessel, J.G.: 1989, *Astrophys. J.* **336**, 655.

Barnes, J.: 1988, *Astrophys. J.* , **331**, 699.

Barnes, J.: 1990, in *"Dynamics and Interactions of Galaxies"*, ed. R. Wielen (Springer-Verlag) Heidelberg, p. 186.

Binney, J.: 1990, in *"Dynamics and Interactions of Galaxies"*, ed. R. Wielen (Springer-Verlag) Heidelberg, p. 328.

Bontekoe, T.R., van Albada, T.S.: 1987, *Mon. Not. R. Astr. Soc.* **224**, 349.

Borne, K.D., Balcells, M., Hoessel, T.G.: 1988, *Astrophys. J.* **333**, 567.

Borne, K.D., Hoessel, T.G.: 1988, *Astrophys. J.* **330**, 51.

Brouillet, N., Combes, F., Baudry, A.: 1990, in *"Dynamics of Galaxies and their Molecular Cloud Distributions"*, ed. F. Combes and F. Casoli (Reidel), in press.

Byrd, G.G., Saarinen, S., Valtonen, M.J.: 1986, *Mon. Not. R. Astr. Soc.* **220**, 619.

Clutton-Brock, M.: 1972, Astrophys. Space Sci. **17**, 292.

Combes, F.: 1990a, in *"Dynamics of Galaxies and their Molecular Cloud Distributions"*, ed. F. Combes and F. Casoli (Reidel), in press.

Combes F.: 1990b, *Astron. Astrophys.* in press.

Combes, F., Dupraz, C., Casoli, F., Pagani, L.: 1988, *Astron. Astrophys. Letters* **203**, L9.

Combes, F., Dupraz, C., Gerin, M.: 1990, in *"Dynamics and Interactions of Galaxies"*, ed. R. Wielen (Springer-Verlag) Heidelberg, p. 205.

Chandrasekhar, S.: 1943, *Astrophys. J.* **97**, 251.

Davoust, E., Prugniel, P.: 1988, *Astron. Astrophys. Letters* **201**, L30.

Farouki, R.,Shapiro, S.L.: 1981, *Astrophys. J.* **243**, 32.

Gerhard O.E.: 1981, *Mon. Not. R. Astr. Soc.* **197**, 179.

Gerin, M., Combes, F., Athanassoula, E.: 1990, *Astron. Astrophys.* **230**, 37.

Gingold, R.A., Monaghan, J.J.: 1982, J. Comput. Phys. **46**, 429.

Hernquist L.: 1990, in *"Dynamics and Interactions of Galaxies"*, ed. R. Wielen (Springer-Verlag) Heidelberg, p. 108.

Hernquist L., Katz, N.: 1989, *Astrophys. J. Supp. Ser.* **70**, 419.

Hernquist L., Weinberg, M.D.: 1989, *Mon. Not. R. Astr. Soc.* **238**, 407.

King, I.R.: 1962, *Astron. J.* **67**, 471.

Komerdy, J.: 1977, *Astrophys. J.* **218**, 333.

Komerdy, J.: 1989, *Astrophys. J. Letters* **342**, L63.

Larson, R.: 1988 in *"Starbursts and Galaxy Evolution"*, Rencontres de Moriond, ed T.X. Thuan, T. Montmerle, J. Tran Thanh Van, p. 467.

Lin, D.N.C., Tremaine, S.: 183, *Astrophys. J.* **264**, 364.

Lynds, R., Toomre, A.: 1976, *Astrophys. J.* **209**, 382.

McGlynn, T.A.: 1990, *Astrophys. J.* **348**, 515.

Merritt, D.: 1984, *Astrophys. J.* **276**, 26.

Miller, R.H.: 1986, *Astron. Astrophys.* **167**, 41.

Miller, R.H., Prendergast, K.H., Quirk, W.J.: 1970, *Astrophys. J.* **161**, 903.

Mirable, I.F.: 1982, *Astrophys. J.* **256**, 112.

Mulder, W.A.: 1983, *Astron. Astrophys.* **117**, 9.

Negroponte, J., White, S.D.M.: 1983, *Mon. Not. R. Astr. Soc.* **205**, 1009.

Nieto, J.L., Prugniel, P.: 1987, *Astron. Astrophys.* **186**, 30.

Noguchi, M.: 1987, *Mon. Not. R. Astr. Soc.* **228**, 635.

Noguchi, M.: 1988, *Astron. Astrophys.* **203**, 259.

Pagel, B.E.H., Edmunds, M.G.: 1981, *Ann. Rev. Astron. Astrophys.* **19**, 77.

Palmer, P.L., Papaloizou, J.: 1982, *Mon. Not. R. Astr. Soc.* **199**, 869.

Palmer, P.L., Papoloizou, J.: 1985, *Mon. Not. R. Astr. Soc.* **215**, 691.

Prieur, J.L.: 1990, in *"Dynamics and Interactions of Galaxies"*, ed. R. Wielen (Springer-Verlag) Heidelberg, p. 72.

Prugniel, P., Davoust, E., Nieto, J.L.: 1989, *Astron. Astrophys.* **222**, 5.

Prugniel, P., Combes, F.: 1990, in *"Paired and Interacting Galaxies"*, ed. J.W. Sulentic, IAU **124**, p.

Quinn, P.J., Goodman, J.: 1986, *Astrophys. J.* **309**, 472.

Rix, H.W., White, S.D.M.: 1989, *Mon. Not. R. Astr. Soc.* **240**, 941.

Rood, H.J., Leir, A.A.: 1979, *Astrophys. J. Letters* **231**, L3.

Sancisi, R., Broeils, A., Kamphuis, J., van der Hulst, T.: 1990, in *"Dynamics and Interactions of Galaxies"* ed. R. Wielen (Springer-Verlag) Heidelberg, p. 304.

Struck-Marcell, C., Lotan, P.: 1990, *Astrophys. J.* **358**, 99.

Tagger, M., Sygnet, J.F., Athanassoula, E., Pellat, R.: 1987, *Astrophys. J. Letters* **318**, L43.

Toomre, A.: 1981, in *"The Structure and Evolution of Normal Galaxies"*, ed. S.M. Fall and D. Lynden-Bell (Cambridge University Press), p. 111.

Toomre, A.: 1990, in *"Dynamics and Interactions of Galaxies"*, ed. R. Wielen (Springer-Verlag) Heidelberg, p. 292.

Toomre, A., Toomre, J.: 1972 *Astrophys. J.* **178**, 623.

Tremaine, S.: 1990, in *"Dynamics and Interactions of Galaxies"*, ed. R. Wielen (Springer-Verlag) Heidelberg, p. 394.

Valtaoja, L.: 1990, *Astron. Astrophys.* **228**, 37.

van der Hulst, J.M., Sancisi, R.: 1988, *Astron. J.* **95**, 1354.

Wakker, B., Broeils, A., Tilanus, R., Sancisi, R.: 1989, *Astron. Astrophys.* **226**, 57.

Weinberg, M.D.: 1986, *Astrophys. J.* **300**, 93.

Weinberg, M.D.: 1989, *Mon. Not. R. Astr. Soc.* **239**, 549.

White, S.D.M.: 1978, *Mon. Not. R. Astr. Soc.* **184**, 185.

White, S.D.M.: 1983, *Astrophys. J.* **274**, 53.

White, S.D.M.: 1990, in *"Dynamics and Interactions of Galaxies"*, ed. R. Wielen (Springer-Verlag) Heidelberg, p. 380.

Zaritsky, D., White, S.D.M.: 1988, *Mon. Not. R. Astr. Soc.* **235**, 289.

Discussion

J. Sulentic - Mixed morphology (S+E/S0) pairs are a problem if we assume they are primordial (i.e. that initial environmental conditions determine galaxy morphology). They could be hybrid pairs in the sense that their current types are a product interaction. One possibility could be that some pairs were originally S0+E systems. Given such a pair, could you produce a model where the final product is an S+E system?
F. Combes - It is indeed possible that a tidal interaction produces spiral extensions, i.e. tidal tails in the disk galaxy of the S0+E pair the lenticular would then look spiral for a while.

R. Buta - Does it seem from simulations that, morphologically, we could not distinguish an externally generated bar from an internally generated one? (apart from obvious companions and peculiarities).
F. Combes - Indeed, it is impossible. The bars in the two cases have the same physical characteristics, including the same angular velocity. These depend only on the mass distribution and velocity distribution of the galaxy.

INTERACTIONS: STRONG

Joshua E. Barnes
Canadian Institute for Theoretical Astrophysics
Toronto, Ontario, Canada

ABSTRACT. Strong interactions between galaxies are implicated in the origins of such features as shells and counter-rotating cores; moreover, at least some ellipticals probably formed as a direct consequence of mergers between disk galaxies. At the same time, interactions may also give rise to the large-scale variation of galactic type with environment. I first review recent numerical results on disk galaxy mergers. Models combining N-body and gas-dynamics illustrate the fueling of nuclear starbursts during such mergers; such starbursts may significantly enhance the central metallicity and phase-space density of merger remnants. At larger radii, however, the remnants are dominated by stars formed *before* the merger, which should in some cases exhibit distinctive kinematic signatures pointing to their origin in a pair of disks. In the second part of this review I discuss numerical results bearing on the relationship between galactic morphology and large-scale structure. Models of compact groups yield relatively short coalescence timescales, implying that those groups we observe today must have formed within the past few Gyr. On larger scales, galaxy clusters may evolve from irregular, spiral-rich systems to centrally concentrated objects dominated by elliptical galaxies.

1 Introduction

A physicist might define *strong interactions* as those in which galaxy number is not conserved. This definition is reasonable in that, although not all close encounters lead to merger, most collisions slow and close enough to permit strong interactions are likely to eventually result in the merger of the galaxies involved. In this article I will describe some recent numerical results bearing on the role of strong interactions in determining and modifying galactic morphology. The first part of this review focuses on the small-scale effects of collisions and mergers on the structure of elliptical galaxies. The second part is devoted to the implications of merging for structures on larger scales.

Before embarking on a description of the numerical results, it is worth reviewing the relevance of strong interactions for galactic morphology. Of ~ 4000 systems in the NGC catalog, Toomre & Toomre (1972) and Toomre (1977) list a dozen which clearly involve a pair of strongly interacting disk galaxies, either still directly distinguishable or evidenced by a pair of tidal tails extending from a single body. Timing arguments suggest that each such object represents at least one major disk-disk merger per $\sim 10^9$ yr; extrapolating this rate over a Hubble time, Toomre & Toomre concluded that the NGC catalog could

G. Longo et al. (eds.), Morphological and Physical Classification of Galaxies, 277–292.

well contain as many as 500 merger remnants, most of which formed long ago. Whil this estimate is highly uncertain, there seems to be little way to escape the conclusio that mergers have restructured a significant fraction of all galaxies.

1.1 Ways and means

Thanks to rapid progress in computer hardware and software, numerical simulations o merging galaxies are yielding results merely dreamed of five years ago. Detailed model of merging galaxies with self-consistent gravitational forces *and* gas-dynamical effects ar now available, as are larger-scale simulations of groups and clusters. These developments however, have built on and elaborate more than two decades of numerical work. Man of the basic techniques and results are reviewed by White (1982, 1983).

Encounters and mergers between disk galaxies may be modeled rather straightfor wardly provided only collisionless components (i.e. stars and dark matter) are included The main demand which must be met by a faithful simulation is particle number; roughl 10^4 particles are needed to approximate a self-gravitating disk galaxy. This requiremen may be attained with a variety of numerical methods; however the further demand of high spatial resolution in three dimensions favors hierarchical "tree" methods (e.g. Greengard 1990) over uniform-grid methods. The code used in the collisionless N-body experiment is fully described and tested in Barnes (1990a).

Models including gas dynamics are significantly more complex. The interstellar ma terial in real galaxies is a multiphase brew capable of exploding violently under starburst conditions; even the most ambitious models only treat these effects schematically. The approach adopted here employs "smoothed particle hydrodynamics" or SPH (e.g. Mon aghan, 1985), in which particles representing the fluid respond pressure forces as well as gravity. In principle, SPH can accurately model the hydrodynamics and thermodynam ics of reasonably smooth gas flows, using artificial viscosity to deal with shocks. The code used for the N-body/gas-dynamical simulations is discussed in detail by Hernquist & Katz (1989).

The experiments described here employ composite disk galaxy models, each consist ing of a central spheroidal bulge, a thin exponential disk, and an extended dark halo, with mass ratios of 1: 3: 16 respectively. All three components are modeled as self-consistent N-body systems, following the set-up procedure described by Barnes (1988). Gas, when included, amounts to 10% of the disk mass and is initially distributed just like the disk component. Results are given in an arbitrary system of units with $G \equiv 1$. Each galaxy has a total mass of 1.25 and a binding energy of -1.38; the disks have exponential scale lengths of $\alpha^{-1} = 1/12$ and rotate once every 0.93 time units at a radius of $3/\alpha$. Crudely scaling one of these disks to the Milky Way, the units of length, time, and mass become 40 kpc, 250 Myr, and 2.2×10^{11} $M_\odot$, respectively.

2 Small scales

Many of the morphological features commonly ascribed to strongly interacting galaxies were already apparent in the test-particle experiments of Toomre & Toomre (1972). Bridges, tails, and rings (Lynds & Toomre, 1976) are all well understood as essentially *kinematic* responses to external perturbations, while tidally induced spiral structure (e.g. Toomre, 1981) results when the self-gravity of the disk is included. The most fundamental questions to be asked in numerical studies concern the morphology of merger remnants; to address such questions, fully self-consistent models are required. From the first, it has been apparent that at least some of the gross structural features of elliptical galaxies could result from mergers of disk galaxies (White, 1978, 1979). Recent studies of ellipticals have uncovered features such as shells (e.g. Prieur, 1990) and counter-rotating cores (e.g. Bender, 1990) possibly due to mergers. There is thus considerable point in attempting more detailed simulations to see if such features can indeed be produced when galaxies collide.

Parameters for a small set of disk-disk merger experiments are listed in table 1. These particular examples were selected to illustrate a variety of different encounters and resulting remnants. The first two models, developed in a collaborative project with Lars Hernquist, include gas dynamics. The other three are pure N-body calculations, run for significantly longer times to examine the equilibrium structure of merger remnants.

Table 1. Parameters for parabolic disk-galaxy collisions. R_p is the pericentric separation. The orientation of each disk is specified by the inclination i and pericentric argument ω, following Toomre & Toomre (1972). N is the total particle number, while N_{sph} is the number of SPH particles.

No.	R_p	i_1	ω_1	i_2	ω_2	N	N_{sph}
1	0.2	71	30	0		90112	16384
2	0.4	71	90	109	90	45056	8192
3	0.4	71	30	0		32768	
4	0.4	109	30	180		32768	
5	0.2	71	90	109	90	32768	

2.1 Scenes from a merger

Figure 1 shows four stages in the evolution of model 1, a relatively close parabolic encounter between an exactly prograde disk and an identical disk inclined at an angle of 71°. Each disk throws out a long tail as a result of tidal interaction between them; the prograde one develops a bridge linking it to its inclined partner. The dark halos surrounding these disks are also gravitationally disturbed. Such violent interactions transfer a considerable fraction of the energy of the relative motion of these galaxies to their internal degrees of freedom; this transfer leaves the galaxies on a tightly bound orbit, leading to a second passage and shortly thereafter a merger.

The basic properties of the merger remnants produced by these simulations are discussed by Barnes (1988, 1990b). To summarize: mergers between bulge/disk/halo galaxies of equal mass produce slowly rotating triaxial remnants with $r^{1/4}$ luminosity profiles.

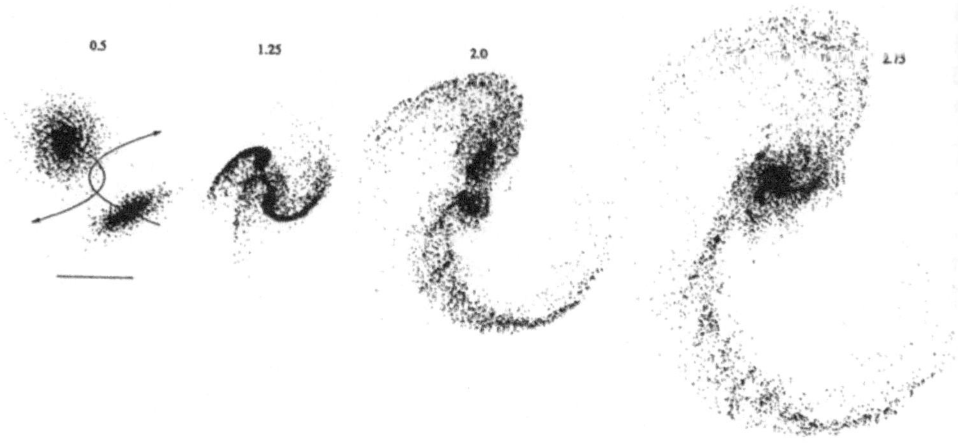

Figure 1. Close parabolic encounter and merger of two disk galaxies, viewed face-on to the orbital plane. Only "luminous" (bulge + disk) particles are plotted. The solid line is one length unit (40 kpc). Numbers give elapsed time, with pericenter at $t = 1$ (250 Myr).

Merging by and large preserves the ordering of the components in binding energy, so the remnants are dominated by bulge and disk particles at small radii and by halo particles at large radii, with a gradual transition at one or two effective radii. The velocity dispersion of the central luminous component climbs by a factor of ~ 1.4 thanks to the transfer of binding energy from the luminous material to the dark halos during the merger. While many details must still be checked, it seems fair to say that the expected properties of disk-galaxy merger remnants are largely consistent with elliptical galaxies.

2.2 Gas dynamics in merging disk galaxies

On large scales the gas and disk stars behave similarly in the early phases of a merging encounter, but on small scales the two develop remarkably different distributions. Some of these differences are illustrated in figure 2, which compares the gas and disk stars in the central region of the prograde disk of figure 1. As in other calculations (e.g. Noguchi 1987, 1988; Barnes, 1988; Gerin et al., 1990) the stellar disk develops a very pronounced bar as a result of the strong tidal field it has been subjected to. The gas initially responds in a similar fashion, but even in the first frame narrow features representing high Mach-number shocks are seen. In later frames the gas bar first becomes narrower, and then shrinks in length as well. A surprising amount of material – over half the gas in the disk – is concentrated in the dense oval ring seen in the last frame. The other disk exhibits imilar behavior, despite its 71° inclination.

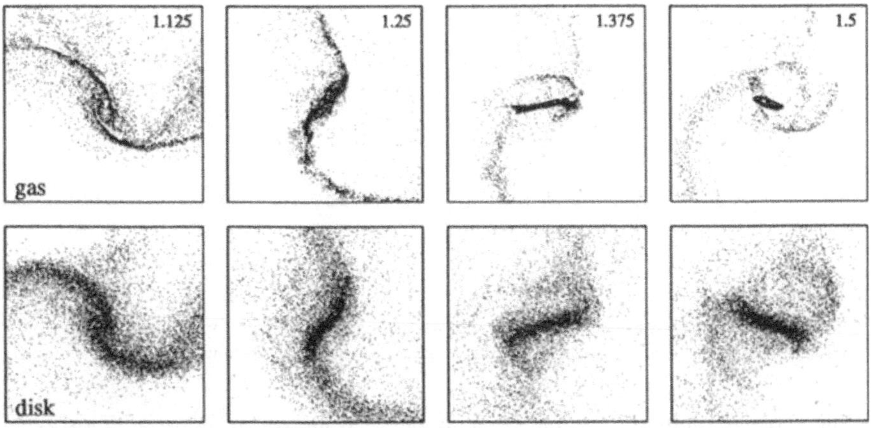

Figure 2. Close-ups of the gas (top) and disk particles (bottom) in the prograde disk from figure 1, plotted at the indicated times.

The torques responsible for this rapid inflow are strictly *gravitational*; hydrodynamic orces play no direct role in driving the gas to the center (e.g. Combes et al., 1990; Barnes & Hernquist, 1990). Physically, this situation is analogous to that found in a tidally driven accretion disk (e.g. Spruit, 1990). In such an accretion disk, an external tidal field sets up a spiral shock wave, which extracts angular momentum from the disk; this angular momentum is transferred to the object generating the tidal field. In the present galactic disk, the gravitational field of the stellar bar sets up two (very open) spiral shocks in the gas; angular momentum flows from the gas disk to the stellar bar. We found that inflow could be shut off by disabling radiative cooling; instead of rapidly cooling to $\sim 10^4$ K (where we placed a floor on the cooling function) the gas is shock-heated to $\gtrsim 10^6$ K and forms an extended cloud around the center of each galaxy. Strong radiative shocks thus appear necessary for the bar to drive a rapid inflow of gas.

The final encounter and merger of these two galaxies leaves 60% of all the gas in the system within a radius of 0.005; in dimensional units, this corresponds to $\sim 5 \times 10^9$ $M_\odot$ within $\sim$ 200 pc. Figure 3 compares the distribution of stars and gas in the merger remnant. The effective radius r_e of the stellar component is 0.1 length units or $\sim$ 4 kpc, a typical scale for a bright elliptical galaxy. The stars have a rather smooth and regular distribution, in marked contrast to the gas. Indeed, the stellar distribution is quite similar to that seen in near-IR images of Arp 220 which disclose an older population with a de Vaucouleurs' profile (Wright et al., 1990).

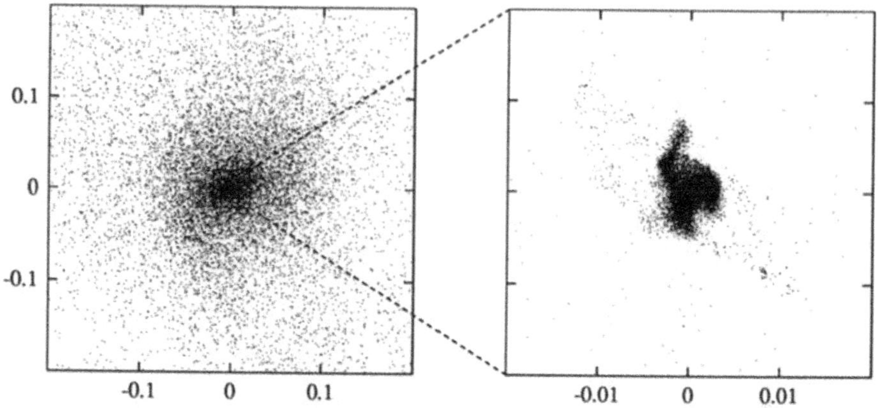

Figure 3. *Left:* luminous particles in the remnant produced by the merger shown in figure 1, projected onto the orbital plane. *Right:* a 10× enlargement showing the nuclear gas cloud.

Yet another remnant, produced by model 2, is shown in figure 4. This experiment involved a wider ($R_p = 0.4$) encounter of two inclined disk galaxies. When the merger is complete, much of the gas within $\sim r_e$ settles into a nested pair of annular disks. As the line-of-sight velocity plot shows, these disks rotate in opposite directions. The kinematics of this gas are somewhat reminiscent of the kinematics of the emission-line gas in NGC 7252 (Schweizer, 1982). Such counter-rotating nuclear gas disks are plausible progenitors of the counter-rotating cores observed in some elliptical galaxies. Counter-rotating cores may also be produced in purely stellar-dynamical mergers between galaxies of unequal mass (e.g. Balcells, these proceedings).

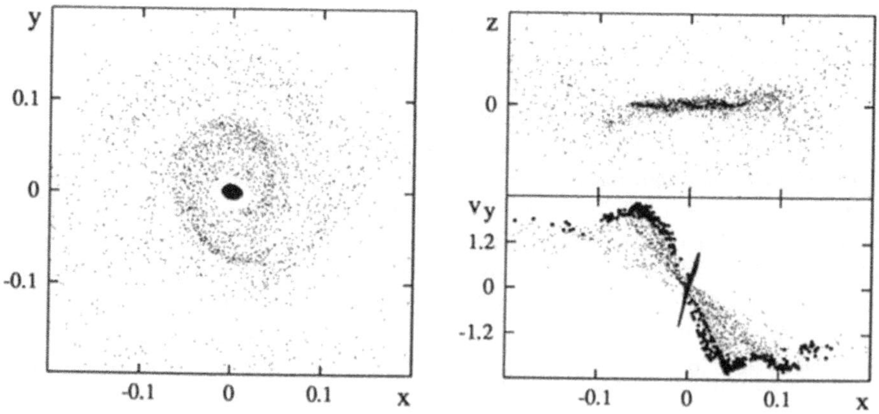

Figure 4. *Left:* face-on view of the nested gas disks produced by encounter 2. *Right, top:* edge-on view of gas disks. *Right, bottom:* line-of-sight velocities for gas particles within $|z| < 0.025$; those also within $|y| < 0.025$ are plotted as small filled circles.

2.3 Orbit structure of merger remnants

Orbit analysis provides a way to "deconstruct" merger remnants and thereby gain a more detailed understanding of their dynamics. In a generic triaxial potential, three major orbit families exist (cf. Binney & Tremaine, 1987). These are:

1. Box orbits: the orbit fills out a box-shaped region passing through the center.
2. Z-tube orbits: the orbit loops around the minor axis without changing direction. Oblate potentials support only this orbit family. A
3. X-tube orbits: the orbit loops around the major axis without changing direction. Two sub-families occur, but will not be distinguished in this simplified analysis.

When this classification scheme is applied to N-body merger remnants which have been allowed to relax for several crossing times, 95% or more of the orbits belong to one of these three families. This may be taken as further evidence that the remnant potentials are approximately triaxial.

The relatively simple structure of the remnant produced by model 3 is nicely elucidated by orbit analysis. In this encounter, which is merely a wider version of encounter 1, both galaxies experienced prograde encounters and responded by flinging out long tidal tails before merging. The final collision of the two central bulges produced a slightly elongated central region surrounded by material from the disks of the two victims. Figure 5 shows the cumulative distribution of orbit class as a function of specific binding energy E for a sample of 2048 luminous particles from this remnant. The most tightly bound particles are almost all on box orbits, helping to support the triaxial shape of the central region. Further out, the remnant is dominated by particles on Z-tube orbits; these tubes give the outer part of this remnant a basically oblate shape. The remnant is significantly flattened by rotation at large radii, a residue of the spins of the original disks and the angular momentum of their original orbit.

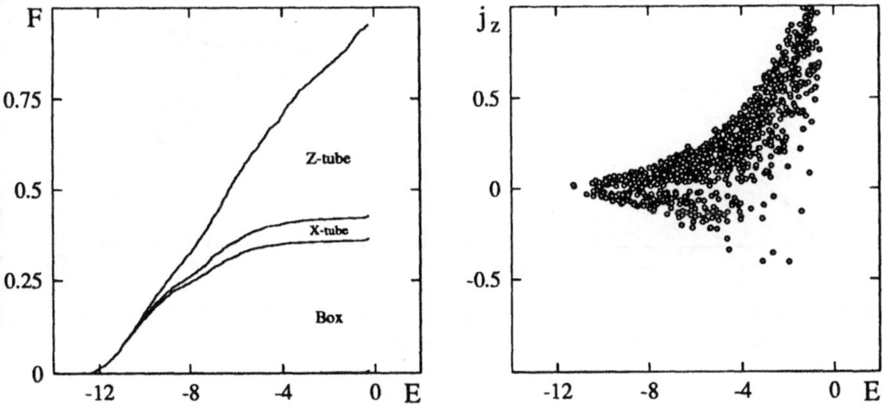

Figure 5. *Left:* Cumulative distribution of orbit class of luminous particles in remnant 3, plotted as a function of binding energy. *Right:* z-component of angular momentum vs. binding energy for the Z-tube orbits.

Simply reversing the spins of both disk, as in encounter 4, produces a very different result. This collision is somewhat less dramatic since the disks do not develop such impressive tidal tails in a retrograde passage. This remnant is nearly oblate at the cente and becomes more triaxial at larger radii. As figure 6 shows, it has a larger populatio of box orbits than the previous example, and a much smaller population of X-tubes. Th properties of the Z-tube orbits are shown in the right-hand panel of figure 6. Here result are presented for the halo (points) as well as the luminous component (circles). Most hal Z-tube orbits have $j_z > 0$, a relic of the orbital angular momentum of the two galaxie The luminous Z-tube orbits, however, tend to travel in the ***opposite*** direction, recallin the spins of the disks they came from. Such counter-streaming of different population is a distinctive signature of the merging process.

Results for model 5 are presented in figure 7. In this case both disks are inclined, bu spin in opposite directions. This encounter is considerably closer than the two previou examples, and the galaxies merge more rapidly. The resulting remnant is triaxial at smal radii but becomes more nearly prolate further out; it supports a large population of X tube orbits. The right-hand panel of figure 7 shows how the opposing spins of the tw original disks have left their mark on the remnant; particles from one disk tend to hav $j_x > 0$, while those from the other move the opposite way. These two sub-population largely cancel each others angular momentum, but had the initial spin of one of the tw disks been reversed, the resulting remnant would probably exhibit significant minor axi rotation.

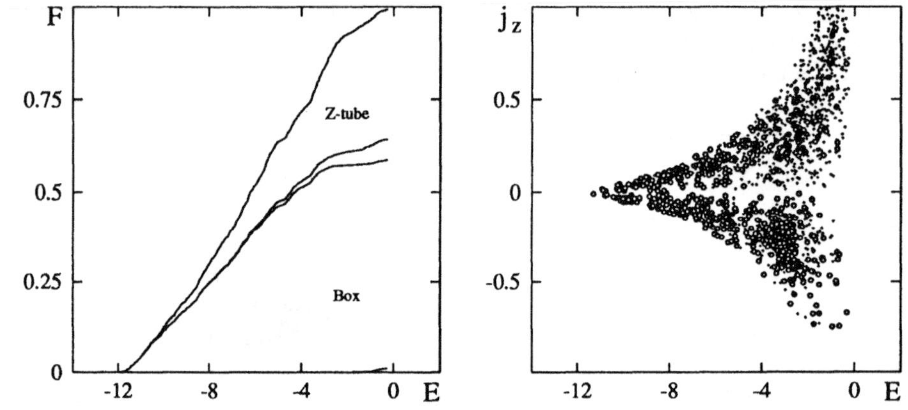

Figure 6. *Left:* Orbit class vs. binding energy for remnant 4. ***Right:*** j_z plotted against E for Z-tube orbits. Circles are luminous particles, points are halo particles.

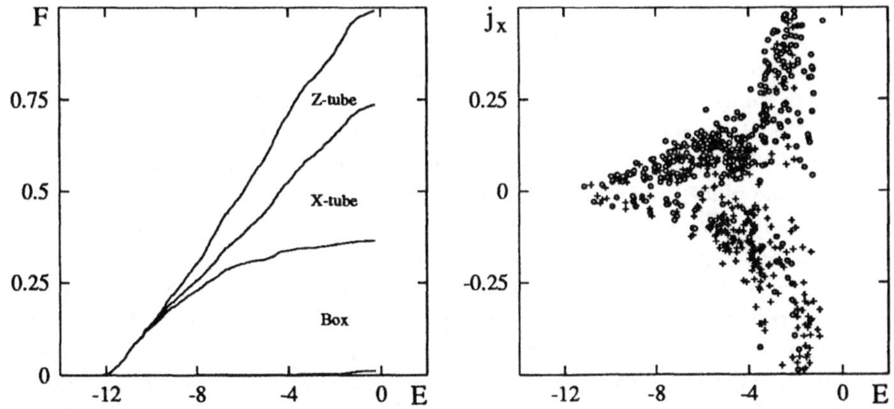

Figure 7. *Left:* Orbit class vs. binding energy for remnant 5. *Right:* j_x plotted against E for X-tube orbits. Circles are luminous particles from galaxy one, crosses from galaxy two.

3 Larger scales

On larger scales, it is well known that the morphology of galaxies varies in a systematic way with environment, or more quantitatively, with local density (e.g. Dressler, 1980). How do large scale conditions influence the outcome of galaxy formation? Features such as the voids, sheets, and filaments of galaxies reported in deep redshift surveys (e.g. de Lapparent et al., 1986) can be reproduced in simulations of gravitational clustering (e.g. White et al., 1987) starting from a spectrum of linear fluctuations predicted in the Cold Dark Matter theory (e.g. Peebles, 1982). As Blumenthal et al., (1984) noted, the CDM fluctuation spectrum is characterized by effective spectral indices $n \simeq -2$ on galactic scales, permitting significant "crosswalk" between a protogalaxy and its environment; in such a situation, galaxy formation is a prolonged process overlapping with the collapse of larger-scales structures. Merging of galaxies is a natural feature in CDM, or indeed in any theory in which structures grow gravitationally (e.g. Efstathiou, 1990). However, recent studies suggest that the Universe has more structure on scales $\gtrsim 10$ Mpc than predicted by CDM (Lynden-Bell et al., 1988; Geller & Huchra, 1990; Maddox et al., 1990). In view of this development, it seems appropriate to study mechanisms which might link galactic morphology with environment in gravitational clustering theories without specifically adopting a CDM model.

Results such as those described in the previous sections lend credence to the hypothesis that mergers of disk galaxies form at least some ellipticals. The techniques used above may also be used to examine the role of merging on larger scales. In the following section I describe models of compact groups, which are the first step from individual pairs of galaxies to larger structures. The section after that presents a preliminary model for the evolution of a rich galaxy cluster.

3.1 Evolution of compact groups

Compact groups are small, dense configurations of galaxies with crossing times of $\lesssim$ 1 Gy A catalog of $\sim$ 100 such groups, chosen by objective criteria, was presented by Hickson (1982). The dynamics of such systems has attracted both observational (e.g. Rose, 1979) and theoretical studies (e.g. Barnes, 1985). The upshot of this work is not completely clear; although many galaxies in compact groups seem to be interacting with each other observational studies tend to find little evidence for extensive evolution. From a theoretical point of view, however, there is little doubt that rapid dynamical evolution and merging will take place in any bound compact group of a few galaxies.

A detailed numerical model of a compact group was reported by Barnes (1989) This purely collisionless simulation, with a total $N = 65536$, followed the evolution of a virialized group of six disk/halo galaxies for a period of $\sim$ 4 Gyr. Figure 8 presents a snapshot of this model after only $\sim$ 400 Myr of evolution; of the six galaxies initially present, two pairs are violently interacting and soon merge. As a rule of thumb, a compact group experiences roughly one close encounter or merger per dynamical time. The group modeled here, in fair agreement with this rule, is reduced to a pair of merger remnants after $\sim$ 3 Gyr; these remnants coalesce after another $\sim$ 800 Myr.

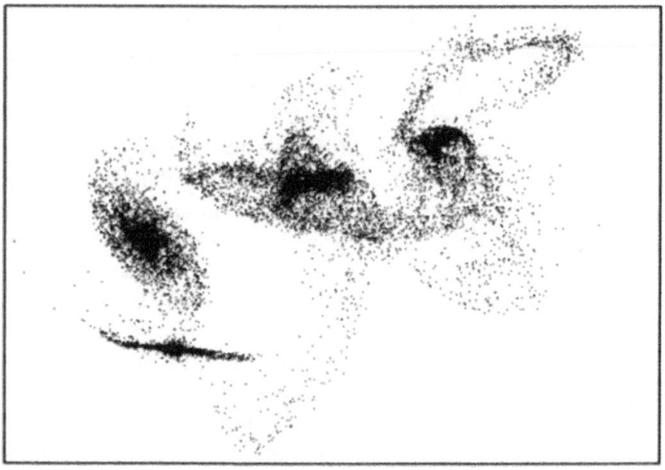

Figure 8. Early stage in the evolution of a compact group of six disk galaxies. Only luminous particles are plotted.

The merger remnants produced as this compact group evolves are similar to those produced in other simulations of disk galaxy mergers; that is, they are slowly rotating ellipsoidal objects with luminosity profiles approximating de Vaucouleurs' law (Barnes, 1989). It is quite likely that group models including gas dynamics would exhibit potentially starburst-feeding nuclear inflows similar to those described in the two-galaxy experiments above. Compact groups should thus be "breeding grounds" for made-by-merger ellipticals. But while some blue ellipticals in compact groups have recently come o light (Zepf & Whitmore, 1990), the brightest group members are not preferentially elipticals (Hickson, 1982). Groups containing recent merger remnants may be rare because ny such group would originally have contained at least one more galaxy, and groups f five galaxies are intrinsically much rarer than those of four; likewise, those mergers ccurring in groups of four must now be sought among the more numerous triples.

Many of the "paradoxes" associated with compact groups have been dispelled by White (1990), although it is still not clear if the compact groups we can observe are ontaminated by chance projections of members of surrounding looser groups (e.g. Manon, 1990). The picture that emerges is one in which compact groups continually form y dynamical evolution of looser groups and rapidly evolve, producing relatively normal lliptical galaxies. Rough demographic arguments indicate that compact groups may ccount for a fraction ~ 0.1 of all mergers between comparable galaxies. However, unless ome yet-unreported dynamical mechanism causes a large fraction of loose groups to volve into compact ones, the relatively small number of compact groups we see could vould nonetheless seem to require very prevalent and general evolution of loose groups nd larger-scale structures. Another example of such general evolution may be involved n the formation of rich clusters of galaxies.

3.2 Formation of rich clusters

Rich regular clusters are in some sense at the opposite extreme from compact groups. Although close encounters between galaxies occur frequently in rich clusters, relative elocities are so high that few such encounters lead to mergers. If this is the case, why do the richest clusters contain the greatest fraction of putative merger remnants? In hierarchical models of galaxy formation it is natural to suppose that the mergers esponsible for the formation of cluster ellipticals might have occurred in smaller groups which then collapsed together to produce a rich cluster. Yet now a further difficulty presents itself: if at each level of the hierarchy the merging timescale is comparable o the dynamical timescale, how did the progenitors of a rich cluster avoid coalescing nto a single remnant containing all the luminosity of the cluster? In other words, what breaks the merger hierarchy? Dissipation provides a partial answer: baryonic clouds of mass $\lesssim 10^{12}$ $M_\odot$ can cool in less than a Hubble time (e.g. Rees & Ostriker, 1977) and thus become less susceptible to merging than the dark halos they collapsed in (White & Rees, 1978). The rest of the answer may hinge on the form of the fluctuation spectrum: the same spectral features which promote crosswalk between galactic and larger scales also favor the survival of individual galaxies in larger systems. Carlberg & Couchman (1989) have illustrated both of these effects with CDM simulations including a dissipative component. In such calculations the role of dissipation is to concentrate baryonic matter within individual dark halos on the scale of galaxies. Alternatively, nitial conditions including individual core/halo galaxies may be crafted "by hand" and volved dissipationlessly to study the galactic survival rate.

Adopting the latter approach, I describe here a pure N-body calculation starting with 128 galaxies, each consisting of a compact core surrounded by an extended halo of 7 times the core mass. The galaxies were laid down in a binary hierarchy (Soneria & Peebles, 1978) and given velocities consistent with virial equilibrium at every level. Such nitial conditions cannot be simply related to a linear fluctuation spectrum since they are non-linear from the very start. In the present calculation, significant coupling betwee levels was obtained by setting the length rescaling parameter to the relatively small valu of $b = 1.6$; this emulated the relatively low contrast "cloud in cloud" structures expecte from CDM-like fluctuation spectra (e.g. Efstathiou, 1990).

White & Negroponte (1982) found that the dynamical evolution of such hierarchie rapidly erases the substructure within them. The present simulation supports their re sult, but also shows that significant numbers of galaxies escape merging as substructur is erased. Figure 9 shows the mass and light distributions toward the end of the simula tion. By this stage the cluster has developed a centralized potential well which contains good deal of dark matter as well as several fairly massive merger remnants; more than 5 distinct galaxies, however, still survive. The largest remnant is built from the wreckage o 12 original galaxies. However, a good 75% percent of the surviving galaxies have partici pated in at most one merger. Dynamical evolution continues in pockets of substructure but the galaxies incorporated into the overall cluster have attained speeds sufficient t protect them from most close encounters. Considerably more detailed calculations ar required to evaluate remnant production and galactic survival rates in realistic cosmo logical models. Nonetheless, results to date suggest that mergers occurring during th collapse of rich clusters can produce the ellipticals we see today without falling afoul o the "overmerging" problem.

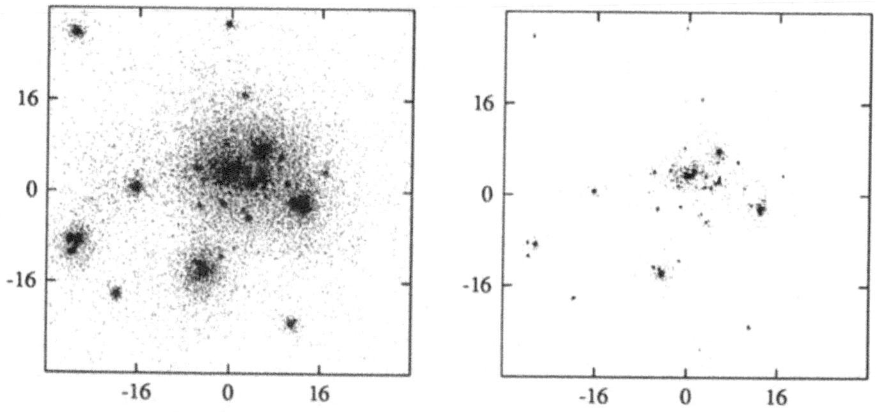

Figure 9. Late stage in the evolution of a cluster of core/halo galaxies. *Left:* mass distribution. *Right:* luminosity distribution.

The model presented here evolves from an irregular configuration of a large number of smaller galaxies into a centralized, roughly spherical system rich in merger remnants. Could the morphology of rich galaxy clusters likewise be viewed as an evolutionary sequence, with irregular, spiral-rich clusters as one extreme and regular, elliptical-rich clusters at the other? Such an interpretation seems consistent with the large numbers of interacting galaxies seen in irregular clusters; many of these interactions will presumably result in mergers. However, those systems now observed as regular clusters would have passed through their irregular phase at redshifts $z \gtrsim 1$, when they would presumably have had much higher densities than irregular clusters at $z \simeq 0$. Furthermore, the galaxies which were merging within proto-clusters at $z \gtrsim 1$ may have been different than those found today in the field and in irregular clusters. It seems plausible that systematic differences between field and cluster ellipticals will be found, and that these will give an indirect indication of the properties of disk galaxies at $z \gtrsim 1$.

4 Problems

In this review I have waxed rather enthusiastic about the role of strong interactions in the formation of elliptical galaxies. Recent numerical results do indeed give considerable cause for optimism on this point. Some caution, however, is in order; the successes of the "merger hypothesis" cannot be used to rule out other formation mechanisms. Only when such other mechanisms have been as well studied as merging will we be in a position to say how elliptical galaxies actually formed. Until then, several potential problems should be kept in mind:

1. Why should faint ellipticals rotate rapidly, when merging is "blind" to overall scale? Are faint E's perhaps continuous with S0's?
2. How do the color-magnitude relations seen among elliptical galaxies arise? On the whole, one expects big ellipticals to form from big spirals. Do properties of spiral galaxies vary systematically with luminosity in such a way as to induce significant trends among their merger products?
3. How are color gradients in ellipticals set up? Are abundance gradients in disk galaxies sufficient? Do metals produced during nuclear starbursts significantly pollute the cores of merger remnants?
4. Why are globular clusters so much more common in cluster ellipticals and especially in cD galaxies? Perhaps high-z disk galaxies had more globular clusters per unit of luminosity?

It is a pleasure to thank Lars Hernquist for permission to present results from our work in advance of publication. I also wish to thank Marijn Franx, Scott Tremaine, Alar Toomre, and Simon White for many valuable discussions.

References

Barnes, J.E.: 1985, *Mon. Not. R. Astr. Soc.* **215**, 517.

Barnes, J.E.: 1988, *Astrophys. J.* **331**, 699.

Barnes, J.E.: 1989, *Nature* **338**, 132.

Barnes, J.E.: 1990a, *J. Comp. Phys.* **87**, 161.

Barnes, J.E.: 1990b, in *Dynamics and Interactions of Galaxies*, ed. R. Wielen (Berlin: Springer-Verlag), p. 186.

Barnes, J.E., Hernquist, L.: 1990, *Astrophys. J. Letters* in press.

Bender, R.: 1990, in *Dynamics and Interactions of Galaxies*, ed. R. Wielen (Berlin: Springer-Verlag), p. 232.

Blumenthal, G.R., Faber, S.M., Primack, J.R., Rees, M.J.: 1984, *Nature* **311**, 517.

Binney, J., Tremaine, S.: 1987, *Galactic Dynamics* (Princeton: Princeton Univ. Press).

Carlberg, R., Couchman, H.: 1989, *Astrophys. J.* **340**, 47.

Combes, F., Dupraz, C., Gerin, M.: 1990, in *Dynamics and Interactions of Galaxies*, ed. R. Wiele (Berlin: Springer-Verlag), p. 205.

de Lapparent, V., Geller, M.J., Huchra, J.P.: 1986, *Astrophys. J. Letters* **302**, L1.

Dressler, A.: 1980, *Astrophys. J.* **236**, 351.

Efstathiou, G.: 1990, in *Dynamics and Interactions of Galaxies*, ed. R. Wielen (Berlin: Springer-Verlag) p. 2.

Geller, M.J., Huchra, J.P.: 1990, *Science* **246**, 897.

Gerin, M., Combes, F., Athanassoula, A.: 1990, *Astron. Astrophys.* **230**, 37.

Greengard, L.: 1990, *Computers in Physics* **4**, 142.

Hernquist, L., Katz, N.: 1989, *Ap. J. Suppl.* **70**, 419.

Hickson, P.: 1982, *Astrophys. J.* **255**, 382.

Lynden-Bell, D., Faber, S.M., Burstein, D., Davies, R.L., Dressler, A., Terlevich, R.J., Wegner, G.W. 1988, *Astrophys. J.* **326**, 19.

Lynds, R., Toomre, A.: 1976, *Astrophys. J.* **209**, 382.

Maddox, S.J., Efstathiou, G., Sutherland, W., Loveday, J.: 1990, *Mon. Not. R. Astr. Soc.* **242**, 43p.

Mamon, G.A.: 1990, in *Paired and Interacting Galaxies*, eds. J.W. Sulentic, W.C. Keel, in press.

May, A., van Albada, T.: 1984, *Mon. Not. R. Astr. Soc.* **209**, 15.

Menon, T.K., Hickson, P.: 1985, *Astrophys. J.* **296**, 60.

Monaghan, J.: 1985, *Comp. Phys. Rep.* **3**, 71.

Nieto, J.-L., Bender, R.: 1989, *Astron. Astrophys.* **215**, 266.

Noguchi, M.: 1987, *Mon. Not. R. Astr. Soc.* **228**, 635.

Noguchi, M.: 1988, *Astron. Astrophys.* **203**, 259.

Peebles, P.J.E.: 1982, *Astrophys. J.* **258**, 415.

Prieur, J.-L.: 1990, in *Dynamics and Interactions of Galaxies*, ed. R. Wielen (Berlin: Springer-Verlag) p. 72.

Rees, M.J., Ostriker, J.P.: 1977, *Mon. Not. R. Astr. Soc.* **179**, 541.

Rose, J.: 1979, *Astrophys. J.* **231**, 10.

Schweizer, F.: 1982, *Astrophys. J.* **252**, 455.

Soneria, R., Peebles, P.J.E.: 1978, *Astron. J.* **83**, 845.

Spruit, H.C.: 1990, in *Theory of Accretion Disks*, eds. F. Meyer, W.J. Duschl, J. Frank, E. Meyer-Hofmeister (Dordrecht: Kluwer), p. 325.

Toomre, A.: 1977, in *The Evolution of Galaxies and Stellar Populations*, eds. B.M. Tinsley, R.B. Larson (New Haven: Yale Univ. Obs.), p. 401.

Toomre, A.: 1981, in *The Structure and Evolution of Normal Galaxies*, eds. S.M. Fall, D. Lynden-Bel (Cambridge: Cambridge Univ. Press), p. 111.

Toomre, A., Toomre J.: 1972, *Astrophys. J.* **178**, 623.

White, S.D.M.: 1978, *Mon. Not. R. Astr. Soc.* **184**, 185.

White, S.D.M.: 1979, *Mon. Not. R. Astr. Soc.* **189**, 831.

White, S.D.M.: 1982, in *The Morphology and Dynamics of Galaxies*, eds. L. Martinet, M. Mayor (Sauverny: Geneva Obs.), p. 289.

White, S.D.M.: 1983, in *Internal Kinematics and Dynamics of Galaxies*, ed. E. Athanassoula (Dordrecht D. Reidel), p. 337.

White, S.D.M.: 1990, in *Dynamics and Interactions of Galaxies*, ed. R. Wielen (Berlin: Springer-Verlag) p. 380.

White, S.D.M., Rees, M.J.: 1978, *Mon. Not. R. Astr. Soc.* **183**, 341.

White, S.D.M., Negroponte, J.: 1982, *Mon. Not. R. Astr. Soc.* **201**, 401.
White, S.D.M., Frenk, C.S., Davis, M., Efstathiou, G.: 1987, *Astrophys. J.* **313**, 505.
Wright, G.S., James, P.A., Joseph, R.D., McLean, I.S.: 1990, *Nature* **344**, 417.
Zepf, S.E., Whitmore, B.: 1990 in *Dynamics and Interactions of Galaxies*, ed. R. Wielen (Berlin: Springer-Verlag), p. 421.

Discussion

J. Sulentic - 1) Your [group] model began with six galaxies close together. How do you keep the first two or three from merging while you bring in the rest? 2) Assuming there is a way to do the above, you will eventually have partial mergers. Why do we not find any strongly FIR or radio enhanced systems in Hickson groups (like we do in pairs)?
J. Barnes - 1) I set up a group of six galaxies "to illustrate what might happen in an admittedly extreme – but not inconceivable – situation" (Barnes, 1989); most compact groups contain fewer galaxies. In any case, one can't "keep galaxies from merging" – try it sometime – so the formation of a compact group presumably requires that *several* galaxies become bound together more or less *simultaneously*. 2) While no ultra-luminous IRAS sources reside in these systems, compact groups as a whole *are* IR-loud (Menon & Hickson, 1985). We see plenty of strongly interacting galaxies in compact groups, but as far as I know, nobody has checked to see if these galaxies are any *less* likely to be strong FIR or radio sources than an equivalent sample of isolated pairs.

I. King - Can you give us an idea of the economics of this business? Is the computing time so great that you are severely limited in the number of cases you can look at?
J. Barnes - I selected the models described here from a pool of about twenty; each required between ten and one hundred hours of supercomputer time. One such calculation might thus be the economic equivalent of a deep exposure on a large telescope. As workstations get faster, however, it is becoming increasingly possible to run significant calculations on a local machine – the final cluster simulation is an example.

D. Burstein - Are you developing a feeling for what happens during a merger in the real world? Gas turns into stars on a hydrodynamical time scale, those stars return gas to the process on similar time scales, and supernovae provide energy on local scales comparable to the tidal fields. How far are the models from looking at these processes?
J. Barnes - My gut feeling is that effects associated with star formation only perturb the gas-dynamics responsible for the nuclear inflows Hernquist and I found in our models. After all, massive nuclear molecular clouds *are* observed in a number of merging galaxies. As the gas density increases, star formation and supernovae probably become more effective, and nuclear starburst systems may represent the dividing line where explosively driven outflow overcomes tidally driven inflow. As for the models, they are limited largely by our poor understanding of star formation; observational measurements of star formation rates in interacting systems should be very useful in constraining the simulations.

D. Burstein - I think we need to proceed in an iterative manner: theorists provide us with hints of what to look at in the *early* stages of the merger process. We go look, see what we can find, [and the] models are modified until we converge at a consisten picture.

J.–L. Nieto - 1) Your results on the properties of the merger remnants (triaxialit slow rotation, $r^{1/4}$ profile) and the questions you are rasing about faint E's rotatin rapidly (and thus possibly being S0-like systems) are quite consistent with the dichotom hypothesis and its interpretation in terms of different evolutionary tracks (see, e.g., Niet & Bender, 1989). 2) The same question that I had to M. Balcells: are your simulation consistent with the cores of bright E's and cD's (namely merger remnants) being flatte than those of the original galaxies?
J. Barnes - 1) I'm not sure if the evidence points to a true dichotomy rather than smooth continuum. By and large, the differences between "disky" and "boxy" elliptical are pretty subtle – and it seems remarkable that two distinct evolutionary tracks (on of which, perhaps, is merging) should nonetheless yield such nearly indistinguishabl products. Moreover, mergers including gas dynamics *can* make disk-like structures, a seen here in figure 4. I'm inclined to view purely stellar-dynamical merging as on endpoint in a continuum of formation processes. 2) The pure N-body experiments I'v done yield remnants with core radii comparable to the core radii of the original galaxies roughly as expected from phase-space density arguments (e.g. May & van Albada, 1984) Star formation could even *increase* the maximum phase-space density and thereby reduc the core radius. The properties of the original galaxies making up a given merger remnan are in any case a matter of speculation.

P. Pişmiş - In your computations you have also considered the merging of two disks. In such [a] case the orientation of the two disks with respect to one another must result in differences, probably quite significant, in the morphological and other physical propertie of the merger. Have you studied different orientations of the disks?
J. Barnes - Yes indeed. The results described in § 2.3 show how the orientation o the disks with respect to each other *and* to the orbital plane of the pair can influence the structure of the remnant. In general the shape, dynamics, and especially the degre of rotational support depend on the orientations of the initial galaxies as well as the pericentric separation of their initial orbit. On the other hand the velocity dispersion half-light radius, and luminosity profile seem insensitive to such variables.

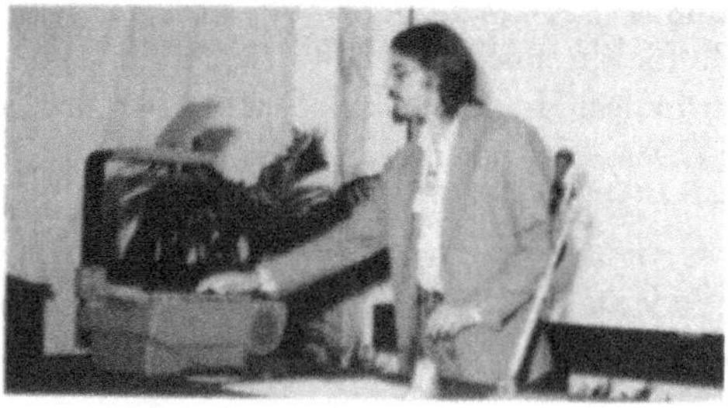

Joshua Barnes swearing on the overhead projector.

THE MORPHOLOGY OF BINARY GALAXIES

J.W. Sulentic
Department of Physics and Astronomy
University of Alabama, USA

ABSTRACT. We review the existing catalogs and major sources of data relevant to the morphology of pairs. Emphasis is on results from the Catalog of Isolated Pairs (CPG) authored by Karachentsev (1972). Sources of bias and uncertainty that are present in pair catalogs are discussed. We consider both: 1) Hubble component morphology and 2) pair interaction morphology. The results on the Holmberg effect and morphological concordance in pairs, as well as evidence for interaction-induced enhancements, are discussed.

1 Introduction

There are many reasons for studying pairs of galaxies that extend beyond their attraction as peculiar systems. Historically the interest in pairs has focused around equation 1 which provides a means of deriving statistical masses for galaxies from their measured properties.

$$\Delta V \times r = GM \sin^3\theta \sin^2\phi \tag{1}$$

Here ΔV represents the component velocity difference, r the physical separation and $\sin^3\theta \sin^2\phi$ the orbital projection. In the past ten years interest has shifted from a simple determination of statistical masses to the search for dark matter. The procedure is to compare estimated pair masses with the sum of the individual masses in order to derive the mass of the unseen component. In a particularly useful review of the sources of uncertainty in this procedure, van Moorsel (1982) has identified four potential problems: 1) sample selection, 2) accuracy of redshift measures, 3) relation of pairs to their environment, and 4) the adopted model of the interaction potential. The importance of these considerations is made clear when one encounters the disagreement in conclusions among recent derivations (van Moorsel, 1982; Karachentsev, 1985; Schweizer, 1987). We will not attempt any detailed discussion or citation of the dynamical studies of pairs.

Moving beyond the use of pairs for mass estimation, we are reminded of the quote in the Atlas of Peculiar Galaxies (Arp, 1967) that "If we could analyze a galaxy in the laboratory we would deform it, shock it, probe it in order to discern its properties." In other words, pairs as nonequilibrium systems are useful for studying the structure of galaxies by observing the response of that structure to an external stimulus. If pairs are primordial and if they retain their basic forms from birth, they can provide clues about the role

G. Longo et al. (eds.), *Morphological and Physical Classification of Galaxies*, 293–307.

of environment in galaxy formation. They can also provide information about galax evolution since they apparently experience accelerated evolution—if evolved component survive long enough (i.e., avoid merging). In this context, pair statistics may also giv useful information on the role and frequency of mergers. Finally, pairs provide valuabl insights into the physics of interaction and its effects, including dynamics, hydrodynamics, star formation, mass transfer, plasma physics, and possibly AGN stimulation.

We will focus upon the morphology of binary galaxies in this review after considering available sources of pairs and attendant biases. We will first consider the Hubbl morphology of the component galaxies in the pairs. This will be followed by a discussio of the interaction morphology of the pairs. The latter discussion will involve all forms o evidence that might be used to deduce the existence of direct interaction in individual or samples of, pairs.

2 History of pair selection. The catalogs

The first pair list was assembled well before the coming of the Palomar Sky Survey. Th pioneering studies of Holmberg (1937, 1958) and Page (1952, 1961) selected pairs wit component separations less than 2 (3 for Page) times the sum of the component majo axis diameters. The modern age can be said to begin with the publication of the Catalo of Isolated Pairs (CPG) by Karachentsev (1972) which was based upon the Palomar Sky Survey. The CPG contains 603 entries north of $\delta = 0°$ and brighter than $m_{pg} = 15.7$. The sample was assembled with an isolation criterion expressed by the following relation:

$$\frac{x_{ji}}{x_{12}} \geq \frac{5a_i}{a_1} \quad \text{where} \quad 0.5a_1 \leq a_i \leq 4a_1 \quad j = 1, 2$$

where x corresponds to separation and a to major axis diameter with subscripts $j = 1, 2$ representing the pair components and i the nearest neighbor. "Harder" selection criteria are also indicated for subsample in the CPG defined by replacing 5, 0.5 and 4 with 10, 0.5, 4 or 5, 0.25, 4 in the relation. A matching catalog of isolated galaxies (CIG) that facilitates comparisons between field and pair properties was subsequently published by Karachentseva (1973). The size of the CPG sample, the complete redshift information, and the existence of a control sample combine to make it the most useful for morphological analysis. Hence the emphasis on CPG results in this review. A Russian language monograph titled "Double Galaxies" (Karachentsev, 1987: hereafter CPG*) catalogs and discusses all CPG data.

The first attempt at a completely objective (i.e., automated) pair survey was published by Turner (1976ab) who used the CGCG (Zwicky et al., 1961–68) as a source. The survey departed from the CPG in selecting pairs using only magnitudes and positions. Pairs were selected as being isolated with respect to all galaxies brighter than m_{pg}=15.0. The original redshift data and subsequent dynamical analyses from this sample have been critically analyzed by Rood (1982). The use of a magnitude cutoff resulted in problems for the fainter pairs in the sample which were not isolated from galaxies below the cutoff. A refinement of this approach was made by Peterson (1979ab). His UGC (Nilson, 1973) based list restricted the isolation requirement to galaxies with apparent magnitudes similar to those of the pair components (as in the CPG, but with a weaker solation criterion). The result was the inclusion of more wide pairs but at the cost of a higher fraction of nonphysical systems.

Further progress in the construction of binary galaxy catalogs has involved attempts at increasing the number of wide pairs and obtaining better estimates of the contamination by nonphysical systems. The UGC-based Catalog of Multiple Galaxies (CMG) was assembled by T. S. van Albada. It has previously been used in part for a dark matter search (van Moorsel, 1982) but was only recently published in full (Soares, 1989). The catalog was assembled by assigning a probability of association to the nearest neighbor for each UGC galaxy brighter than m_{pg}=14.2. This was done for a number of different apparent magnitude intervals. The probability is derived from an estimate of the local surface density, estimated using the 10th nearest neighbor in each of the adopted apparent magnitude intervals. Galaxies that occur more than once (in more than one magnitude bin) in this analysis are merged so that the resulting catalog contains pairs and multiplets with up to 11 components.

Lists for the southern sky have appeared in the past five years. Schweizer (1987) assembled a small (n=48) list of southern pairs using selection procedures based on a dynamical model that assumed bound pairs with Keplerian orbits. Effects of the nearest neighbor were estimated with a perturbation analysis. Other recent pair lists include 1) the Arp and Madore (1985) catalog based upon a visual search (with a size limit of about 10 arcsec) of the SRC–J plates of the southern sky survey and 2) a compilation of 630 pairs brighter than B$\sim$17.0 in the south galactic cap by Zhen-Long et al. (1989).

3 Pair samples are biased

Rather than present a detailed critique for each of the pair compilations referenced in the previous section, it was decided to summarize the possible sources of bias in a separate section. In this way we avoid a tremendous amount of duplication since all of the catalogs suffer from these sources of bias or incompleteness at some level. This summary is intended to alert the potential user to the considerations that should go into adopting a given list for more detailed study. We focus our attention on the CPG since it provides the largest and deepest pair sample with complete redshift information.

1) Pair samples selected with an isolation criterion are biased toward binaries with certain orbital phases (specifically, pairs near pericenter) and viewing angle. In the case of the CPG the mean physical separation $< r_{12} >=$ 83 kpc. The CMG most effectively complements the lack of wide pairs in the (bright end of the) CPG. We find that $< r_{12} >=$ 146 kpc for 230 pairs from the CMG that were studied by Soares (1989).

2) Samples with an isolation criterion reject nearby pairs and pairs near unrelated bright galaxies. The former apparently bright pairs have such wide separations that they are rejected while the latter effect is independent of apparent magnitude. A recent V/V_m test for the CPG sample (Xu and Sulentic, 1991) shows that these effects produce incompleteness at all magnitude levels. Most of the pairs near bright foreground galaxies that are rejected from the CPG were subsequently cataloged (Sulentic et al., 1978).

3) All samples favor size and magnitude concordant pairs. The mean difference in apparent magnitude between components of CPG pairs is $< \Delta m >=$ 0.7$\pm$0.3. Less than 9% of the CPG pairs have a major axis component ratio A/B> 2.0. NGC5194/5 with

A/B= 1.7 is one of the few hierarchical pairs that was included in the CPG. Catalog that followed the CPG often did not use diameter as a selection parameter which reduce somewhat the bias against uneven pairs. Whether one employs an isolation criterion o a more elaborate selection process, the inclusion of uneven pairs greatly complicates th exclusion of nonphysical pairs. No effective catalogs of hierarchical pairs exist excep for a compilation of 160 "M51 type" systems discussed, but not listed, by Vorontsov-Velyaminov (1975). This work and a spectroscopic study by Arp (1969) suggests tha many uneven pairs are physical binaries.

4) Samples without an isolation criterion contain discordant redshift pairs. Surpris ingly, even restrictive catalogs like the CPG contain a significant number of such pair (57/603=0.09 with $\Delta V > 1000$ km s^{-1}). It is interesting to note that recent sample with more elaborate selection procedures (e.g., Schweizer, 1987) are contaminated a about the same level. Some of these discordant pairs also show morphological evidenc for peculiarities usually attributed to interaction (Sulentic et al., 1985ab).

5) Samples also contain significant numbers of accordant redshift optical pairs. Thi is the largest and most insidious form of contamination in pair samples. In the CPC there are 42 accordant redshift pairs that have derived M/L > 100 (CPG*), raisin suspicions about their nature as physical binaries. Most such optical pairs, however, show measured properties completely within the domain of expectation for physical systems Karachentsev (1982) performed Monte Carlo simulations that modeled the observec clustering properties of galaxies. This led to the conclusion that 32% of the CPG pair are apparent alignments in larger groupings rather than isolated binaries. A simila estimate for the contamination was derived by van Moorsel (1983ab). Attempts to remove the accordant contamination (e.g. White et al., 1983) have met with limitec success (Picchio and Tanzella-Nitti, 1983). Physical pairs are apparently very difficult to identify unless the components show direct signs of interaction (see later discussion of interaction criteria).

6) Samples will often reject mergers in process. This stems from the selection requirement that two galaxies be discernable. It is somewhat dependent on the data source employed for assembling the catalog. Only one CPG entry (Arp220) shows properties generally acknowledged as merger-related.

7) Finally, single objects (like Arp220!) are sometimes misidentified as binaries. The CPG lists 16–18 pairs misclassified in this way (CPG*). "Companions" in some hierarchical pairs can be misidentified stellar objects. This problem is largely avoided in lists assembled from pre-existing galaxy catalogs.

4 Pair component morphology

The study of binary galaxy morphology is still in a primitive state. The main reasons for this are 1) the aforementioned biases resulting in statistically incomplete samples and 2) the poor quality and inhomogeneity of the imaging data for all but the least distant pairs. The size and relative completeness of the CPG mandates its use as our principal data source. After exclusion of 57 discordant and 16 misidentified single galaxies, we are left with a sample in excess of 500 pairs.

The most elementary comparison of pair component morphology is given in Table

1 where we distinguish between pairs composed of early type E/S0 galaxies (EE), late type S/Irr galaxies (SS), and those with a mixture of the two (ES). We compare the observed fraction of the three types relative to expectations based upon "field" galaxy samples. We derive expectations from: a) the CIG sample (Karachentseva, 1973), b) estimates given by Noerdlinger (1979) which are the same as the E/S fraction derived from the lowest density measure on the edge of the Hydra cluster (Wirth and Gallagher, 1980), and c) the field values given in Gisler (1980). The second reference presents a particularly useful and relevant discussion of the dependence of the E/S ratio upon the source material used to derive it. The EE, SS and ES expectations are derived using the following formulae from binomial statistics: $P(EE)= P(E)^2$; $P(SS)= [1-P(E)]^2$; $P(ES)= 2P(E)[1-P(E)]$. Uncertainties for the expectations based upon $\sqrt{n}$ statistics range from 0.01 for EX<0.10 to 0.03 for EX>0.60. The CIG expectation was the most appropriate source of comparison for the original CPG types. The present compilation of CPG types is taken from CPG* which makes use of considerable 6m data as well as other published sources. Thus the CPG values are now better compared to a mean of the expectations. The most significant result that emerges is the ***overrepresentation of early type galaxies in binary samples.*** The observed number of SS and ES pairs are near the expected values. Thus the only significant evidence for "morphological concordance" (the excess over expected population) in pairs is found in the EE class. This suggests that the morphology–density relation for galaxies (Dressler, 1980) persists at the lowest order level of galaxy clustering. These are all luminous binaries because a recent calculation of the optical luminosity function (OLF) for the CPG reveals a complete absence of dwarf EE pairs (Xu and Sulentic, 1991).

Table 1. Morphological concordance.

PAIR TYPE	CPG TOTAL	CPG FRACT.	CIG EXPEC.	NOERD EXPEC.	GISLER EXPEC.
SS	303	0.60	0.54	0.65	0.64
ES	152	0.30	0.39	0.31	0.32
EE	51	0.10	0.07	0.04	0.04

One of the big surprises to emerge from statistically useful catalogs like the CPG is the large number of ES pairs. They seem to fly directly in the face of current ideas about galaxy formation where local environment is seen as determining galaxy morphology. This is the theoretical motivation for the belief that most pairs have morphologically concordant components. This general belief has led some (Faber and Gallagher, 1979; Noerdlinger, 1979) to argue that all ES pairs must be discordant or accordant optical alignments. We find, however, that ES pairs are well–represented in the CPG, Accounting for 30% of the sample. There are now several forms of evidence that point toward the existence of a considerable population of physical ES binaries. Many show obvious signs of interaction (see e.g. Arp, 1966). They show a distribution of M/L values that is very similar to the distribution for SS pairs (CPG*). They show the distribution of velocity differences expected for bound pairs (Schweizer, 1987). An imaging survey has recently been completed for 150 CPG pairs designated ES or related type (Sharp and Sulentic 1991). Preliminary classifications reveal at least 35 certain and 35 probable ES pairs. Rampazzo and Sulentic (1990, 1991) have obtained CCD observations for 22 southern (ES candidate) pairs. They find that 13 are ES systems while the remainder are "disky" (S+S0 or S0+S0) and "early type" (E+S0) pairs. They find two kinds of evidence that suggest some ES pairs are products of secular evolution: 1) three S components show tidally generated arms (integral sign) without any sign of a disk structure and 2) at least two E components show shell structure. The former may indicate that some ES pairs are created by interaction between E and S0 galaxies while the latter may indicate that some ES pairs are partially merged higher–order multiplets. These studies suggest that between 100–200 ES pairs brighter than m_{pg}=15.7 exist on the sky. They also raise questions about whether the Hubble types for pairs are primordial or direct products of interaction events.

It is possible with the CPG sample to look more closely at the distribution of Hubble types in pairs. Table 2 presents this study in tabular form where again expectations are presented for each Hubble class based upon available field morphology studies. Figure presents a graphical depiction of the different Hubble populations for CPG pairs. The horizontal and vertical axes represent the two components of each pair. The boxes contain the observed (O) number of CPG pairs. Three expectations are calculated in Table 2 based upon : a) the populations of Hubble types in the UGC as presented by van den Bergh (1980), b) the Gisler (1980) field galaxy sample, and c) the Peterson (1979ab) binary sample as presented in van den Bergh (1980). Figure 1 expectations (E) represent averages of the results from sources a and b. They are given only if they differ from observed values by 3σ or more. Several comments are necessary before any results can be considered: 1) the CPG morphology data is heterogeneous representing previously published as well as considerable 6m data as summarized in CPG*. 2) the UGC estimate for the Sc fraction is probably too high. 3) the Gisler value for Sm/I is too high because it includes all Markarian and otherwise peculiar galaxies. 4) the Peterson values for E/S0 galaxies are probably too high due to misclassifications and bias—their selection criteria accepted the brightest members of some galaxy clusters as binaries (Schweizer, 1987).

Table 2. Component Hubble types.

HUBBLE TYPE	NILSON #	NILSON FRACT.	GISLER #	GISLER FRACT.	PETERSON #	PETERSON FRACT.	CPG #	CPG FRACT.
E	123	0.06	67	0.08	48	0.18	205	0.20
S0	240	0.11	103	0.12	54	0.20	71	0.07
Sa	196	0.09	77	0.09	43	0.16	204	0.20
Sb	468	0.22	194	0.22	59.5	0.23	274	0.27
Sc	962	0.46	250	0.29	50.5	0.19	199	0.20
Sm/I	98	0.05	171	0.20	9	0.03	59	0.06

We find several results worth mentioning in this comparison. Figure 1 suggests that the excess of early–type galaxies in CPG binaries manifests itself both in the EE and ES pair amples. There is an apparent deficit of S0 galaxies in the CPG. This probably indicates hat S0's have been frequently misclassified as E or early type spiral galaxies. This may xplain part of the apparent excess in both of those classes. Another possibility is that ome S0's have acquired spiral arms through the interaction process as was suggested y the ES pair studies. Among the spiral classes there is an apparent excess of Sa and b type galaxies which is almost certainly too large to be due entirely to misclassified nticulars. We may be seeing evidence for secular evolution in the pairs. There might be tendency for dynamical effects to enhance the bulge component of spiral galaxies, espeially in the presence of a bar (see e.g., Noguchi, 1988). The evidence for a corresponding xcess or deficit of Sc galaxies is still very uncertain. A deficit among physical pairs, due o secular evolutionary effects, might be masked by an excess from the accordant optical airs present in the CPG sample.

	E	S0	Sa	Sb	Sc	Sm
E	E= 2.4 O= 49					
S0	E= 7.9 O= 22	E= 6.6 O= 2				
Sa	E= 6.1 O= 27	O= 17	E= 4.0 O= 33			
Sb	E= 15.4 O= 37	O= 17	E= 20.1 O= 66	E= 25.0 O= 47		
Sc	O= 19	E= 43.0 O= 9	O= 26	O= 51	O= 39	
Sm	O= 2	O= 2	O= 2	O= 9	O= 16	O= 14

FIGURE 1
HUBBLE PAIRING
CPG DATA

An interesting question concerning the spiral components of pairs involves the frequency of bars, rings, and grand design morphologies. Karachentsev and Karachentseva (1975) argued that early type galaxies and barred galaxies were overrepresented in the CPG. Very little followup work has been done in order to check these conclusions. We confirm he general excess of early type systems. We are unable, however, to find much support in the literature for the excess of barred structure. The review of van den Bergh (1980), based on a UGC control sample, found evidence for a small deficit of barred members in binaries and clusters. A quick check of the 31 and 38 SS pairs found in the Schweizer (1987) and van Moorsel (1982) samples shows bar fractions of 18% and 33%, respectively—commensurate with no excess. The problems with all such investigations are the quality and degree of homogeneity of the data. R or I band images are probably the most effective for detecting stellar bars in spirals while most existing data is sensitive to shorter wavelengths. A study of the Turner sample (Elmegreen and Elmegreen, 1982) produced evidence that most grand design spirals are barred members of pairs—but little evidence for an excess of barred galaxies. Very recent work (Elmegreen et al. 1990), based upon RC2 and UGC morphology information, indicates that early–type spirals (Sa,b) in binary galaxies show a barred fraction of 0.50±0.05, while field and group systems show 0.32±0.09 and 0.28±0.02, respectively. No excess of barred galaxies is found for the later–type spirals in binaries. It is also found that 77±5% of SB members of binaries are early type while 44±2% of the entire SA sample are early type. The conclusion is that spirals in binaries are preferentially barred and early type. The new work is based upon the Turner (1976ab) and Peterson (1979ab) samples which are subject to considerable contamination from nonphysical pairs. It is clear that homogeneous imaging surveys using catalogs such as the CPG and CMG will be needed in order to further clarify the morphology of binary galaxies.

4.1 The Holmberg effect

An appropriate bridge between a discussion of Hubble component and binary interaction morphologies is provided by a discussion of the Holmberg effect. We refer here to the apparent correlation between the integrated colors of binary components first suggested by Holmberg (1958). This color correlation is a quantitative measure of the tendency for galaxies of similar type to pair. If earlier discussion is accurate, the existence of a Holmberg effect is not necessarily proof that there is strong morphological concordance in complete pair samples. It is apparently more significant than that. In fact, dissimilar (mixed) pairs exist in significant numbers. While the existence of the Holmberg effect is widely accepted, there is little high quality confirmatory data in the literature. Sharp and Jones (1980) presented evidence for a correlation between the B–V colors in 36 pairs. Most existing data refers to pairs in the CPG. Tomov (1979) has published UBV photoelectric colors for more than 100 CPG pairs. A catalog of recalibrated colors for 105 pairs was subsequently published (Demin et al., 1982; 1984). Unfortunately, full details of the calibration are not given, but comparison with limited independent data suggests an rms on the order of 0.1 for B–V. Taken at face value, this data suggests that a Holmberg effect exists for EE, SS and ES pairs. The B–V and U–B correlation coefficients for a subsample of 40 SS pairs are 0.65±0.14 and 0.77±0.07, respectively. Evidence is presented for an increase in the color difference between components with both increasing separation and velocity difference. Early type spirals (Sa-b) appear bluer than normal while the later types have near normal colors. A sample of 13 EE pairs shows an almost normal range of colors but an apparently significant correlation of component colors within that range. Twelve ES pairs show a B–V correlation coefficient of 0.88±0.06 with indications that the "E" components are bluer than normal. This may represent evidence for cross fueling in these pairs (Demin, 1984). The early–type components in 13 S+S0 pairs also appear abnormally blue. There is a great need for more accurate colors and magnitudes in order to expand upon the interesting correlations suggested in previous work.

5 Pair interaction morphology

A discussion of pair interaction morphology raises the difficult issue of defining "interaction" objectively. It is perhaps interaction criteria which provide the only means for dentifying individual physical binaries. This becomes particularly important in view of the difficulty in excluding large numbers of accordant redshift optical pairs from the catalogs. We will define interaction morphology in its broadest sense and will attempt to list all measures that have emerged with an apparent connection to interaction. Thus we will be, at the same time, assembling a list of environmental effects that paired galaxies manifest.

5.1 Optical morphology

We begin with the most subjective criterion for assessing the presence of interaction. Despite the difficulty of quantification, it remains important because of the information it provides about the effects of interaction on galaxy forms. Vorontsov–Velyaminov (1959, 1970) and Arp (1966) provided the first large compilations of objects that include the prototypes of most of the peculiar galaxy classes that are studied today. A simple and useful set of descriptors of the optical peculiarities in pairs were provided with the CPG. Pairs with peculiarities were distinguished in three classes: **LIN**– those with evidence of bridges [LIN(br)], tails [LIN(ta)], or both [LIN(br+ta)]; **ATM**– those with components in a common envelope of symmetrical amorphous [ATM(am)] or asymmetrical shredded character [ATM(sh)]; and **DIS**– those with one [DIS(1)] or both [DIS(2)] components showing signs of distortion. The DIS class is clearly the most subjective. The three types represent approximately 18%, 15% and 23% of the CPG sample, respectively, leaving almost 45% of the pairs without visible signs of interaction. Clearly many physical binaries do not show structural peculiarities that can be recognized 1) before redshift measures are available and 2) from a survey based on a uniform dataset.

Several samples of binaries have been studied in greater detail. Davoust and Prugniel (1988) find that the ATM haloes around EE pairs are often symmetrically distorted. A sample of dumbbell galaxies (EE pairs in clusters) not represented in the CPG was recently studied by Valentijn et al. (1988). Their analysis of 44 pairs suggests that they are dynamically distinct binaries with near circular orbits. They find $\sigma_V \approx 438\pm88$ km/s while $\sigma_V \approx 200\pm30$ km/s for CPG–EE pairs and $\approx 110\pm10$ km/s for CPG–SS and ES pairs. A study of ES and SS pairs by Schombert et al. (1990) shows that tidal features contain a significant fraction of the luminosity in some systems (25% on average). Tails and bridges are found to be bluer than parent galaxies which suggests that they may be the sites of excess star formation.

5.2 21cm profile morphology

The global 21cm profile is a potentially useful and sensitive measure of the dynamical state of the gas component in a galaxy. Only the wider pairs can be studied with large single–dish radio telescopes such as Arecibo. A considerable number of CPG pairs were included in a 21cm survey (Davis and Seaquist, 1983) but no profiles were presented (see also Peterson, 1979a). Gallagher et al. (1981) published 21cm data for four strongly interacting systems and noted profile peculiarities. Sulentic and Arp (1983) proposed a profile classification scheme that might prove useful as an interaction diagnostic. Tifft (1978) had previously proposed a classification for the line profiles of dwarf galaxies. Van Moorsel (1982) published 21 cm profiles derived from Westerbork synthesis data for a sample of SS pairs taken from the CMG. The 21cm profile has not yet been adequatel exploited as a tool for refining optically selected binary samples.

5.3 Optical spectroscopy. Hα imaging

Narrow band imaging of interacting galaxies has produced less than clearcut evidenc that it is useful as an interaction diagnostic. Hodge (1975) found examples of interactin systems where the distribution of HII regions (on Hα images) was strikingly asymmetri both towards and away from the companion. Most of the 14 pairs studied, howeve showed no obvious effect. There is some evidence that HII regions might be more centrall concentrated in interacting systems (Bushouse, 1987; Kennicutt et al., 1987).

Recent work suggests that the Hα equivalent width (EW) may be a useful interactio diagnostic. This results because it is an effective measure of the star formation activit that is induced by gravitational perturbation. The distribution of integrated (Hα+ [NII]) EW for paired galaxies is skewed towards larger values for pairs compared t control samples. Median EW=21 for control sample and 27 and 46, respectively, fo complete and Arp (1966) Atlas pair samples. They also report a Holmberg effect in th EW and especially in a measure of the EW excess for the binary components. The wor of Kennicutt et al. (1987) has provided evidence for enhanced star formation activity i both the nucleus and disk of paired galaxies.

Two large compilations of spectroscopic data exist for the CPG (Tifft, 1982 and CPG*). The latter reference and Tifft (1985) present discussions of the correlation o emission line properties with morphology.

5.4 Optical. FIR. Radio luminosity enhancement

Paired galaxies are apparently more intrinsically luminous than galaxies in the field. This has now been established in at least three regions of the electromagnetic spectrum. We consider this effect here because it is both an indicator of physical interaction and because it affects the results of all comparisons between magnitude limited paired and nonpaired samples. Complete reviews of the nuclear (Heckman, 1990) and disk (Kennicutt, 1990) enhancement in pairs can be found in the Proceedings of IAU Colloquium # 124, *Paired and Interacting Galaxies*, (Sulentic et al., 1990).

Tovmassian (1968) first proposed that there was an enhancement in the radio emission from pairs. The issue was debated for a number of years (see e.g., Wright, 1974) Finally, a series of complementary surveys established the statistical validity of the claim (Sulentic, 1976; Stocke et al., 1978; Hummel, 1981). Stocke (1978) argued that the effect was correlated with the average galaxian density, but subsequent studies (e.g. Kotanyi 1981; Altschuler et al., 1984) suggested that the enhancement was due to individual companions. In other words, isolated pairs show evidence of an enhancement. Sulentic (1976) argued, on the basis of spectral indices derived from single–dish measures, that the radio enhancement was primarily a disk phenomenon. Later surveys (e.g., Hummel, 1981) provided evidence that nuclear emission dominated the enhancement. More recently (see e.g., Condon et al. 1982) it has been established that disk emission also contributes to the phenomenon. Apparently both nuclear and disk emission related to starburst activity can be enhanced in binaries. The relationship of a more energetic form of nuclear activity (AGN) to interaction is still being debated (see e.g., Dahari 1985, Fuentes–Williams and Stocke, 1988) although many are convinced that there is a direct link.

The evidence for an optical enhancement could be said to originate with the work of Larson and Tinsley (1978) which showed an unusual dispersion of colors for a sample of objects taken from the Arp Atlas. White and Valdes (1980) used a binary sample from Turner (1976ab) to propose that binary components were about one magnitude brighter than field galaxies. Sharp and Jones (1980) argued that the brightness enhancement was due to interaction induced star formation. These results did not gain widespread acceptance because of the aforementioned problems with the Turner sample. Several studies involving CPG pairs have shown, through comparisons with simulations (Karachentsev, 1982) and control samples (Sulentic, 1989), that binary components are between a factor of 2 and 3 brighter (~one magnitude) than galaxies in the field. A recent calculation of the OLF for CPG pairs (Xu and Sulentic, 1991) shows that all three morphological classes of pairs (SS, ES and EE) show this enhancement. It was also found that 10% of field galaxies congregate in (physical and nonphysical) isolated pairs.

Recent infrared and mm wave (e.g. Young et al., 1984) observations have expanded our perspective on interaction enhancement by monitoring the response of dust, young stars and molecular gas. The IRAS database has played a leading role in the many studies of FIR emission from paired galaxies. Interest was sparked when it was shown that galaxies from infrared selected samples have more near neighbors than expected (Soifer et al., 1984; see also Lawrence et al., 1986 for a counter view). Subsequent studies have shown evidence for enhanced FIR emission in interacting galaxies (Lonsdale et al., 1984; Cutri and McAlary, 1985; Joseph and Wright, 1985). Some of the FIR results have been far from overwhelming (e.g. Haynes and Herter, 1988). Apparently many binaries show little or no FIR enhancement. It is probable that many are nonphysical pairs but it is difficult to establish this objectively. A recent analysis of the dust temperatures (through the IRAS 60/100μm ratio) for various classes from the Arp and Madore (1985) catalog suggests that only the closest pairs exhibit enhanced emission (Telesco et al., 1988).

Evidence for an enhanced FIR detection fraction among SS and ES pairs in the CPG has been found (Sulentic, 1989). EE pairs are rarely detected at FIR wavelengths. Evidence for a weak enhancement is even found for pairs with little or no signs of structural perturbation. This study showed that the results from all previous FIR studies of pairs could be reconciled if proper attention was paid to the biases discussed in III (along with pair morphology and the depth of the sample). A recent calculation of the FIR luminosity function (Xu and Sulentic, 1991) shows that SS pairs are dominant at the bright end ($L_{fir} > 2\times10^{11}\ L_{\odot}$). It was found that the SS pairs with LIN, ATM and DIS interaction descriptors, as well as separations less than the diameter of the primary, show clearly enhanced FIR properties compared to other CPG pairs. It is not yet possible to compare the strength of the enhancement in the individual components due to the resolution limits of IRAS. However, the presence of a Holmberg effect in the broad band optical colors, Hα emission measures, and apparent magnitudes (Karachentsev, 1982), suggests that both components of many pairs are enhanced.

6 Summary

Our knowledge of binary galaxy morphology is still in a primitive state. The preliminar evidence, however, points toward an excess of early type galaxies in pairs. All or part c this excess may be a direct result of the interaction process. The considerable numbe of mixed morphology pairs may arise from the same processes. It has become apparen that it is difficult to isolate pure samples of physical pairs. The aspects of interactio morphology discussed here may be the only means of isolating physical systems.

References

Altschuler, D., et al. 1984, *Astron. J.* **89**, 1531.

Arp, H.: 1966, *Atlas of Peculiar Galaxies*, (California Institute of Technology, Pasadena).

Arp, H.: 1969, *Astron. Astrophys.* **3**, 418.

Arp, H. and Madore, B.: 1985, *A Catalog of Southern Peculiar Galaxies and Associations: I-II*, Cambridg University Press.

Bergh, S. van den: 1980, *Phil. Trans. Roy. Soc. London, A* **296**, 319.

Bushouse, H.: 1987, *Astrophys. J.* **320**, 49.

Condon, J., et al. 1982, *Astrophys. J.* **252**, 102.

Cutri, R. and McAlary, C.: 1985, *Astrophys. J.* **296**, 90.

Dahari, O.: 1985, *Astrophys. J. Suppl. Ser.*, **57**, 643.

Davis, L. and Seaquist, E.: 1983, *Astrophys. J. Suppl. Ser.* **53**, 269.

Davoust, E. and Prugniel, P.: 1988, *Astron. Astrophys.* **201**, L30.

Dëmin, V., Dibaĭ, É., and Tomov, A.: 1982, *Soviet Astron.* **25**, 528.

Dëmin, V., Zasov, A., Dibaĭ, É., and Tomov, A.: 1984, *Soviet Astron.* **28**, 367.

Dëmin, V.: 1984, *Soviet Astron.* **28**, 622.

Dressler, A.: 1980, *Astrophys. J.* **236**, 351.

Elmegreen, D. and Elmegreen, B.: 1982, *Mon. Not. R. Astr. Soc.* **201**, 1021.

Elmegreen, D., Elmegreen, B., and Bellin, A.: 1990, *Astrophys. J.* December 1990.

Faber, S. and Gallagher, J.: 1979, *Ann. Rev. Astron. Astrophys.* **17**, 135.

Fuentes-Williams, T. and Stocke, J.: 1988, *Astron. J. 96*, 1235.

Gallagher, J., Knapp, G., and Faber, S.: 1981, *Astron. J.* **86**, 1781.

Gisler, G.: 1980, *Astron. J.* **85**, 623.

Heckman, T.: 1990, In *Paired and Interacting Galaxies* Proceedings of IAU Colloquium #124, eds. J Sulentic, W. Keel, and C. Telesco, C., NASA.

Haynes, M. and Herter, T.: 1988, *Astron. J.* **96**, 504.

Hodge, P.: 1975, *Astrophys. J.* **202**, 619.

Holmberg, E.: 1937, *Annals Lunds Astr. Obs.* **No. 6**.

Holmberg, E.: 1958, *Medd. Lunds Astr. Obs. Ser. II* **136**, 1.

Hummel, E.: 1981, *Astron. Astrophys.* **96**, 111.

Joseph, R. and Wright, G.: 1985, *Mon. Not. R. Astr. Soc.* **214**, 87.

Karachentsev, I.: 1972, *Comm. Spec. Astrophys. Obs.* **7**, 1 (CPG).

Karachentsev, I.: 1982, *Astrophysics* **17**, 135.

Karachentsev, I.: 1985, *Soviet Astron.* **29**, 243.

Karachentsev, I.: 1987, *Double Galaxies* (Nauka: Moscow) (CPG*).

Karachentseva, V.: 1973, *Comm. Spec. Astrophys. Obs.* **8**, 1.

Kennicutt, R., Keel, W., van der Hulst, J., Hummel, E. and Roettiger, K.: 1987, *Astron. J.* **93**, 1011.

Kennicutt, R.: 1990, In *Paired and Interacting Galaxies* Proceedings of IAU Colloquium #124, eds. J. Sulentic, W. Keel, and C. Telesco, C., NASA.

Kotanyi, C.: 1981, thesis, University of Groningen.

Larson, R. and Tinsley, B.: 1978, *Astrophys. J.* **219**, 46.

Lawrence, A., Walker, D., Rowan–Robinson, M., Leech, K. and Penston, M.: 1986, *Mon. Not. R. Astr. Soc.*, **219**, 687.

Lonsdale, C., Persson, S. and Matthews, K.: 1984, *Astrophys. J.* **287**, 1009.

Moorsel, G. van: 1982, thesis, University of Groningen.

Moorsel, G. van: 1983a, *Astron. Astrophys. Suppl. Ser.* **53**, 271.

Moorsel, G. van: 1983b, *Astron. Astrophys. Suppl. Ser.* **54**, 1.

Nilson, P.: 1973, *Uppsala Astr. Obs. Ann.* **6**.

Noerdlinger, P.: 1979, *Astrophys. J.* **229**, 470.

Noguchi, M.: 1988, *Astron. Astrophys.* **203**, 259.

Page, T.: 1952, *Astrophys. J.* **116**, 63.

Page, T.: 1961, *Astron. J.* **66**, 614.

Peterson, S.: 1979a, *Astrophys. J. Suppl. Ser.* **40**, 527.

Peterson, S.: 1979b, *Astrophys. J.* **232**, 20.

Picchio, G., and Tanzella–Nitti, G.: 1985, *Astron. Astrophys.* **142**, 21.

Rampazzo, R., and Sulentic, J.: 1990, In *Paired and Interacting Galaxies*, Proceedings of IAU Colloquium 124, eds, J. Sulentic, W. Keel and C. Telesco, NASA.

Rampazzo, R., and Sulentic, J.: 1991, in preparation.

Rood, H.: 1982, *Astrophys. J. Suppl. Ser.* **49**, 111.

Schombert, J., Wallin, J., and Struck–Marcell, C.: 1990, *Astron. J.* **99**, 497.

Schweizer, L.: 1987, *Astrophys. J. Suppl. Ser.* **64**, 411.

Sharp, N. and Jones, B.: 1980, *Nature* **283**, 275.

Sharp, N. and Sulentic, J.: 1991, in preparation.

Soares, D.: 1989, thesis, University of Groningen.

Soifer, B. et al.: 1984, *Astrophys. J. Letters* **278**, L71.

Sulentic, J.: 1976, *Astrophys. J. Suppl. Ser.* **32**, 171.

Sulentic, J.: 1989, *Astron. J.* **98**, 2066.

Sulentic, J., Arp, H. and di Tullio, G.: 1978, *Astrophys. J.* **220**, 47.

Sulentic, J. and Arp, H.: 1983, *Astron. J.* **88**, 489.

Sulentic, J., Arp, H. and Lorre, J.: 1985a, *Astron. J.* **90**, 522.

Sulentic, J. and Arp, H.: 1985b, *Astrophys. J.* **297**, 572.

Sulentic, J., Keel, W. and Telesco, C.: 1990, *Paired and Interacting Galaxies*, Proceedings of IAU Colloquium #124, NASA.

Stocke, J.: 1978, *Astron. J.* **83**, 348.

Stocke, J., Tifft, W. and Kaftan–Kassim, M.: 1978, *Astron. J.* **83**, 322.

Telesco, C., Wolstencroft, R. and Done, C.: 1988, *Astrophys. J.* **329**, 174.

Tifft, W.: 1978, *Astrophys. J.* **221**, 449.

Tifft, W.: 1982, *Astrophys. J. Suppl. Ser.* **50**, 319.

Tifft, W.: 1985, *Astrophys. J.* **228**, 65.

Tomov, A.: 1979, *Soviet Astron.* **22**, 540.

Tovmassian, H.: 1968, *Australian J. Phys.* **29**, 231.

Turner, E.: 1976a, *Astrophys. J.* **208**, 20.

Turner, E.: 1976b, *Astrophys. J.* **208**, 304.

Valentijn, E., and Casertano, S.: 1988, *Astron. Astrophys.* **206**, 27.

Vorontsov–Velyaminov, B.: 1959, 1970, *Atlas and Catalog of Interacting Galaxies*, (Sternberg Institute Moscow State University).

Vorontsov–Velyaminov, B.: 1975, *Soviet Astron.* **19**, 422.

White, S. and Valdes, F.: 1980, *Mon. Not. R. Astr. Soc.* **190**, 55.

White, S., et al.: 1983, *Mon. Not. R. Astr. Soc.* **203**, 701.

Wirth, A., and Gallagher, J.: 1980, *Astrophys. J.* **242**, 469.

Wright, A.: 1974, *Mon. Not. R. Astr. Soc.* **167**, 251.

Xu, C., and Sulentic, J.: 1991, *Astrophys. J.* in press.

Young, J., Kenney, J., Lord, S. and Schloerb, F.: 1984, *Astrophys. J. Letters* **287**, L65.

Zhenlong, Z. et al.: 1989, *Publ. Beijing Astr. Obs.* **12**, 8.

Zwicky, F., et al.: 1961–1968, *Catalog of Galaxies and of Clusters of Galaxies*, **1–6**, Pasadena, California Institute of Technology.

Discussion

M. Persic - The classical procedure to estimate mass density enhancements involves measurements of the velocity dispersion within pairs of galaxies (e.g. Davis and Peebles 1983). To what extent do such measurements depend on the choice of a particular catalog?

J. Sulentic - All else being equal, the accuracy of the velocity measures will be the most important factor. One also has to watch out for the bias toward pericentric pairs in some of the catalogs. The other significant problem involves differences in sample selection criteria that affect the number of false pairs (accordant redshift optical alignments in groups and clusters).

L. Buson - There exist some EE pairs, whose components are comparable in luminosity and where only one component shows strong twisting effects which appear to be absent in the other one. What could be the origin of such a different behaviour?

J. Sulentic - Our experience is that the most convincing evidence for peculiarity in EE systems involves both components. I refer to the work of Prugniel and Davoust which has shown that the halo isophotes in EE pairs are distorted symmetrically relative to the major axis of the pair. Since isophotal twisting is observed in galaxies without obvious companions, it is not clear that it relates directly to interaction.

R. Bender - Since dE's have low surface brightness, doesn't this explain why many dE pairs have not been found up to now?

J. Sulentic - The LF for EE pairs falls so rapidly near m≈ -18 that it is unlikely to be due to sample bias. Instead, my experience suggests that dwarf E's are frequently paired with much larger E or S galaxies (in hierarchical pairs). Many of these dwarfs are bright enough to have been selected for the CPG if they existed as isolated pairs.

P. Pişmiş - If you intercompared the different catalogs of pairs of galaxies, how many of the objects would be common to all the catalogs? The effective criterion for selection of duplicity in this case would be very stringent.

J. Sulentic - The principal range of overlap would involve the close pairs with separations less than 2–3 arcmin. The principal range of complementation would involve combining samples with relaxed isolation criteria (such as van Albada's CMG) and the CPG. The relaxed samples would extend the range of physical separation covered in the CPG. The price, however, would be an increase in contamination of the combined sample by discordant and accordant redshift optical pairs.

D. Burstein - It is sobering to realize that the one spiral–spiral pair we do know exists—The Galaxy and M31—are separated by 700 kpc. I am pessimistic that we will be able to understand galaxy pairs in a statistical sense until: a) we have a well-defined sample of galaxies selected optically, b) radial velocities for all galaxies, and c) distance estimates for most galaxies.

Jack Sulentic mimiking binary galaxies.

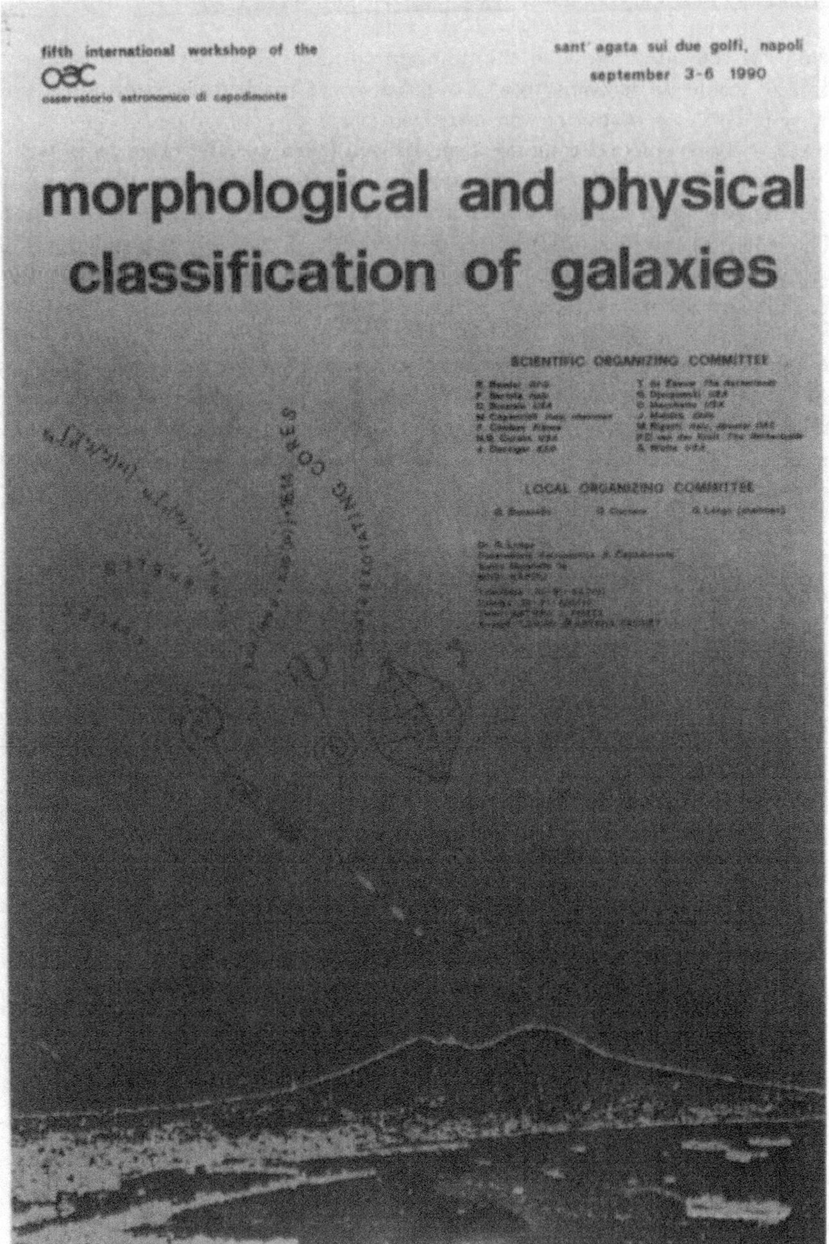

The poster of the meeting.

INTRINSIC SHAPES OF ELLIPTICAL GALAXIES

D. Merritt
Department of Physics and Astronomy
Rutgers University Piscataway, NJ, USA

ABSTRACT. Tests for the intrinsic shape of the luminosity distribution in elliptical galaxies are discussed, with an emphasis on the uncertainties. Recent determinations of the ellipticity frequency function imply a paucity of nearly spherical galaxies, and may be inconsistent with the oblate hypothesis. Statistical tests based on the correlation of surface brightness, isophotal twisting, and minor axis rotation with ellipticity have so far not provided strong evidence in favor of the nearly oblate or nearly prolate hypotheses, but are at least qualitatively consistent with triaxiality. The possibility that the observed deviations of elliptical galaxy isophotes from ellipses are due to projection effects is evaluated. Dynamical instabilities may explain the absence of elliptical galaxies flatter than about E6, and may also play a role in the (apparent) lack of nearly-spherical galaxies.

1 Introduction

About ten years ago, a number of observational programs designed to elucidate the systematics of elliptical galaxy morphology and kinematics were begun. The new data generated a flurry of papers in the early 1980's on the subject of elliptical galaxy intrinsic shapes. The question was usually phrased in one of two ways: either "Are elliptical galaxies more nearly oblate or prolate?," or "Is there strong evidence for triaxiality?". While the answer to the second question was generally felt to be "yes," none of the studies provided convincing evidence of a preference in nature for oblate or prolate forms (perhaps because most elliptical galaxies are maximally triaxial). After a hiatus of several years, the topic has once again begun to attract attention; one of the new developments has been the widespread use of CCD's, which have permitted much more accurate measurements of isophotal shapes than in the past. Rather than attempt a comprehensive review of the subject—Schecter's summary of 1987 is still essentially up to date—I will focus on the uncertainties in this game, and on possible alternative interpretations. While the problem of intrinsic shapes is a difficult one, I will argue that the accumulation of high quality data over the last few years puts us in an excellent position to improve on the earlier work.

G. Longo et al. (eds.), Morphological and Physical Classification of Galaxies, 309–320.

2 Ellipticity Distributions

The relation between the observed and intrinsic distributions of elliptical galaxy axis ratios has been discussed many times since the classic paper of Hubble (1926), who interpreted the observations in terms of a simple oblate model. It is often stated that the frequency function of apparent axis ratios cannot be used to distinguish between different hypotheses for the intrinsic shape, since the observed distribution (a function of *one* variable) is generally consistent with an infinite set of functions of the *two* axis ratios that define a triaxial galaxy. Strictly speaking, this is not true, since an inferred distribution may be negative for certain axis ratios. For instance, a distribution of apparent axis ratios $f(q_{app})$ ($q_{app} \leq 1$) is consistent with the oblate hypothesis only if the quantity

$$\int_0^{q_{app}} \frac{f(x)dx}{\sqrt{q_{app}^2 - x^2}}$$

is an increasing function of q_{app}; violation of this condition means that—for instance—the observed number of nearly round galaxies is too few to be produced by random orientations of oblate spheroids, which have a tendency (stronger than that of prolate spheroids) to appear nearly round in projection.

Until now, the positivity of the intrinsic distribution has never been an important constraint: deconvolutions based on the oblate, prolate or triaxial hypotheses have always turned out to be non-negative for all axis ratios (e.g. Noerdlinger, 1979; Binggeli, 1980; Binney and de Vaucouleurs, 1981). This is partly because, in the past, the distribution of apparent ellipticities was thought to be a smoothly increasing function of axis ratio, with E0 galaxies the most common (e.g. Sandage, Freeman and Stokes, 1970). The ellipticities in these studies were usually taken from estimates in the *First* and *Second Reference Catalogues*, based on the appearance of galaxies on photographic plates. However more recent determinations of $f(q_{app})$, based on fitting of ellipses to CCD intensity data for small and homogeneous samples, tend to favor a rather different frequency function with a pronounced *minimum* at $q_{app} = 1$. Such a distribution has been found, for instance, by Djorgovski (1986); by R. Bender, and T. Lauer (private communications) and, in an earlier study of cluster ellipticals, by Benacchio and Galletta (1980). Figure 1 shows a second-order polynomial fit to the Djorgovski (1986) data, and the inferred intrinsic distribution $N(q)$ under the oblate and prolate hypotheses. The paucity of nearly-round galaxies appears to cause problems for the oblate deconvolution: $N(q) < 0$ for $q > 0.93$, although the small number of galaxies in this sample would probably permit an acceptable fit to $f(q_{app})$ with a positive $N(q)$ (as in Franx, 1988). Nevertheless the shortage of nearly-round galaxies, if it persists in larger samples, might eventually be shown to be inconsistent with the oblate hypothesis. (Deconvolution algorithms like Lucy's (1974), which *guarantee* positive-definite solutions, should be avoided in this business, since they may "force" a positive solution where a negative one is implied by the data.)

It is interesting to speculate what sort of galaxy formation process would prefer elongated galaxies to nearly spherical ones. A recent simulation of the formation of halos in the "cold dark matter" universe (Frenk et al., 1988) predicts a frequency function that s probably too strongly weighted toward spherical systems to be consistent with figure 1.

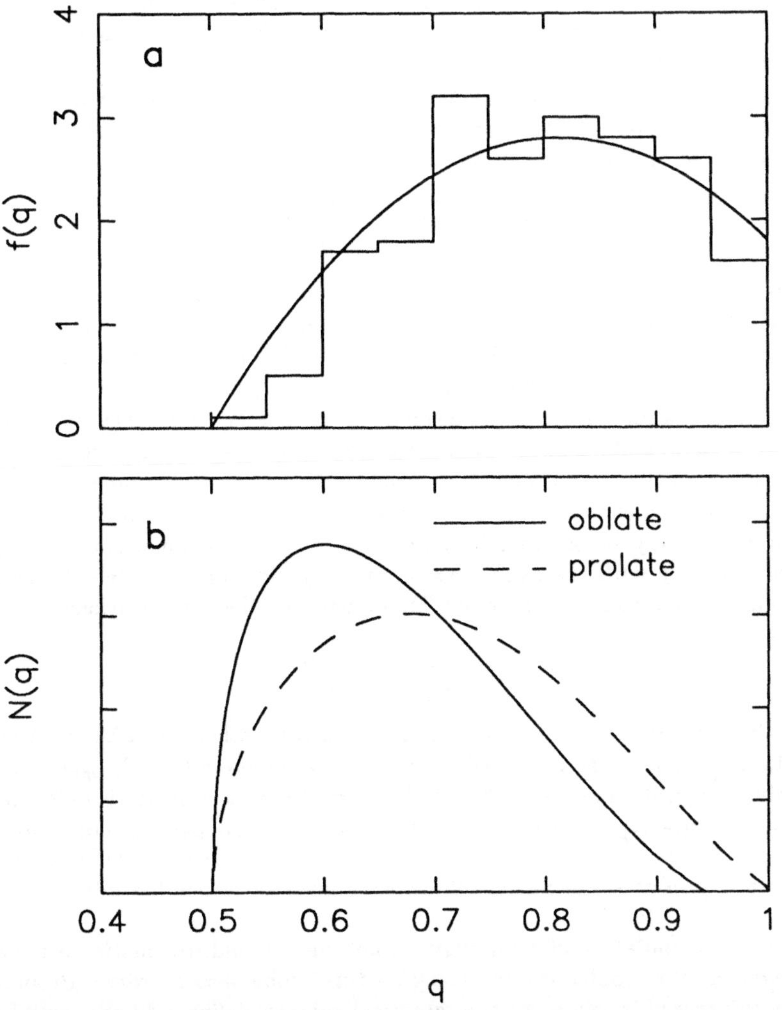

Figure 1 (a) Observed apparent ellipticity distribution, from Djorgovski (1986). (b) Inferred true ellipticity distributions under the oblate and prolate hypotheses.

Simple collapse naturally produces elongated (~E4), prolate/triaxial galaxies (Aguilar and Merritt, 1990), but only from initial conditions that are unphysically smooth. Mergers of disk systems can easily result in elongated final states, especially if dissipation is important. Whatever the actual formation mechanism, it would be ironic if the spherical galaxies so loved by theorists were never present in nature.

3 Correlations with ellipticity

The correlation of apparent ellipticity with surface brightness was first used as a test t distinguish between the oblate and prolate hypotheses by Marchant and Olson (1979 and Richstone (1979). The surface brightness of an oblate galaxy is highest when viewe edge-on, and t therefore when the galaxy appears most highly flattened; the convers is true for a prolate galaxy. The dependence of the central surface brightness Σ on th apparent axis ratio q_{app} may be expressed simply as

$$\Sigma = \begin{cases} q_{app}^{-1}\Sigma_p & \text{(oblate)} \\ q_{app}\Sigma_p & \text{(prolate)} \end{cases}$$

where Σ_p is the polar surface brightness. For a sample of *identical* galaxies with randon orientations, the observed correlation of surface brightness with axis ratio would imme diately reveal the intrinsic shape. Similar tests have been proposed based on velocit dispersion (Lake, 1979) or isophotal radius (Fasano, 1987) instead of surface brightness

Unfortunately, what works well for a sample of *identical* galaxies does not necessaril work well for a sample of galaxies with different axis ratios and luminosities: even afte eliminating the (strong) dependence of surface brightness on total luminosity, one has t allow for a possible dependence of polar surface brightness on *intrinsic* ellipticity. Ther is no reason why a very elongated galaxy should have the same polar surface brightnes as a spherical galaxy with the same total luminosity. If—for instance—the *equatoria* surface brightness is assumed to be more "fundamental," then the relations given abov should be written

$$\Sigma = \begin{cases} q q_{app}^{-1}\Sigma_{eq} & \text{(oblate)} \\ q^{-1} q_{app}\Sigma_{eq} & \text{(prolate)} \end{cases}$$

where q is the *intrinsic* axis ratio. For constant Σ_{eq} (or, rather, for a Σ_{eq} that depends only on total luminosity), the predicted distribution of points in the (Σ, q_{app}) plane turns out to be almost correlation-free (although the *form* of the distribution is different unde the oblate and oblate hypotheses). Studies which allow for a dependence of polar surface brightness or velocity dispersion on intrinsic axis ratio (Merritt, 1982; Fasano, 1990) have so far not produced conclusive evidence in favor of either the oblate or prolate hypothesis.

Given the accumulation of high-quality photometric and kinematic data over the last few years, a more sophisticated attack on this problem is overdue. Probably the correct way to proceed is to assume that elliptical galaxies define a "fundamental plane" something like

$$L \propto \Sigma_p^l \sigma_p^m q^n,$$

with L the luminosity, Σ_p and σ_p the polar surface brightness and velocity dispersion, and q the intrinsic axis ratio. This "fundamental plane" differs from the one discussed by e. g. Djorgovski and Davis (1987) in that (a) the number of significant variables is increased by one; (b) the variables defining the plane are all *intrinsic*, and can only be related to *observed* quantities after making some assumption about intrinsic shapes. Presumably, either the oblate or prolate hypothesis will lead to a better match between the predicted and observed distribution of galaxies in the space of observables $\{L, \Sigma, \sigma, q_{app}\}$. The high dimensionality of the fit makes this a tough problem, but definitely worth attempting.

4 Isophotal Twists

Of course, we now believe that elliptical galaxies are neither oblate nor prolate, but triaxial. The strongest evidence for triaxiality is probably still the prevalence of isophotal twists, which are a common feature of triaxial models seen in projection if the axis ratios vary with radius. The interpretation of twists in terms of triaxiality is not airtight: first, because galaxies may be intrinsically twisted, and certainly are in some cases of recent close interaction; and second, because some triaxial models with ellipticity gradients exhibit no twists at any viewing angle (the best examples being models with separable potentials; see Franx, 1988). Nevertheless the abundance of significantly twisted galaxies (estimates range as high as 50% or 60%), and the expectation that intrinsic twists are short-lived, suggest that triaxiality is common.

Little progress has been made on unravelling twists since the early papers of Carter (1978), Benacchio and Galletta (1980), and Leach (1981). These studies seemed to show that the frequency and amplitude of the twists was consistent with the observed ellipticity gradients, assuming either a constant degree of triaxiality (e.g. Benacchio and Galletta, 1980), or a mixture of triaxial and axisymmetric galaxies (e.g. Leach, 1981). With much larger samples, one could imagine inferring the distribution of axis ratios and their gradients $N(q_1, q_2, dq_1/dr, dq_2/dr)$ from the observed twists. Such an analysis would be formidable, given the fact that both ellipticities and position angles often seem to vary in a complicated way with radius.

Perhaps a more fundamental question—not really addressed in the early studies—is to what degree triaxiality (in the commonly accepted sense) is *required* to explain the twisted isophotes. Fasano and Bonoli (1989) have recently attempted to answer this question by studying twists in a sample of isolated elliptical galaxies, taking care to understand all the effects (dust lanes, improper flat fielding, misidentification of lenticular galaxies) that might result in spurious twists. They find a frequency of twisting in their "unperturbed" sample that is comparable to the highest estimates given by earlier workers for more randomly selected samples; thus they conclude that tidal interactions do not make a dominant contribution to the twists. In addition, however, they find that significant twisting is only seen in galaxies that deviate from a de Vaucouleurs luminosity profile. This they interpret as the signature of a disk superimposed on a spheroid, and they suggest that some fraction of the twisted ellipticals may be SB0 galaxies, or else an intermediate class of elliptical galaxies with a disklike component. It would be interesting to determine whether there is any kinematical evidence for this interpretation.

5 Non-elliptical isophotes

Up to now I have avoided the question of why or whether elliptical galaxies are precisely ellipsoidal. The standard answer to these questions is, of course, that they are not: deviations of the isophote shapes from perfect ellipses, both in the directions "boxiness" and "diskiness", are now routinely measured at the $\sim 0.5\%$ level (e.g. Bender, Döbereiner and Möllenhoff, 1988; Michard and Simien, 1988). Nevertheless it is remarkable that elliptical galaxies are so *well* described by ellipsoids. The first fully self-consistent model for a pressure-supported triaxial galaxy (Schwarzschild, 1979) was decidedly "peanut-shaped," and the non-classical integrals that are thought to maintain triaxial figures in the absence of rotation are present in many models with strongly non-ellipsoidal isodensity surfaces (de Zeeuw, Peletier, and Franx, 1986). Furthermore, the bulges of disk galaxies are often extremely boxy (e.g. Kormendy and Illingworth, 1982). Apparently some physical mechanism, as yet unguessed, guarantees that the isodensity surfaces of elliptical galaxies never deviate very strongly from perfect ellipsoids.

To what extent could the observed deviations—like the isophotal twists discussed above, and the kinematic misalignments discussed below—be an artifact of projection? Such a possibility was raised by several of the early workers in this field (e.g. Leach, 1981; Williams, 1981), but apparently never worked out in quantitative detail. The basic idea is that—whereas similar, coaxial ellipsoids always project to perfectly elliptical isophotes—the isophotes of a galaxy in which the ellipticity, orientation, or center of the isodensity surfaces varies with radius are *not* ellipses. Consider the first of these cases, i.e. a galaxy with ellipticity gradients, but without any intrinsic twists or offsets. Suppose that the luminosity is stratified on spheroidal surfaces of constant m, where $m^2 = x^2+y^2+h(m)z^2$. The simplest choice for $h(m)$ is $t_0 + am^2$, where $t_0^{-1/2}$ is the central axis ratio and a is a constant related to the ellipticity gradient; for this choice of $h(m)$, m may be expressed simply in terms of the coordinates, i.e. $m^2 = (x^2 + y^2 + t_0z^2)/(1 - az^2)$. The resulting model is bounded by the cylinder $x^2 + y^2 = -t_0/a$ for $a < 0$, and by the slab $z = \pm a^{-1}$ for $a > 0$; however for small a the bounding surfaces lie at very large radii. For a power-law luminosity profile $\rho \propto m^{-3}$ (a good fit to a de Vaucouleurs law near r_e), the surface brightness, seen along the y axis, is

$$\Sigma(x,z) \propto \begin{cases} (1-az^2)\sqrt{1+ax^2/t_0}/\left(x^2+t_0z^2\right), & a<0 \\ (1-az^2)^{3/2}/\left(x^2+t_0z^2\right), & a>0. \end{cases}$$

Define the "best-fit" ellipse passing through $(x,0)$ to be the ellipse around which the surface brightness, expanded in a Fourier series

$$\Sigma(\theta) = \Sigma_0\left[1 + a_2\cos(2\theta) + 2a_4\cos(4\theta) + \ldots\right]$$

in the angle $\theta = \tan^{-1}(\sqrt{t}z/x)$, has no $\cos(2\theta)$ component. (The factor of two in front of the $\cos(4\theta)$ term is due to the fact that observers tend to quote a_4 in terms of the *radial* deviation of the best-fit ellipse from the true isophote, rather than the intensity variation around the best-fit ellipse; the difference is a factor of $|d\log\Sigma/d\log r| = 2$ in the present case.) Setting $a_2 = 0$ fixes the axis ratio of the best-fit ellipse passing through $(x,0)$:

$$t = t_0 + \frac{3}{2}ax^2 + \ldots$$

Around this ellipse, the $\cos(4\theta)$ term has coefficient

$$2a_4 = \frac{3}{64}\frac{a^2x^4}{t_0^2} + \ldots$$

Note that, to lowest order in ax^2, the ellipticity gradient just results in a change in the axis ratio of the best-fit ellipse; the deviations from "ellipticality" are of order a^2x^4, and always (at least in this restricted model) go in the direction of making the isophotes more disky ($a_4 > 0$). In terms of the ellipticity gradient $\delta = dq_{app}/d\log x$,

$$2a_4 = \frac{t_0}{48}\delta^2 + \ldots$$

Figure 2 shows a_4 as a function of $dq_{app}/d\log r$ in a model with $t_0 = 1$ (spherical at the center) and $a > 0$ (increasingly oblate). Diskiness at the observed level evidently requires an ellipticity gradient $dq_{app}/d\log x$ greater than 0.3. Bender, Döbereiner and Möllenhoff (1988) find a few elliptical galaxies with $dq_{app}/d\log x$ as large as 0.15, but for the majority of galaxies in their sample, the ellipticity gradient is much less. Therefore I conclude that the observed ellipticity gradients fail by a factor of several to explain the observed deviations of the isophotes from ellipses.

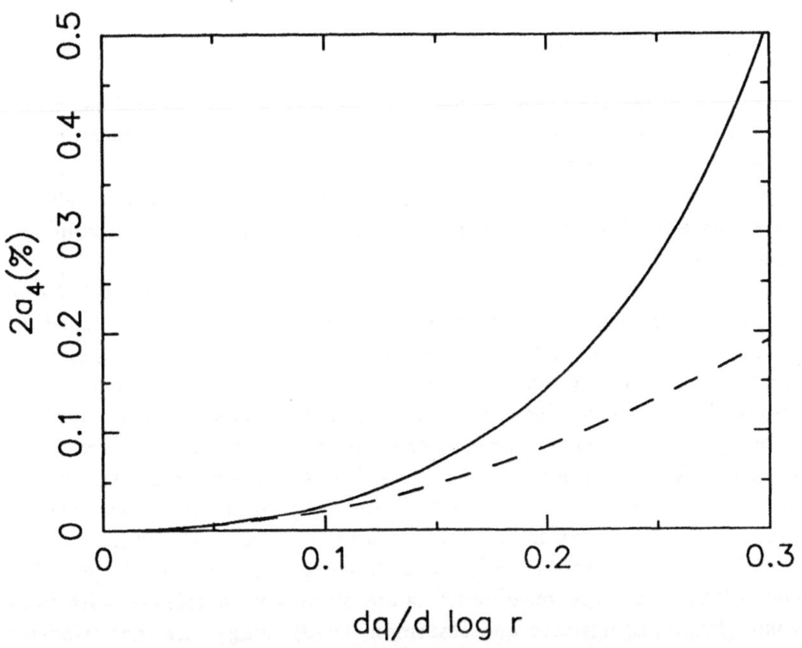

Figure 2. Variation of a_4 with ellipticity gradient in the model discussed in the text. Solid line: major axis; dashed line: minor axis.

Although this simple analysis appears to rule out ellipticity gradients as an important contributor to non-elliptical isophotes, it would be worthwhile to extend the calculation to models with non-aligned or non-concentric isodensity surfaces. It would also be helpful to understand how one goes about formally deprojecting an observed intensity distribution to obtain the three-dimensional luminosity distribution, *without* assuming that the isodensity surfaces are ellipsoids, and without making an *ad hoc* assumption (as above) about the radial dependence of the shape parameters. Rybicki (1987) has carried out some preliminary work in this direction; the problem is inherently underdetermined in the absence of information about the inclination angle.

The next simplest interpretation of the non-elliptical isophotes is a model in which the underlying elliptical galaxy is, again, accurately ellipsoidal, but with a second component (presumably different kinematically) that has a very different morphology. "Disky" isophotes are naturally modelled by superimposing an exponential disk on an otherwise normal spheroid (e.g. Carter, 1987; Jedrzejewski, Davies, and Illingworth, 1987; Rix and White, 1990). Disky ellipticals are often found to be rotationally supported (Bender, 1988), and many such galaxies may simply be misclassified S0's (e.g. Capaccioli, Piotto and Rampazzo, 1988). No comparably simple interpretation of "boxy" isophotes has so far been suggested. Two conclusions may be drawn from these studies: first, that many elliptical galaxies probably contain a more or less significant disk; and second, to the extent that the non-elliptical isophotes are due to photometric superposition of a spheroid and a disk, the underlying elliptical galaxies must be even *more* accurately ellipsoidal than the measured values of a_4 would suggest.

6 Kinematic Tests

Broadly speaking, these are of two kinds. First, the rings of dust and gas seen in a few elliptical galaxies can sometimes be identified with periodic orbits. Given an assumption about the radial form of the underlying potential (which may or may not trace the light) it is then possible to infer the intrinsic shape of the potential from the behavior of the orbit. F. Bertola and T. de Zeeuw discuss this sort of analysis in their contributions to these proceedings.

The second kind of kinematical test is based on a purely *geometrical* property of triaxial ellipsoids. Because the *apparent* minor axis of a triaxial galaxy need not be coincident with the projection of the *intrinsic* minor axis, it follows that stellar streaming around the intrinsic minor axis will generally result in so me component of line-of-sight velocity along the apparent minor axis (Binney, 1985). Thus the observation of "minor axis rotation" provides a certain amount of evidence in favor of the triaxial hypothesis. As always, the inference is not airtight, since (a) stellar streaming may occur around the intrinsic *major* axis, i.e. a prolate "spindle"; (b) elliptical galaxies certainly have rotating figures, and the rotation axis may even be inclined to the symmetry axis. However observations show a preference for small kinematic misalignments (e.g. Davies and Birkinshaw, 1988), and large misalignments are often seen in galaxies with large photometric twists (Franx, Illingworth and Heckman, 1989), suggesting that triaxiality is the main culprit.

As in the case of isophotal twists, making detailed inferences about the shapes of galaxies from statistics of the observed rotation fields is extremely difficult. Since a major study of this problem is nearing completion (Franx, Illingworth and de Zeeuw, 1990), I will limit myself to an outline of the main points. In its simplest form, the problem is very similar to the surface brightness/ellipticity test described above: given some measure of the kinematic misalignment for a sample of galaxies (e.g., the ratio of minor to major axis rotation velocities on some isophote), what hypothesis for the distribution of intrinsic shapes is most consistent with the observed correlation of this parameter with apparent axis ratio, assuming that the streaming is about the intrinsic minor axis? Franx (1988) finds that a triaxiality $Z = (1-q_2)/(1-q_1) \approx 0.5$ fits the data well, except for a possible bundance peak at misalignments of $\sim 90^\circ$, indicative of rotation around the intrinsic ajor axis.

As Franx points out, a major uncertainty in his analysis is the likelihood of *intrinsic* isalignments: since triaxial galaxies contain families of tube orbits that circulate both bout the major and minor axes, there is no good reason to expect streaming around nly one of these axes. Furthermore, the intrinsic misalignment may itself be correlated ith intrinsic axis ratios, in a way that masks the expected correlation between observed uantities (just as in the surface brightness/ellipticity test discussed above). Work in rogress should elucidate to what extent these difficulties can be overcome.

Stability

ne of the reasons that the oblate hypothesis persisted so long in the minds of galactic ynamicists is that the number of orbital integrals in axisymmetric potentials (always wo, sometimes three) is clearly sufficient to allow the existence of self-consistent models f any axis ratio. Schwarzschild's (1979) construction of an equilibrium *triaxial* model ut an end to this mind set: clearly it was now impossible to prefer axisymmetric over on-axisymmetric models on the basis of equilibrium arguments alone. Statler (1987) vent on to demonstrate that essentially no point in the plane of model axis ratios could e excluded by the requirement of equilibrium, at least for models based on the so-called perfect" potentials in which all parts of phase space are characterized by three isolating ntegrals. But it has recently become clear that many of these equilibrium models are ynamically unstable.

Most of the work to date on stability of nonrotating models has focused, for obvious easons, on the spherical case. While isotropic spherical models are generally stable Antonov, 1962), surprisingly small amounts of radial anisotropy can induce instability o an "$m = 2$" or bar mode, which causes the model to evolve into a triaxial bar. A recent tudy of this "radial-orbit instability" in a particular family of spherical isochrone models Saha, 1990) finds instability for $\sigma_r/\sigma_t > 1.2$, where σ_r and σ_t are the mean radial and one-component) tangential velocity dispersions. Work in progress should soon reveal vhether this result holds for a wider class of anisotropic spherical models. Although it vould probably be extreme to blame this instability alone for the apparent absence of early-spherical galaxies (Fig. 1), it is likely that radial collapse sometimes produces ar-unstable final states (Aguilar and Merritt, 1990), thus leading to a preference for onspherical (even non-axisymmetric) galaxies over spherical (axisymmetric) ones.

At the other extreme of the ellipticity distribution, it appears more and more likely hat the absence of elliptical galaxies flatter than about E6 can be ascribed to instailities. Fridman and Polyachenko (1984) suggested that oblate and prolate models vith axis ratios more extreme than about 2 : 5 might be generally unstable to bending nodes, a prediction that has now been verified in one family of prolate models (Merritt nd Hernquist, 1991). Figure 3 shows the instability acting on an initially E9 model. emonstrating the existence of this instability in a range of prolate and oblate models vill require considerably more work, but may solve once and for all the classic problem of why highly flattened elliptical galaxies do not exist.

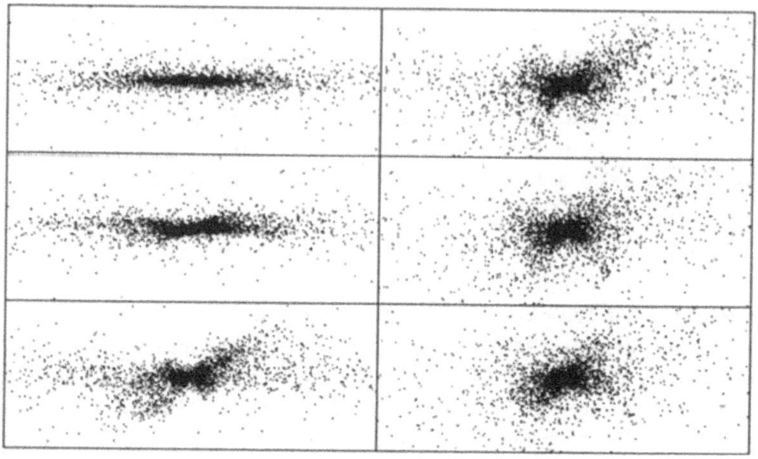

Figure 3 Evolution of a prolate model with initial axis ratio 1:10 (from Merritt and Hernquist, 1991).

Perhaps the ultimate lesson to be learned from these stability studies will be simply that nature prefers triaxial equilibria to axisymmetric ones, since the latter often evolve spontaneously into the former.

Acknowledgements

I am indebted to Ted Williams for bringing to my attention the influence of ellipticity gradients on isophotal shapes.

References

Aguilar, L.A. and Merritt, D.: 1990, *Astrophys. J.* **354**, 33.

Antonov, V.A.: 1962, *Vestnik Leningrad Univ.* **19**, 96.

Benacchio, L., Galletta, G.: 1980, *Mon. Not. R. Astr. Soc.* **193**, 885.

Bender, R.: 1988, *Astron. Astrophys.* **193**, L7.

Bender, R., Döbereiner, S., Möllenhoff, C.: 1988, *Astron. Astrophys. Suppl. Ser.* **74**, 385.

Binggeli, B.: 1980, *Astron. Astrophys.* **82**, 289.

Binney, J.: 1985, *Mon. Not. R. Astr. Soc.* **212**, 767.

Binney, J., de Vaucouleurs, G.: 1981, *Mon. Not. R. Astr. Soc.* **194**, 679.

Capaccioli, M., Piotto, G., Rampazzo, R.: 1988, *Astron. J.* **96**, 487.

Carter, D.: 1978, *Mon. Not. R. Astr. Soc.* **182**, 797.

Carter, D.: 1987, *Astrophys. J.* **312**, 514.

de Zeeuw, T., Peletier, R., Franx, M.: 1986, *Mon. Not. R. Astr. Soc.* **221**, 1001.

Davies, R.L., Birkinshaw, M.: 1988, *Astrophys. J. Suppl. Ser.* **68**, 409.

Djorgovski, S.: 1986, PhD thesis, University of California, Berkeley.

Djorgovski, S., Davis, M.: 1987, *Astrophys. J.* **313**, 59.

Fasano, G.: 1987, in *IAU Symposium 127, Structure and Dynamics of Elliptical Galaxies* (T. de Zeeuw, Ed.), 395, Reidel, Dordrecht.

Fasano, G., Bonoli, C.: 1989, *Astron. Astrophys. Suppl. Ser.* **79**, 291.

Fasano, G.:1990, preprint.

Franx, M.: 1988, PhD thesis, University of Leiden.

Franx, M., Illingworth, G., Heckman, T.: 1989, *Astrophys. J.* **344**, 613.

Franx, M., Illingworth, G., de Zeeuw, T.: 1990, in preparation.

Frenk, C.S., White, S.D.M., Davis, M., Efstathiou, G.: 1988, *Astrophys. J.* **327**, 507.

Fridman, A.M., Polyachenko, V. L.: 1984, *Physics of Gravitating Systems* Springer, New York.

Hubble, E.: 1926, *Astrophys. J.* **64**, 321.

Jedrzejewski, R.F., Davies, R.L., Illingworth, G.D.: 1987, *Astron. J.* **94**, 1508.

Kormendy, J., Illingworth, G.: 1982, *Astrophys. J.* **256**, 460.

Lake, G.: 1979, in *Photometry, Kinematics and Dynamics of Galaxies* (D. S. Evans, Ed.), 381, University of Texas Press.

Leach, R.: 1981, *Astrophys. J.* **248**, 485.

Lucy, L. B.: 1974, *Astron. J.* **79**, 745.

Marchant, A.B., Olson, D.W.: 1979, *Astrophys. J.* **230**, L157.

Merritt, D.: 1982, *Astron. J.* **87**, 1279.

Merritt, D. and Hernquist, L.: 1991, preprint.

Michard, R., Simien, F.: 1988, *Astron. Astrophys. Suppl. Ser.* **74**, 25.

Noerdlinger, P.D.: 1979, *Astrophys. J.* **234**, 802.

Richstone, D.O.: 1979, *Astrophys. J.* **234**, 825.

Rix, H.-W., White, S.D.M.: 1990, *Astrophys. J.* **362**, 52.

Rybicki, G.B.: 1987, in *IAU Symposium 127, Structure and Dynamics of Elliptical Galaxies* (T. de Zeeuw, Ed.), 397, Reidel, Dordrecht.

Saha, P.: 1990, preprint.

Sandage, A., Freeman, K.C., Stokes, N. R.: 1970, *Astrophys. J.* **160**, 831.

Schwarzschild, M.: 1979, *Astrophys. J.* **232**, 236.

Statler, T. S.: 1987, *Astrophys. J.* **321**, 113.

Williams, T.: 1981, *Astrophys. J.* **244**, 458.

Discussion

R. Buta - In your plots comparing $N(\varepsilon)$ vs ε_{app} for visual and photometric samples, it would seem that the excess of round E's in the visual samples is an artifact of the tendency of observers or classifiers to measure a nearly round galaxy as exactly round.
D. Merritt - Yes, a number of observers have pointed out this effect. There is also a tendency to exaggerate the ellipticities of very flat galaxies.

R. van der Marel - You mentioned the $m = 1$ instability, which manifests itself in a center shift of the isophotes. Such shifts are not easy to detect observationally, since flat fielding problems generally also manifest themselves as a center shift of the isophotes. This is why observers generally do not publish the centers of the isophotes. However,

it may be a good ideas if they would do so in the future. One could then compare the results of different authors, and may be separate isophotal shifts from flat fielding problems.
D. Merritt - I agree.

G. Djorgovski - Two comments: the reason why the simple tests for oblateness/prolaten• failed is that there is zero correlation between the ellipticity and either surface brightness or velocity dispersion. It is striking how ellipticity, which serves as the basis of Hubble's classification of E's, does not correlate at all with any physical property. Second, I think that we shouldn't worry too much about the stability of E's; for all we know, they may be oscillating or changing their shapes on a secular time scale.
D. Merritt - The correlation coefficient by itself is *never* sufficient to distinguish between oblate and prolate hypotheses, regardless of its value. It is the *distribution* in the surface brightness/ellipticity plane, and not just the shape, that is a useful discriminant, as I have argued. Certainly we should not rule out slow oscillations or gradual evolution of elliptical galaxies. However any model that evolves strongly on a few dynamical time scales can be securely excluded.

T. de Zeeuw - Do you think that elliptical galaxies fill the whole space of *stable* solution or that processes during galaxy formation reduce the solution space even further?
D. Merritt - I think the proper way to answer this question is to thoroughly map out the stable region of model space, then to see whether observed galaxies fill this space, or whether they are confined to some smaller region. Certainly stability arguments on their own will never predict the detailed distribution of galaxies in model space.

David Merritt

ON UNDERSTANDING THE GLOBAL PROPERTIES OF GALAXIES

D. Burstein
Department of Physics and Astronomy
Arizona State University, USA

ABSTRACT. The global properties of galaxies must necessarily be determined from large samples of galaxies. Yet, the means by which astronomers can choose such large samples yield data that are biased in ways known and sometimes unknown. Problems in selecting galaxy samples are reviewed. Two general techniques for overcoming these biases are discussed. One of these techniques is applied to the vexing problem of determining the average change of spiral galaxy diameter and magnitude with inclination.

1 Introduction

There are many questions concerning the global properties of galaxies that we would like to be able to answer:

What is the volume density of galaxies by Hubble type? By absolute luminosity? By absolute size? What is the relationship of size to luminosity among galaxies? What are the physical relationships of the stellar populations of galaxies to the size of the galaxy? The density of the galaxies? The environments in which the galaxies are found? What are the relationships of dark matter to luminous matter in galaxies?

The above list of questions is not intended to be exhaustive, but simply illustrative. On each of the above questions much has been written by many researchers, often with contradictory conclusions. It is the opinion of this writer that, in most cases of disagreement, the principal source of disagreement is the manner in which the galaxies were selected for study, not the accuracy of the data *per se*.

Unfortunately, the very act of choosing a galaxy sample is biased, often in more ways than one. If our eventual goal is to understand the structure and evolution of galaxies, these biases must be identified, and analysis techniques developed to overcome their influence. This review summarizes the most important biases known to this writer. Two methods are discussed that can be used to overcome some of these biases. As an illustration of its utility, one of these methods is applied to the much-discussed problem of determining the inclination dependence of the optical diameters and magnitudes of galaxies.

G. Longo et al. (eds.), Morphological and Physical Classification of Galaxies, 321–336.

2 The kinds of biases that affect galaxy samples

By definition, in order to 'study' a galaxy, one must be able to 'see' the galaxy, wher the term 'see' can be applied to any wavelength region. The very act of 'seeing' a galax is a function of the galaxy's distance, surface brightness, wavelength region in which i observed and inclination of the galaxy to the line of sight. Each of these dependencies ca introduce different biases for different kinds of galaxies, and separate biases can combin to produce even more complicated problems with galaxy samples. To be more specific:

Distance-dependent bias (often called Malmquist [1920, 1922] bias): As discussed b this writer (Burstein, 1990), Malmquist bias is one of the most misunderstood concept in observational astronomy. Briefly, this bias originates from the fact that any sample o galaxies has an observed dispersion of absolute luminosity or size, due to a combinatio of cosmic dispersion of these properties and observational error. Rather than try in th space here to give a summary of a rather complicated subject, the reader is referred t the original papers of Malmquist for the most comprehensive description, to Lynden-Bel et al. (1988), and to Burstein (1990) for this writer's discussion of this subject.

It is important to emphasize here that the sense, and size, of errors introduced int a galaxy sample by Malmquist bias depend on the true density distribution of galaxies i space. In a relatively homogeneous distribution of galaxies (but not necessarily uniforn or smooth), galaxies selected by a distance-dependent parameter will be intrinsicall brighter than one would estimate based strictly on predicted distances. The increase i absolute brightness caused by this bias is proportional to the square of the measure dispersion in absolute magnitude of the galaxy sample. Thus, a galaxy sample with a small measured dispersion in absolute magnitude is much less affected than one with a large dispersion.

Surface-Brightness bias: As has been emphasized by Disney and others (e.g., Disney 1976; Bothun et al., 1990), the selection of galaxies is often made as much by apparen surface brightness as by apparent magnitude or diameter. The surface brightness of a galaxy is plausibly related to other galaxian physical properties, such as inclination of a galaxy to the line of sight, level of star formation, kind of stellar population or amoun of obscuring dust. Attempts to determine the inclination dependence of magnitude and diameter for spiral galaxies are particularly affected by surface-brightness dependen biases, as discussed in section 4.1.

Wavelength-dependent bias: The physical mechanisms that produce radiation emission in galaxies can vary systematically as a function of galaxy type. For example, selection of galaxies in the optical discriminates against galaxies that have large amounts o dust, while selection of galaxies in the far-infrared (e.g., by IRAS) discriminates against galaxies with little or no dust. Similar problems exist when selecting galaxies via emissior in X-rays, far-ultraviolet light, HI 21 cm line or radio continuum.

Inclination-dependent bias: The three previous biases could just as well have been applied to the study of stars as to the study of galaxies (indeed, the papers of Malmquist discuss stars, not galaxies). Only galaxies have three-dimensional forms that are not necessarily spheroidal in shape, are spatially-resolved on the sky, and which present themselves to us at all viewing angles. Ideally, one would like to choose samples of galaxies based on an inclination-invariant property of the galaxy. Conversely, samples of galaxies chosen on the basis on an inclination-*variant* property will have measured global roperties that are biased as a function of inclination. To further complicate matters, ne inclination-invariant property for galaxies can, in principle, itself be a function of ubble type, absolute luminosity, wavelength and surface brightness.

The above kinds of biases can act together to create even more intricate selection iases in galaxy samples. One example of how such biases can be inadvertently combined one galaxy sample is discussed in section 4.2. In addition, the relative faintness of alaxies ensures that measurement of their physical properties often involves separating low flux signal (from the galaxy) from a large flux signal (due to background emission). s a result, available data on galaxies span a wide range of observational accuracy. Since ne size of a bias often is a function of observational errors, catalogs that assemble data om a variety of sources (e.g., de Vaucouleurs, de Vaucouleurs and Corwin, 1976; RC2) nay be affected by biases too interconnected to reliably disentangle.

Two general methods to overcome these biases

.1 Using malmquist bias to our advantage

ranted one must be able to 'see' a galaxy to study it, the effects of Malmquist bias are nportant for most galaxy samples. A method to make practical use of Malmquist bias vas introduced by Burstein and Lebofsky (1986), in studying the inclination-dependent roperties of far-infrared emission from spiral galaxies: One should first select a sample f galaxies strictly according to a distance-dependent parameter (e.g., apparent luminosty), to a well-defined limiting value. One can then correlate any distance-*independent* arameter (e.g., inclination) with the absolute value of the distance-dependent parameer, as calculated by an appropriate model of the large-scale velocity field (e.g., a smooth Iubble flow) and measured redshifts. The variation of distance-dependent parameter as function of distance-independent parameter (e.g., absolute magnitude as a function of nclination) will give the relationship between these two parameters.

For example, to test whether the blue isophotal diameter of a type of galaxy increases vith inclination, choose a sample of this kind of galaxy that is strictly diameter-limited nd free of other biases (e.g., surface brightness bias). If edge-on galaxies have larger sophotal diameters than face-on galaxies, one would be able to see edge-on galaxies to reater distances than face-on galaxies. The mean absolute isophotal diameter of this ind of galaxy would increase with increasing axial ratio in a diameter-limited sample. The same kind of test, termed a 'Malmquist bias' test, can be made for any measurements f flux, or size, of galaxies as a function of inclination.

If galaxies are chosen strictly by the distance-dependent quantity, the Malmquist bias est should yield the correct correlation of distance-independent parameter with distancelependent parameter, providing other selection effects are not important. Malmquist ias tests made with more heterogeneous data sets are also informative, as such tests an discover the form and sense of the correlation, if not the correct slope.

The Malmquist bias test has been applied to the question of inclination dependence f diameter and magnitude in spiral galaxies by Burstein and Lebofsky (1986) for a mall sample of galaxies, and will be applied by Haynes and Burstein (1991) and de Vaucouleurs et al. (1991, Third Reference Catalog of Galaxies; RC3) for large samples f galaxies. The results of these tests are given in section 4.

3.2 Double-Blind samples of galaxies

Most galaxies emit much, if not most of their energy at wavelengths in or near th visible part of the spectrum. Our eyes, photographic plates and photoelectric devices all being sensitive to visual radiation, can therefore readily detect galaxies (the principl exception being very low surface brightness galaxies). In contrast, detection of galaxie in other passbands (X-rays, far-ultraviolet, far-infrared, radio) has generally been mor difficult (HI 21 cm observations excepted), owing both to relatively fainter emission a these wavelengths, and the need to use space-based observatories with small aperture for many of these wavelength regions.

Determination of both the absolute amplitude and the spectral energy distribution o emergent flux in different wavelength regions is required, if we are to eventually construc a complete and realistic picture of galaxies. The most reliable means by which sample of galaxies can be assembled is through comprehensive sky surveys with well-define sensitivity limits, preferably of the whole sky. Maps of the sky in the far-infrared (IRAS COBE), in the radio continuum (e.g., Condon et al., 1989), in X-rays (ROSAT), an in yet-to-be-observed passbands each give us a new picture of galaxies which has to b merged with our old picture of galaxies.

Malmquist bias selection effects are present in each survey, such that the intrinsicall brightest objects in each passband (both stars and galaxies) will be the objects mos readily detected in all-sky surveys. In passbands other than the visual, the intrinsicall brightest galaxies are often the most active galaxies, generating flux in a given passband greatly disproportionate to the flux from the stellar populations that dominates th baryonic masses of galaxies.

The only set of galaxies for which all passbands will detect some flux are the neares 150 or so galaxies in the universe. Detailed study of these nearest galaxies can give of examples of what kinds of physical properties galaxies *can* have. Unfortunately, the exten to which the nearest galaxies are a representative set is *a priori* unknown, analogous to the problems in stellar astronomy created by studying only the nearest 150 stars.

Due to the different selection effects that operate in each passband, it is therefore necessary to perform what will be termed here 'double-blind' surveys of galaxies. By 'double-blind' it is meant that one first selects galaxies in passband A according to well-defined, objective criteria, and measures the observed flux from those galaxies in passbands A and B. One then reverses the process, selects galaxies from passband B by equally well-defined criteria, and measures the flux from these galaxies in both passbands. The combined data set should yield a sample of galaxies significantly less biased than either data set individually. Indeed, in the absence of samples that are selected in this double-blind manner, it is very hard *not* to have biased comparisons of galaxian properties in different passbands.

4 Inclination-dependence of magnitude and diameter for spiral galaxies

4.1 A statement of the basic problem

In the opinion of this writer, the question of how to correct the intrinsic properties of galaxies for inclination to the line of sight is one of the most confused subjects in extragalactic astronomy. The principal reasons for the existing confusion are the use of implicit assumptions and the biased selection of galaxies. Of the two major classes of galaxies — those without much visible dust (E and S0) and those with visible dust (spiral and irregular) — the literature on the latter is the most confusing to read.

The modern discussion of the inclination-dependent properties of spiral galaxies began with Holmberg (1958). In that classic study, Holmberg assumed only magnitude changed with inclination; the possibility that diameter changes with inclination was not discussed. In contrast, de Vaucouleurs (1960) concluded from the analysis of magnitude-selected samples of galaxies that diameter changed with inclination, with magnitude not changing, a conclusion incorporated into the prescriptions given in the first *Reference Catalog of Bright Galaxies* (de Vaucouleurs and de Vaucouleurs, 1964).

These two opposing views of the inclination-dependent properties of galaxies have dominated the discussion of this problem, with some modification, up until the present. If diameters do not change with inclination, spiral galaxies behave as if their disks are optically-thick. If diameters do change with inclination, spirals behave as if their disks are optically-thin. Magnitude can still vary with inclination in the optically-thin case, albeit not as much as in the optically-thick case. For a fine recent discussion of optical depth effects in galaxy magnitudes, the reader is referred to Bruzual et al. (1988).

In the discussion of this problem, a very basic empirical point should be made clear. The samples of galaxies studied by many workers (e.g., Holmberg, 1958; Heidemann, Heidemann, and de Vaucouleurs, 1972, [H^2V]; RC2; Valentijn, 1990) show that the apparent B magnitude surface brightnesses of spiral galaxies, $L_B/(\pi ab)$, are independent of inclination. Edge-on spirals have the same mean surface brightness as face-on spirals, and a large intrinsic scatter in B mag surface brightness exists at each inclination.

As a direct result, a magnitude-selected sample of galaxies will select galaxies according to a minimum surface area criterion: Only galaxies which present a minimum surface area to us will have a bright-enough magnitude to make it into our sample. As the apparent size of galaxies in these samples becomes smaller, the axial ratios of the galaxies become, on average, rounder. At the limiting magnitude, galaxies are viewed primarily face-on. Alternately, a diameter-selected sample selects spirals independently of apparent magnitude, yielding a sample in which the faintest galaxies in the sample are dominated by edge-on galaxies.

The question, "Which manner of selecting galaxies is correct?" is analogous to asking the question "Which parameter, if either, is the inclination-invariant?" To answer either question, distance information is needed, as Malmquist bias effects are plausibly present in both magnitude-limited and diameter-limited samples. Are the small, round spirals in the magnitude-selected sample at the same distance as the edge-on galaxies, or are they more-distant interlopes? Are the faint, edge-on galaxies in the diameter-limited sample at the same distance as the face-on galaxies, or are *they* the more-distant interlopers? It was to answer these questions that the Malmquist bias test was developed by this write and Marcia Rieke (Burstein and Lebofsky, 1986).

Before using the Malmquist bias test on appropriate samples of spirals to answer th above questions, it is illustrative to discuss how biases in galaxy samples have affecte two of the more significant attempts to determine the inclination properties of galaxies

4.2 An example of connected biases in a galaxy sample

Both the RC2 and H^2V employ a test of the inclination dependence of diameter fo spirals which reasoned that the HI 21 cm flux from galaxies was not optically thick a any inclination. Under such an assumption (which is not necessarily valid; Sec. 4.4) if the isophotal diameters of spirals become larger as the galaxies become more edge on, the effective HI 'surface brightness,' formed from combining HI flux and B isophota diameter, would become fainter. (Note that this does not require that HI flux be confine to the optical disk.) This test was first applied to a sample of 180 galaxies with measure HI flux by H^2V, and then to a sample of 270 galaxies for the RC2. Both tests conclude that HI surface brightness became fainter with increasing inclination, implying tha the diameters of spirals increased with increasing inclination. The form of increase o isophotal diameter with inclination given in the RC2 is $\delta \log D = 0.235 \log(a/b)$.

A combination of Malmquist bias effect and the surface brightness selection effec discussed in section 4.1 affected both the H^2V and RC2 studies. In a sample that i HI flux-limited, the effect of Malmquist bias is that the faintest-appearing spirals in HI flux will be preferentially HI-rich galaxies. This same sample has a B mag surface brightness dependent selection effect that results in the faintest-appearing galaxies being preferentially seen face-on. Thus, if one combines both systematic biases, the resulting sample has the most distant galaxies being small, round, and HI-rich; i.e., with high HI surface brightnesses as defined by H^2V. Galaxies that are more edge-on would have a range of somewhat lower HI surface brightness.

In the samples of H^2V and the RC2, galaxies that have the smallest apparent optica diameters also are among the galaxies with the highest HI surface brightnesses. Galaxies with large apparent optical diameters show a range of HI surface brightnesses that is lower in the mean than that of the small, round galaxies in the sample. Thus, the HI surface brightness test used by H^2V and the RC2 cannot be used to determine change of isophotal diameter with inclination for spiral galaxies.

4.3 A second example: size of galaxy sample does not matter

Valentijn (1990) examines the dependence of diameter with inclination for a *diameter-selected* sample of over 9000 spiral galaxies for which luminosity distributions were determined from the ESO Southern Sky Survey plates (Lauberts and Valentijn, 1989). Three surface brightness parameters are used in his analysis: the observed central surface brightness (μ_0), surface brightness at the effective radius (μ_e), and extrapolated central surface brightness from an exponential fit to the luminosity profile (μ_0^{Bfit}). For each kind of surface brightness, Valentijn examines whether it is a function of axial ratio for his sample. Since he finds no dependence of these surface brightness parameters on axial ratio, he reasons that galaxies must be, in his terms, 'opaque' to B magnitude light.

However, as Valentijn himself points out, these three surface brightness parameters are strongly related to each other for exponential luminosity distributions. What he did not point out is that both the mean effective surface brightness within an effective radius (SB_e) and the mean surface brightness within an isophotal radius (SB_{25}) are also strongly correlated with these three other surface brightness measures. These strong relationships exist because an exponential luminosity distribution has only two independent parameters — a scale length and an amplitude. All surface brightness measures of galaxies with exponential luminosity distributions will be strongly related to each other, as all such measures are derived from these two parameters.

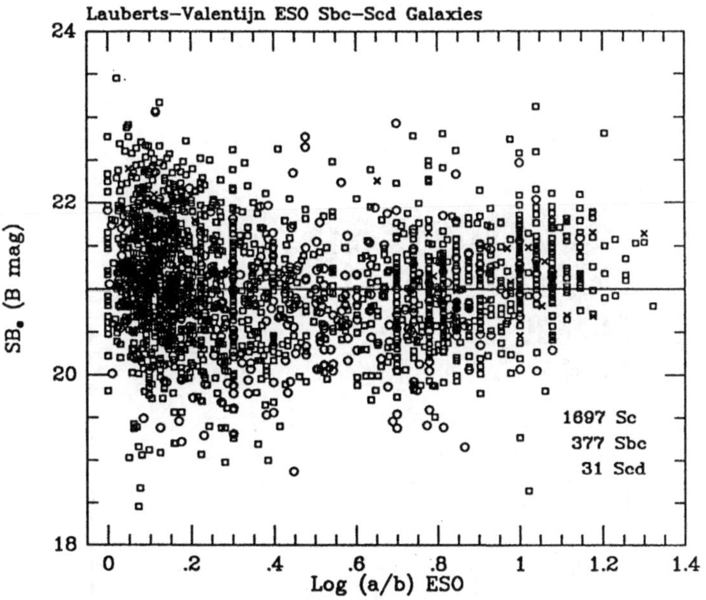

Figure 1. Mean effective B magnitude surface brightness, as given by Lauberts–Valentijn surface photometry, plotted vs. the log of the axial ratio, as tabulated in the original ESO catalog (Lauberts, 1982), for 2105 Sbc–Scd galaxies. The number of galaxies of each type is given in the figure. Data for Sc galaxies are plotted as open squares, for Sbc galaxies as open circles and for Scd galaxies as crosses. Mean effective surface brightness shows no net trend with axial ratio, although there is evidence of some difference between galaxies with $Log(a/b) = 0.4 \sim 0.8$ and the rest of the sample. The lack of a surface brightness trend from face–on to edge–on galaxies in this diagram is in agreement with the discussion of Valentijn (1990) for a larger set of galaxies, using other definitions of surface brightness. A line of constant surface brightness is drawn for visual reference.

That one can reproduce Valentijn's results using SB_e instead of μ_0 or μ_e is shown in figure 1 for a subset of Valentijn's sample of spirals that is disk-dominated — 2105 Sbc–Scd galaxies. Four points about figure 1 are worth noting. a) this figure uses $\log(a/b)$ rather than a/b as used by Valentijn. b) Valentijn's figures 1 and 2 cut off at $a/b = 6.5$ ($\log(a/b) = 0.8$) for his sample of 9000 galaxies, over a factor of two smaller than the maximum value of $\log(a/b) = 1.25$ that exist in the Sbc-Scd sample alone. c) the apparent 'quantized' nature of axial ratio in figure 1 is due to the fact that the axial ratios used by Valentijn are eyeball estimates given in the original ESO catalo (Lauberts, 1982), *not* photometrically-determined axial ratios. d) the relative paucity c galaxies with $0.4 < \log(a/b) < 0.8$ is puzzling, especially as these galaxies have surfac brightnesses somewhat brighter than face-on *and* edge-on galaxies.

Valentijn supports his conclusion that galaxy diameters are independent of inclina tion by plotting one of the surface brightness parameters versus apparent magnitude, B_T and showing that the magnitudes of galaxies behave as if they are optically-thick. A discussed in section 4.1, this conclusion is a direct consequence of the use of a diameter limited sample of galaxies. To illustrate this point explicitly, figures 2 and 3 plot th relationships of B_T and $\log D_{25}$ with axial ratio for subsets of the Sbc-Scd galaxies use in figure 1, objectively selected in two ways: figure 2, galaxies are selected by apparen magnitude; figure 3, galaxies are selected by apparent diameter.

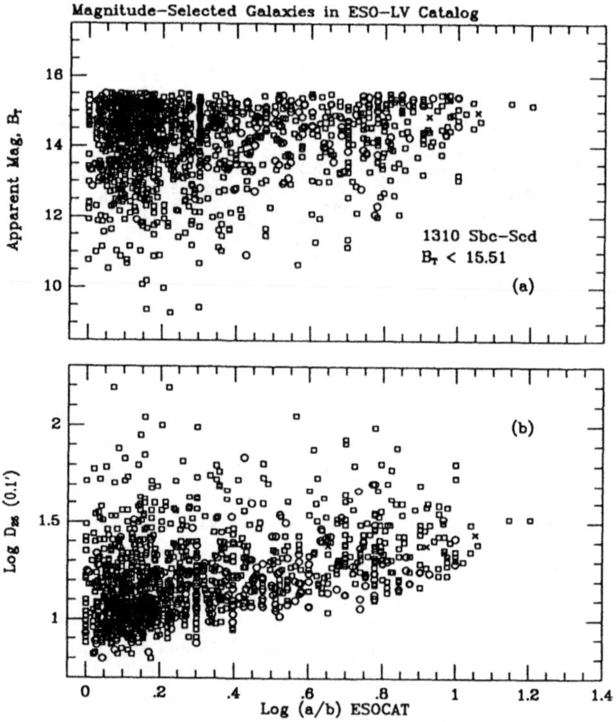

Figure 2. Inclination-dependence of apparent magnitude and diameter for a magnitude-limited sample selected from the Sbc-Scd galaxies in the Lauberts-Valentijn catalog. a) Apparent B mag, B_T plotted vs. axial ratio, here expressed as $Log(a/b)$. b) Apparent isophotal diameter $Log\ D_{25}$ (in units of 0.1′) plotted vs. $Log(a/b)$. Note how diameter appears to increase with axial ratio, while magnitude remains approximately constant for this sample. Symbols for Sbc, Sc and Scd galaxies are as in figure 1.

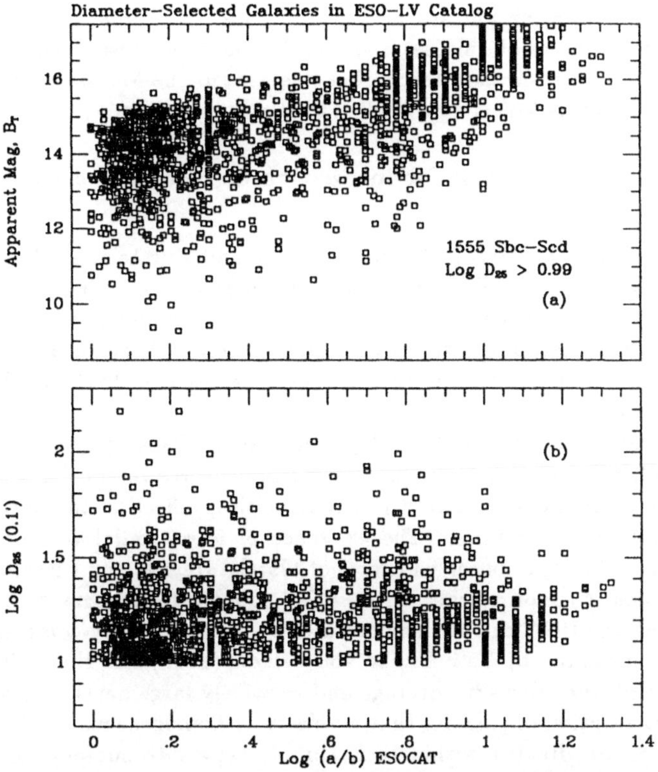

Figure 3. Inclination-dependence of apparent magnitude and diameter for a diameter-limited sample selected from the Sbc-Scd galaxies in the Lauberts-Valentijn catalog. a) B_T vs. axial ratio; b) *Log* D_{25} vs. axial ratio. In contradiction to the magnitude-selected sample, for this sample diameter appears to be independent of axial ratio, while apparent magnitude becomes fainter with increasing axial ratio. All galaxies are given the same plotting symbol in this diagram.

In the diameter-selected sample ($D_{25} \geq 1.0'$), the mean apparent magnitude of the spirals becomes fainter with increasing inclination, while mean apparent diameter does not change with inclination. This is in accordance with the result of Valentijn. In contradiction, in the *magnitude*-selected sample ($B_T \leq 15.5$) mean apparent magnitude changes relatively little with inclination, while mean apparent diameter becomes larger with increasing inclination. (The rather 'scalloped' distribution of points in figure 3b for log$(a/b) > 0.8$ is probably due to the use of visually-determined axial ratios.)

Simply choosing galaxies by apparent diameter or apparent magnitude will dictate the answer one will obtain about the inclination-dependence of magnitudes and diameters of spirals (Sec. 4.1). It is not that either Valentijn's conclusion, based on 9000 galaxies, or the RC2 conclusion, based on 300 galaxies, is wrong *per se*, it is just that their analyses cannot prove their conclusions. It is *how* galaxies are selected that is of paramount importance for this problem, not the size of the sample.

Most, if not all published analyses that have attempted to do tests of inclination dependence of magnitude and diameter are similarly affected by selection biases. Tw additional pieces of information are needed to break this circle of confusion — a objectively-selected sample of galaxies and the distances of these galaxies. It was fo this problem that this writer and Marcia Rieke developed the Malmquist bias test. It i for this problem that this test is now applied in the next section.

4.4 The Malmquist bias test for inclination-dependence of galaxy parameters

In collaboration with both Martha Haynes (Haynes and Burstein, 1991) and the RC consortium (de Vaucouleurs et al., 1991), Malmquist bias tests have been made for galax diameter, B magnitude and HI 21 cm flux. Of the three applications of this method only one, that of diameter, uses a sample of galaxies that is demonstrably selected onl by that parameter. It is this sample that will be here used to illustrate the use of th Malmquist bias test.

The galaxy catalog of Nilson (1973, Uppsala Galaxy Catalog; UGC) is the catalo of choice for application of this Malmquist bias test for three reasons. First, a majorit of galaxies in the UGC have measured radial velocities, reflecting the northern bias o telescope observations of galaxies. The existence of these radial velocities is due t many workers, including the concerted efforts of Martha Haynes and Riccardo Giovanell to observe the most edge-on spirals in the UGC (obviously important to the presen problem). Second, the UGC can be demonstrated to be statistically-complete to one ar minute diameter, as found by both Thuan and Seitzer (1979) and this writer using th V/V_{max} test. Third, the wide sky coverage and relatively large depth in distance of th UGC yields a distribution of galaxies that is relatively homogeneous on the sky.

UGC galaxies are divided within each Hubble type into subsets: four ranges o apparent UGC diameter ($D_{UGC} \geq 3.0'$, $3.0' < D_{UGC} \geq 2.0'$, $2.0' < D_{UGC} \geq 1.5$ and $1.5' < D_{UGC} \geq 1.0'$) and four ranges of UGC axial ratio (here expressed as b/a $0.0 \geq b/a < 0.25$, $0.25 \geq b/a < 0.50$, $0.50 \geq b/a < 0.75$, $0.75 \geq b/a \leq 1.00$). The resulting 4×4 matrix divides galaxies within a single Hubble type into bins defined by these ranges in diameter and axial ratio. The product $D_{UGC} \times V_R$, termed the 'diameter-distance product', is calculated for all galaxies with measured redshifts. (The diameter-distance product of a galaxy is clearly proportional to the absolute size of the galaxy, modulo a Hubble constant. Local group velocities are used here, but the result would be the same if Cosmic Background Radiation velocities were used.)

Median values for the following parameters are calculated for galaxies in each bin of (Hubble type, diameter, axial ratio): axial ratio, radial velocity and diameter-distance product. Owing to different numbers of galaxies in each of the 16 bins, the median values of diameter-distance product are normalized to unity for all axial ratio bins in each range of (Hubble type, diameter). This normalization procedure permits comparison of the relation between diameter-distance product and axial ratio for all galaxies in all diameter ranges. The normalization constants for each Hubble type contain information about the absolute sizes of the galaxies as a function of apparent diameter.

Figures 4a,b,c plot the median value of normalized diameter-distance product versus median value of axial ratio for the four axial ratio ranges and four diameter ranges within

Iubble types Sb, Sbc and Sc, respectively (Haynes and Burstein, 1991). Rather than calculate errors for each median value, the scatter in the median values for the four diameter ranges in each axial ratio bin is more representative of the real errors of this analysis. The number of galaxies included within each axial ratio range is given in each figure by Hubble type, together with the percentage this number represents of all UGC galaxies that could have qualified in that range. As of 1987, when the data for this figure were assembled by Martha Haynes and this writer, redshifts for 60–70% of the UGC Sb, Sbc and Sc galaxies were available, more for face-on galaxies than edge-on galaxies. Today, that figure is undoubtedly higher, and this test will be updated in Haynes and Burstein (1991).

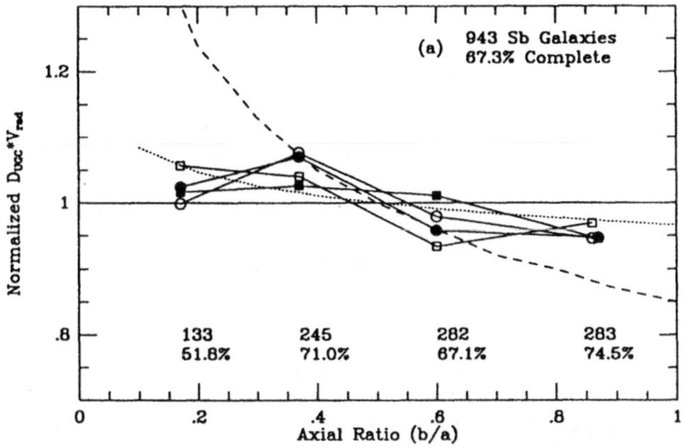

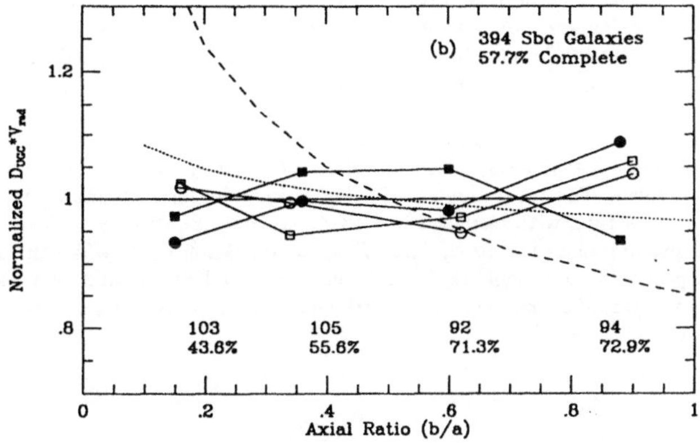

Figure 4. Malmquist bias test for inclination dependence of diameter. a) UGC Sb galaxies; b) Sbc galaxies, c) Sc galaxies (next page). A more detailed explanation of this figure is given on the next page.

If these spiral galaxies increase their isophotal diameters as they become more edge-

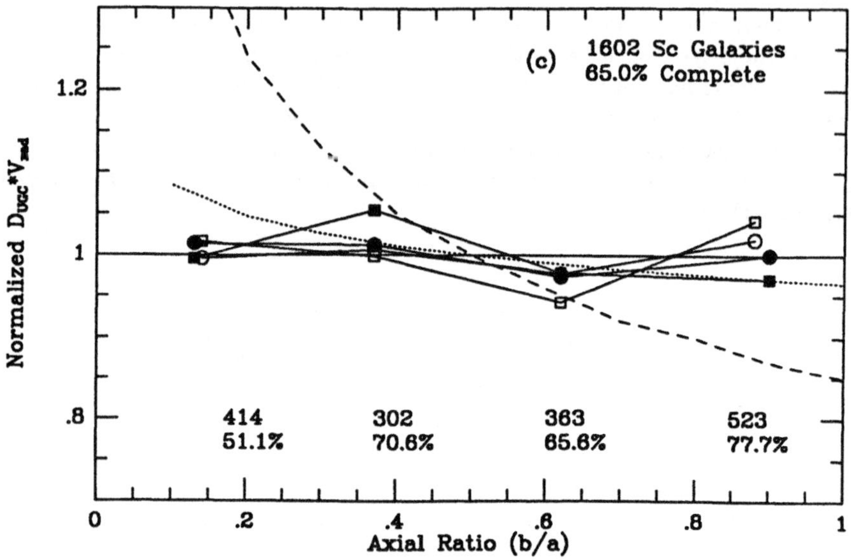

Figure 4. (continuation) The diameter–selected galaxies in the UGC are used for a Malmquist bias test of inclination–dependence of isophotal diameter for a) Sb, b) Sbc and c) Sc galaxies. UGC galaxies are divided into four ranges of apparent diameter: ($\geq 3.0'$: closed squares; $2.0' \sim 3.0'$: closed circles; $1.5' \sim 2.0'$: open squares; and $< 1.5'$: open circles), as well as four bins of axial ratio, b/a (note different definition from figures 1–3). Median values are calculated for both axial ratio and the product $D_{UGC} \times$ *radial velocity* (diameter–distance product). The diameter–distance product is normalized to unity in each diameter range for compatibility in plotting. Median values of $D_{UGC} \times V_{rad}$ are plotted vs. median values of b/a for each (diameter, axial ratio) bin, with points in the same diameter bin connected by a line. The number of galaxies used in each axial ratio bin (all diameters) and the completeness of each axial ratio bin, relative to UGC galaxies of this type and axial ratio range, are given. The total number of galaxies included in each galaxy type, and the completeness of each type with respect to all UGC galaxies are also given. The dashed line represents the prediction of the RC2, $\delta Log\ D_{25} = 0.235 Log(a/b)$, while the dotted line represents $\delta Log\ D_{25} = 0.05 Log(a/b)$ for reference. Sbc and Sc galaxies show no evidence of change of diameter with inclination. Sb galaxies show a marginal increase of diameter with inclination.

on, the Malmquist bias test would show the diameter-distance product getting larger with decreasing b/a. The Sc galaxy sample has the largest number of galaxies, both in absolute terms and in the 1602 galaxies with measured redshifts. For Sc's the diameter-distance product is independent of axial ratio for galaxies in all four diameter ranges (Fig. 4c). The consistent result for all diameter ranges is important because a) the sample is nearly 100% complete for the two largest diameter bins and b) each diameter bin is a separate Malmquist bias test.

Could Nilson have preferentially chosen edge-on galaxies with small absolute diameters? Highly unlikely, as all of the known biases would lead one to select edge-on galaxies with *large* absolute diameters in a diameter-limited sample. Both the so-called 'Holmberg' (1946) effect of visually measuring edge-on galaxies to fainter isophote levels than face-on galaxies, and possible discrimination against galaxies of low surface brightness, also work in the wrong direction. Could the incompleteness in the available radial velocities affect the result? Consistency of result from all diameter ranges strongly argues against this possibility, as well as the use of median values for all parameters.

A similar lack of inclination-dependence of diameter-distance product is seen for 394 Sbc galaxies (Fig. 4b), while 943 Sb galaxies (Fig. 4a) show a slight increase of diameter-distance product with axial ratio. In each figure, two models of change of diameter with inclination are drawn: the RC2 model, $\delta \log D = 0.235 \log(a/b)$ (note use of a/b here), and a model that fits the trend for Sb galaxies, $\delta \log D = 0.05 \log(a/b)$. Possible inhomogeneity in Nilson's type classification between edge-on galaxies and face-on galaxies would lead one to suspect that the derived change of diameter with inclination for any one Hubble type is uncertain by ~ 0.05 dex.

From these Malmquist bias tests, one concludes that the diameters of Sbc and Sc galaxies are independent of inclination, while the diameters of Sb galaxies may have a mild dependence on inclination. Thus, Valentijn is more correct than the RC2 in how diameter changes with inclination for spiral galaxies. The fact that Valentijn obtained the correct result was a direct consequence of his studying a diameter-limited sample. If he had studied a magnitude-limited sample, he would have arrived at a result closer to that of the RC2 (Sec. 4.3).

Valentijn is *not* correct in concluding that, if the diameters of spirals are independent of inclination, spirals must be 'opaque' to optical radiation. On the contrary, we know that we can see through parts of galaxies (e.g. M31, M33), and, more importantly, we can easily see through our own galaxy (Burstein and Heiles, 1982). For spiral galaxies not to change blue isophotal diameter with inclination, they must have disks that are 'optically-thick' on average.

That spirals are optically-thick does not necessarily imply that the mass associated with luminosity increases in direct proportion to the level of dust extinction. Dust also reddens (albeit not directly proportional to extinction in the optically-thick case), so that when one corrects apparent luminosity for extinction, one also must make the stellar population bluer. The decrease in predicted $(M/L)_*$ that results offsets the increase in L_*, yielding no net increase in predicted stellar mass.

The term 'optically-thick', as used here, is meant to describe patchy dust distribution in a spiral that manages to produce optical depths of unity when averaged over the whole disk. In this respect, the idea of 'optically-thick' should be viewed as analogous to that of exponential luminosity distributions in spirals, since the latter is only seen when average over the whole disks of spiral galaxies.

It is also important to note that, as evident in figure 1 of this review, spiral galaxie of the same Hubble type display a very wide range of effective surface brightness. Is thi range due to a range of intrinsic dust properties, such that some spirals could really b optically-thin? Is it due to an intrinsic range of star formation rates in different galaxies Or is it due to a more complex combination of differences in both dust content and sta formation rate? These questions still need to be answered.

The consortium producing the *Third Reference Catalog of Bright Galaxies* (de Vau couleurs et al., 1991; RC3) have performed Malmquist bias tests, in collaboration wit this writer, on blue isophotal diameter, blue apparent magnitude and HI flux for th galaxies in the RC3. Those tests: a) confirm the lack of inclination-dependence of di ameter for spiral galaxies; b) show that inclination-dependence for irregular, S0 and E galaxies are uncertain due to a combination of other factors; c) show a change in blu magnitude with inclination of the form $\sim 1.5 \log(a/b)$, twice the value used in the RC2 but consistent with the recent analysis of Bruzual et al. (1988); and d) yield evidenc that HI 21 cm flux becomes optically-thick as galaxies are seen at higher inclinations These results will be detailed in the introduction to the RC3 and subsequent papers.

5 Conclusions

Our understanding of the global properties of galaxies is as good as the sample of galaxie we select for study. Selection of galaxies is an inherently biased process, so it is necessar to construct methods to overcome those biases. Two such methods are discussed — usin Malmquist bias effects to our advantage (Burstein and Lebofsky, 1986), and double-blinc surveys to compare the properties of galaxies in two different wavelength regions.

To illustrate the utility of the Malmquist bias test, the thorny problem of how spira galaxies change diameter and magnitude with inclination is addressed. Two example (out of many that could have been chosen) of how previous studies of this problem are affected by selection biases are discussed — the analyses made in the RC2 and by Valentijn (1990).

The Malmquist bias test is applied here to the Sb, Sbc and Sc galaxies of the diameter-selected Uppsala General Catalog (Nilson, 1973). The result of this test is that the diameters of Sbc and Sc galaxies are found to be independent of inclination while diameters of Sb galaxies may have a slight dependence on inclination. The in- terpretation of this result is that spiral galaxies are, on average, optically-thick at B magnitude, but with an intrinsically wide range in either amount of dust, level of sta formation, or both.

If one's goal is to select spiral galaxies by the observational parameter that is inclination-invariant, one should choose galaxies based on their *apparent diameter*. Choos ing galaxies by apparent magnitude means that one will exclude edge-on galaxies which if observed face-on, would be included in the sample, starting at ~ 1.5 mag brighter than the magnitude limit of the survey.

Finally, this writer notes that one can now, or will soon be able to, select galaxies from digital all-sky surveys in the infrared, X-rays and radio continuum. Is it not time hat a digital all-sky survey be done that will produce optical data of comparable quality?

\cknowledgements

t is a pleasure to thank Marcia Rieke, Martha Haynes, and Gerard de Vaucouleurs for nany stimulating discussions on the topic of inclination corrections for galaxies.

References

othun, G.D., Schombert, J.M., Impey, C.D. and Schneider, S.E.: 1990, *Astrophys. J.* **360**, 427.

ruzual, G.A., Gladis Madris, C. and Calvet, N.: 1988, *Astrophys. J.* **333**, 673.

urstein, D.: 1990, *Rep. Prog. Phys.* **53**, 421.

urstein, D. and Heiles, C.: 1982, *Astron. J.* **87**, 1165.

urstein, D. and Lebofsky, M.J.: 1986, *Astroph. J.* **300**, 683.

ondon, J.J., Broderick, J.J. and Sielstad, G.A.: 1989, *Astron. J.* **97**, 1064.

)isney, M.: 1976, *Nature* **243**, 573.

e Vaucouleurs, G.: 1960, *Astrophys. J.* **131**, 265.

e Vaucouleurs, G. et al.: 1991, *Third Reference Catalog of Bright Galaxies*, in preparation. (RC3).

e Vaucouleurs, G. and de Vaucouleurs, A.: 1964, *Reference Catalog of Bright Galaxies*, Univ. of Texas Press, Austin, Texas.

e Vaucouleurs, G., de Vaucouleurs, A. and Corwin, H.G.: 1976, *Second Reference Catalog of Bright Galaxies*, Univ. of Texas Press, Austin, Texas. (RC2).

laynes, M.P. and Burstein, D.: 1991, in preparation.

leidemann, J., Heidemann, N. and de Vaucouleurs, G.: 1972, *Mem. R.A.S.* 75, Part 4–6, 85. (H^2V)

lolmberg, E.: 1958, *Medd. Lund. Obs.* **II**, no. 136.

lolmberg, E.: 1946, *Medd. Lund. Obs.* **II**, no. 117.

auberts, A.: 1982, *The ESO/Uppsala Survey of the ESO(B) Atlas*, European Southern Observatory. (ESO).

auberts, A. and Valentijn, E.A.: 1989, *The Surface Photometry Catalogue of the ESO-Uppsala Galaxies*, European Southern Observatory.

ynden-Bell, D., Faber, S.M., Burstein, D., Davies, R.L., Dressler, A., Terlevich, R. J., and Wegner, G.A.: 1988, *Astrophys. J.* **326**, 19.

Aalmquist, K.G.: 1920, *Medd. Lund.* Ser. II, Band 3, no. 22.

Aalmquist, K.G.: 1922, *Medd. Lund.* Band 5, no. 100.

Jilson, P.N.: 1973, *Uppsala General Catalogue of Galaxies*, Uppsala Obs. Ann. **6**. (UGC).

'huan, T.X. and Seitzer, P.O.: 1979, *Astrophys. J.* **231**, 680.

'alentijn, E.A.: 1990, *Nature* **346**, 153.

David Burstein at the corner waiting for the "bell".

GALAXY MANIFOLDS AND GALAXY FORMATION

S. Djorgovski
Palomar Observatory
Caltech, Pasadena, CA, USA

ABSTRACT. We are witnessing the emergence of a physical classification of galaxies. New methods of inquiry, including multivariate statistical analysis of galaxy manifolds, scaling laws and correlations provide important clues towards understanding the formative mechanisms and evolutionary histories of galaxies. Galaxies of different families form two-dimensional sequences in the parameter spaces of their many (but not all) global observable properties; this relative simplicity probably reflects the regularities and feedbacks in the processes of galaxy formation. A minimal three-dimensional parameter space, whose axes include measures of galaxy size, density, and kinetic temperature promises to become the equivalent of an H-R diagram for galaxies.

1 The taxonomy of galactic species

Every young empirical science begins its task by describing systematically the objects of its study; at first, by their outward appearance, i.e., the general morphology. The abundance of empirical facts soon gives birth to morphological classification schemes and systems. These are initially seldom meant to be anything more than organizing frameworks, to serve as playgrounds for developments of interpretations, and as such, they channel the growth of their subjects. As they age, such early schemes are often burdened by various extensions, subclassifications, and other impedimenta, in order to cope with the new data which do not fit simply into the original framework. If taken too seriously for too long, such purely phenomenological structures can hamper, rather than further, progress. Ultimately, if there is enough understanding of the underlying principles and mechanisms, new organizing frameworks arise.

In extragalactic astronomy, which certainly is a young science, comparisons with biology (and even some of the language) come naturally. In the words of Lewontin (1990): "*... Precisely because biology has no general theory of what organisms ought to look like, we have nothing but our accumulated historically contingent experience about what they do look like as a guide for ordering the world. The divisions that we regard as basic are simply the result of what has been left over or dug up from the extinctions of the past...*" Like paleontologists and evolutionary biologists trying to understand the origin and evolution of life, we try to understand and map the formation and evolution of galaxies; like them, we cannot do direct experiments on the extinct species (for us,

G. Longo et al. (eds.), Morphological and Physical Classification of Galaxies, 337–356.

young and primeval galaxies), but only study the fossil traces and carefully piece to gether the scarce bits of evidence. However, just as molecular biology and evolutionar genetics have surpassed the old taxonomical biology, while at the same time providin fundamental explanations for its many empirical facts, we now have an incipient physi cal classification and understanding of the diversity of galactic species and their origins which transcends and explains much of the morphological schemes pioneered by Hubble de Vaucouleurs, and others. We can now proceed with our real task: understandin galaxies and galaxy families as physical systems, and the formative and evolutionar processes which produced them. In an approach complementary to a direct search fo forming and evolving galaxies at large redshifts, we can use the systematics of galax properties at $z \sim 0$ (galaxy manifolds, scaling laws, etc.) as fossils of galaxy formatio and evolution.

Our basic classification scheme is surely the Hubble sequence, and its many deriva tives and refinements. It is a tribute to the physical intuition of Hubble that, at least fo the spirals, the Hubble type does indeed correlate with many of the physical properties e.g., the bulge-to-disk ratio (B/D), color (i.e., the relative star formation rate), gas frac tion, the ratio of the kinetic energy in the rotational and random components, average luminosity density, etc. On the other hand, Hubble's classification of early-type galaxie on the basis of their apparent ellipticity was a less lucky guess: as it turns out, ellipticity is remarkable in that it does not seem to correlate significantly with *any* other properties of elliptical galaxies, and never mind the projection effects. The Hubble sequence can be plausibly interpreted as a sequence in the history of star formation rate (Sandage, 1986) This accounts partly, but not entirely, for the variation of the global mass-to-light ratio (M/L) along the sequence (Faber and Gallagher, 1979). Significantly, the Hubble type does *not* correlate well, or at all, with the total mass, or even the luminosity, or many other basic observables, e.g., the maximum rotational velocity (Rubin, 1987). There is a significant spread of physical properties at any given Hubble type, suggesting that the Hubble classification alone is an incomplete descriptor of galaxies.

Perhaps inevitably, many classifications in the past fixed their attentions on the morphological superficialities, typically as seen in the blue light: the prominence of spira arms (or their flocculent absence), bars, rings, etc. Whereas such ornamental details represent interesting phenomena worthy of study in their own right, they usually involve only a small fraction of the total, or even of the visible mass, and tell us little about the fundamental properties of galaxies. Recall also that the stars which produce most of the blue light live for only a couple of percent of the Hubble time, and that density waves and other disk instabilities may be transient in nature. We need a physical classificatior of galaxies based on their fundamental properties, such as the mass, density, energy angular momentum, etc., or the nearest observable equivalents thereof, rather than or the esthetics of star formation patterns and density waves. There are also biases and selection effects due to the historically preferred wavelengths and technical limitations of our observations, e.g., surface brightness (Bothun, 1990).

A basic methodological problem is that we can study only the visible parts of galaxies whereas about 90% of the total mass is dark, and thus very poorly understood. The dark matter, being the gravitationally dominant component, may control many or all of the galaxian properties; for example, Lake and Carlberg (1988) suggested that the velocity dispersion of the dark matter determines the eventual Hubble type of a forming galaxy. In many galaxy formation models which involve some form of biasing, including the CDM scenario (Silk, 1983; Blumenthal et al., 1984; etc.), the density contrast of the protogalactic peaks, i.e., the dark matter, determines the galaxy type and other properties. We may be able to use scaling laws for the dynamical and other observable parameters to deduce the properties of dark halos (Athanassoula et al., 1987; Kormendy, 1990).

Figure 1 attempts to summarize roughly a contemporary picture of galaxy taxonomy, or at least my refracted view thereof. First, we should realize that here and now we are dealing with galaxy survivors: there may be as-yet undiscovered types of galaxies nearby (remember the low surface brightness disks!), and there may have been galaxy types at large redshifts which did not evolve simply into any known type at $z \sim 0$. Also, the relation of the Lyα absorption clouds to the galaxies we know is still quite uncertain.

One of the major discoveries of extragalactic astronomy in the last few years is that there are two fundamentally different kinds of galaxies, loosely called the dwarfs and the giants (Wirth and Gallagher, 1984; Kormendy, 1985; Sandage et al., 1985). This fundamental dichotomy was completely missed in the old classifications: gas-poor dwarfs, including the dwarf spheroidals (dSph's) were shoehorned with ellipticals, despite obvious and dramatic differences (compare, for example, M32 and NGC 205 or NGC 185); while gas-rich dwarfs were (perhaps not too wrongly) binned with irregulars at the end of the Hubble sequence. Giants are the classical, ordinary galaxies, characterized by relatively high luminosity densities and relatively low (M/L); the dwarfs are the opposite. The two types of galaxies have very different luminosity functions (Sandage et al., 1985), and differ in density and other properties at luminosities where both types coexist (Kormendy, 1985). The dwarf family may represent galaxies swept by galactic winds during their initial starburst phase; the mass loss drives an adiabatic expansion, leading to the low-density remnants we see today (Dekel and Silk, 1986; Yoshii and Arimoto, 1987; etc.) Whereas galactic winds could drive out baryons by pressure, they would not affect any nonbaryonic, electrically neutral material, thus perhaps accounting for the large (M/L) values seen in at least some dSph's. The evolution of dwarfs' gas content is less well understood (for a brief discussion and references, see Kormendy and Djorgovski, 1989, or Davies and Phillipps, 1988). It is possible, but by no means clear, that the gas-rich, luminous dwarfs tie in with the tail end of the Hubble sequence. This family may even include some genuinely young, late-forming galaxies, such as I Zw 18 (Kunth and Sargent, 1986), or H I 1225+0146 (Giovanelli and Haynes, 1989). Progenitors of the present-day dwarfs (and possibly some extinct genera as well) may have been the original building blocks of giant galaxies in an early era of vigorous and dissipative merging.

Giants are thought to be descendants of protogalaxies massive enough to retain their processed material, and to accrete smaller fragments and gas. Dissipative collapse, be it smooth, or by gradual merging, leads naturally to high luminosity densities in them. The basic distinguishing characteristic is the relative prominence of an actively star-forming disk. There are several reasons for this. First, the youngest (massive) stars have the highest (L/M), and ionize the gas, and thus dominate many optical/IR variables. Second, the very presence of a star-forming disk is contingent on the existence of a substantial gaseous disk, and therefore all ISM-related variables will be closely coupled to the relative prominence of a young stellar disk.

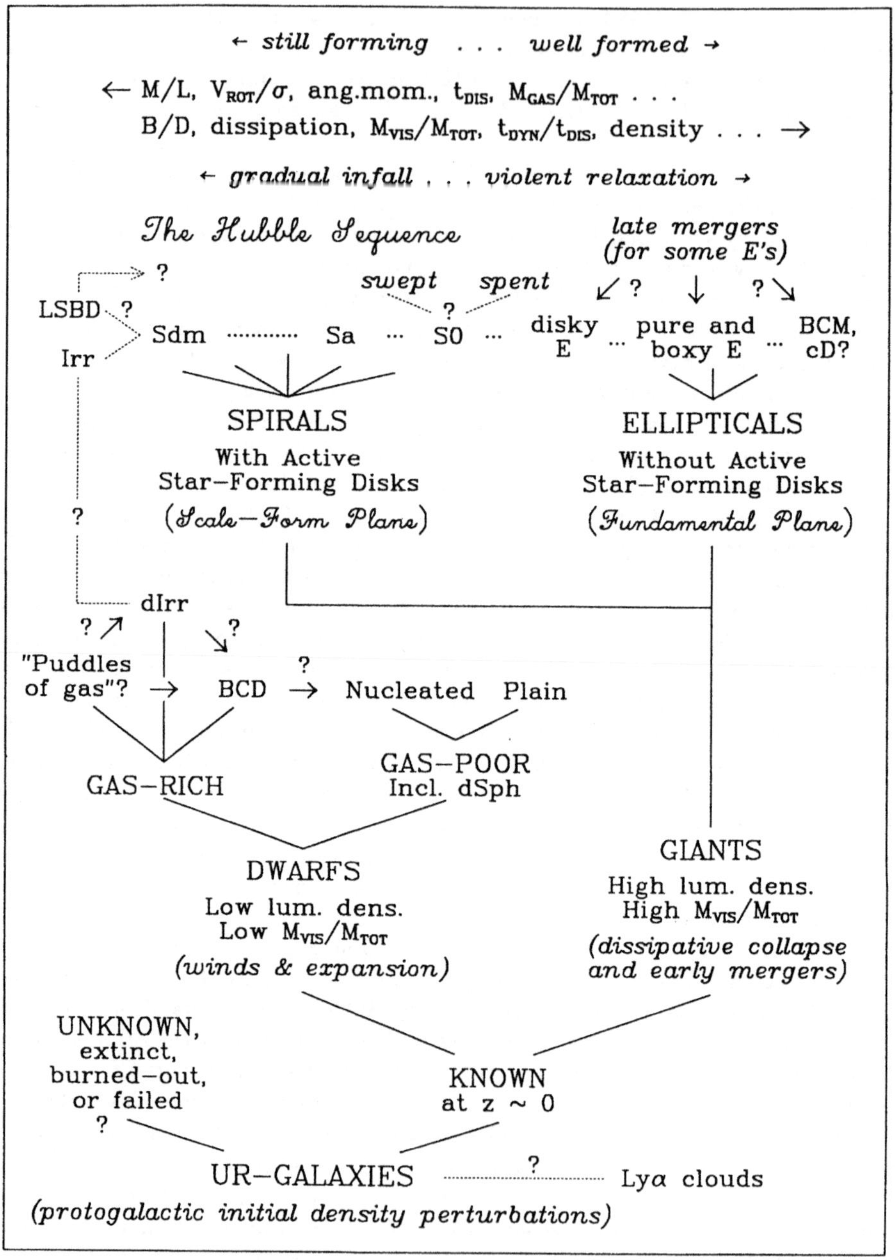

Figure 1. A modern view of galaxy taxonomy.

Finally, the existence of thin (i.e., dynamically cold, rotationally supported) disks implies formative history with no major mergers for the last $\sim 5-10$ Gyr (Ostriker, 1990). If a alaxy is granted a massive gaseous disk during its inception (i.e., a supply of gradually ccreting gas, which settles to a principal dynamical plane before forming stars), and ie chance to keep it (no disruptive mergers), then its subsequent properties, including s eventual Hubble type, are governed largely by its star formation history (Sandage, 986; Tinsley, 1982). Both factors depend on the environment, thus giving rise to the iorphology-density relation. The former, the propensity to form a disk, may also depend n the details of protogalactic cooling and statistics of peaks in the primordial density eld (Evrard, 1989; Lake, 1990), which may well depend on the environment, through iasing and possibly other mechanisms.

Towards a physical classification: the galaxy parameter space

heuristically motivated playground, such as depicted in figure 1, can thus contain or rovoke some understanding of galaxy formation. Short of wonderful surprises, it appears hat we have identified the basic galaxy families, and have at least plausible explanations f how they came to be and branch out. But there is more to this forest than family trees, norphological bins, and qualitative trends. We can *measure* many important quantities bout galaxies, their luminosities, radii, internal velocities, etc., and these quantitative, ontinuous variables can be used to form an empirical basis for a physical classification nd understanding of galaxies and their families. Whereas much of the recent research in his field concentrated on understanding the origin of basic galaxian species (e.g., dwarfs s. giants, spirals vs. ellipticals), in order to learn more about their origins we need to tudy the systematics of physical properties within individual galaxy families. For that ask, a new and different approach is needed.

We have at least the initial necessary methodology and tools for this purpose: the alaxy parameter space (hereafter the G-space, for short), galaxy manifolds, and multi-ariate statistical techniques. The subject of multivariate analysis of galaxy properties nd galaxy manifolds was pioneered in extragalactic astronomy by Brosche (1973), Bu-arrabal et al.(1981), Brosche and Lentes (1983, 1989), Lentes (1983), Efstathiou and 'all (1984), Whitmore (1984), Okamura et al.(1984), and others. From the galaxy lassification viewpoint, Okamura et al.(1990) provide a good review.

Briefly, consider the parameter space defined by several independent measurements or each object in a data set. If there are any genuine correlations present, the statisti-al dimensionality of the data hyperellipsoid (i.e., the number of independent quantities vhich can fully specify the data points) will be smaller than the number of input param-ters. The problem of the minimal manifold is to find this dimensionality, and identify he eigenvectors of the data ellipsoid. This solution at once identifies and optimizes *all* enuine correlations which may be present in the data. Generally, oblique projections of he data onto just the planes defined by pairs of observables involve a loss of information. The full information, and thus the optimal correlations, are recovered in a multivariate it. Derived parameters, such as the mass, mean density, etc., can also be considered in his framework, and all correlations involving them immediately follow.

Now consider a parameter space whose axes are a measure of galaxy size (e.g., mass, luminosity, or radius), density (or surface density), and kinetic temperature of stars (kinetic energy per unit mass, e.g., rotational V_{max} or velocity dispersion, depending on the dominant support component), as schematically shown in figure 2. Individual coordinate planes of this G-space are some of the familiar diagrams in extragalactic astronomy and cosmology: the Tully-Fisher (T-F) or Faber-Jackson (F-J) relations (luminosity vs. velocity scale plane), the cooling diagram (projected density - kinetic temperature), or the radius - surface brightness (the Kormendy relation) or mass-density plane (which describes the primordial density fluctuation spectrum).

Remarkably, galaxies of all families (e.g., spirals, ellipticals) do not fill the G-space but sit on two-dimensional surfaces embedded in this three-dimensional space, which are in general tilted with respect to all of its axes. Projecting these actual distributions of data points onto any of the coordinate planes smears the intrinsic bivariate correlation: that is why G-space needs three dimensions. Just as stars represent a one-dimensional sequence of mass embedded in the space defined by the luminosity and temperature (the H-R diagram), galaxies form two-dimensional sequences embedded in the G-space. Two statistical dimensions are enough: most global parameters of galaxies can be derived within the accuracy of measurement errors from any two others.

The low dimensionality of galaxy manifolds suggests that a small number of physical processes were operating in forming galaxies, and/or that they were well-regulated. The resulting bivariate correlations (typically power-law scaling relations) may contain some useful information about them. The eigenvectors of the data hyperellipsoid form the natural coordinate system, given by the data, i.e., the properties of galaxies themselves. This system may be more physically meaningful than the historically accidental coordinates representing our customary observables. Deciphering the physical meaning of the new coordinate axes may lead to the unravelling of the processes which reduce the dimensionality of the manifolds, by imposing correlations between physical properties.

3 The manifold of elliptical galaxies

Studies of the manifold of elliptical galaxies include those by Brosche, Lentes (1983), Lentes (1983), Efstathiou and Fall (1984), Lauer (1985), Dressler et al.(1987), Djorgovski and Davis (1987). All of them independently found that two statistically significant dimensions are sufficient to account for most of the observed dispersion of properties of elliptical galaxies, but only the more recent works made some astronomical use of this fact, and had enough data of a sufficient quality to apply it. Briefly, the global parameters of elliptical galaxies, which describe their overall dynamical structure and chemical composition, such as the luminosity, radius, surface brightness, velocity dispersion, colors, metallicity, etc., are unified in a two-dimensional manifold, the so-called "fundamental plane" (hereafter FP). A number of important derived parameters, such as the mass, luminosity or mass density, (M/L), etc., are also contained in the FP. The old Faber-Jackson $(L - \sigma)$, luminosity-color, mass-metallicity, and Kormendy's $R - \mu$ relations are simply its oblique projections. The tilt of the FP with respect to the planes of observables produces the correlated intrinsic scatter seen in these projected relations. However, whereas many of the global properties are unified in the FP, the variables which describe the *shape* of the light distribution, be it radial or azimuthal, do not correlate with the FP, or even mutually; thus, the *entire* manifold of ellipticals is more accurately described as $(2+n)$-dimensional. Recent work on the subject has been reviewed by Djorgovski (1987), Kormendy and Djorgovski (1989); Djorgovski and de Carvalho (1990).

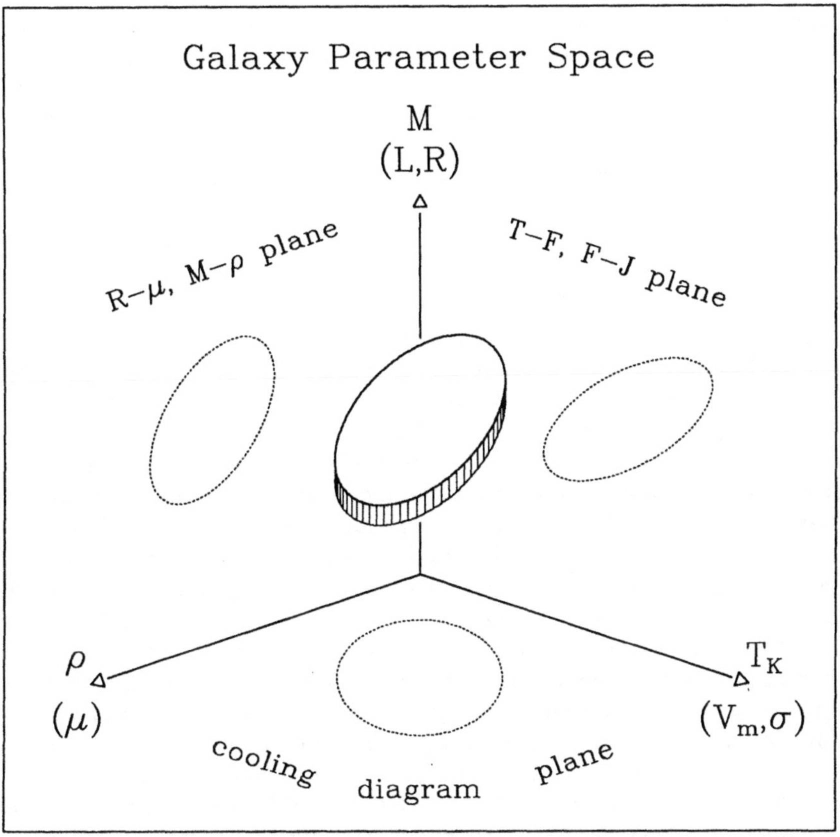

Figure 2. A schematic representation of the galaxy parameter space. Galaxies of a given family (ellipticals or spirals, and probably dwarfs as well) occupy two-dimensional regions (thickened in the third dimension mainly by the measurement errors) in a parameter space whose axes can be called size (mass, luminosity, or radius), density (or surface brightness), and temperature (i.e., kinetic energy per unit mass, typically the maximum rotational velocity for cold disks, or the central velocity dispersion for pressure-supported systems). The particular choice of axes depends on the application and available observables, but the basic picture remains unchanged. The coordinate planes thus defined are some of the well-known diagrams in extragalactic astronomy and cosmology; however, none of them contains all the information, only the oblique projections of the galaxy manifolds.

The FP implies a strong regularity among the global properties of elliptical galaxies, which may be a product of the physics of their formation and evolution. Its low dimensionality suggests self-regulation mechanisms in the process of formation of ellipticals: if a single process was dominant, the dimensionality of the manifold would have been 1;

if many different processes were equally important, the dimensionality might have been higher. The principal eigenvector of the manifold is identifiable with the mass, but we still do not understand what physical variable or process causes the spread of properties at a fixed mass. On the other hand, the lack of correlations among the "shape" variables (ellipticity, isophotal twist rate, radial profile shape, etc.) suggests that the details of dynamical structure are determined by some stochastic process (e.g., mergers), which may operate separately from whatever processes correlate the global properties.

The two-dimensionality of the manifold of global properties implies that any three of them are connected by a bivariate correlation, and that any residual scatter is due to the measurement errors alone. Observationally the best determined of these is the scaling law between radius, R (defined in some consistent way, but not isophotally, e.g., the r_e) velocity dispersion, σ, and surface brightness, I (this notation implying a measurement in linear flux units, rather than magnitudes): $R \sim \sigma^A I^B$, where the observed power law slopes are $A = 1.4 \pm 0.15$ and $B = -0.9 \pm 0.1$. The $D_n - \sigma$ relation of Dressler et al.(1987) is equivalent to this expression, with the modified isophotal diameter D_n containing both the radius and surface brightness terms, provided that all galaxies have the same surface brightness profiles (which is roughly, but not exactly true). We have found that, using the same data, the $R - \sigma - \mu$ formulation is slightly more robust and shows a slightly less scatter than the $D_n - \sigma$ formulation. The equivalent relation using luminosity instead of radius is operationally more difficult to evaluate, due to the correlated errors between the luminosity and surface brightness. Instead, assuming homology, i.e., that all galaxies have the same luminous and dynamical structure, and that the observed velocity dispersion is directly proportional to the kinetic energy per unit mass, $\sigma = CE_{kin}/M$, where $C = const.$, one can derive the following scaling relations luminosity $L \sim \sigma^{2A} I^{2B+1}$, mass $M \sim \sigma^2 R \sim \sigma^{A+2} I^B$, and $(M/L) \sim \sigma^{2-A} I^{-1-B}$. Note that for a pure virial theorem, assuming the same homology, $R \sim \sigma^2 I^{-1} (M/L)^{-1}$.

Given the observed values of the coefficients A and B, it then follows that $(M/L) \sim M^{1/6}$ (Faber et al., 1987; Djorgovski, 1988; Djorgovski and de Carvalho, 1990). An early suggestion of a slightly steeper relation was already made by Faber and Jackson (1976). If this is true, then the FP contains two additional clues about the formation of ellipticals: (1) the formation of elliptical galaxies and bulges must have been at least partly dissipative, otherwise (M/L) would not be a function of mass (cf. also Carlberg, 1986); (2) the more massive galaxies were less efficient in making stars per unit mass, or at least the progenitors of the present day red giants ($M_\star \sim 0.8 - 1 M_\odot$). This may be achieved either by modifying the IMF (e.g., the more massive galaxies would retain more of their metals produced in the early star formation, which could affect the protostellar cooling and thus the stellar mass fragmentation spectrum), or by a less efficient star formation overall (e.g., the protostellar cloud collisions in the potential wells of more massive dark halos may be too fast, disruptive rather than simply compressive), etc. In any case, this relation is an important empirical constraint for theories of galaxy formation.

The observed values of the coefficients A and B are about 3-σ away from the "pure virial" values, 2 and -1. But, could the FP be merely the virial theorem in disguise, i.e., is the (M/L) dependence on the mass real? What can go wrong?

First, the observed surface brightness may be affected by the metallicity. This could manifest itself as a dependence of the FP coefficients on the bandpass. The fact that the same solution is obtained in the B band (Faber et al., 1987) and the red bands (Lauer, 1985; Djorgovski and Davis, 1987; Djorgovski 1988), suggests that the metallicity effects cannot be very important. On the other hand, Recillas-Cruz et al.(1990) find a trend going to the infrared bands. Following the early suggestion by Faber, Jackson (1976) that (M/L) scales with mass, the question of metallicity effects was examined using stellar population synthesis models by Smith and Tinsley (1976), who found that the metallicity is unlikely to account for the entire effect. Finally, if higher metallicity causes a higher apparent (M/L), then it should also cause the surface brightness to be lower; that is contrary to the observed relation (de Carvalho and Djorgovski, 1989): at a fixed luminosity or radius, galaxies with a higher metallicity have ***higher*** surface brightness. Therefore, it appears unlikely that the relation $(M/L) \sim M^{1/6}$ is due to the metallicity effects alone.

Second, the assumption of homology is only approximate. The more luminous ellipticals tend to be more supported by velocity anisotropy (Davies et al., 1983), and thus the observed central velocity dispersion reflects a different fraction of the kinetic energy per unit mass, as a function of luminosity or radius. Indeed, the coefficients A and B do depend on the mean luminosity in the sample, and possibly on the environment as well (Djorgovski and de Carvalho, 1990). However, the more luminous objects (and BCMs in particular) tend to have *lower* values of A, i.e., the FP solutions are further away from the pure virial theorem. It is not obvious which luminosity range is more relevant: whereas the lower luminosity objects are less affected by the caprices of velocity anisotropy, they also have a portion of their kinetic energy siphoned off to rotation, which is not accounted for in the standard FP fits. There is also a weak and noisy trend towards shallower surface brightness profiles for the more luminous ellipticals (Schombert, 1986), so that the observed r_e relates to some physically relevant radius (e.g., the half-mass radius) in a way which depends slightly on the luminosity. It is thus not yet clear how the departures from homology will affect the $(M/L) \sim M^{\alpha}$ relation.

The FP contains stellar population variables as well as the global and dynamical ones. A relation between radius, surface brightness, and a metallicity indicator (color or Mg_2 index) has been obtained by de Carvalho and Djorgovski (1989, 1990). The implication is that there is a second parameter in the relation between mass and metallicity, which is identified as the luminosity density. This scaling relation can be understood in the context of dissipative galaxy formation. If the initial enrichment is regulated by galactic winds (energy input by supernovæ), the final metallicity will depend on the escape velocity, and thus the mass. The new term, luminosity density, implies an additional dependence at a fixed mass, similar to the Schmidt law for star formation in protoellipticals, SFR $\sim \rho_{gas}^{n}$.

We also find that various FP solutions for the field and for the cluster ellipticals separately differ, on a marginal level, possibly reflecting some systematic difference in their formative histories, or in the ages; this is especially evident in correlations involving colors or metallicities. Brightest cluster members also obey a different scaling laws from normal ellipticals. This probably reflects the relative importance and history of mergers and tidal interactions in different luminosity ranges and environments (Djorgovski and de Carvalho, 1990). We are probing the "fine structure" of the FP: subtle differences i solutions for differently selected samples.

Related to this is the question of distinction between the "boxy" and "disky" el lipticals. The isophotal shape parameter a_4/a correlates or shows trends with severa important quantities, such as the degree of velocity anisotropy, $(V_m/\sigma)^*$, x-ray or radi luminosities (e.g., Bender et al.1989; Bender, this volume, and references therein), bu it does *not* correlate at all with such basic quantities as the luminosity, radius, surfac brightness, velocity dispersion, metallicity, etc. This is not too surprizing, since the $a_4/$ parameter typically reflects isophotal deviations at a level of only a few percent of th total light. Djorgovski and de Carvalho (1990) find only a very mild difference in th FP solutions for the boxy and disky ellipticals, which can be understood in terms c increased velocity anisotropy for the boxy objects. There is no sign of any bimodality i the properties of elliptical galaxies which would correlate with a_4/a. To make the thing more complicated, individual ellipticals sometimes show both boxy and disky isophote at different radii. This suggests that a_4/a represents a measure of the details of dynam ical structure, rather than a fundamental global property of elliptical galaxies. Ther is thus no firm basis for the suggestion that a_4/a distinguishes two different kinds o ellipticals (e.g., Nieto and Bender, 1989); the boxy objects may have undergone mor mergers or tidal encounters, but there is no evidence for a dichotomy.

The existence of disks in elliptical galaxies is not difficult to understand: given modest delay ($\sim$ 1 Gyr?) in the initial period of star formation, and absence of majo mergers thereafter, any galaxy can grow and keep a disk. One naturally expects continuity in the (B/D) ratios, rather than a clear-cut separation of disky and non disky objects. The "pure" and boxy ellipticals may have suffered more (or more recent mergers, but do not seem to be fundamentally distinct from their disky relatives. Th (B/D) sequence of S0's can be extended into the realm of ellipticals, although, as th relative unimportance of ellipticity and a_4/a shows, small disks do not matter for mos properties of ellipticals. Operationally, the classical S0's are galaxies with distinct, bu inactive disks. There may be more than one way to skin a disk: we know of at leas two mechanisms by which they could have reached such a state, simply by running ou of gas through star formation, or by being swept (van den Bergh, 1990, and reference therein).

It may be possible to use the scaling relations for galaxies to probe their formation ir another way. Gott and Rees (1975) derived a scaling relation between mass and density for protogalactic clumps which separate from an evolving density field, whose powe spectrum is a standard power law, $|\delta_k|^2 \sim k^n$, in the form: $\rho \sim M^{-3/2-n/2}$. Scalin relations for other quantities can be derived by an application of the virial theorem, fo example, the proto-Tully-Fisher-Faber-Jackson relation, $M \sim \sigma^{12/(1-n)}$, where σ is characteristic velocity. Faber (1982) pointed out that the observed T-F or F-J slope $\sim$ 4 implies $n \simeq -2$, close to the value predicted by the CDM scenario on the mass scale o galaxies. On the other hand, if we identify surface brightness with surface density, the Gott-Rees hierarchy gives the scaling relation: $I \sim R^{-2(2+n)/(5+n)}$. The average observed scaling relation for ellipticals is: $I \sim R^{-1.2\pm0.1}$ (Kormendy and Djorgovski, 1989, and references therein); thus, $n \sim +2.5 \pm 0.8$, very different from Faber's result.

There are several problems with this argument. First, we do not observe mass, but ght. Second, we know that both F-J and Kormendy relations are just projections of e FP, and that we should use a bivariate scaling relation. This is a problem, since e Gott-Rees hierarchy describes a single-parameter family of objects, parameterized y the mass. Thus, let us introduce the conversion relation $(M/L) \sim M^{\alpha}$ in the Gott-ees hierarchy, where the observed slope value is $\alpha = 1/(6 \pm 2)$. Using the FP relation, $\sim \sigma^{A} I^{B}$, one can derive the condition equation: $2n + 10 = A(1-n) - B(12\alpha + 4n + 8)$. rom the observed values of A, B, and α, we find a solution which is degenerate in : the observed scaling relations for elliptical galaxies thus contain no memory of the itial density perturbation spectrum. This information may have been erased by the iolent relaxation and dissipative formation processes, which convert bound protogalactic umps, which the Gott-Rees theory describes, into galaxies which we observe today.

The observed properties of elliptical galaxies indicate a spread in density at any given ass, even though there is some anticorrelation between ρ and M. However, this spread relatively small. If we identify the turnover epoch, $(1+z_{turn})$, and the protogalaxy ollapse factor R_{init}/R_{final}, as the factors which determine the density of ellipticals oday, $\rho^{1/3} \sim (1+z_{turn}) R_{init}/R_{final}$, the data indicate a very narrow dynamical range of hat product, only $\sim 50\%$ after accounting for the measurement errors, even assuming no ariations in the collapse process. Explaining this puzzlingly narrow range of densities ay teach us something nontrivial about the formation of elliptical galaxies.

The FP may also provide some constraints on biased galaxy formation. Kaiser 1988) proposed a biasing model in which the luminosity of a galaxy scales with its mass nd turnover redshift as: $L \sim M^{\alpha}(1 + z_{turn})^{\beta}$, where coefficients α and β are related o the biasing parameter b; $\beta = 0$ is an unbiased model; $\beta = 2$, $\alpha = 4/3$ is the so-alled natural or autonomous bias, preferred by n-body simulators. If there is a T-F r F-J relation of the form $L \sim \sigma^{\gamma}$, the following condition applies: $\alpha - \beta(n+3)/6 = (1-n)/12$, where n is the standard primordial density spectrum index; for $\beta = \gamma/2$, ne gets a scatterless F-J relation. Now assume that ellipticals form a homologous amily, with $\sigma_{stars} \sim \sigma_{halo}$, $M_{baryonic}/M_{total} = const.$, and $R_{init}/R_{final} = const.$ (a easonable assumption, as shown above). Then the mean halo density at the turnover s a function of the redshift alone, which gives the connection with β. Using the FP elation $R \sim \sigma^{A} I^{B}$, after some algebra one obtains: $\alpha = (1 - A + 2B)/3B = 0.82 \pm 0.08$, nd $\beta = -(2 + A + 4B)/2B = -0.11 \pm 0.3$. Thus, the observed scaling relations for lliptical galaxies, under the assumptions, indicate a weak or no bias; the autonomous ias parameters are off by several sigma. A similar discussion and conclusions were resented independently by Peacock (1990).

The FP also contains optimal correlations between distance-dependent and distance-ndependent quantities, which can be used for mapping of the large-scale structure and elocity fields. The residual scatter about the FP is $\sim 15 - 20$ % per galaxy (given as he relative error of distance or radius), and it is mostly or entirely due to the mea-urement errors. The FP is thus a potentially powerful distance indicator for early-type alaxies. Using it in its $D_n - \sigma$ form, Dressler et al.(1987) and Lynden-Bell et al.(1988) laimed large-scale galaxy streaming motions toward the Hydra-Centaurus Supercluster the "Great Attractor" model). However, the reliability of the peculiar velocity mea-urements is still a subject of some disputes (Djorgovski and de Carvalho, Han, 1988 vs.

Burstein and Faber and Dressler, 1990). We contend that the distance indicator relations, being a product of formative processes of early-type galaxies and their dynamical evolution, may vary with the environment at a level of $\sim 10-20\%$, and could then masquerade as fake peculiar velocities. Recent results from other groups have raised some doubts about the accuracy and reality of at least some of the previously claimed peculiar velocities (e.g., Rowan-Robinson et al., 1990; Lucey et al.1991). The systematic variations we see are near the level of accuracy of the available photometric and spectroscopic data, and more and better data are needed in order to resolve this problem.

4 The manifold of spiral galaxies

The reason why disks exist at all is that energy can be dissipated, but angular momentum cannot. If for some reason star formation in a collapsing protogalaxy is delayed until the gas settles to the principal dynamical plane of the system (determined by its net angular momentum), a stellar disk will ultimately result. This requires a relatively quiet infall, since disks are dynamically cold. Such a situation may be provided by a gradual infall in the potential well of a gravitationally dominant dark halo. Numerical simulations have by now well established that objects formed through violent relaxation or rapid collapse resemble ellipticals, whether or not a massive halo is present, and whether or not dissipation is involved (Carlberg et al., 1984; Barnes, 1988). Disk galaxies must form slowly, require a supply of bound gas, and a quiescent neighborhood: thus the morphology-density relation. Vigorous star formation rates, and gas masses which cannot supply such extravagance for much longer (Sandage, 1986) suggest that the infall continues to the present day, and that formation of disks is an ongoing process. Wyse and Silk (1989) and White (1991) review some theoretical issues of disk formation. Structural properties of disk galaxies were reviewed recently by van der Kruit (1989) and star formation and chemical evolution issues by Kennicutt (1989, 1990) and Dopita (1990). Disks seem to form from the inside out, as suggested by their finite optical radii which are smaller than the H I radii, and by the radial metallicity gradients.

Despite their apparent complexity, the manifold of disk galaxies is not any more complicated than that of ellipticals. Whitmore (1984) found two dominant dimensions: the "form", which is a combination of the (B/D) ratio and color, and the "scale", which is a combination of the luminosity and radius. The former quantifies the Hubble sequence, the later reflects the luminosity spread at a given Hubble type (Rubin, 1987). While the Hubble type represents one of the two principal axes of the scale-form plane, it cannot be used to predict galaxy luminosity. Whitmore's results were confirmed and extended by Staveley-Smith and Davies (1988), and Magri (1990), who found that if FIR data are added to the optical and dynamical parameters, the dimensionality of the manifold increases to 3. The most extensive study of the manifold of spirals to date is that by Magri (1990). Among the other things, he finds that the spiral arm class (e.g., grand design vs. flocculent), which a superficial morphologist would likely be tempted to use in classifying spirals, does not correlate with any other physical quantities of interest.

The most important scaling relation for disk galaxies is the T-F relation, $L \sim V_m^\gamma$, where $\gamma \simeq 4$, and the residual intrinsic scatter (at least for the near-IR bands) is less than $\sim 0.2^m$ in L, or $\sim 20\%$ per galaxy (expressed as a relative error of distance).

Multivariate analysis indicates that a second parameter is present, since the V_m axis is located between the scale and form axes of the manifold; introducing the $(B-H)$ color, surface brightness, (B/D) ratio, or the FIR power as the second parameter improves the scatter of the T-F relation in the B band; however, the improvement is small in the IR, where the effects of obscuration and recent massive star formation are weaker. The slope and possibly the intercept of the relation seem to depend on the Hubble type (Djorgovski et al., 1988, and references therein), but within a narrow range of types, with good CCD photometry in the I band, the intrinsic scatter may be as low as $\sim 10\%$, or even less.

Using the virial theorem, and the tidal torque spinup theory of Fall, Efstathiou (1980), it can be shown that the following scaling relation should apply: $L \sim V_m^4 X$, where $X \sim (M/L)^{-1}\Sigma_h^{-1} \sim (M/L)^{-2} I_0^{-1} \lambda^{-2}$, where Σ_h is the halo surface mass density, I_0 the disk central surface brightness, and $\lambda = G^{-1}|J|E^{1/2}M^{-5/2}$ is Peebles's spinup parameter, which in turn is related to the collapse factor, by $\lambda = \sqrt{2}R_{disk}/R_{halo}$. The existence of a good T-F relation implies $X \simeq const$. Each of the quantities which compose the function X is known or suspected to vary by at least a factor of 2; yet, their products are constant to better than 20%, the scatter of the T-F relation. Thus, not only should the quantities composing X be connected by the appropriate scaling relations, but their scatter should also be correlated. Some coupling mechanism must be at work.

The small scatter of the T-F relation is quite remarkable, since it connects a halo-driven property (V_m) with a disk property (L, which is roughly proportional to the integrated SFR over the Hubble time). This points to a halo-regulated disk formation, as proposed by Silk and Wyse (1989; Wyse and Silk, 1989). In their model, the SFR is regulated both by the gas density (the Schmidt law), and the local velocity shear rate: SFR $= \epsilon \mu_{HI}^n \Omega$, where ϵ is the efficiency factor (a few % in the normal disks), μ_{HI} the gas surface density, $n \simeq 1-2$, and $\Omega = V(r)/r$ the local angular frequency; Ω reflects the rotation curve and is directly proportional to its amplitude, which in turn is largely determined by the dark halo. This model is in a good agreement with the data, and the heuristic model proposed by Kennicutt (1989). He found that the SFR depends on the local gas density, but only above a given threshold, which in turn depends on the local angular frequency, i.e., the velocity shear. These models predict an increasingly good correlation between L and V_m, but do not yet guarantee the right slope and intercept.

Silk and Wyse (1989) suggest that their SFR formula should apply in *all* star forming systems, from protogalaxies on, modulo a different efficiency factor (cf. also Brosche and Lentes, 1985). A similar bivariate dependence of SFR was suggested for protoelliptical galaxies on the basis of mass - metallicity - luminosity density relation (de Carvalho and Djorgovski, 1989, 1990; Djorgovski and de Carvalho, 1990). It is possible that star formation in galaxies is a universally bivariate process, which along with the virial theorem may be the ultimate causes of the two-dimensionality of galaxy manifolds.

Finally, the low surface brightness disks (LSBD) present an interesting problem (Impey and Bothun, 1989; Schombert et al., 1990). It is not clear whether LSBD are the extreme end of the Hubble sequence instead of irregulars, or a parallel sequence, distinguished by a different, super-regulated mode of star formation, biased against massive stars. Their relation to the scale-form plane is unknown at this time.

5 Concluding remarks

The family of dwarf galaxies was neglected in this review, partly for the lack of spac and partly because the situation there is still quite murky. At least at the low lumino ity end, the gas-poor dwarfs show quite different scaling solutions from those of norma bright ellipticals. Their manifold is almost one-dimensional: the principal eigenvecto identifiable with the mass, carries more than 90% of the total variance (Djorgovski an de Carvalho, 1990; and this volume). This again points to a fundamental difference i formation histories of dwarfs and normal galaxies: dwarfs were probably a product of single, highly self-regulated formation process, driven by the mass as the dominant pa rameter (i.e., galactic wind sweeping); giants may have experienced at least two differen formative processes, e.g., the initial dissipative collapse, and the post-formation mergers It is not clear at this point whether the dimensionality of the manifold of dwarfs increase to two at higher luminosities, nor what is their relation to the normal ellipticals in th regime where the two families overlap in luminosity (Bender and Nieto, 1990). As fo the gas-rich dwarfs, virtually no multivariate analysis work has been done so far.

One outstanding methodological problem is that we cannot compare very easily th manifolds of spirals and ellipticals, since we measure different things for them, or compar either with the theoretical quantities. The problem is operational, rather than funda mental, and a unified description of all galaxy manifolds in terms of physical quantitie (e.g., mass, density, etc.), rather than in terms of raw observables, should be possible.

The field of paleocosmology is gaining interest. The G-space and galaxy manifold provide a potentially very useful and powerful new framework for studies of systemati properties of galaxies and their formation. Different formative and evolutionary processe (dissipation, collapse, merging, galactic wind sweeping, etc.) can obviously move galaxie in the G-space, in the ways which are only partially and qualitatively understood at thi point, and remain yet to be fully explored theoretically. This multivariate approach ca also lead to a better understanding of the distance-indicator relations and the limits o their accuracy. At long last, we may have an analog of the H-R diagram for galaxies We are only beginning to use it, but at least the initial framework is in place.

It is a pleasure to express appreciation and thanks to the organizers of this workshop Drs. G. Longo, G. Busarello, and M. Capaccioli, whose efforts made it one of thos memorable conferences everybody enjoyed, and to thank them in their editorial rol for their great and abused patience. I also wish to thank Dr. A. Meiksin for usefu comments on the manuscript. Much of the work and ideas described here were develope in collaboration with Dr. R. de Carvalho. I acknowledge a travel grant from the AAS and partial support from the Alfred P. Sloan Foundation.

References

Athanassoula, E., Bosma, A., Papaioannou, S.: 1987, *Astron. Astrophys.* **179**, 23.

Barnes, J.: 1988, *Astrophys. J.* **331**, 699.

Bender, R., Nieto, J.-L.: 1990, *Astron. Astrophys.* **239**, 97.

Bender, R., Surma, P., Döbereiner, S., Möllenhoff, C., Madejsky, R.: 1989, *Astron. Astrophys.* **217**, 35.

Blumenthal, G.R., Faber, S.M., Primack, J.R., Rees, M.J.:1984, *Nature* **311**, 517.

Bothun, G.: 1990, in Kron, R.G. (ed.), *A.S.P. Conf. Ser.* **10**, 54.

Brosche, P.: 1973, *Astron. Astrophys.* **23**, 259.

Brosche, P., Lentes, F.T.: 1983, in IAU Symp. #100, ed. E. Athanassoula, p. 377. Dordrecht, Reidel.

Brosche, P., Lentes, F.T.: 1985, *Astron. Astrophys.* **153**, 157.

Brosche, P., Lentes, F.T.: 1989, in *The World of Galaxies*, eds. H. Corwin and L. Bottinelli, p. 50. NY: Springer Verlag.

Bujarrabal, V., Guibert, J., Balkowski, C.: 1981, *Astron. Astrophys.* **104**, 1.

Burstein, D., Faber, S., Dressler, A.: 1990, *Astrophys. J.* **354**, 18.

Carlberg, R.G., Lake, G., Norman, C.: 1984, *Astrophys. J.* **300**, L1.

Carlberg, R.G.: 1986, *Astrophys. J.* **310**, 593.

Davies, J., Phillipps, S.: 1988, *Mon. Not. R. Astr. Soc.* **233**, 553.

Davies, R.L., Efstathiou, G., Fall, S.M., Illingworth, G., Schechter, P.L.: 1983, *Astrophys. J.* **266**, 41.

de Carvalho, R., Djorgovski, S.: 1989, *Astrophys. J. Letters* **341**, L37.

de Carvalho, R., Djorgovski, S.: 1990, in Kron, R.G. (ed.), *A.S.P. Conf. Ser.* **10**, 106.

Dekel, A., Silk, J.: 1986, *Astrophys. J.* **303**, 39.

Djorgovski, S.: 1987, in IAU Symp. #127, ed. T. de Zeeuw, p. 79. Dordrecht, Reidel.

Djorgovski, S., Davis, M.: 1987, *Astrophys. J.* **313**, 59.

Djorgovski, S.: 1988, in *Starbursts and Galaxy Evolution*, Moriond Astrophysics Workshop, eds. Thuan, T.X. et al., p. 549. Gif sur Yvette: Eds. Frontières.

Djorgovski, S., de Carvalho, R., Han, M.-S.: 1988, in *The Extragalactic Distance Scale*, ed. S. van den Bergh, C.J. Pritchet, p. 329. *A.S.P. Conf. Ser.* **4**, 329.

Djorgovski, S., de Carvalho, R.: 1990, in Fabbiano, G. et al. (eds.), *Windows on Galaxies*, Erice Astrophysics Workshop, p. 9. Dordrecht, Kluwer.

Dopita, M.: 1990, in *The Interstellar Medium in Galaxies*, eds. H. Thorson and J.M. Shull, p. 437. Dordrecht: Kluwer.

Dressler, A., et al.: 1987, *Astrophys. J.* **313**, 42.

Efstathiou, G., Fall, S.M.: 1984, *Mon. Not. R. Astr. Soc.* **206**, 453.

Evrard, A.: 1989, *Astrophys. J.* **341**, 26.

Faber, S.: 1982, in *Astrophysical Cosmology, Pontif. Acad. Scripta Varia* #48, pp. 191 and 219.

Faber, S.M., Jackson, R.E.: 1976, *Astrophys. J.* **204**, 668.

Faber, S., Gallagher, J.: 1979, *Ann. Rev. Astron. Astrophys.* **17**, 135.

Faber, S.M. et al.: 1987, in *Nearly Normal Galaxies*, ed. S. M. Faber, p. 175. NY, Springer Verlag.

Fall, S.M., Efstathiou, G.: 1980, *Mon. Not. R. Astr. Soc.* **193**, 189.

Giovanelli, R., Haynes, M.: 1989, *Astrophys. J. Letters* **346**, L5.

Gott, J.R., Rees, M.: 1975 *Astron. Astrophys.* **45**, 365.

Impey, C., Bothun, G.: 1989, *Astrophys. J.* **341**, 89.

Kaiser, N.: 1988, in IAU Symp. #130, eds. J. Audouze et al., p. 43. Dordrecht, Kluwer.

Kennicutt, R.: 1989, *Astrophys.J.* **344** 685.

Kennicutt, R.: 1990, in *The Interstellar Medium in Galaxies*, eds. H. Thorson and J.M. Shull, p. 405. Dordrecht: Kluwer.

Kormendy, J.: 1985, *Astrophys. J.* **295**, 73.

Kormendy, J., Djorgovski, S.: 1989, *Ann. Rev. Astron. Astrophys.* **27**, 235.

Kormendy, J.: 1990, in Kron, R.G. (ed.), *A.S.P. Conf. Ser.* **10**, 33.

Kunth, D., Sargent, W.L.W.: 1986, *Astrophys. J.* **300**, 496.

Lake, G., Carlberg, R.: 1988 *Astron. J.* **96**, 1581.

Lake, G.: 1990, *Astrophys. J.* **364**, L1.

Lauer, T.R.: 1985, *Astrophys. J.* **292**, 104.

Lentes, F.T. :1983, in *Proc. Statistical Methods in Astronomy*, ed. E. Wolfe, p. 73. ESA SP-201.

Lewontin, R.C.: 1990, *The New York Review of Books* XXXVII, #10, p. 3.

Lucey, J., Gray, P., Carter, D., Terlevich, R.: 1991, *Mon. Not. R. Astr. Soc.* **248**, 804.

Lynden-Bell, D., et al.: 1988, *Astrophys. J.* **326**, 19.

Magri, C.: 1990, Ph.D. thesis, Cornell University.

Nieto, J. L., Bender, R.: 1989, *Astron. Astrophys.* **215**, 266.

Okamura, S., Kodaira, K., Watanabe, M.: 1984, *Astrophys. J.* **280**, 7.

Okamura, S., Watanabe, M., Kodaira, K.: 1989, in *The World of Galaxies*, eds. H. Corwin and L. Bottinelli, p. 75. NY, Springer Verlag.

Ostriker, J.: 1990, in Kron, R. G. (ed.), *A.S.P. Conf. Ser.* **10**, 25.

Peacock, J.: 1990, *Mon. Not. R. Astr. Soc.* **243**, 517.

Recillas-Cruz, E., Carrasco, L., Serrano, A., Cruz-Gonzales, I.: 1990, *Astron. Astrophys.* **229**, 64.

Rowan-Robinson, M., et al.: 1990, *Mon. Not. R. Astr. Soc.* **247**, 1.

Rubin, V.: 1987, in IAU Symp. #117, eds. J. Kormendy and G. Knapp, p. 51. Dordrecht, Reidel.

Sandage, A.: 1986 *Astron. Astrophys.* **161**, 89.

Sandage, A., Binggeli, B., Tammann, G. A.: 1985, *Astron. J.* **90**, 1759.

Schombert, J.M.: 1986, *Astrophys. J. Suppl.* **60**, 603.

Schombert, J., Bothun, G., Impey, C., Mundy, L.: 1990, *Astron. J.* **100**, 1523.

Silk, J.: 1983, *Nature* **301**, 574.

Silk, J., Wyse, R.: 1989, in *The Epoch of Galaxy Formation*, eds. C. Frenk *et al.*, p. 285. Kluwer.

Smith, H., Tinsley, B.: 1976, *P.A.S.P.* **88**, 370.

Staveley-Smith, L., Davies, R.: 1988, *Mon. Not. R. Astr. Soc.* **231**, 833.

Tinsley, B.: 1982 *Mon. Not. R. Astr. Soc.* **194**, 63.

van den Bergh, S.: 1990, in Kron, R. G. (ed.), *A.S.P. Conf. Ser.* **10**, 70.

van der Kruit, P.: 1989, in *The World of Galaxies*, eds. H. Corwin, L. Bottinelli, p. 256. NY, Springer.

White, S.: 1991, in IAU Symp. #146, eds. F. Combes and F. Casoli, p. 383. Dordrecht, Kluwer.

Wirth, A., Gallagher, J.S.: 1984, *Astrophys. J.* **282**, 85.

Whitmore, B.C.: 1984, *Astrophys. J.* **278**, 61.

Wyse, R., Silk, J.: 1989, *Astrophys. J.* **339**, 700.

Yoshii, Y., Arimoto, N.: 1987, *Astron. Astrophys.* **188**, 13.

Discussion

I. King - You began by making a strong distinction between giants and dwarfs; then you introduced the fundamental plane for giant ellipticals. What about the dwarf ellipticals? Do they have a different fundamental plane?

G. Djorgovski - Low surface brightness dwarfs are not "dwarfs ellipticals", but an entirely different family of galaxies; that old name is very much a misnomer. Indeed they follow different scaling laws from those of the ellipticals family. For single-parameter correlations, this was clearly demonstrated by Kormendy, and in particular the behaviour of (M/L) with L is strikingly different. Using multivariate analysis on the sample of gas-poor dwarfs in the Virgo and Fornax clusters (from Caldwell, Bothun and others), and the local group dSph's, De Carvalho and myself found that their manifold is nearly one-dimensional; however, there is a "second parameter". Gas-rich LSB dwarfs have at least one additional parameter, but I cannot say much more about them.

. Sulentic - I began agreeing with your statement that our current galaxy classification scheme may be obsolete. Our inability to quantify that scheme suggests that we need to try something different. Your analogy to the H-R diagram however is not quite true. We still use it and we understand which physical parameters the "measures" refer to. We further have a physical model for what it is telling us. We have no H-R for galaxies (it will no doubt involve more dimensions). It represents a direct interface between theory and observation. Do you seriously advocate discarding our present classification scheme in favor of one based upon models assuming the existence of dark matter? Can you propose an alternate scheme or set of quantities that we should work with, a set that is model independent?

G. Djorgovski - Certainly we should not discard the good old Hubble classification. What I am proposing is that once we divided galaxies into obvious families (e.g., LSB dwarfs, ellipticals, star-forming, spirals, etc.), we use the multivariate approach and define the manifolds and physical correlations and scaling laws for each family. Then, some day, we might be able to unify them all in the "theoretical" galaxy parameter space of mass, temperature and density. Such studies can and should be model independent. I only mentioned the Lake-Carlberg models to indicate how the properties of the dark halos my substantially affect the morphology of the resulting visible galaxy; however, this is by no means a complete model.

R. Bender - Do you think the fundamental plane for dwarf is 1-dimensional in all environments? As far as I remember your dwarf sample is taken from the Local Group and these seem to have a one-dimensional relation. However Binggeli, Sandage, Tammann as well as Bothun, Impey and others found very low surface brightness objects. So, to me it seems that the dwarf sequence may also populate a 2D plane.

G. Djorgovski - We considered three samples of LSB Dwarfs: those in the Virgo and Fornax clusters from the CCD photometry by Caldwell, Bothun and collaborators, and the LG dSph's. The large spread of surface brightness at a fixed luminosity, as seen in the Impey et al. and Binggeli et al. samples may be either indicative of the selection effects in our samples, or of the errors in their photographic photometry, or be due to the presence of gas rich dwarfs in their samples, or to a combination of the three. I should emphasize that our results are clearly very preliminary.

J.L. Nieto - Essentially, the fundamental plane is informative about the formation of galaxies because of the deviation that it presents with respect to the virial theorem. Each sample of objects, sorted out because they have a physical property in common, shows its own amount of deviation, but deviations from sample to sample are small. If this physical property has been recognized to define a sample and to specify it as a "family", doesn't it mean in a sense that this property tells you already very much about formation and evolution of its family in a more informative way than its deviation from the virial theorem.

G. Djorgovski - It is informative to consider the differences in the fundamental plane solutions with respect to an additional parameter, such as a_4/a, as one might learn about the effects of velocity anisotropy, etc. However, such variations are typically a ~ 10% (or 1-2 σ) level, at the limit of our data. I don't think that such slight effects can b used to divide ellipticals into distinct families, e.g. "Disky" vs. "Boxy". They are, i my opinion, just a small perturbation on the overall manifold. Recall, for example, tha the fundamental plane spans some 3 orders of magnitude in luminosity or radius, which is far more than the 10% effects.

A. Renzini - Have you got your fundamental plane constructed using some IR (e.g. K luminosity? and if so, is the trend in M/L_K shallower than that in M/L_B?
G. Djorgovski - There just isn't enough data to do this well quite yet. However, th preliminary results from Recillas-Cruz et al. (1990, A. A.) using K-band photometry, an by De Carvalho and myself using bolometric corrections from Buzzoni's models indicat that indeed the slope of the (M/L) vs. L relation is shallower at longer wavelength and/or for the bolometric luminosities. However, the relation does not disappear entirely I think.

R. Buta - Just a philosophical note. I think if Hubble were here today, he would probably be very surprised that we still use his classification system so religiously. Hubble knew what he was doing, that morphology could be a starting point and that the key to understanding galaxies lays in physical measurements. Whether morphological classification has impeded progress in extragalactic astronomy, as Van Maanen's work rotation of spirals impeded the final understanding of nebulae, or has served a useful purpose, is something each of us has to judge. But I am certain that most of us would agree that morphological classification has served a useful purpose. Now we are finally approaching a true physical understanding of galaxies, and Hubble no doubts would have been pleased to hear the wonderful work you presented today.
G. Djorgovski - Of course. It is a real tribute to Hubble that we still use his basic classification, and profitably at that. What I am proposing is that the basic morphological classification as we have it today is probably as good as it needs to be, and the time is now to do something else.

D. Burstein - George, I think you are probably ten years ahead of your time, in that neither the quality of the data, nor the manner in which they are selected, permit us yet to reliably derive global properties of galaxies, let alone as a function of environment density, etc. For example, from the magnitude limited subset of the 7 *Samurai* sample, we find no improvement of distance estimates for E's by incorporating stellar population information (Mg_2). I greatly appreciate the ESO key projects in E's that are ongoing as they hold the hope for the assemblage of good data on well selected samples.
G. Djorgovski - We find that the typical uncertainties of the power-law coefficients in the scaling laws are of the order of 10–20%. Certainly, if these relations can be used to map the large scale velocity field, for example, they can be used with exactly the same degree of accuracy and reliability to study the systematics of galaxy properties. As for the versions of the mass-metallicity-luminosity density relation which De Carvalho and myself proposed as new distance indicator relations, we said that a metallicity indicator (a suitable chosen color or the Mg index) can be used instead, not in addition to the velocity dispersion in the fundamental plane solutions. These relations do not improve on the $R - \delta - M$ or the $D_\eta - \sigma$ relations, rather, they are alternative to them.

P. Pişmiş - I shall consider only the spiral branch of the Hubble sequence. It is an empirical relation as we all know. But Hubble hit the right thing. Many of us, if not all, are intrigued by it and want to know why these different types of galaxies arranged in a neat sequence do exist. It is my conviction that the underlying cause must be sought for in the global primordial parameters with which a mass of galactic size gets started. In other words, the primordial parameters shape the destiny of a galaxy. With this in mind we looked into the total mass of spirals within the different Hubble types. The average total mass from Sa through Sc and irregulars shows a well defined decrease (Pişmiş, et al., 1978). However, the scatter around the mean mass is large, and this suggests that there is another global parameter independent of mass that differentiates the types (at least one more independent parameter). All the remaining physical properties should be derivable from the fundamental parameters if one knows the dynamical state of a galaxy at least in principle. **G. Djorgovski** - No answer needed.

S. Djorgovski listening to King's question.

Discussions during the coffee break.

SUB-COMPONENTS IN ELLIPTICAL GALAXIES AND THEIR FORMATION

R. Bender[1,2]
[1]*Landessternwarte Königstuhl, Germany*
[2]*Department of Physics, University of Oxford, England*

1 Introduction

Until recently it was widespread belief that elliptical galaxies, beyond their global scaling relations (Djorgovski, this conference, Faber et al., 1987), were essentially individuals reflecting in their detailed properties very individual formation histories. Weak trends between the various parameters describing the detailed properties of ellipticals had been observed but none of these appeared suited to arrange this class of galaxies in a meaningful sequence of physical properties, e.g. similar to the late-type Hubble-sequence. In analogy to spiral galaxies, mass also appeared to be the primary parameter not only for the global scaling relations, but also for the detailed properties of ellipticals; however, in the case of elliptical galaxies no parameter equivalent to the disk-to-bulge ratio of the late-type sequence was known (Tremaine, 1987).

Then, in the last few years, it became evident that isophote shapes of elliptical galaxies show surprisingly good correlations with a variety of other properties of these objects. Most interestingly, ellipticals with pointed isophotes have properties like S0 galaxies which led to the conclusion that pointed isophotes most likely are due to the presence of weak stellar disks, and furthermore, that the 'disky' E's may form a continuous sequence in disk-to-bulge ratio with S0 galaxies (for a recent summary, see Bender, 1990). These findings not only have the potential to provide a better *ansatz* for a physical classification of elliptical galaxies, but moreover, tell us that faint stellar disks, despite of being dynamically unimportant for the galaxy as a whole, represent important 'genetical' information about the structure and formation of the ellipticals in which they reside.

Another long-standing problem, or matter of debate, is whether elliptical galaxies can be formed by merging from evolved galaxies (Toomre, 1977, for a recent summary: Schweizer, 1990) and how many of these objects were actually formed in this way. Also in this field, a significant step ahead towards a better understanding was achieved recently. The detection of counter-rotating cores and other kinematical peculiarities (see below and Franx, Balcells, this conference) showed convincingly that at least some *bona fide*, luminous ellipticals have experienced violent merging processes (although these may have

G. Longo et al. (eds.), Morphological and Physical Classification of Galaxies, 357–370.

happened quite early). Detailed investigation of some of these cores has shown that th peculiar kinematics may be due to *cold* stellar sub-components superimposed on th otherwise hot cores of these objects (Franx and Illingworth, 1988; Bender, 1988b, 1990) So, kinematically peculiar cores indicate the presence of the smallest disks so far detecte in ellipticals (see also Nieto, this conference).

This review will mainly deal with isophote shapes, peculiar cores, and stellar disk in elliptical galaxies and their relation to other properties of these objects; it will updat and specify what I presented recently at another occasion (Bender, 1990). In addition t the mentioned sub-components I will also briefly discuss radio emission and X-ray halo as far as they are relevant in this context. However, I will not deal with cold and warn gas. This has two reasons. First, this topic has been covered in fair detail by severa other reviews presented at this conference (Bertola, van Gorkom, Macchetto, Renzini) Second, as is evident from these reviews, the mass, spatial distribution, and kinematic of warm and cold gas in ellipticals are in general not well correlated with the othe properties of these galaxies; hence, as has been argued convincingly, warm and cold ga most likely result from secular accretion events and do not trace the primary formatio history of an elliptical galaxy.

Although I tried to keep this paper self-contained, I was unable to cover all aspect in similar detail and, therefore, especially for a comprehensive account to previous work refer the reader to other recent papers (e.g. Nieto, 1988; Davies, 1989; Kormendy an Djorgovski, 1990; Bender, 1990).

2 Isophote shapes and faint stellar disks: general aspects

Massimo Capaccioli summarized in his review the evidence why disks may be frequentl hidden in the spheroids of elliptical galaxies (see also Capaccioli, 1987). At present, th most reliable and systematic way to search disks in ellipticals is by means of isophota shape analysis (Carter, 1978; Lauer, 1985; Jedrzejewski, 1987; Bender et al., 1988, 1989a Michard and Simien, 1988; Franx et al., 1989; Peletier et al., 1990). Since sufficientl inclined disks cause otherwise elliptical isophotes to appear pointed along their majo axes, the presence of a disk is easily revealed by a Fourier series expansion of the isophota deviations from ellipses. Indeed, surveys of large numbers of elliptical galaxies (Bender e al., 1989a) have shown that systematic deviations from elliptical isophotes are frequentl present and that the 4th cosine coefficient (named a_4/a hereafter) in most cases describe the strongest deviations (with negative a_4/a indicating boxy, and positive a_4/a pointec isophotes).

A complication in the search for disks via isophotal shape analysis is given by th fact that we cannot *a priori* be sure that the spheroids *themselves* always show elliptica isophotes in projection and that positive a_4/a may therefore not always indicate th presence of a disk (e.g. many Staeckel models show non-ellipsoidal isodensity surfaces de Zeeuw et al., 1986; in addition, triaxiality alone may cause non-elliptical isophotes) Consequently, in order to be confident that positive a_4/a, at least in general, trace disk similar to those in S0 galaxies, it is essential (a) to compare the statistical properties o ellipticals with pointed isophotes to those of other ellipticals and S0 galaxies, and (b) to study single characteristic cases of disky ellipticals in more detail (i.e. to derive disk arameters, spheroid kinematics etc.). Detailed studies of single disky elliptical are under 'ay and examples for photometric disentangling of disk and spheroid are described in :apaccioli et al. (1987), Simien and Michard (1990), and Scorza & Bender (1990, and nis conference). In this paper, I will mainly concentrate on statistical aspects.

The study of statistical correlations between isophote shapes and other galaxian roperties requires a reasonable way to select characteristic values for radially varying arameters in a distance-independent way. Here I follow the definition of Bender et al. 1989a)[1]. Furthermore, only galaxies with dominating a_4/a coefficients are considered, e. all ambiguous cases with strong sine and odd cosine deviations are excluded from he analysis. This point needs careful inspection of each object but is important since iclusion of dubious cases can spoil possibly existing correlations considerably. In this 'ay, some 20% to 30% of ellipticals are excluded from the analysis, but that procedure ppears acceptable until a more sophisticated investigation can be carried out (this could, .g., include, besides a_4/a, also the rms-values of some sine coefficients which could serve s a measure for the irregularity of the isophotes).

Although it is the purpose of the following sections to show that pointed isophotes 1 ellipticals are in fact due to faint stellar disks, I will, for the sake of simplicity, already uring the course of my argumentation address these objects as disky ellipticals.

3 Photometric parameters *vs.* isophote shapes

'he isophotal shapes of elliptical galaxies do not correlate strongly with any one of the tandard photometric parameters, like luminosity, effective radius, surface brightness, llipticity, or color. However, interesting trends are present:

There is a tendency that boxy ellipticals are on average more luminous than ellipticals vith pointed (disky) isophotes (figure 1, see also Bender et al., 1989a). If we ignore he four boxy low-luminosity objects in the lower left of figure 1, this trend becomes specially evident. There is probably good reason not to consider these four objects. 'hey are all companions to more massive galaxies and therefore may be exceptional in he sense that their shape is strongly influenced or even changed due to their extreme nvironment (see Nieto and Bender (1989), Peletier et al. (1990)).

Deviations from elliptical isophotes are on average stronger in more elongated objects, e. the largest a_4/a values occur in 'edge-on' systems. Furthermore, boxy ellipticals are n average rounder than disky ellipticals and there exist only few boxy E's which have llipticity higher than 0.5 (Bender et al., 1989a). Recent investigations have confirmed nd extended this result. Nieto (1990) investigated a large sample of elongated ellipticals rawn from the RC2 (de Vaucouleurs et al., 1976) and the RSA (Sandage and Tammann,

[1]Different definitions only marginally influence whether a galaxy is classified as boxy or disky since in most galaxies the isophote shapes change only slowly with radius. However, cores and outer parts of the galaxies should always be excluded from the radius range from which the characteristic values are derived because cores may show properties decoupled from the main bodies of the galaxies (see below) and the outer parts may be influenced by interactions or the tidal field of neighbours. A useful radius range to derive characteristic values is therefore the interval between a few tenths of the effective radius and 1-2 effective radii (or an equivalent range in surface brightness). Such a definition also avoids some ambiguities caused by changing isophote shapes observed especially in the outer parts of some galaxies and described by Nieto and Bender (1989)

1981). He found that about 80% of ellipticals of ellipticity class E5 or flatter have pointed isophotes and, also, that in his sample boxy E's are on average more luminous than pointed E's. Similar results were obtained by Capaccioli et al. (1990) on a sample of Virgo ellipticals. A comparison of the ellipticity distributions of boxy and disky E's by Fasano (1990) suggests that boxy E's most likely have mildly triaxial figures while pointed E's are probably oblate (see also Merritt, this conference).

• Djorgovski and de Carvalho (1990) investigated whether boxy and disky ellipticals show different relations between their 'Fundamental Plane' parameters (effective radius, surface brightness and velocity dispersion). However, only marginal evidence was found.

• Peletier et al. (1990), Longo et al. (1989), and Kormendy and Djorgovski (1990) checked whether isophotal shapes correlate with colors or color gradients. Some trends have been detected which are however small compared to the spread between individual objects. Also, the still small size of CCD's prevents an accurate measurement of color gradients over a large radius range (a way to cope with this problem is the use of focal reducers, see Surma et al., this conference).

The last two points in this list still suffer from a too small number of objects or insufficient data and, therefore, more significant conclusions require further observations.

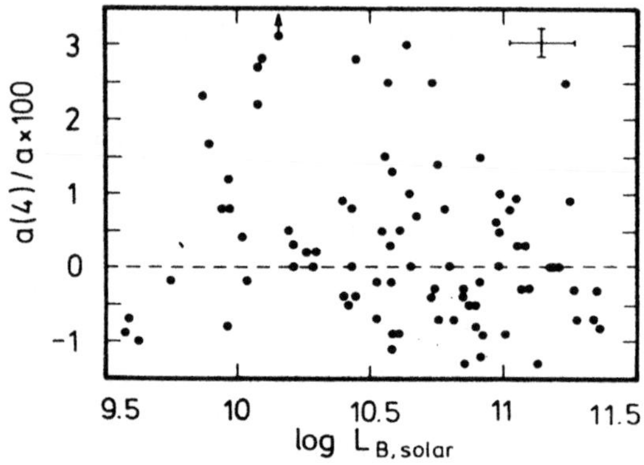

Figure 1. Isophotal shape parameter a_4/a against solar blue luminosity of elliptical galaxies (cf. Bender et al., 1989a). Positive a_4/a describe pointed (disky) isophotes, negative a_4/a indicate boxy isophotes. $H_o = 50$ km/s/Mpc is assumed here and also in the following figures.

4 Radio and X-ray emission *vs.* isophote shape

Figures 2 and 3 show the radio and X-ray luminosities of elliptical galaxies against their blue luminosities (diagrams adapted from Bender et al., 1989a). In these figures disky ellipticals (selected to have $a_4/a \geq 0.5\%$) are characterized by open circles while boxy ellipticals and ellipticals with no significant deviations from ellipticity are represented by filled circles. It is again evident that disky ellipticals are on average less luminous than boxy/perfect ellipticals. However, in the luminosity range where their luminosity

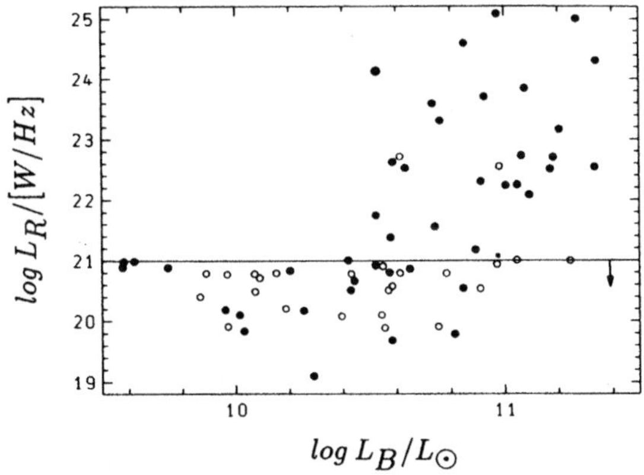

Figure 2. The radio-luminosity per frequency interval L_R at 1.4 GHz of elliptical galaxies against their blue luminosity L_B in solar units. Open symbols refer to disky ellipticals ($a_4/a \geq 0.5\%$), filled symbols to boxy E's and E's with no significant deviations from elliptical isophotes. Most points below the horizontal line are upper limits. References to the origin of the data can be found in Bender et al. (1989a).

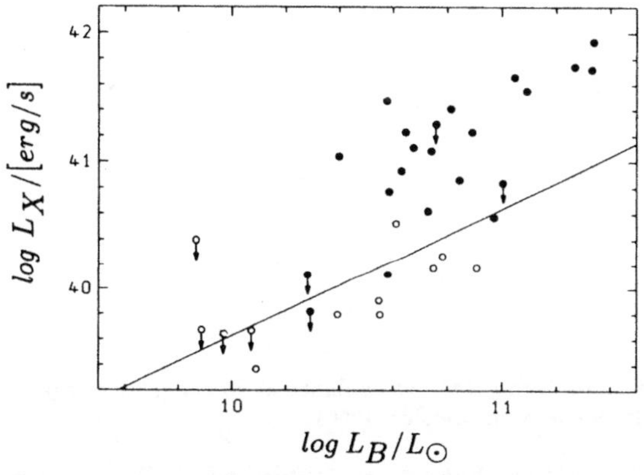

Figure 3. The X-ray luminosity L_X between 0.5 keV and 4.5 keV of elliptical galaxies against their blue luminosity L_B in solar units. Open symbols refer to disky ellipticals ($a_4/a \geq 0.5\%$), filled symbols to boxy E's and E's with no significant deviations from elliptical isophotes. The line indicates the presumed contribution of discrete sources (according to Canizares et al., 1987). References to the origin of the data can be found in Bender et al. (1989a).

functions overlap, a quite clear difference between the two types is obvious. Disky E's are in general radioquiet and also show no or only little X-ray emission in excess of their discrete sources (full line in figure 3, cf. Canizares et al., 1987). These properties put disky ellipticals in one line with S0 galaxies and spirals which are as well weak radi and X-ray emitters when compared to ellipticals on average (e.g. Hummel and Kotany 1982, Fabbiano and Trinchieri, 1985, Forman et al., 1985)[2]. The dependence of radi and X-ray properties not only on luminosity but also on structural parameters may hav interesting implications on our understanding of galaxy formation. Fortunately, ROSA will soon provide us more data to investigate this relationship in better detail.

The differences between pointed and boxy ellipticals in their radio and X-ray prop erties are also of importance in another respect. This distinction definitely rules out tha boxy and pointed ellipticals are intrinsically the same objects just seen from differen viewing angles.

5 Kinematics *vs.* isophote shape

Figure 4 shows anisotropy $(v/\sigma)^*$ against blue luminosity L_B for luminous elliptica galaxies ($\log L_B/L_\odot > 10.2$ or $M_T < -20.1$ mag, $H_\circ = 50$ km/s/Mpc); only objects wit sufficiently well measured kinematics and known isophotal shape have been included (fo references, see Bender, 1990).

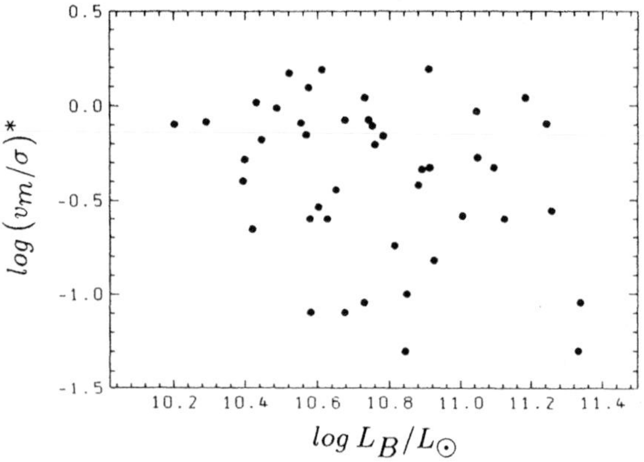

Figure 4. The anisotropy $(v/\sigma)^*$ of luminous elliptical galaxies against their blue luminosity L_B in solar units (from Bender, 1990).

This diagram is equivalent to the one first published by Davies et al. (1983) and demonstrates that luminous ellipticals show a large variety of kinematical properties but ar mostly supported by anisotropic velocity dispersion, i.e. they have $(v/\sigma)^* < 1$ (Binney 1982). No clear correlation between $(v/\sigma)^*$ and L_B is observed for luminous E's. Only

[2]Donnelly et al. (1990) speculated whether the large scatter in the L_X *vs.* L_B diagram as well as th different X-ray properties of disky and boxy ellipticals could be attributed to bad distances. The concluded that using distances derived from the $D_n - \sigma$ - relation (Dressler et al., 1987) reduces th scatter significantly and, consequently, also the relevance of second parameters (like isophotal shapes) In my view, such an explanation is probably premature and more data are certainly necessary befor final conclusions can be drawn (there is also another problem outlined by Djorgovski and de Carvalho 1990: disky and boxy ellipticals may follow slightly different $D_n - \sigma$ - relations).

when less luminous ellipticals are included in this diagram, a trend between luminosity and anisotropy becomes evident showing that less luminous ellipticals tend to be more supported by rotation than luminous E's (Davies et al., 1983, see also Bender and Nieto, 1990).

Figure 5 shows anisotropy $(v/\sigma)^*$ against isophote shape a_4/a for the same sample of objects as in figure 4 (Bender, 1988a). Although a large scatter is present, boxy ellipticals in general appear to be anisotropic systems. On the other hand, nearly all disky ellipticals are more or less flattened by rotation, i.e. they have $(v/\sigma)^* > 0.7$ (dashed horizontal line in figure 5; 0.7 is a somewhat arbitrary but reasonable value to separate rotationally flattened from anisotropic systems). The rapid rotation of disky ellipticals has been taken as evidence for their similarity to S0 galaxies (Carter, 1987; Bender, 1988a; Nieto et al., 1988). Indeed, as van der Marel et al. (1990) recently demonstrated, the $(v/\sigma)^*$ *vs.* a_4/a relation cannot be accounted for by a simple deformation of the isodensity contours towards pointedness or boxiness. Consequently, the rapid rotation of disky ellipticals is not just accidental and the different kinematics of disky and boxy E's really indicate significant differences in their internal structures.

The generally slow rotation of boxy ellipticals shows that their boxiness must be of a different origin than the boxiness of the bulges in late-type galaxies (e.g. Shaw, 1987), which show rapid rotation (Kormendy and Illingworth, 1982). It has been discussed by several authors that boxiness in slowly rotating ellipticals may result from interaction or merging (for references, see Nieto, 1988; Bender, 1990).

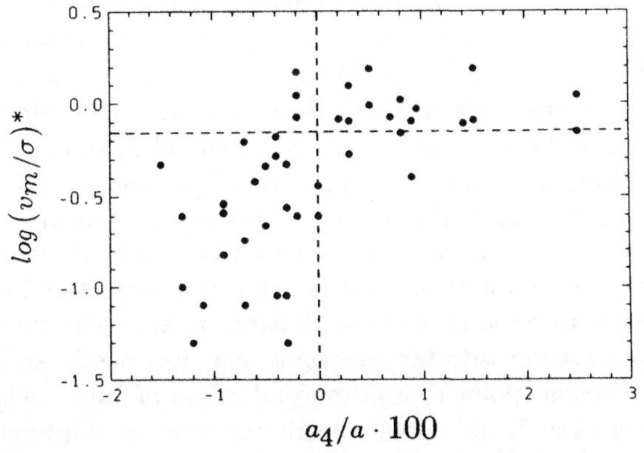

Figure 5. The anisotropy $(v/\sigma)^*$ of luminous elliptical galaxies against their isophotal shape a_4/a. The objects are the same as in Figure 4. Above the dashed horizontal line (corresponding to $(v/\sigma)^* > 0.7$) the galaxies are mainly supported by rotation.

Accepting the presence of faint stellar disks in disky ellipticals poses however another interesting question concerning their internal kinematics: to what extent are the measured rotation velocities of these objects influenced by the presence of weak, but rapidly rotating disks? There are two straightforward ways to analyze this question: (a) if the disk is sufficiently inclined, velocity curves measured parallel to the major axis of the object easily reveal the bulge kinematics, and (b) since the disks are small, the kinematics of the bulge can be derived from large enough radii where the line-of-sigh ratio of disk-to-bulge light becomes negligible[3]. So far, only few pointed E's have bee investigated in sufficient detail to apply these tests. I will mention only two example here. In NGC 4697 (Binney et al., 1990), the decrease of rotation velocity perpendicula to the major axis indicates that the bulge presumably rotates significantly slower tha suggested by the measured rotation along the major axis (where the disk mainly con tributes). In NGC 3610, the maximum rotation velocity along the major axis is hardl altered when the disk's surface brightness becomes negligible compared to the one of th bulge. This indicates that the bulge itself rotates quite rapidly, which is also confirme by velocity curves measured parallel to the major axis (Bender et al., in preparation) So, apparently, we may have to expect some variety in the kinematics and shapes c the spheroids of disky E's (a similar variety has been claimed for S0's and early spiral by Whitmore et al., 1984; Gerhard and Vietri; 1990, but also see Kormendy and Illing worth, 1982, Jarvis and Freeman, 1985). An interesting side aspect of the presence c faint stellar disks in ellipticals is that modelling of the isophotal shapes can be used t put relatively strong constraints on the inclination of the objects (see Scorza and Ben der, 1990) and in turn on dynamical models. The systematic analysis of the kinematica properties of pointed E's is part of a larger project already under way (Bender et al. 1989b) and will hopefully soon help to answer our questions concerning the kinematica structure of disky ellipticals.

The detailed investigation of the kinematical properties of spheroids and weak disks i also of importance for our understanding of the formation processes of disky ellipticals. I the faint stellar disks were acquired by secular evolution, then the disks' angular moment should not be correlated with those of the spheroids: some disks should extend along th minor axes of the spheroids, others should be found counter-rotating in respect to th spheroid; i.e. the faint stellar disks should show similar behaviour as the dust/gas ring frequently observed in ellipticals (see e.g. Bertola et al., 1990, Möllenhoff and Bender 1990: the kinematics of the warm gas in E's is in general decoupled from the kinematic of the underlying ellipticals and counter-rotation of gas and stars is indeed observed ir some cases). On the other hand, if the faint stellar disks are primordial and formed in a similar way to the disks in S0 galaxies, then we would expect that the angular moment of spheroid and disk are always parallel (although counter-rotation between gas and star has also been detected in some S0 galaxies, Bettoni et al., 1990, the stellar disks of S0' always appear in co-rotation with the spheroids, e.g. Kormendy and Illingworth, 1982)

So far, all observations point to a primordial origin of faint stellar disks in elliptica galaxies: none have been found along the minor axes of ellipticals, counter-rotation between disk and mainbody in 50% of all disky E's would presumably result in, or average, lower rotation velocities of these objects than is observed, and, finally, counter-rotation has not been found in any of the half dozen objects investigated in more detai until now. However, further analysis certainly is desirable.

[3]A possible third method makes use of the line-of-sight velocity distributions in pointed E's which contain information about both the bulge and disk rotation velocity. However this method is not yet well enough tested and also requires a modelling of the galaxies' mass distribution.

Frequency of disks in ellipticals, disk-to-bulge ratios

resently, only one unbiased sample to test the frequency of disks in ellipticals is available Bender et al., 1988)[4]. This sample is a nearly complete sub-sample of the ellipticals sted in the Revised Shapley Ames catalog (Sandage and Tammann, 1981) and comprises early all E's brighter than B=12.4 *mag* and having $\delta > -10°$. In this sample, 12% of all bjects have complex (irregular) isophote shapes and are not classifiable in an easy way, 9% have boxy isophotes, 31% show significantly pointed isophotes, and 35% exhibit nly small deviations from perfectly elliptical isophotes ($|a_4/a| < 0.5\%$).

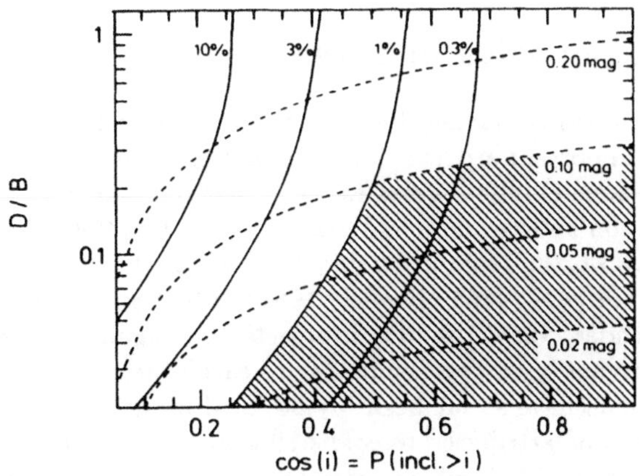

Figure 6. The diagram shows the influence of a disk on the isophote shapes and the surface brightness profile of an axisymmetric elliptical galaxy with inclination i (diagram adapted from Rix and White, 1990). The disk model has an exponential surface brightness profile, the underlying spheroid model follows an $r^{1/4}$-law. D/B denotes the disk-to-spheroid luminosity ratio. The inclination axis is parametrized by $\cos(i)$ which is equivalent to the probability to observe an inclination larger than i. The dashed lines correspond to a constant excess in surface brightness compared to the $r^{1/4}$ profile of the spheroid, the solid lines indicate constant a_4/a values. The disks having D/B and i values in the shaded area are undetectable by photometric means, i.e. 50% of all disks with $D/B < 0.3$ can be hidden in ellipticals.

Rix and White (1990) analysed the detectability of stellar disks in ellipticals in detail. They found that disks even as bright as 1/3 of the total luminosity can be easily hidden n ellipticals (see also Capaccioli, this conference) and can only be detected by isophotal hape analysis if their inclination is higher than $\approx 60°$ (Fig. 6). Furthermore, Rix and White found that the cumulative distribution of the positive a_4/a values of the Bender t al. (1988) sample is consistent with being due to disks and that presumably 50% of ll disks with disk-to-bulge ratios below 0.25 cannot be detected by photometric means. They concluded that possibly most of the radioquiet ellipticals may contain faint stellar lisks.

[4] This situation will improve in the near future when a complete, apparent magnitude selected sample from the southern hemisphere becomes available, see Bender et al (1989b).

In a few disky ellipticals disks and spheroids have been modelled (Simien and Michar 1990; Rix and White, 1990; Scorza and Bender, 1990, and this conference). The derive disk-to-bulge ratios range between 0.05 and 0.3. Central surface brightness values wer found to be similar to or brighter than the canonical value of 21.6 Bmag/sq.arcsec foun in spiral galaxies (see e.g. van der Kruit, this conference).

So far, when dealing with isophote shapes, stellar disks and their correlations wit other properties of ellipticals, we considered exclusively the main bodies of these objects The following section of this review will briefly discuss the kinematics of the innermos regions of ellipticals and how these show us further evidence for the existence of col sub-components, now on the smallest scales.

7 Peculiar kinematics in cores

This topic has been addressed during this conference by several speakers and I will onl summarize the most important aspects. Peculiar core kinematics has now been detecte in about 1/3 of luminous elliptical galaxies. The first detected and still most strikin cases of counter-rotation between core and mainbody are IC 1459, NGC 3608, and NGC 5322 (Franx and Illingworth, 1988; Jedrzejewski and Schechter, 1988; Bender, 1988b) Even more peculiar are NGC 4365 (Bender, 1988b, 1990) and NGC 4406 (Illingwort and Franx, 1989; Bender, 1988b) in which objects the angular momentum vectors o cores and main bodies are nearly perpendicular to each other.

The few cases which have so far been investigated in more detail indicate that th peculiar kinematics can be attributed to *cold* stellar sub-components extending to up t 4 photometric core radii and being kinematically decoupled from the rest of the galax (Franx and Illingworth, 1988; Bender, 1988b, 1990). The suggestion that we observe col components is mainly due to the analysis of line-of-sight velocity distributions which ca plausibly be separated in a fast, low-velocity dispersion component and an underlying non- or slowly rotating hot component (for a possible alternative explanation, see Bal-cells, this conference). Until recently, only one case was known (NGC 5322) where th cold component also showed its presence by pointed isophotes inside of 3 core radii. A discussed by Nieto (this conference), high resolution imaging has now revealed man more objects showing pointed isophotes in their core regions. In the case of NGC 5322 photometric modelling of the central isophotes by superimposing a flat component on a elliptical core gives a result consistent with the decomposition of the line-of-sight velocit distribution into two components: the rapidly counter-rotating component contribute about 1/4 to the luminosity inside of $\approx 1.5\,\mathrm{kpc}$ and about 2% to the total luminosity o the galaxy (Bender et al., in preparation). The mass of the counter-rotating component in NGC 5322 can be estimated to $\approx 10^{10} M_\odot$ (assuming an M/L of 6) which is similar to the mass in molecular gas observed in the centers of some starburst galaxies (e.g. Dupraz et al., 1990) and also the 'standard merger' NGC 7252 (Schweizer, 1990). No color difference between the flat component and the main body of NGC 5322 has been detected (Bender et al., in preparation) indicating that the counter-rotating component cannot be a very young phenomenon, i.e. much younger than $\approx 2\,\mathrm{Gyr}$.

The process of galaxy merging leading to end products similar to NGC 5322 has been modelled extensively (see Balcells' and Barnes' contributions to this conference) giving results consistent with the observational material. So, there is little doubt left that elliptical galaxies can be formed from mergers of evolved galaxies (although some questions still await more careful investigation, like e.g. metallicities and globular cluster frequencies).

It is important to note that nearly all ellipticals with peculiar cores have slowly rotating main bodies and show irregular or boxy isophotes. Also, ellipticals with peculiar cores have been found in different environments: NGC 5322 and IC 1459 are 'field' ellipticals, NGC 4365 and NGC 4406 are members of the Virgo cluster. This indicates that these objects can be formed under very different environmental conditions and from very different progenitors, as is indeed also evident from the numerical simulations (e.g Barnes, 1989; Balcells and Quinn, 1990). For a more detailed discussion, see Bender (1990).

3 Concluding remarks

Cold stellar sub-components (disks) are hidden in many, if not most elliptical galaxies and show their presence in pointed isophotes or in peculiar core kinematics.

The disks detected in isophotal shape analysis can be quite luminous and can contribute between 5% and 30% to the total light of disky elliptical galaxies, i.e. the disk-to-bulge ratios of disky E's are lower than or sometimes similar (!) to those of S0 galaxies. The central surface brightness values of the disks analysed so far are roughly equal to or brighter than 21.6 Bmag/sq.arcsec.

The kinematically decoupled disks/cold stellar components found in the cores of ellipticals with boxy/irregular main bodies contribute in general presumably less than a few per cent to the total luminosity of the galaxies. However, their surface brightnesses can be expected to be considerably higher than those of the larger disks detected in isophotal shape analysis because they otherwise would not have such a strong influence on the observed velocity fields. This also suggests that at least some sub-systems may so far have not been detected because they just have too low surface brightness.

It is certainly too early to discuss whether there is some kind of continuity in disk to total luminosity ratio (and possibly other properties) between boxy/irregular E's with peculiar cores and disky E's. The rather clear differences in their non-optical properties rather suggest the contrary.

Acknowledgements. I am grateful to my colleagues in Heidelberg and to all with whom I had the possibility to discuss during other recent conferences, notably at the 1990's Astrophysics Workshop in Aspen. This work was partially supported by the Deutsche Forschungsgemeinschaft (SFB 328) and by a NATO grant provided by the DAAD.

References

Balcells, M., Quinn, P.J.: 1990, *Astrophys. J.* **361**, 381.

Barnes, J.: 1989, *Nature* **338**, 123.

Bender, R.: 1988a, *Astron. Astrophys. Letters* **193**, L7.

Bender, R.: 1988b, *Astron. Astrophys. Letters* **202**, L5.

Bender, R.: 1990, in *Heidelberg Conference on Dynamics and Interactions of Galaxies*, p. 232, ed. R. Wielen, Springer Verlag Heidelberg.

Bender, R., Döbereiner, S., Möllenhoff, C.: 1988, *Astron. Astrophys. Suppl. Ser.* **74**, 385.

Bender, R., Nieto, J.-L.: 1990, *Astron. Astrophys.* , in press.

Bender, R., Surma, P., Döbereiner, S., Möllenhoff, C., Madejsky, R.: 1989a, *Astron. Astrophys.* **217**, 35.

Bender, R., Capaccioli, M., Macchetto, D., Nieto, J.-L.: 1989b, *ESO The Messenger* **55**, 6.

Bertola, F., Bettoni, D., Buson, L.M., Zeilinger, W.W.: 1990, in *Heidelberg Conference on Dynamics and Interactions of Galaxies*, p. 249, ed. R. Wielen, Springer Verlag Heidelberg.

Bettoni, D., Fasano, G., Galletta, G.: 1990, in *Heidelberg Conference on Dynamics and Interactions of Galaxies*, p. 84, ed. R. Wielen, Springer Verlag Heidelberg.

Binney, J.: 1982, *Ann. Rev. Astron. Astrophys.* **20**, 399.

Binney, J., Davies, R.L., Illingworth, G.: 1990, *Astrophys. J.* **361**, 78.

Canizares, C.R., Fabbiano, G., Trinchieri, G.: 1987, *Astrophys. J.* **312**, 503.

Capaccioli, M.: 1987, in IAU Symp. 127, *Structure and Dynamics of Elliptical Galaxies*, ed. T. de Zeeuw, p. 47, Reidel, Dordrecht.

Capaccioli, M., Held, E.V., Nieto, J.-L.: 1987, *Astron. J.* **94**, 1519.

Capaccioli, M., Caon, N., Rampazzo, R.: 1990, *Mon. Not. R. Astr. Soc.* **242**, 24P.

Carter, D.: 1978, *Mon. Not. R. Astr. Soc.* **182**, 797.

Carter, D.: 1987, *Astrophys. J.* **312**, 514.

Davies, R.L., Illingworth, G.D.: 1983, *Astrophys. J.* **266**, 516.

Davies, R.L.: 1989, in *Les Mondes des Galaxies*, p. 312, eds. L. Bottinelli and H.G. Corwin, Springer Verlag, New York.

Davies, R.L. Efstathiou, G., Fall, S.M., Illingworth, G.D., Schechter, P.: 1983, *Astrophys. J.* **266**, 41.

Djorgovski, S., de Carvalho, R.: 1990, in *Windows on Galaxies*, eds. G. Fabbiano et al., Kluwer, Dordrecht.

Donnelly, R.H., Faber, S.M., O'Connell, R.M.: 1990, *Astrophys. J.* , in press.

Dressler, A., Lynden-Bell, D., Burstein, D., Davies, R.L., Faber, S.M., Terlevich, R.J., Wegner, G.: 1987, *Astrophys. J.* **313**, 42.

Dupraz, C., Casoli, F., Combes, F., Kazes, I.: 1990, *Astron. Astrophys. Letters* **228**, L5.

Faber, S.M., Dressler, A., Davies, R.L., Burstein, D., Lynden-Bell, D., Terlevich, R., Wegner, G.: 1987, in *Nearly Normal Galaxies*, ed. S. Faber, Springer Verlag, New York.

Fabbiano, G., Trinchieri, G.: 1985, *Astrophys. J.* **296**, 430.

Fasano, G.: 1990, *Mon. Not. R. Astr. Soc.* , submitted.

Forman, W., Jones, C., Tucker, W.: 1985, *Astrophys. J.* **293**, 102.

Franx, M., Illingworth, G.: 1988, *Astrophys. J.* Letters **327**, L55.

Franx, M., Illingworth, G., Heckman, T.M.: 1989, *Astron. J.* **98**, 538.

Gerhard, O.E., Vietri, M.: 1990, in *Heidelberg Conference on Dynamics and Interactions of Galaxies*, p. 342, ed. R. Wielen, Springer Verlag Heidelberg.

Hummel, E., Kotanyi, C.: *Astron. Astrophys.* **106**, 183.

Illingworth, G.D., Franx, M.: 1989, in *Dynamics of Dense Stellar Systems*, ed. D. Merritt, Cambridge University Press, Cambridge.

Jarvis, B., Freeman, K.: 1985, *Astrophys. J.* **295**, 324.

Jedrzejewski, R.I.: 1987, *Mon. Not. R. Astr. Soc.* **226**, 747.

Jedrzejewski, R.I., Schechter, P.L.: 1988, *Astrophys. J.* Letters **330**, L87.

Kormendy, J., Illingworth, G.: 1982, *Astrophys. J.* **256**, 460.

Kormendy, J., Djorgovski, S.: 1989, *Ann. Rev. Astron. Astrophys.* **27**, 235.

Lauer, T.R.: 1985, *Mon. Not. R. Astr. Soc.* **216**, 429.

Longo, G., Capaccioli, M., Bender, R., Busarello, G.: 1989, *Astron. Astrophys. Letters* **225**, L17.

van der Marel, R., Binney, J., Davies, R.L.: 1990, *Mon. Not. R. Astr. Soc.* **245, 582**.

Michard, R., Simien, F.: 1988, *Astron. Astrophys. Suppl. Ser.* **74**, 25.

Möllenhoof, C., Bender, R.: 1990, in *Heidelberg Conference on Dynamics and Interactions of Galaxies*, p. 272, ed. R. Wielen, Springer Verlag Heidelberg.

Nieto, J.-L.: 1988, 2da *Reunion de Astronomia Extragalactica*, p. 239, Academia Nacional de Ciencias de Cordoba, Cordoba.

Nieto, J.-L.: 1990, *Astron. Astrophys.* , submitted.

Nieto, J.-L., Bender, R.: 1989, *Astron. Astrophys.* **215**, 266.

Nieto, J.-L., Capaccioli, M., Held, E.V.: 1988, *Astron. Astrophys. Letters* **195**, L1.

Peletier, R.F., Davies, R.L., Illingworth, G.D., Davis, L.E., Cawson, M.: 1990, *Astron. J.* **100**, 1091.

Rix, H.-W., White, S.D.M.: 1990, *Astrophys. J.* **362**, 52.

Sandage, A., Tammann, G.A.: 1981, *A Revised Shapley-Ames Catalog of Bright Galaxies*, Carnegie Institution of Washington, Pub. 635, Washington.

Schweizer, F.: 1990, in *Heidelberg Conference on Dynamics and Interactions of Galaxies*, p. 60, ed. R. Wielen, Springer Verlag Heidelberg.

Scorza, C., Bender, R.: 1990, *Astron. Astrophys.* **235**, 49.

Shaw, M.A.: 1987, *Mon. Not. R. Astr. Soc.* **229**, 691.

Simien, F., Michard, R.: 1990, *Astron. Astrophys.* **227**, 11.

Toomre, A.: 1977, in *The Evolution of Galaxies and Stellar Populations*, eds. B.M. Tinsley and R.B. Larson, p.401, Yale University Observatory.

Tremaine, S.: 1987, in IAU Symp. 127, *Structure and Dynamics of Elliptical Galaxies*, ed. T. de Zeeuw, p. 367, Reidel, Dordrecht.

de Vaucouleurs, G., de Vaucouleurs, A., Corwin, H.G.: 1976, *Second Reference Catalog of Bright Galaxies*, University of Texas Press, Austin.

Whitmore, B.C., Rubin, V.C., Ford, W.K.: 1984, *Astrophys. J.* **287**, 66.

de Zeeuw, T., Peletier, R., Franx, M.: 1986, *Mon. Not. R. Astr. Soc.* **221**, 1001.

Ralf Bender answering questions happily.

MORPHOLOGY AND CLASSIFICATION OF GALAXIES. WHERE DO WE STAND?

I.R. King
Astronomy Department
University of California, Berkeley, USA

These concluding remarks note some of the high points of the workshop and offer some comments about the future of the morphology of galaxies.

It is of course a requirement that any summary of a conference begin with a statement that this is a hard job to do. In this case the statement is particularly true, however, because it has been more than ten years since I worked actively in the field of galaxies.

It is really impressive to see how much has changed since, for example, the Yale meeting of thirteen years ago (Tinsley and Larson, 1977). For one thing, we have infinitely better data to work with. A large part of the improvement has been in instrumentation—particularly the widespread availability of CCD's—and in computing power—both for image processing and for simulation of the dynamics of stellar systems. But we also have new types of observations: the IRAS data have literally opened a new eye on the Universe; CO surveys have greatly improved our knowledge of the interstellar (and particularly the star-forming) component; and it is my impression that the radio maps—particularly in HI—have increased in quality as well as in quantity. And in addition, we should take important note of the improvement of the quality of optical imaging, with the sharpening of seeing that has taken place at so many observatories. The widespread recognition that an important part of the seeing arises in the dome rather than in the atmosphere has made a real difference both to limiting resolutions and to limiting magnitudes.

Along with the better data, we now have a lot of new understanding. Here I would single out two areas. One is of course the elliptical galaxies. In the mid-seventies they looked terribly simple; a few years later they turned out to be simply terrible. The other area that has changed so much is the recognition of the significance—no, I ought rather to say the importance—of mergers. At Yale I was a complete skeptic; but I was converted to the faith at the Heidelberg meeting last year (Wielen, 1990).

And here we are, taking a broad look at the whole range of galaxies. At first it seemed to me that this meeting had a very dull title. And in fact, if we had talked only about morphology and classification it might have been a very dull meeting indeed. But the word "physical" is in the title too, and when we think about the physical basis of our astronomy, things tend to get a lot more exciting. Besides, a lot of astronomers simply behave like astronomers and talk about what they are interested in and about what they

G. Longo et al. (eds.), *Morphological and Physical Classification of Galaxies*, 371–376.

are doing. As a result we have had a wide range of talks, from reviews to reports of nev science. In fact, the nature of any meeting depends on the individuals who participate and the people in this room have had a lot of interesting things to say, both on and of the platform.

This very diversity of the nature of the contributions makes it impossible to giv anything that could be called a summary. Just as I went through reading the progran to the participants, I now ask the reader to scan the table of contents once again. I wil merely comment on a few of the presentations.

Ron Buta set the tone very well in his introductory talk on classification. And Gusta Tammann made very clear the distinction between luminosity functions of galaxies o different types. As for his sartorial splendor in spite of this southern heat, I have no wa of conveying it in words.

I got a lot out of the talk on ellipticals by Marijn Franx, as well as the somewha related discussion by Tim de Zeeuw, who could have contributed so much more ha his time not been so limited. Perhaps after a dozen more such talks I will begin t understand something about these difficult dynamical problems.

I hope that Lia Athanassoula's written paper is as clear as her oral presentation was Theories of spiral structure are difficult and often contradictory, and she made a lot o complicated things seem clear. (And what would a meeting be without her laugh?)

This is probably just exposing my own ignorance, but I learned a lot about molecula clouds from the very clear review by Michel Guelin. Similarly Jacqueline van Gorkom on an afternoon when I desperately wanted to sleep through all the papers, she kept m awake for every moment of her talk about gas in early-type galaxies. And Renzo Sancis fascinated me with his description of superbubbles in the HI of nearby galaxies.

As usual, we were treated to a fascinating update on Josh Barnes' simulations. was sorry to hear him say that he can afford to do only a few runs per year, but perhap that means that we can keep current by going to one meeting per year.

In addition to the reviews, we had a number of contributions of original work. Amon those, I was impressed with that from Francesca Matteucci, on the evolution of chemica abundances. I must admit, however, that part of my reaction was, "Se non è vero, è ber trovato." The presentation by George Djorgovski, on the fundamental plane of elliptica galaxies, had so much in it that it went by me too fast; I look forward to studying th printed manuscript.

I don't mean to slight the presentations that I haven't mentioned, but I am sure tha I would bore you to tears if I went over every paper.

A better thing to do, I am sure, is to reflect on our purpose in this meeting. A some point in the discussions—I can't remember whether it shows up in the printed dis cussions or whether it was dinner-table conversation—some one suggested that galax classification was an obsolete idea. I think that this is quite wrong. We need a classifi cation scheme, because it is both a language and a description. But at the same time w should be careful not to overclassify. At the extreme, each individual galaxy would ge a different classification—the ultimate *reductio ad absurdum*. It is well to remember th aphorism of our mentor Gérard de Vaucouleurs, "After a collision a car is a wreck, no a new type of car!"

As for specific schemes, I believe that the world is moving toward the classificatio scheme of the *Revised Shapley–Ames Catalog* (Sandage and Tammann, 1981). Although Sandage reveres Hubble, he has in fact been very wisely eclectic (as Hubble would also have been if he had lived into these recent decades), and the RSA has pleasant flavors of de Vaucouleurs and van den Bergh in it. The first essential of a classification scheme is to be well illustrated, and the RSA has done just that. But what is likely to cast the system in bronze is Sandage's plan to publish, with NASA support, reproductions of images of all 1300 of the Shapley–Ames galaxies.

But this is just an opinion, not an edict. The impending publication of the third edition of the de Vaucouleurs *Reference Catalog* will give classifications, in his system, of a much longer list of galaxies.

And there are some shortcomings in the RSA that have not yet been remedied: it fails to recognize the cD's as a separate class from the ellipticals, nor does it distinguish the dwarf spheroidals (perhaps because none of these low-surface-brightness objects met the selection criteria of Shapley and Ames [1932]).

These objections may be irrelevant, however. It is becoming increasingly clear that elliptical galaxies are a totally different sort of beast from the spirals, or even from the S0's. (The distinction can be confused by failures to distinguish between E's and S0's, for lack of adequate quantitative photometry, but it is a sharp physical distinction nevertheless.)

A worse problem—and one that increasing numbers of astronomers believe is related—is that of interactions and mergers. Here especially we need to take heed of de Vaucouleurs' aphorism; indeed, we are seeing the wrecks in the process of happening. These systems include many, if not all, of the systems classified by Hubble and Sandage as Irr II. I would suggest, with all due respect for my betters, that Irr I become just plain Irr—or perhaps Si, since these objects are clearly the extrapolation of the spiral sequence, even though they are not literally spiral. As for Irr II, why not call them Int, for interacting or products of interactions?

A final remark about classification: this meeting has had a lot of discussion of *physical* properties of galaxies. In nearly all cases the galaxies concerned have been identified by the classification scheme to which I have referred, and meaningful statements have been made about the classes. This should call our attention to the fact that our classification efforts must be doing something right.

But back to the ellipticals now. I would really like to suggest that they be excluded from the classification schemes that we have been talking about, except for the mere fact of being noted as elliptical rather than disk. Yes, the ellipticity digit is useful, as a description, but let us not attribute physical significance to it. It is clear that the giant ellipticals are triaxial systems, whose true shape we can only faintly hope to divine from detailed observations of their velocity fields. But the essential point is that the "tuning-fork diagram" is dead. The ellipticals are a different type of galaxy, and their superficial resemblance to the bulges of spirals can only mislead us.

Again I have to hedge my bets. Low-luminosity ellipticals have a considerable resemblance to the bulges of disks. The resemblance may be coincidental—but who knows what will turn up in extragalactic research?

The ellipticals do bother me. They really are terrible. Along with triaxiality, we have to deal with such horrors as counter-rotating cores and out-of-plane gas disks.

Paradoxically, it is a blessing that these systems do not conform to Hubble's ideal c similar ellipsoidal surfaces. He showed (1926) that in that case we would never be abl to tell the true orientation and shape of any individual elliptical galaxy. Fortunately, th ellipticities change outward (in the twisted-isophote phenomenon), so that the triaxi systems with which we now know we must deal give us some evidence of their tru nature; and it may be that in conjunction with an observed velocity field we may be abl to decipher the real shape of an individual elliptical galaxy. In the dynamical analysi we have another terrible problem, however. Not knowing the analytic nature of the thir integral, we have to resort to Schwarzschild's method. Bless it, for it is all that we have but it gives a multiplicity of possible solutions. From another point of view, we can ge insight into the orbits through study of Stäckel potentials (de Zeeuw, 1985); but the are only an approximation, and real galaxies have orbit families that are outside tha complicated but clear realm.

Indeed, discussion of elliptical galaxies has dominated a large part of this meeting as it does in all galaxy conferences nowadays. I think that this is appropriate, becaus at this moment ellipticals are "where the action is." But we can hope that the dynamic of disks, and of spiral structure, will regain the place that they deserve.

So, where are we now? I feel very hopeful. Even in our general field of astronomy where the *e*-folding time of knowledge tends to be a decade or two, understanding of th structure of galaxies seems to me to be advancing at least as fast as that of any othe area.

It is interesting to speculate about another meeting on this subject, ten years fron now. First, we can hope that Dave Burstein and George Djorgovski will not have eate each other up totally, so that we will again be privileged to witness their encounter. Mor seriously, however, I feel that progress—observational, theoretical, and interpretative (le us never forget that joint process that really *does* it)—is rapid today, and that ten year from now we can look forward to solutions to many (but not all!) of the problems tha plague us today—but that we can also look forward to facing new problems of which w cannot conceive today. Such has always been the history of astronomy.

Finally, I must remark on the circumstances of the meeting itself. We have beer extremely well supported and entertained here in Sant'Agata, in all the senses of thos two words, and we can thank our hosts in a way that can only be inadequate. Perhaps th most meaningful statement that I can make is that astronomers who consider attendanc at future workshops of the Osservatorio Astronomico di Capodimonte should look mos favorably on the opportunity.

References

Hubble, E.: 1926, "Extra-Galactic Nebulae." *Astrophys. J.* **64**, 321–369.

Sandage, A., and Tammann, G.A.: 1981, *A Revised Shapley-Ames Catalog of Bright Galaxies*, Carnegie Institution of Washington, Washington, DC.

Shapley, H., and Ames, A.: 1932, "A Survey of the External Galaxies Brighter than the Thirteentl Magnitude", in *Annals of Harvard College Obs.* **88**, no. 2.

Tinsley, B.M, and Larson, R.B. (eds.): 1977, *The Evolution of Galaxies and Stellar Populations*, Yal Univ. Press, New Haven, CT.

Vaucouleurs, G. de: 1974, "Structure, Dynamics and Statistical Properties of Galaxies", in *The Formation and Dynamics of Galaxies*, (IAU Symposium no. 58), ed. J.R. Shakeshaft, Reidel, Dordrecht NL, pp. 1–53.

Vielen, R. (ed.): 1990, *Dynamics and Interactions of Galaxies*, Springer-Verlag, Berlin.

eeuw, T. de: 1985, "Elliptical Galaxies with Separable Potentials." *Mon. Not. R. Astr. Soc.* **216**, 273–334.

Ivan King thanks the local organizing committee during the social banquet.

George Djorgovski presents a "boxy" box harboring a "lemon-shaped" lemon. Massimo Capaccioli expresse some concerns about its position in the fundamental plane.

DETAILED MORPHOLOGICAL CLASSIFICATION OF EARLY-TYPE GALAXIES IN THE PERSEUS CLUSTER[1]

P. Poulain, J.-L. Nieto, E. Davoust
Observatoire Midi-Pyrénées, Toulouse, France

We have tackled a long-term observational project to derive the photometric and morphological parameters of a large number of early-type galaxies in clusters. Our aim is to identify the effects of environment on the detailed morphology of early-type galaxies. For this purpose, we have obtained V and R images of 90 early-type galaxies belonging to the Perseus Cluster, selected from their redshift and their morphological classification in Kent and Sargent (1983) and in the *Catalogue of Principal Galaxies* (Paturel et al., 1990).

The observations were made at the 2m telescope of Pic-du-Midi Observatory (scale = 0.32 arcsec/px) and at the Cassegrain focus of CFHT (scale = 0.11 arcsec/px), most in excellent seeing conditions (see figure 1). We therefore reclassified these galaxies from a careful inspection of the images and the isophotal shape analysis technique presented in Bender et al. (1988).

We present in figure 2 the results of this detailed classification, where we extracted from the E class only the most obvious disky E's (E's presenting more subtle deviations are left among the E class). Figure 2 shows that the improved resolution produces a shift toward later types along the Hubble sequence, by revealing fine structures and reducing to a minimum the disturbing effects of foreground stars.

At this stage, we would like to mention the large variety of morphological properties found in this sample: E's presenting the two classical types of boxy and disky deviations ("boxy lemons"; see Nieto and Bender, 1989 and Nieto et al., 1990), SO's with boxy or X-shaped bulges, one ringed barred spiral, etc. The complete set of data is presented in Poulain et al. (1991), and its analysis is underway.

References

Bender, R., Döbereiner, S., Möllenhoff, C.: 1988, *Astron. Astrophys. Suppl. Ser.* **74**, 385.

[1]Based on observations made with the 2-m telescope of Pic-du-Midi Observatory and the Canada-France-Hawaï telescope.

G. Longo et al. (eds.), *Morphological and Physical Classification of Galaxies*, 377–378.

Kent, S.M., Sargent, W.L.W.: 1983, *Astron. J.* **88**, 697.

Nieto, J.-L., Bender, R.: 1989, *Astron. Astrophys.* **215**, 66.

Nieto, J.-L., Poulain, P., Rosenblatt, P., Davoust, E.: 1990, "Isophotal shapes of early-type galaxies: Elongated ellipticals", *Astron. Astrophys. Suppl. Ser.*, submitted.

Paturel, G., Fouqué, P., Botinelli, L., Gouguenheim, L.: 1989, *Catalogue of Principal Galaxies*, Monographies de la base de données extragalactiques, 1.

Poulain, P., Nieto, J.-L., Davoust, E.: 1991, "Isophotal shapes of early-type galaxies: II. The Perseus cluster", *Astron. Astrophys. Suppl. Ser.*, submitted.

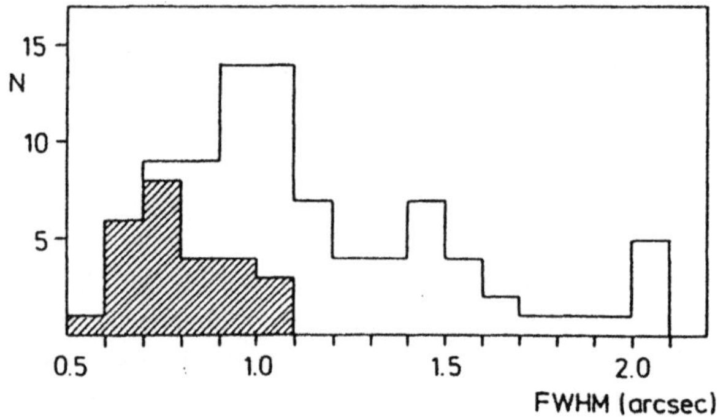

Figure 1. Histogram of the image qualities (FWHM) of our observations (essentially R band). The shaded area defines the histogram from CFHT images.

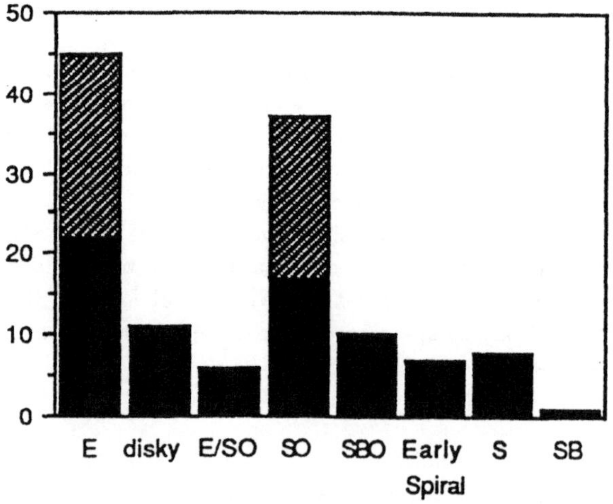

Figure 2. Histogram of the morphological classifications: earlier classification (PGC, shaded area) and detailed classification (this poster, black area).

PROPERTIES OF DWARF SPHEROIDALS

S. Djorgovski[1], R. de Carvalho[1,2]
[1] *Palomar Observatory, Caltech, Pasadena, USA*
[2] *Observatorio Nacional, CNPq, Rio de Janeiro, Brasil*

Low surface brightness dwarf galaxies (at least the gas-poor variety) represent a fundamentally distinct family of objects, rather than just a low luminosity extension of the Hubble sequence; they are *not* "dwarf ellipticals" (Wirth and Gallagher, 1984; Kormendy, 1985; see Kormendy and Djorgovski, 1989 for a brief discussion and further references). Kormendy (1985) has demonstrated that dwarfs obey simple scaling laws and correlations which are quite different from those of normal ellipticals, pointing towards different formation histories. We extended this line of inquiry to multivariate correlations, using the sample of Local Group dwarf spheroidals (dSph), and samples of low surface brightness dwarfs (LSBD) in the Virgo and Fornax clusters. Whereas normal ellipticals form a statistically two-dimensional family in the parameter space of observables (Djorgovski, this volume, and references therein), dwarfs form nearly one-dimensional families; i.e., all of their properties seem to be well correlated (Djorgovski and de Carvalho, 1990). The only exception are velocity dispersion (only 5 available), which do not correlate with *any* other observed properties. Furthermore, the corresponding correlations for dSph's and ellipticals often have the opposite signs. Some examples are shown in figures 1 and 2. These results are only preliminary: selection effects are notoriously difficult.

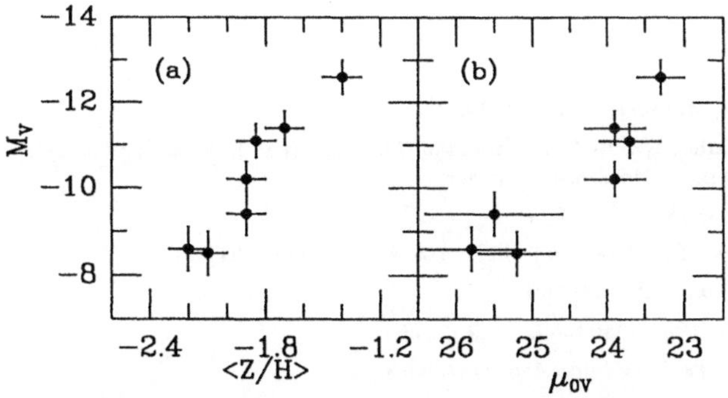

Figure 1. Correlations between the metallicity and luminosity (a), and central surface brightness and luminosity (b), for the seven dSph satellites of our Galaxy. The excellent correlation (a) would be difficult to understand if dSph's were simply faded tidal interaction fragments.

G. Longo et al. (eds.), Morphological and Physical Classification of Galaxies, 379–380.

The correlations presented here, and other observed properties of LSBD's can b interpreted very naturally within the galactic wind and adiabatic expansion scenario fo their formation (Dekel and Silk, 1986; Vader, 1986; Yoshii and Arimoto, 1987; etc.).

More details will be presented in a forthcoming paper. SD acknowledges a trave grant from the AAS, and a partial support from the Alfred P. Sloan Foundation.

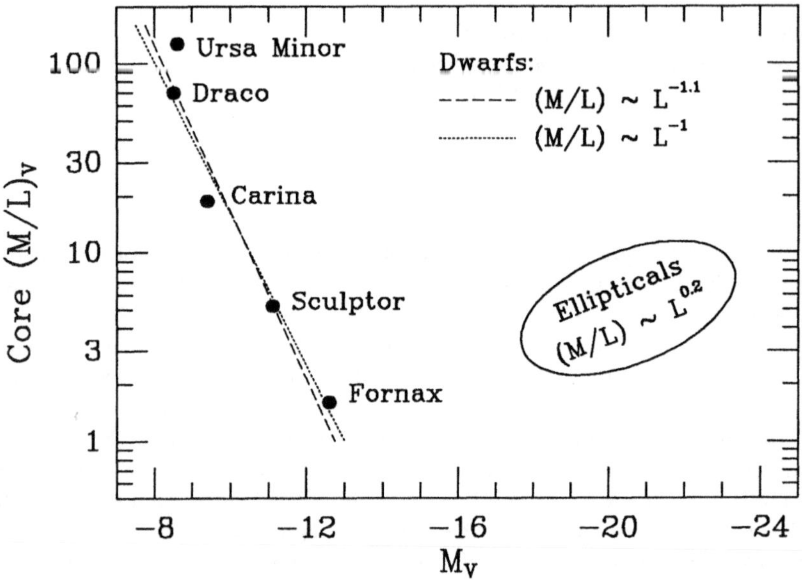

Figure 2. The scaling of the core (M/L), computed using the King-Minkowski formula, with the total luminosity, for the five dSph's with available velocity dispersions. The observed slope indicates that *the mass and luminosity are uncorrelated* for these objects. The sign of the (M/L) vs. L correlation is opposite of that for the normal E's, which occupy the region indicated with the ellipse. This shows the dramatic difference between the physical properties and formation of these two galaxy families.

References

Dekel, A., Silk, J.: 1986, *Astrophys. J.* **303**, 39.

Djorgovski, S., de Carvalho, R.: 1990, in Fabbiano, G. et al. (eds.), *Windows on Galaxies*, Erice Astrophysics Workshop, p. 9. Dordrecht, Kluwer.

Kormendy, J.: 1985, *Astrophys. J.* **295**, 73.

Kormendy, J., Djorgovski, S.: 1989, *Ann. Rev. Astron. Astrophys.* **27**, 235.

Vader, J.P.: 1986, *Astrophys. J.* **305**, 669.

Wirth, A., Gallagher, J.: 1984, *Astrophys. J.* **282**, 85.

Yoshii, Y., Arimoto, N.: 1987, *Astron. Astrophys.* **188**, 13.

BRIGHT CORES IN NEARBY SOUTHERN GALAXIES

O.R. Hainaut, B.J. Jarvis
European Southern Observatory
Santiago, Chile

1 Introduction

The presence of bright cores in galaxies is of considerable interest since they may be the site of current star formation, radio activity, black holes etc.. Probably the best known case is M 87 (NGC 4486)(Young et al., 1978). Recent observations of the core of M 87 in the near IR by Jarvis and Melnick (1990) convincingly show that there is a massive young star cluster as revealed by a large increase in the strength of the Ca triplet lines (a strong indicator of supergiants) and velocity dispersion.

An important question is whether this bright core is exceptional. If bright cores are widespread, how do they correlate with radio activity, morphological type etc.. For example, bright cores in early-type galaxies due to localized star formation could oblige us to continue revising our traditional view in which elliptical galaxies contain little or no interstellar material and consequently are not expected to show evidence for star formation.

To start to answer this question, we have embarked on a program of looking at the cores of suitable galaxies searching for significant enhancement in core luminosities over what may be expected from "normal" ellipticals. Our sample of galaxies was selected from the Revised Shapley-Ames Catalog (Sandage et al., 1987) according to the following criteria: 1) radial velocity less than 2000 km s^{-1}; 2) Hubble type Sb or earlier. Forty eight galaxies have already been observed in an ongoing program using 2.2m ESO/MPI and the New Technology Telescope (NTT) at La Silla, Chile. From these, we show preliminary results for 15 of these galaxies.

2 Description of fitting

For each galaxy, a Hubble law (Hubble, 1930) was fitted to the outer part of the photometric profile *NOT including data points interior to* 2″ to exclude a possible bright core from the fit. Four parameters I_0, R_h, R_0 and I_{sky} were extracted through a least-square

G. Longo et al. (eds.), Morphological and Physical Classification of Galaxies, 381–384.

minimization of the function

$$I_{Model} = \frac{I_0 R_h^2}{(|X - R_0| + R_h)^2} + I_{sky}$$

where I_{Model} is the intensity of the model at position X. On the plots given at figure 1 the residual (in magnitude) of the Hubble law is plotted against the radius, up to 20 from the center.

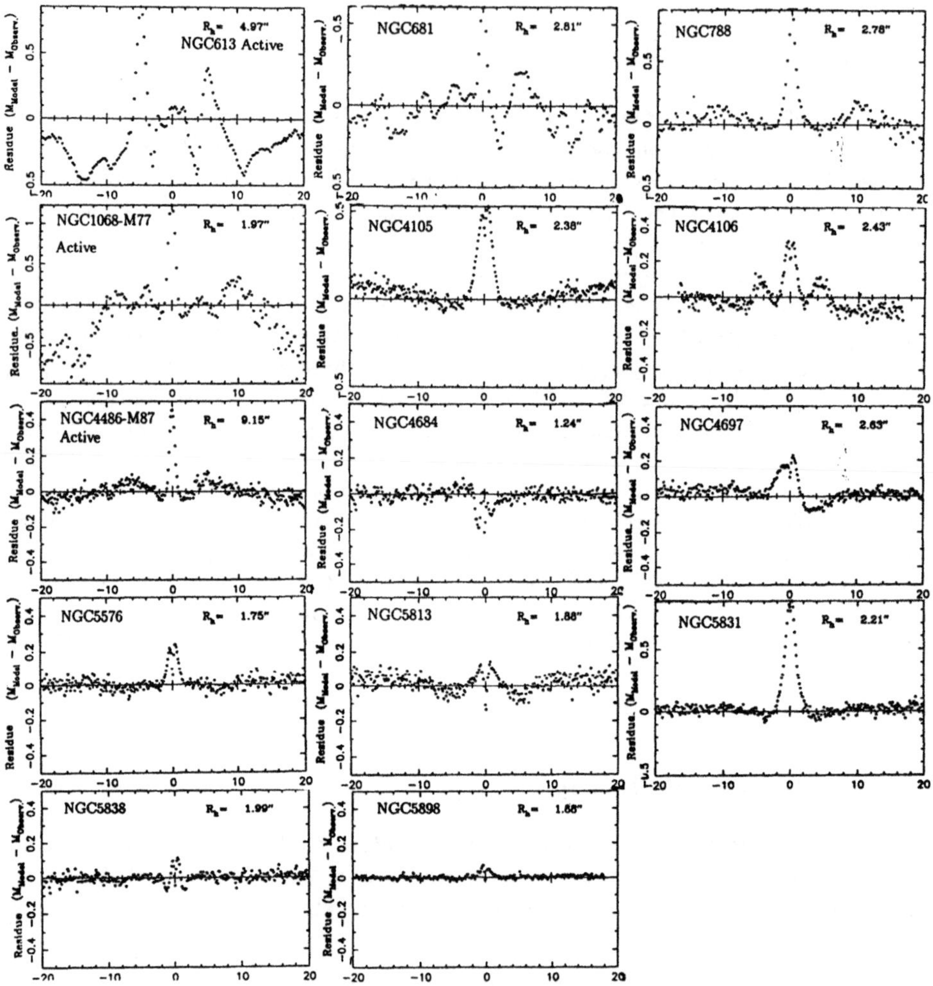

Figure 1. The residual in magnitudes of the observed luminosity profile in R with a subtracted least-square fitted Hubble law. The galaxies with active nuclei are also noted.

For an example galaxy (NGC 1052), we give also the isophotal contour and the magnitude vs. $\log R$ profile with the fitted Hubble law. We have similar graphs for all the galaxies ut no space to give them here. It is important to note that even if the residuals are iverging at the edges of the graph, they have to be very small at large radii since this imposed by the fit of the I_{sky} parameter on the sky level.

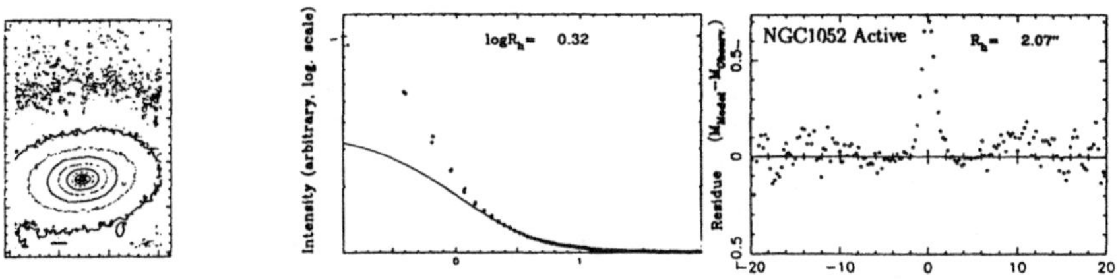

Figure 2. Example of (a) the isophotal contour; (b) the magnitude vs. $\log R$ profile (dots) and the Hubble law fit (solid line), and (c) the residual in magnitude (like in figure 1 for each other galaxy).

3 Results

rom this initial sample of 15 galaxies, we found an unexpectedly large number of bright ores with several clear cases (residuals larger than 0.5 magnitude): NGC's 788, 1052, 068, 4105, 4486 (M 87) and 5831. Some galaxies show a weaker peak (0.2–0.5 magitude) and *may* have a bright core: these are NGC's 681, 4106, 4697. The remaining alaxies show no significant evidence for a bright core (with a residual of less than 0.2 agnitude).

4 Conclusion

'rom a preliminary analysis, we find that bright (M 87 type) cores are not exceptional. 'he analysis of the remaining observed galaxies is being performed, and we also plan to bserve more galaxies, to complete our sample and to perform some statistical analysis n the results. We will also observe spectroscopically those galaxies showing enhanced ore brightness. This we hope will tell us something about the change in stellar content 1 these region.

We are grateful to J. Melnick, who suggested this project.

References

Iubble, E.P.: 1930, *Astrophys. J.* **71**, 231.

arvis, B.J., Melnick, J.: 1990, submitted *Astron. Astrophys. Letters* .

andage, A., Tammann, G.A.: 1987, *A Revised Shapley-Ames Catalog of Bright Galaxies*, Carnegie Institution of Washington Publication 635.

'oung, P.J., Westphal, J.A., Kristian, J., Wilson, C.P., Landauer, F.P.: 1978, *Astrophys. J.* **221**, 721.

Brian Jarvis (left) and Françoise Combes (right) in the poster room.

SHELL FORMATION AND THICK DISKS IN ELLIPTICAL GALAXIES

R.C. Thomson
Institute of Astronomy, Cambridge, UK

Numerical simulations by Quinn (1984), Dupraz and Combes (1986), and Hernquist and Quinn (1988, 1989) have shown that regular shell structures can be formed when a small secondary galaxy collides nearly head-on with a larger elliptical galaxy to form a merged system. Thomson and Wright (1990 - hereafter TW) have discussed several problems associated with this merger model and have proposed a viable alternative model for shell formation wherein shell structures are spiral density waves induced in a thick disk population of dynamically cold stars by a weak interaction (but not a merger) with another galaxy. Numerical simulations by TW supported this weak interaction model, but were limited to spherical potentials only. This work has now been extended to include non-spherical potentials and the results are found to further support the weak interaction model of TW.

The numerical simulations shown here are the results of restricted three body calculations with a primary galaxy (of unit mass) modelled by a Kuzmin-de Zeeuw oblate spheroidal mass distribution:

$$\rho(m) = \frac{\rho_0}{(1+m^2)^2}$$

where $m^2 = 1 + R^2 + z^2/c^2$ in cylindrical polar coordinates, and c is a measure of the flattening of the system.

An analytic expression for the associated gravitational field was used to integrate the equations of motion for test particles with short axis tube orbits and zero radial thickness (de Zeeuw, 1985; Bishop, 1987; Merritt and Stiavelli, 1990). The initial phase space distribution of the test particles was calculated according to the method described by Bishop (1987), and by Merritt and Stiavelli (1990). This enabled an equilibrium model of a dynamically cold population ('thick disk') to be constructed.

Figure 1 shows the effect of a parabolic encounter with a secondary galaxy of 1/10th the mass of the primary and perigalactic distance of about $2r_e$. Note the formation of a predominantly one armed spiral density wave post-encounter which appears as interleaved shells when the system is viewed from the side or front (i.e., edge-on). Note also the near-circular appearance of the shells when the system is viewed edge-on, whereas

G. Longo et al. (eds.), Morphological and Physical Classification of Galaxies, 385–386.

the underlying galaxy is E4. This is consistent with the low ellipticity of shell structure found in the interleaved shell galaxy NGC3923 as described by Prieur (1988).

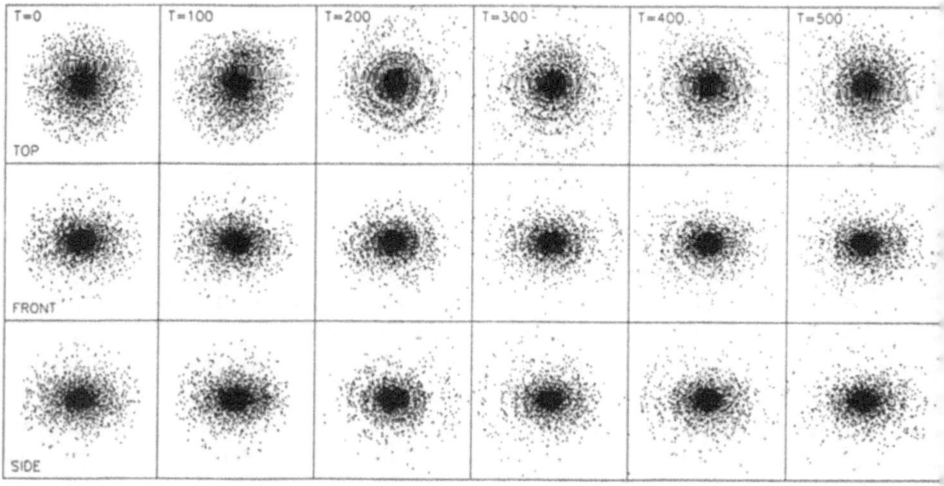

Figure 1. Effect of interaction with a secondary galaxy of mass 1/10th that of the primary galaxy. The secondary's orbit is parabolic and reaches perigalacticon at time T=100 units (each time unit represents about 3×10^7 years).

Numerical simulations of weakly interacting systems which contain a significant popula- tion of dynamically cold stars ('shell orbits') have provided further evidence to suppor this mechanism of shell formation in elliptical/S0 galaxies.

References

Bishop, J.L.: 1987, *Astrophys. J.*, **322**, 618.

de Zeeuw, T.: 1985, *Mon. Not. R. Astr. Soc.*, **216**, 273.

Dupraz, C., Combes, F.: 1986, *Astron. Astrophys.*, **166**, 53.

Hernquist, L., Quinn, P.J.: 1988, *Astrophys. J.*, **331**, 682.

Hernquist, L., Quinn, P.J.: 1989, *Astrophys. J.*, **342**, 1.

Merritt, D., Stiavelli, M.: 1990, *Astrophys. J.*, **358**, 400.

Quinn, P.J.: 1984, *Astrophys. J.*, **279**, 596.

Thomson, R.C., Wright, A.E.: 1990, *Mon. Not. R. Astr. Soc.*, **247**, 122.

THE ORIGIN OF STRONG ISOPHOTAL TWISTS IN ELLIPTICAL GALAXIES[1]

J.-L. Nieto and Ch. Lagoute
Observatoire Midi-Pyrénées
Toulouse, France

Strong isophotal twists in elliptical galaxies have been one of the major clues to our knowledge of the structure and the formation of elliptical galaxies. Many papers addressing this question have concluded that isophotal twists are evidence for ellipticals being triaxial systems (e.g., Williams, 1981).

Recently, the analysis of isophotal shapes (e.g. Bender et al., 1988) has revealed interesting insights on the existence of subtle systems, such as disks: the physical similarity of disk E's with S0's indeed suggests that disk E's are S0-like systems at the end of the disk/bulge luminosity sequence (Bender et al., 1989; Nieto and Bender, 1989).

In this context, we have noticed morphological analogies presented by E's having strong isophotal twists (e.g. NGC 596, NGC 1549) with S0 systems known to contain a small inner bar such as NGC 3384 (Davoust et al., 1984; Ferrario and Capaccioli, this workshop) or SB0's remote enough for the bar to appear as a small deviation from ellipticity on low-resolution images. This analogy is exemplified in figure 1 which shows results from CCD images of NGC 3384 and NGC 596 obtained at the 2m telescope of Pic-du-Midi Observatory; the profiles of the position angle (arbitrary zero point) and a_4, the $\cos 4\theta$ coefficient of the Fourier series fitting best the isophotal contours, are plotted against the semi major axis, a. In NGC 3384, the signature of the inner bar, which is almost perpendicular to the major axis of the system, is clearly visible. It is apparent that: (*i*) the position angle (as well as the ellipticity) experience strong variations near 15", and (*ii*) the a_4 term becomes strongly negative at that distance, while it is positive elsewhere (as expected from a disk system). In NGC 596, the E galaxy with the most spectacular isophotal twist ($\simeq$ 90 degrees), the geometry behaves just like in NGC 3384, although in a more moderate way. In the outer part($r > 8$"), several sharp geometrical features are present at the same distance from the center (15"). In particular, there is a strong change in orientation (40 degrees), a sudden increase in the ellipticity profile, a change

[1]Observations were made with the 2m Telescope of Observatoire du Pic du Midi

G. Longo et al. (eds.), Morphological and Physical Classification of Galaxies, 387–388.

in the sign of the a_4 term (boxy inwards, disky outwards), and the sin 4θ coefficien becoming negative. This description is very consistent with an intricate SB0 structur such as a bar + disk system crossing at a 40 degree angle the main outer structure (lens?). In the inner part ($r < 5"$), similarly, the data suggest a small-size disk oriente 50 degrees away from the bar.

Similar ways of reasoning applies to several other E galaxies having the largest inne isophotal twists, such as, e.g. NGC 636, NGC 1549. The strong inner isophotal twis of those galaxies can thus be considered as signatures of imbricated substructures rath than more complex geometries. In other words, in the same way as S0-like system SB0-like systems would exist among E's. They could be recognized from their sma boxy-type deviations, indicative of bars, and providing the orientation permits it, fro their large isophote twists.

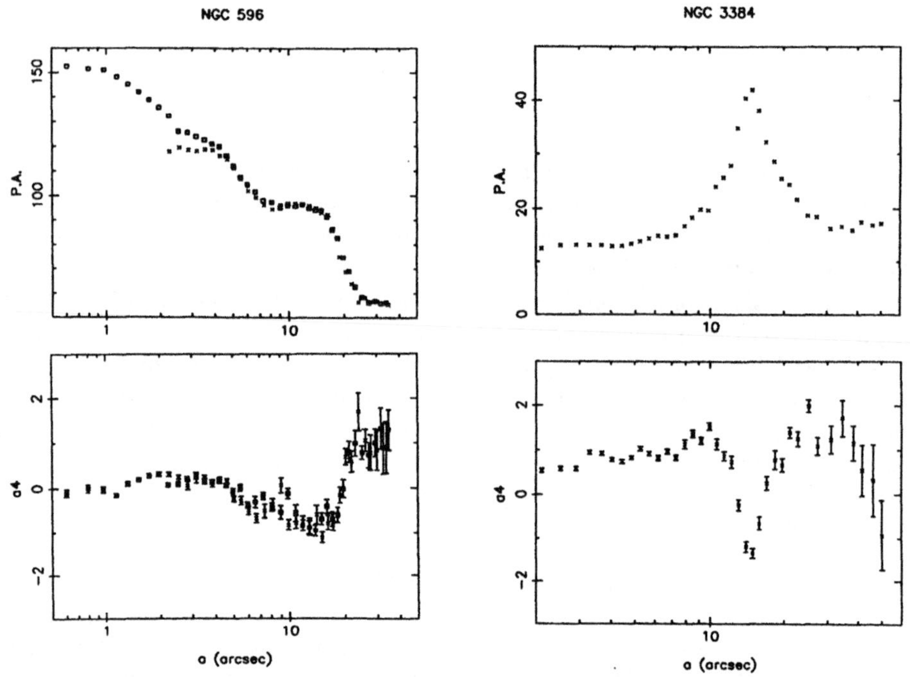

Figure 1. Profiles of position angle and a_4 coefficient for NGC 3384 et NGC 596.

References

Bender, R., Döbereiner, S., Möllenhoff, C.: 1988, *Astron. Astrophys. Suppl. Ser.* **74**, 385.

Bender, R., Surma, P., Döbereiner, S., Möllenhoff, C, Madejski, R.: 1989, *Astron. Astrophys.* **217**, 35.

Davoust, E., Lelièvre, G., Maury, A., Nieto, J.-L.: 1984, *Mon. Not. R. Astr. Soc.* **209**, 503.

Nieto, J.-L., Bender, R.: 1989, *Astron. Astrophys.* **215**, 266.

Williams, T.B.: 1981, *Astrophys. J.* **244**, 458.

HOTOMETRIC PROPERTIES OF DISKS IN ELLIPTICAL GALAXIES

C. Scorza[1], R. Bender[1]
[1]*Landessternwarte Königstuhl, Heidelberg, Germany*

Introduction

ointed isophotes in elliptical galaxies are thought to reflect the presence of a faint early edge-on disk imbedded in the spheroid of the galaxies. This hypothesis has been uggested by the analysis of surface brightness profiles (e.g. Capaccioli, 1987) and by the act that ellipticals with pointed isophotes show kinematical, radio and X-ray properties imilar to those of S0-galaxies (e.g. Bender, 1990). Especially, disky ellipticals generally ppear to be isotropic oblate rotators (Carter, 1987; Bender, 1988 and Nieto et al., 1988).

Here we report on the modeling of the isophotal shapes of three disky ellipticals ssuming perfectly elliptical spheroids and exponential thin disks.

The method

he isophotal shapes of the studied galaxies were parametrized by the coefficients of the ourier expansion of the isophotal deviations from perfect ellipses.

We started our disk modeling by assuming exponential profiles. In these models, he central surface brightness I_0, the radius scale length r_0, the inclination angle i, the osition angle P.A and the center coordinates of the disk are free parameters. The isk models were then subtracted iteratively from the galaxy images until the isophotes f the remaining spheroids appeared closest to ellipses and therefore showed near to ero Fourier-coefficient profiles. We found that the parameters of the disk models are vell constrained because different disk parameters influence in general different Fourier oefficients. For details see Scorza and Bender, 1990.

Results

n NGC 821 and NGC 3377 the superposition of a single exponential thin disk with an lliptical bulge can satisfactorily explain the isophotal shapes of these galaxies. In case of IGC 4697 it was necessary to subtract two disks models in order to obtain zero Fourier oefficients over the whole radius range. The a_4 and the ellipticity profiles of this galaxy how double peaks indicating the presence of these two disk components.

. Longo et al. (eds.), Morphological and Physical Classification of Galaxies, 389–390.

The figures below show the a_4 coefficient profiles of the studied galaxies togeth with their disk and bulge surface-brightness profiles. The table gives the estimated D/ ratios and the inclinations and scale parameters of the disks.

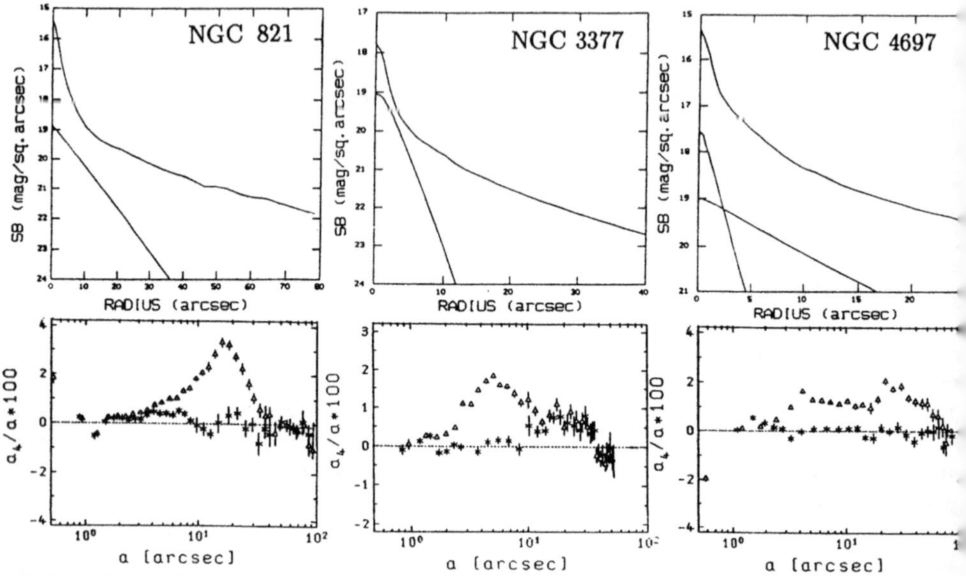

Objekt	M_t	SB(V)	r_0(pc)	D/B	i(grad)
NGC 821	10.8	22.1	1288	0.03	70°
NGC 3377	10.2	19.4	153	0.02	71°
NGC 4697	9.27	19.7	306	0.02	64°
NGC 4697*	9.27	21.9	2025	0.008	67°

* NGC 4697 outer disk parameters.

References

Bender, R.: 1988, *Astron. Astrophys. Letters* **193**, L7.

Bender, R.: 1990, *Heidelberg Conference on Dynamics and Interaction of Galaxies* , Ed. R. Wielen, Springer Verlag, p. 232.

Capaccioli, M.: 1987, in *Structure and Dynamics of Elliptical Galaxies*, IAU Symposium no. 127, Ed. T. de Zeeuw, D. Reidel, p. 47.

Carter, D.: 1987, *Astrophys. J* **312**, 514.

Nieto, J-L., Capaccioli, M., Held, E.V.: 1988, *Astron. Astrophys. Letters* **195**, L1.

Scorza, C., Bender, R: 1990, *Astron. Astrophys.* **235**, 49.

ELLIPTICAL GALAXIES WITH DARK MATTER

G. Bertin[1], R.P. Saglia[1,2,†], M. Stiavelli[3]
[1]*Scuola Normale Superiore, Pisa, Italy*
[2]*Dept. of Astronomy, Keble Road, Oxford, England*
[3]*European Southern Observatory, Germany*
[†] affiliated to the Astrophysics Division of the Space Science Department of ESA

One component stellar dynamical models can provide excellent zeroth order description of currently available data for elliptical galaxies (see, e.g., Bertin, Saglia, Stiavelli, 1988; van der Marel, Binney and Davies, 1990). However, some non–stellar dynamical indicators tend to support the view that dark matter is present in these systems. We now probe the presence and extent of massive dark halos around elliptical galaxies by fitting in detail the observed photometric and kinematical data with spherical, self–consistent models based on two f_∞ distribution functions (Bertin and Stiavelli, 1984). We concentrate our efforts on a sample of bright galaxies which do not show evidence of significant rotation, have small apparent flattening, and for which accurate data are available. A more extended description of the adopted theoretical framework is given in Bertin, Saglia, and Stiavelli (1990).

In order to assess the liability of our "dynamical diagnostics" we have performed several tests on simulated observations. They are described in detail in Saglia, Bertin, and Stiavelli (1990). The results of these tests have convinced us that the fitting procedure that we apply is indeed able to recognize the presence of dark matter in a given system and set limits on the amount present, provided that the system under study does not depart significantly from the physical picture that we have assumed, i.e. that it is characterized by a moderately radial velocity dispersion as observed in the end-products of dissipationless collapse (van Albada, 1982).

In table 1 we summarize our preliminary results. In the table, $(M/L_B)^1$ refers to one component fits (i.e. those performed with models without dark matter), all the other values apply to two-component fits. In figure 1 we show, as an example of the quality of the fits which we are able to obtain, the best–fitting model for NGC 4374. We wish to emphasize that measuring velocity dispersion profiles at larger distances from the center is crucial in order to establish more firmly the presence of dark matter in these regions.

G. Longo et al. (eds.), Morphological and Physical Classification of Galaxies, 391–392.

Table 1. Derived values for the best-fit models ($H_o = 50\ km\ s^{-1} Mpc^{-1}$)

Object	$\frac{r_L}{r_D}$	$\frac{M_L}{L_B}$	$(\frac{M_T}{L_B})_{R_e}$	$\frac{M_T}{L_B}$	$(\frac{M}{L_B})^1$
NGC 1399	0.16	8	12.9	58.1	6.6
	$\leq$ 0.21	6.4-8	9.7-13.2	50.1-58.1	
NGC 1404	0.98	4.6	10.9	12	8.8
		3.3-6.7	6.8-12.4	11-18.8	
NGC 1540	0.27	3.5	8.1	16.2	5.5
		3.5-4.9	7.5-9.2	7.7-30.7	
NGC 3379	0.49	5	8.5	11.7	6.7
		4.6-6	8-9.2	9.7-22.2	
NGC 4278	0.16	12.9	16.1	37.1	17.2
		3.1-14.1	16.1-25.8	18.5-39.3	
NGC 4374	0.96	7.5	18.4	19.8	8.6
		1.1-8.4	11.9-25.3	16.8-30.1	
NGC 4472	0.36	6.7	10.2	18.1	7.2
	$\leq$ 0.52	5.9-7.5	8.6-13	17.3-30.4	
NGC 4486	0.38	7.9	13.4	23.6	9.1
	$\leq$ 0.41	6.1-8.3	13.2-15.3	23.3-34.8	
NGC 4636	0.17	10.8	14.8	52.2	9.9
	$\leq$ 0.3	8.9-12.1	14.8-19.8	32-79.8	
NGC 7626	0.18	5.4	7.8	28	5.3
	$\leq$ 0.24	5.2-5.6	7.4-8.8	22.8-37.6	

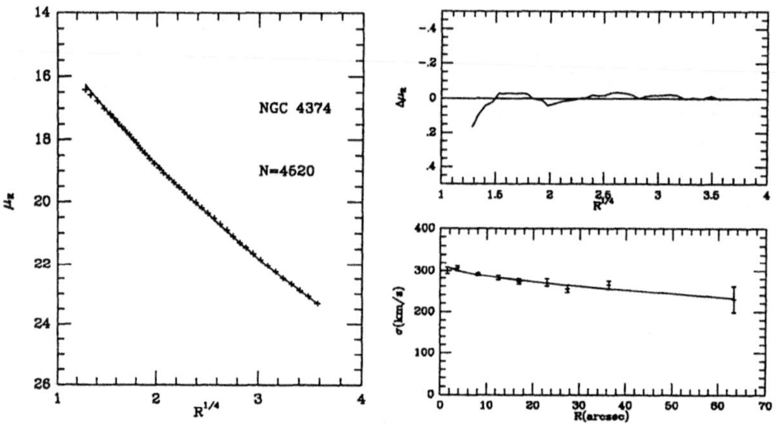

Figure 1.

References

Bertin, G., Stiavelli, M.: 1984, *Astron. Astrophys.* **137**, 26.

Bertin, G., Saglia, R.P., Stiavelli, M.: 1988, *Astrophys. J.* **330**, 78.

Bertin, G., Saglia, R.P., Stiavelli, M.: 1990, preprint.

Saglia, R.P., Bertin, G., Stiavelli, M.: 1990, preprint.

van Albada, T.S.: 1982, *Mon. Not. R. Astr. Soc.* **201**, 939.

van der Marel, R.P., Binney, J., Davies, R.L.: 1990, *Mon. Not. R. Astr. Soc.* **245**, 582.

NUCLEAR DUST-DISCS IN RADIO ELLIPTICALS

W.B. Sparks[1], F. Macchetto[1], D. Fraix-Burnet[1], R. Bender[2], M. Capaccioli[3], J.-L. Nieto[4]
[1]*Space Telescope Science Institute, Baltimore, USA*
[2]*Astronomical Obs., Heidelberg, Germany*
[3]*Dept. of Astronomy, Univ. of Padova, Italy*
[4]*Astronomical Obs., Toulouse, France*

1 Introduction

As part of the ESO Key Program "Towards a Physical Classification of Galaxies", Bender et al., NTT images using narrow-band and broad-band filters are being obtained. The exceptional seeing (0.7–1.0 arcsec) has resulted in the detection of several new dust-lanes, for the most part in galaxies known to be active. Where radio position angle information is available, in one case the radio jet is exactly orthogonal to the nuclear dust (NGC 5090) but in another, (NGC 3557) it lies at about 40 deg to the dust.

2 Observations and results

The primary sample targeted by the emission-line imaging program is those elliptical galaxies which have known X–ray properties. Exposures in the V band, shown here, are mostly 300 sec integrations using the low resolution RCA CCD with a pixel size of 0.26 arcsec per pixel. Models of the unobscured light distribution were constructed using ellipse fits to the data and making images whose isophotes are exactly the fitted ellipses. Division by the model gives effective optical depth. The images presented are selected from about 15 galaxies observed during March 1990.

NGC 5090 with NGC 5091 as the close companion, figure 1. Smith and Bicknell (1986) present radio observations showing a jet in position angle $\approx$ 42 deg, which is perpendicular to the dust-lane to within the uncertainty. The radial filament seen at some distance from the nucleus lies along the radio axis and will provide an interesting target for detailed observation at a later date.

NGC 3268 a cD at the centre of a modest cluster, found to contain a symmetric dust ring located close to the nucleus. The galaxy has a weakly active radio core, Disney and Wall (1977).

NGC 3557 a fairly well-studied galaxy in the optical, e.g. Carter, 1988. A radio map is presented in Ekers et al. (1989) which has an elongation in position angle $\approx$ 75 deg, nowhere near perpendicular to the dust lane! Higher resolution radio observations

G. Longo et al. (eds.), *Morphological and Physical Classification of Galaxies*, 393–394.

would be interesting since this contradicts one of the better correlations that has been established so far for active ellipticals.

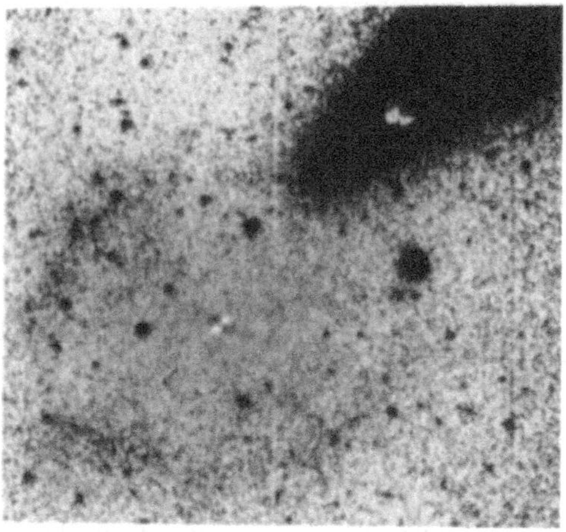

Figure 1. NGC 5090 - Direct CCD image (right) and processed to show the dust lane (below).

3 Discussion

The presence of well-ordered presumably high angular momentum gas discs in giant elliptical galaxies is of considerable interest in the context of galaxy classification. If stellar discs are to form, then such gas would provide a natural source of material. Questions which remain to be answered include, what is the origin of such gas? — is it the relic of a past accretion event, short lived and related to the nuclear activity? — is it a remnant of the galaxy formation process? Are such dust lanes responsible for the far infra-red radiation now found commonly in elliptical galaxies? And is this dust related to the ionized component of the ISM?

References

Carter, D.: 1988, *Astrophys. J.* **312**, 514.

Disney, M.J., Wall, J.V.: 1977, *Mon. Not. R. Astr. Soc.* **179**, 235.

Ekers, R. et al.: 1989, *Mon. Not. R. Astr. Soc.* **236**, 737.

Smith, R.M., Bicknell, G.: 1986, *Astrophys. J.* **308**, 36.

XTENDED Hα FILAMENTS IN ELLIPTICAL GALAXIES

D. Fraix-Burnet[1], F. Macchetto[1], W. Sparks
Space Telescope Science Institute, Baltimore, USA
[1]Affiliated to the Astrophysics Division, Space Science Department, European Space Agency

Ve have undertaken a statistical observational study of elliptical galaxies with narrow and imaging, in order to investigate possible correlations between the presence of exnded Hα filaments and the activity of elliptical galaxies (X-ray, infrared, radio, dust,...).

Sixty-two galaxies have been observed so far and here we present preliminary results btained using the ESO 3.6 meter telescope at La Silla (Chile) in October 1989. (See the oster "Nuclear dust-discs in radio elliptical galaxies" by Sparks et al., for dust lanes ound in galaxies observed during a later ESO NTT run in April 1990.)

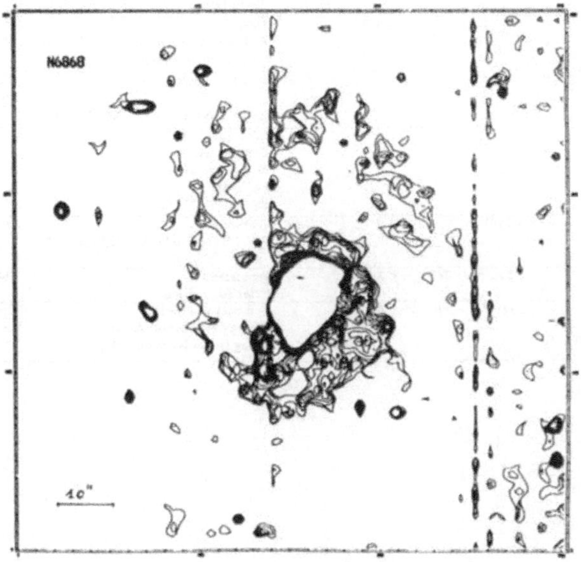

Figure 1

. Longo et al. (eds.), Morphological and Physical Classification of Galaxies, 395–396.

We report in the table some properties of the galaxies studied and a concise descrip[-]tion of the structure of the Hα filaments when present.

It is very striking from even these early results that a large proportion (8 out of 1[…] of these elliptical galaxies show extended ionized filaments. Of particular interest is th[e] relatively common occurrence of very extended one-armed spiral (!) structure, possibl[y] indicative of infalling gas. We also find examples of arc-like features whose origin ma[y] be related to shock waves of cooling gas. See for example NGC 1533, NGC 6868 an[d] NGC 584.

Table 1

Name	M_B M/L	L_{5GHz} (erg/s/Hz)	L_{IR}	L_X (erg/s)	σ (km/s)	HI ($M_\odot$)	EW(Hα) (Å)	Comments (Literature, spectro,...)	H_α (this work)
N 584	-21.06 7-10	< 27.22	12.8 (mJy)	$1.5\,10^{40}$	226	3.89	< 5	Disk component No dust, no spiral S0, no H_α, no NII	Filaments near the nucleus Arc
N 720	-20.98 6-12	$< 1.3\,10^{27}$	< 5.5 (mJy)	$1.1\,10^{41}$	228	< 2.88	–	Perfect ellipses Broad red nucleus No feature, no dust E5, no H_α, no NII	Nothing detected
N 1172	-18.79 7.5			$< 10^{40}$	98	< 2.96	–	Disk? Bar? S-shaped dust lane $M_{dust} \simeq 5\,10^4\ M_\odot$	S-shaped filament crossing the nucleus
N 1316	-22.44	$5.5\,10^{31}$	$1.3\,10^{43}$ erg/s	$1.6\,10^{42}$	239		2.5	Seyfert-like Irregular dust lanes H_α/NII(6583)<1.2 FWHM(H_α)=7.5Å	Lot of filaments Arc
N 1399	-20.83	$2\,10^{29}$		$1.3\,10^{42}$			–	E1, featureless No dust No H_α, no NII	Nothing detected
N 1407	-21.29 14.5	$4.9\,10^{28}$	–	$1.3\,10^{41}$	275	< 1.98	–	E0, featureless No H_α, no NII No dust EW(OII)=1.47 Å	Nothing detected
N 1453	-21.39	$5.8\,10^{21}$				< 5.60		Dust? Red elongated nucleus OII, OIII, NII detected	Elongated structure at the center
N 1533	-17.79	$< 1.5\,10^{28}$	$9.8\,10^{41}$	$5.8\,10^{39}$		7.1		$H_\alpha=10^{38}$ erg/s EW(NII6584)=1.15 Å; NII/H_α=0.99	Blobs Spiral arm
N 6868	-21.59	$1.8\,10^{22}$					5.1	E3, Seyfert-like Compact red nucleus M(dust)=0.4$M_\odot$ FWHM(H_α)=10.3 Å H_α/NII(6583)<1.2	Spiral arm
N 6876	-21.68	$< 9.1\,10^{28}$		$3.5\,10^{41}$				Featureless EW(NII6584)-0.85 Å; NII/H_α=1.18	Small filaments
N7562	-21.25 13.9	$< 5.8\,10^{27}$		$1.3\,10^{41}$	277	< 3.99		Featureless Extended red nucleus	Nothing detected
N 7626	-21.31 22.5	$1.3\,10^{30}$		$2.0\,10^{41}$	292	< 3.2		Compact red nucleus Strong dust absorption or strong isophotal asymmetry	Filaments near the center ?

THE ORIGIN OF THE IONIZED GAS IN S0 GALAXIES

F. Bertola[1], L.M. Buson[2], W.W. Zeilinger[3]
[1]*Dipartimento di Astronomia, Padova, Italy*
[2]*Osservatorio Astronomico, Padova, Italy*
[3]*European Southern Observatory, Garching, Germany*

Several pieces of evidence indicate that the ionized gas present in elliptical galaxies is in the majority of the cases of external origin. In particular, the counter-rotation of the gas with respect to the stars constitutes the most direct evidence of the acquisition phenomenon (see Caldwell, Kirshner & Richstone 1986 *Astrophys. J.*, **305**, 136; Bertola & Bettoni 1988 *Astrophys. J.*, **329**, 102; Bertola, Buson & Zeilinger 1988 *Nature*, **335**, 705). There is *a priori* no reason why such accretion events should be limited only to elliptical galaxies. While in the case of spirals the pre–existing gas may hide the process, S0 galaxies seem to be the best candidates, a part from ellipticals, for detecting these events. Indeed, the acquisition has certainly taken place in those S0s which possess polar rings and in a few galaxies where the gas is found to be in counter–rotation with respect to the stars (Galletta 1987 *Astrophys. J.*, **318**, 531; Dettmar 1989 *Proc. 2nd Wyoming Conf. "The interstellar medium in external galaxies"*; Rubin 1988 in *"Large Scale Motions in the Universe"* p. 541). On the basis of the assumption that the gas in S0 galaxies is of external origin, we expect to observe counter–rotation phenomena in 50% of the objects, where ionized gas has been detected.

In order to investigate if this assumption holds true we have analyzed from the kinematical point of view a sample of 8 S0 galaxies with emission lines selected from the Revised Shapley–Ames Catalog (Sandage & Tammann 1987). The galaxies were observed with the Red Channel spectrograph mounted at the *Multiple Mirror Telescope*. The wavelength region covered was $\lambda\lambda$ 3700 ÷ 4400 Å including the [O II] doublet (λ 3727-29 Å), the Ca II H & K lines (λ 3933 Å, λ 3968 Å) and the G–band (λ 4300 Å). After standard calibration of the CCD frames, the absorption lines were analyzed with a Fourier Quotient technique (Bertola *et al.* 1984. *Astron, J.*, **89**, 356) yielding radial velocities and velocity dispersion. The [O II] doublet was measured using a Gaussian fitting algorithm.

The preliminary analysis yields that 4 galaxies possess a gas component which is kinematically decoupled from the global galaxy kinematics. In particular, the gas com-

G. Longo et al. (eds.), Morphological and Physical Classification of Galaxies, 397–398.

ponent of NGC 2768 is found to be counter–rotating with respect to the stars. I NGC 3414 the gas in the outer parts of the galaxy is in counter–rotation with the inne gas core. NGC 4036 and NGC 4111 on the contrary possess counter–rotating gas core These results are fully consistent with the hypothesis that the ionized gas in S0 galaxie is accreted from outside. Using surface photometry published in the literature in orde to derive the density profile of the galaxies, we corrected the velocity data for the asym metric drift. We noted that the stellar velocities remain significantly higher than the ga velocities. This fact may indicate that the ionized gas component is not in equilibrium

LINE-STRENGTH GRADIENTS IN S0 GALAXIES

A. Paquet, R. Bender, W. Seifert
Landessternwarte Königstuhl, Heidelberg, Germany

1 Introduction

S0 galaxies are key objects for the understanding of galaxy formation not only because they form the link between elliptical and spiral galaxies in the Hubble sequence but, also, because they allow to study the properties of disks unobscured by dust and gas.

Here, we present measurements of Mg_2 and Fe line-strength gradients in 10 edge-on S0 galaxies. Both line-strengths are related to metallicity but are affected differently by stellar parameters (Faber, 1973; Mould, 1978).

Observations and Reductions: The data were obtained at the 2.2 m telescope of the German-Spanish Astronomical Center, Calar Alto with a B&C spectrograph and a CCD camera (Seifert, 1990). The line strengths given here were derived using the definitions by Burstein et al. (1984). The equivalent widths of $Fe_{\lambda_{5270}}$ and $Fe_{\lambda_{5335}}$ were averaged to give a mean <Fe> index. All measured line strengths were corrected to zero velocity dispersion.

2 Results

The S0's presented here have disk-to-bulge ratios between 0.2 and 1 and in about half of our objects the disks dominate the light beyond a certain radius along the major axes. However, the line-strength gradients are always smooth which indicates similar metallicity for disk and bulge at a given radius.

Figure 1 shows the Mg_2 gradients against the absolute magnitudes for our S0 sample (open squares); the filled circles show mean gradients in elliptical galaxies as given by Peletier (1989) with each point representing the average of 5 ellipticals. Between $M_T = -19.0$ mag and $M_T = -20.5$ mag ($H_\circ = 50$ km/s/Mpc) the gradients observed in S0's and E's are equal within the scatter. This result supports the notion that elliptical galaxies and S0 galaxies have similar properties in this magnitude range (especially, ellipticals are also rotationally flattened at these luminosities; Davies et al., 1983; Bender and Nieto, 1990). However, although this result is not very significant, there seem to exist different trends in the Mg_2-gradients vs. luminosity relation for S0's and E's when considering a larger luminosity range. If further measurements confirm this difference, it is a strong

G. Longo et al. (eds.), Morphological and Physical Classification of Galaxies, 399–400.

indication that luminous S0's have a star formation history significantly different fro the one of luminous ellipticals.

Figure 2 presents gradients in <Fe> vs. the Mg_2 gradients; no correlation betwee the two is apparent. The fact that in some objects, despite of a strong Mg_2 gradient, th <Fe> gradient is very small or even absent indicates that the slope of the IMF may var with distance from the center. However, *on average* the gradients relative to the centr line-strength values are similar ($\frac{1}{<Fe>}\frac{d<Fe>}{d\log r} \approx \frac{1}{Mg_2}\frac{dMg_2}{d\log r}$).

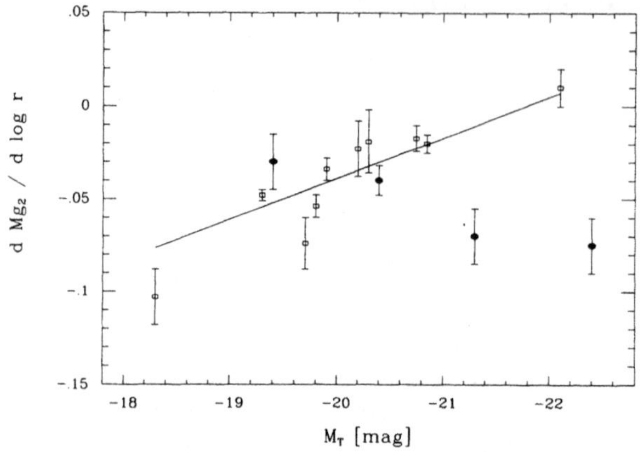

Figure 1

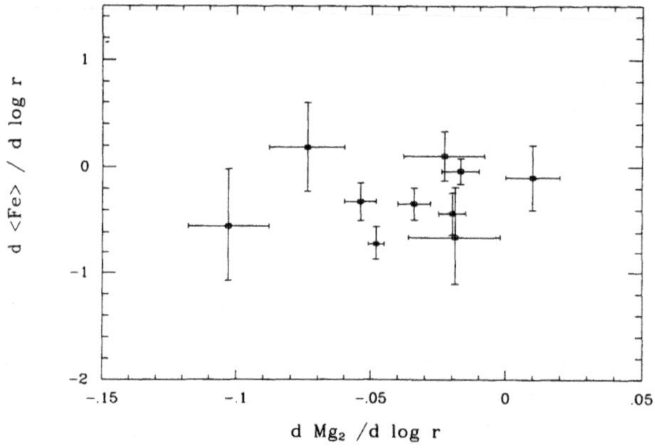

Figure 2

References

Bender, R., Nieto, J.-L.: 1990, *Astron. Astrophys.* in press.

Burstein, D., Faber, S.M., Gaskell, C.M., Krumm, N.: 1984, *Astrophys. J.* **287**, 586.

Davies, R.L., Efstathiou, G., Fall, S.M., Illingworth, G., Schechter, P.L.: 1983, *Astrophys. J.* **266**, 41.

Faber, S.M.: 1973, *Astrophys. J.* **179**, 731.

Mould, J.R.: 1978, *Astrophys. J.* **220**, 434.

Peletier, R.F.: 1989, Ph.D. Thesis, University of Groningen.

Seifert, W.: 1990, Ph.D. Thesis, University of Heidelberg.

'LTRAVIOLET SPECTRA OF EARLY-TYPE GALAXIES

A. Danks[1], M. Pérez[1,2]
[1]*ST System Corporation, Lanhan, USA*
[2]*IUE Observatory, NASA-GSFC, Greenbelt, USA*

Introduction

'he Hubble sequence which is determined from optical images is accompanied by a clear rogression in the manifestation of the ISM and star formation rates. At one extreme liptical galaxies present visible spectra predominantly characteristic of late type stars, nd at the other, spirals show young blue stars with evidence of dust, neutral and ionized as. The spectra of *normal* ellipticals in the visible are remarkably uniform (Gregg 1989) nd differ only in details, but the discovery of significant UV flux in these seemingly ormal galaxies is one of the outstanding contributions of space astronomy (UV rockets, 'opernicus, TD1, ANS and most recently IUE).

The UV flux may arise from a variety of sources, localized starburst episodes, white warf binaries, blue giants on the AGB or massive stars on the main sequence. However, ecause of the brief ($<$ 15 Myr) lifetimes of the latter stars, the UV is a *direct tracer f recent star formation*. In many examples, E and S0 galaxies show evidence of ISM ırough the presence of weak Hα emission (Sadler and Gerhard, 1985) or dust lanes. owever, only integrated stellar flux is observed and the stellar population can only be educed or inferred through spectral synthesis models. Within our galaxy distinct stellar opulations correlate with age, metallicity and position although there is considerable vidence from the visible that this is not quite so simple (Burstein, 1985). The star rmation rate is tied inextricably to the IMF which, in itself, may be complex. Theory as evolved away from the simple Salpeter (1955) IMF and Larson's (1986) bimodal odel, however, the UV IMF from Lequeux et al. (1981) may be adequate. It is clear ıat as the observations become more detailed it will allow more sophisticated modeling pproaching reality.

Observations and results

Ve have concentrated on the gas rich S0s and spirals from the Southern hemisphere ırvey from Reif et al. (1982). In figure 1 we present new and archival IUE merged SWP + LWP; range 1250–3100 Å) unreddened fluxes for NGC 1326, 5102, 3115, 2768,

. Longo et al. (eds.), Morphological and Physical Classification of Galaxies, 401–402.

4762, 4111 and 4382. The galaxies NGC 1326, 5102 (gas-rich, metal-poor) and 311 (gas-poor, metal-rich) present strong UV fluxes shortward of 2000 Å indicative of recen burst phenomena.

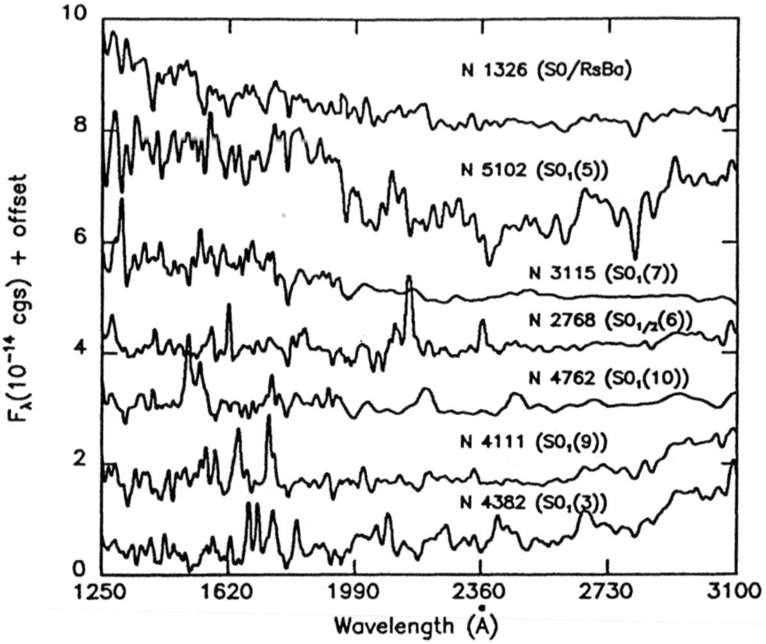

Figure 1. UV unreddened fluxes for a sample of S) galaxies.

Similarly, the somewhat metal-rich galaxies NGC 2768, 4762, 4111, 4382 and also 510 show rising fluxes around 3000 Å representative of a late population. The gradual se quence of increasing fluxes toward the far-UV indicated that there is no correlatio between morphological types, metallicities (as determined by the Mg_2 index, values fron Longo et al., 1989) and their UV slopes or colors (e.g., UV (1550 Å)-V). These U observations have confirmed the individualistic nature of the SO galaxies. We conclud that there is strong new evidence that a large number of SOs appear to simultaneousl support old and late type populations in their bulges.

References

Burstein, D.: 1985, *Astron. Soc. Pac.* **97**, 89.

Gregg, M.D.: 1989, *Astrophys. J. Suppl. Ser.* **69**, 217.

Larson, R.B.: 1986, *Mon. Not. R. Astr. Soc.* **218**, 409.

Lequeux, J., et al.: 1981, *Astron. Astrophys.* **103**, 305.

Longo, G., Capaccioli, M., Bender, R., Busarello, G.: 1989, *Astron. Astrophys. Letters* **225**, L17.

Reif et al.: 1982, *Astrophys. J. Suppl. Ser.* **50**, 451.

Sadler, E.M., Gerhard, O.E.: 1985, *Mon. Not. R. Astr. Soc.* **214**, 177.

Salpeter, E.E.: 1955, *Astrophys. J.* **121**, 161.

ON THE NATURE OF THE COMPACT GALAXIES IN THE UGC

E. Davoust[1], S. Considère[2], Ph. Prugniel[3], N. Denis[1]
[1]*Observatoire Midi-Pyrénées, Toulouse, France*
[2]*Observatoire de Besançon, Besançon, France*
[3]*European Southern Observatory, Garching, Germany*

The very few red compact galaxies that are known have been found near more massive galaxies: M 32 near M 31, NGC 4486B near M 87

- Do these low-luminosity ellipticals owe their very high surface brightness to gravitational interaction, or are they the genuine low-mass end of the family of ellipticals?
- Is it possible to find galaxies that would fill the gap between these compacts and normal ellipticals?

To answer these questions, we have searched for red compact ellipticals in the *Uppsala General Catalogue* (Nilson, 1973). There are 231 objects labelled "compact" in this catalogue.

A detailed analysis of the sample using data in the literature reveals a mixed bag of objects: 21 Markarian galaxies (hereafter Mrk); 26 IRAS galaxies (10 of them also Mrk); 4 Seyferts (545, 3087, 9412, 11763); 2 BL Lac objects (6132, 10599, both also Mrk); 4 Spirals (407, 7778, 12074, 12607); 3 stars (11608, 11668, 12389) and 2 suspected stars (7731, 11761). Furthermore, 42 galaxies have faint outer structure (envelope, halo, ring, or disk), according to the notes of the UGC. Such a structure is incompatible with spatial compactness. Some of the compacts have an asymmetric envelope, distortions, plumes, a jet, a disrupted halo, or streamers. Finally, many of them are not isolated.

We have obtained spectra of 60 compacts at Observatoire de Haute-Provence, and derived 49 new radial velocities. Radial velocities have also been compiled from Bottinelli et al. (1990) and PGC (Paturel et al, 1989). Radial velocities are thus available for 152 galaxies (67% of the sample). 28 galaxies have emission-line spectra (4 Seyferts, 17 Mrk, 16 very or extremely compact) and 26 galaxies have been detected in neutral hydrogen.

We have also obtained CCD images of a subsample of 40 objects with the 2-meter telescope of Pic du Midi. We tried to select only red objects, by comparing the B and R

G. Longo et al. (eds.), Morphological and Physical Classification of Galaxies, 403–404.

diameters given in the UGC. Detailed surface photometry has revealed that most of th galaxies of the subsample are in fact just normal early-type galaxies.

A mean-brightness vs absolute magnitude diagram for images in V (Fig. 1) show that only a few galaxies, such as NGC 2636 and UGC 11546, can be considered a moderately compact. The latter is similar to NGC 5666, the prototype of a class c compact, high surface brightness ellipticals (Lake, Schommer, van Gorkom, 1987). NG 2636 falls just next to NGC 3641, the faintest "compact" companion of the sample c hierarchical pairs (Prugniel et al., 1989). The two dwarf ellipticals below NGC 2636 ar UGC 3654 and UGC 4508.

This study thus shows that red compact galaxies intermediate between NGC 4486 and low-mass ellipticals are rare, and that the "compacts" of the UGC do not belong t this class.

References

Bottinelli, L., Gouguenheim, L., Fouqué, P., Paturel, G.: 1990, Astron. Astrophys. Suppl. Ser. **82**, 39

Lake, G., Schommer, R.A., van Gorkom, J.H.: 1987, Astrophys. J. **314**, 57.

Nilson, P.: 1973, *Uppsala General Catalogue of Galaxies*, Uppsala Astron. Obs. Ann., **6**.

Paturel, G., Fouqué, P., Bottinelli, L., Gouguenheim, L.: 1989, *Catalogue of Principal Galaxies*, Monc graphies de la base de données extragalactiques, **1**.

Prugniel, Ph., Davoust, E., Nieto, J.-L.: 1989, Astron. Astrophys. **222**, 5.

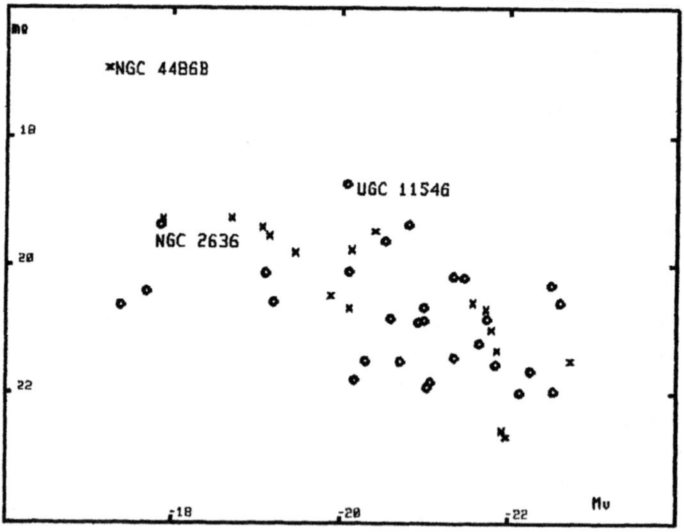

Figure 1. The absolute V magnitudes of the observed sample are plotted versus their mean surface brightness at the effective isophote (open circles), together with those of the sample of hierarchical pairs (Prugniel et al., 1989; crosses) and one previously known red compact (NGC 4486B) which delineate the projection of the Fundamental Plane.

MORPHOLOGY OF GALAXIES SURROUNDING COMPACT GROUPS

G.C. Baiesi-Pillastrini[1], V. Tornatore[2], G.G.C. Palumbo[1,3], P. Hickson [4], C. Mendes de Oliveira [4]
[1]*Department of Astronomy, University of Bologna, Italy*
[2]*Istituto di Radioastronomia, CNR, Bologna, Italy*
[3]*Istituto Te.S.R.E., CNR, Bologna, Italy*
[4]*Department of Geophysics and Astronomy, U.B.C., Canada*

1 Introduction

Compact groups are small, relatively isolated systems of galaxies with projected separations comparable to the diameters of the galaxies themselves. The most recent, and largest, survey of compact groups is that of Hickson (1982) who visually inspected all the red POSS prints, covering about 67% of the sky. These groups have apparent space densities spanning a range of over four orders of magnitude, including the highest density systems known. Extensive observational work has been done on the sample (see Hickson, 1990 for a review). Optical spectra and redshifts are now available for all groups, as well as broad band CCD images and plates. Radio studies include both HI and continuum observations. Infrared and X-Ray studies have also been carried out with the available data. Sulentic (1987) and Rood and Williams (1989) have performed detailed studies on the neighborhoods of these compact groups. Sulentic examined all galaxies within one degree of all HCGs and concluded that the average surface density of galaxies increases slightly within half a degree of a group, where it is comparable to that of the lowest density loose groups. He found that typical HCGs have density enhancements of a factor of 100 or more, and about a third have density enhancements of over 500. Rood and Williams examined galaxies within 10 groups radii of the HCGs and found no significant excess of galaxies in the neighborhood of two thirds of them. They estimated that on the average there is one physical neighbor for every 3.3 group members. With redshifts available for 95% of the galaxies of the HCGs it is now possible to study the environment and its consequences on group structure using physical radii to count galaxies.

2 The data

Kindl (unpublished) examined all galaxies within 1.125 h^{-1} Mpc. of 97 HCGs with known redshift and estimated their magnitude and Hubble types. This is the largest

G. Longo et al. (eds.), *Morphological and Physical Classification of Galaxies*, 405–406.

environment survey so far available. A study of morphological frequency of galaxies each group neighborhood has been performed using a circle of 1 Mpc around the cent of each HCG. The number of galaxies found around each group was plotted vers occurrence. We divided the HCGs into two classes: those with rich environment (hig count fields) i.e. when more than 30 galaxies per 1 Mpc circle were counted (29 group 2102 galaxies) and isolated groups when less then 30 galaxies were counted (66 group 1310 galaxies). In order to investigate the morphological behavior near and away fro the group and compare our results with previous ones we have divided the 1 Mpc circ into two regions. One comprising galaxies found within a 0.5 Mpc circle, the other wit the remaining galaxies contained in the region between 0.5 and 1 Mpc.

3 The results

The most interesting results obtained from the data are: the relatively poor environme near the 66 HCG has less ellipticals (24%) than the equivalent 29 crowded ones (35% This proportion is respected away from the groups, at distances greater then 0.5 Mp 22% and 36% respectively. No appreciable change occurs in lenticulars either close t the groups or far from them. This holds true for both crowded and isolated group The fraction of lenticulars remains, in fact, approximately 15%. Spirals, on the oth hand, are more numerous (63%) far away from isolated groups than for crowded on (52%) while they remain basically constant (52-53%) in their close neighborhood. Th increase in spirals, as one moves away from the group, brings their number close to th 63-75% estimated by Hickson and Sulentic for the field. Spirals in the neighborhood ar however, more than in the groups themselves (41-44%). This result is at variance wit what Sulentic found i.e. that spirals are less in the environment than in the groups. Fro the present results it appears that, in general, we agree with the findings of Rood an Williams (1989) i.e. the fraction of spiral galaxies tend to be smaller and the fraction c E galaxies larger in the HCG than in their neighborhood. Comparisons, however shoul be taken with caution since the Rood and Williams results are based on a total of 21 galaxies and our sample contains 3412 galaxies furthermore both their analysis as we as Sulentic results stem from fixed areas of inspection which only occasionally coincid with ours and the galaxies they consider are the brightest end of the distribution of th galaxies considered in the present study.

References

Hickson, P.: 1983, *Astrophys. J.* **255**, 382.

Hickson, P.: 1990, Review in IAU Colloquium 124.

Rood, H.J., Williams, B.A.: 1989, *Astrophys. J.* **339**, 772.

Sulentic, J.W.: 1987, *Astrophys. J.* **322**, 605.

IEAN SURFACE DENSITY AND MORPHOLOGICAL TYPE IN DISK :ALAXIES

M. Prieto[1], A. Campos Aguilar[2], C. García[3]
[1]*Instituto de Astrofisica de Canarias, Tenerife, Spain*
[2]*Instituto de Astrofisicas de Andalucia, Granada, Spain*
[3]*Observatoire de Marseille, France*

ı Prieto et al. (1989), we have represented the rotational velocity, V, at a radius orresponding to 25 mag/arcsec^2 against this radius, R, for a sample of 56 spiral galaxies. Ve found that galaxies with different morphological types appear in different regions on ıis diagram, with some overlap between Sb and Sc. In the V-R diagram, we have plotted ontours of total mass, angular momentum and "mean surface density" for a simple model f an isolated galaxy, considered as a disc with circular motion, in centrifugal equilibrium, ıowing that, up to a critical mass, "the mean surface density" or its observational arameter V^2/R can represent the morphological type, independent of the mass. We ound too that, for a constant mass, galaxies with different angular momentum are of a ifferent type, such that those with smaller angular momentum are later types.

All these semi-empirical relations found among the global parameters of spiral galax-s are the consequence of the dynamical behaviour of their discs. Obviously, we would ot expect to find similar relations for elliptical galaxies, but we would, for lenticulars nd irregulars. In this contribution, we analyze the behaviour of these galaxies compared o that of spirals.

In figure 1, we represent the linear radius, R, against the rotational velocity V, ıeasured at that radius for a sample of galaxies. For galaxies where V has not been ıeasured as far as R_{25}, it is extrapolated. The spiral sample is that studied by Rubin t al.(1980, 1982 and 1985) in a series of papers relevant to the rotational properties f spirals. The sample of lenticulars and irregulars is chosen from those with rotation urves in the catalogue of Baiesi-Pillastrini et al. (1983). The R_{25} radii for all galaxies re from RCA (de Vaucouleurs et al., 1976). We have used H_o=50 km s^{-1} Mpc^{-1} for the istance scale. As the size of irregulars are almost certainly underestimated, in figure (c) we have adopted a radius of $2R_{25}$ for these objects.

From the figures we can conclude that the lenticular galaxies appear on the V-: diagram, figure 1 (a), superimposed on the spirals, which probably indicates their isc have a similar dynamical behaviour to the spirals. This suggests to us a possible ubclassification of the lenticulars according to their "mean surface density" or V^2/R, erhaps, for example, related to some morphological characteristic such as the size of he bulge.

. *Longo et al. (eds.), Morphological and Physical Classification of Galaxies*, 407–408.

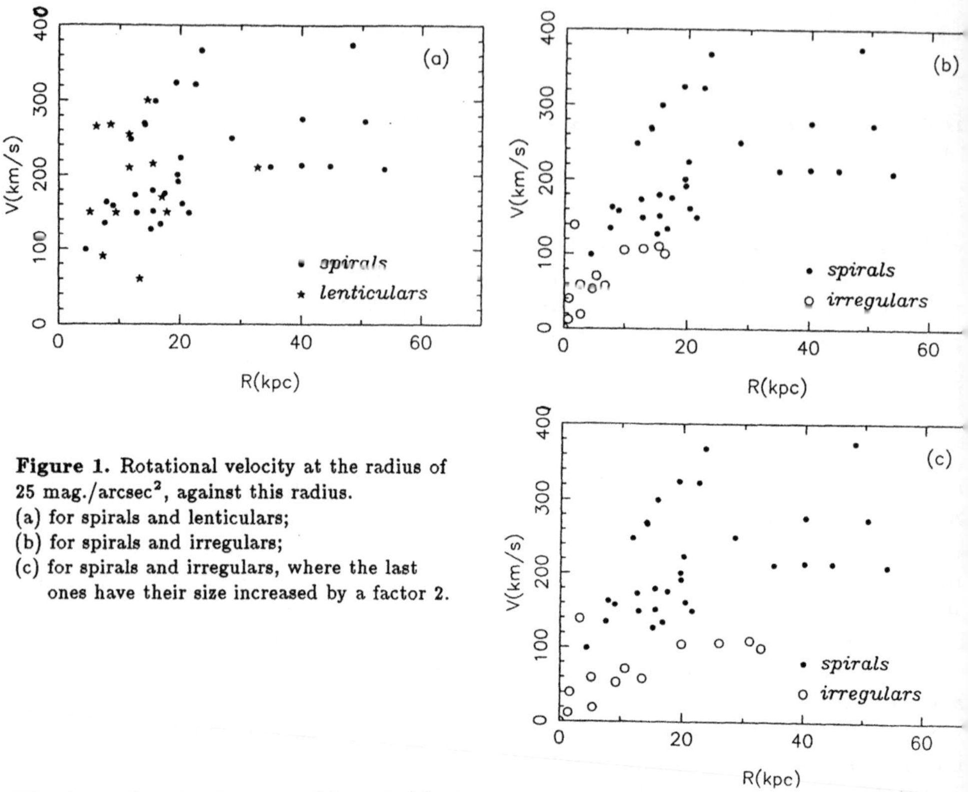

Figure 1. Rotational velocity at the radius of 25 mag./arcsec^2, against this radius.
(a) for spirals and lenticulars;
(b) for spirals and irregulars;
(c) for spirals and irregulars, where the last ones have their size increased by a factor 2.

The irregulars in figure 1 (b) and (c), don't appear superimposed on the spirals, but occupy the regions of the lowest rotational velocity for a constant radius. This suggest that these galaxies form a sequence, together with the spirals (Sa, Sb, Sc and I). If we look at the position of these galaxies on diagrams of figures 2a and 3 by Prieto et al. (1989), we can conclude that these galaxies would be those which, for a constant mass have the lowest angular momentum, the largest radius, the lowest rotational velocity and the smallest "mean surface density". They would be the least evolved galaxies.

Baiesi-Pillastrini, G. C., Palumbo, G. G. C., Vettolani, G.: 1983, *Astron. Astrophys. Suppl. Ser.* **53**, 373.

Prieto, M., Campos Aguilar, A., Garcia, C.: 1989, *Astrophys. Space Sci.* **157**, 321.

Rubin, V. C., Burstein, D., Ford. W. K. Jr., Thonnard, N.: 1980, *Astrophys. J.* **238**, 471.

Rubin, V. C., Burstein, D., Ford. W. K. Jr., Thonnard, N.: 1982, *Astrophys. J.* **261**, 439.

Rubin, V. C., Burstein, D., Ford. W. K. Jr., Thonnard, N.: 1985, *Astrophys. J.* **289**, 81.

Vaucouleurs, G. de, Vaucouleurs, A. de, Corwin, H.: 1976 *Second Reference Catalog of Bright Galaxies* (RC2), University of Texas Press, Austin.

PECULIAR KINEMATICS IN INTERACTING ELLIPTICAL GALAXIES

R. Madejsky
Landessternwarte Königstuhl, Heidelberg, Germany

1 Observations

We have undertaken a morphological investigation of elliptical galaxy pairs using the 1.23 m telescope at the German-Spanish Astronomical Center on Calar Alto, Spain, and the Danish 1.54 m telescope at ESO, La Silla, Chile. Typical projected separations of both components of a galaxy pair are one or two half-light radii, or $20 \cdots 50$ arcsec, corresponding to $10 \cdots 20$ kpc. According to their morphology, several obviously interacting galaxy pairs were selected for spectroscopy to determine their kinematics. All spectroscopic observations presented here were carried out with the B&C-spectrograph at the ESO 3.6 m telescope on La Silla, Chile. The Fourier correlation quotient method described in Bender (1990) was used to determine the kinematical properties.

2 Results and discussion

The presence of tidal tails and major halo distortions in AM 2244-651 indicate efficient tidal interaction (Fig. 1). Both galaxies are separated by 30 arcsec (9 kpc in projection). From the east component a large tidal tail extends to the south. The whole length of the tidal tail amounts to more than 20 kpc and gives an estimate to the time elapsed since closest approach. A minimum time $t = 2 \times 10^8$ yr is required to form the tail if the stars escape with a velocity of 100 km s^{-1}. In the west component, halo distortions are not easily detected. They are only seen at much fainter surface brightness levels and they appear much more diffuse. The morphology resembles much that of the numerical simulation of Barnes (1990, see his figure 1). The reason for the different morphologies presumably is the different kinematical behaviour of both galaxies (Fig. 2). While the east component rotates rapidly, the west component rotates slowly. Stars in the rotating east component having accidentally a velocity vector parallel to the orbit of the other galaxy respond violently to the time-varying potential (the 'perturbing' galaxy in this case is on a prograde orbit) and form the tidal tail. Stars with a velocity vector antiparallel to the perturbing galaxies' orbit (retrograde) experience only small tidal effects. The mean radial velocities of both galaxies differ by less than 50 km s^{-1}. As

G. Longo et al. (eds.), Morphological and Physical Classification of Galaxies, 409–410.

a consequence of the low relative velocity, resonance effects due to the high intern rotation of the east galaxy play an important role.

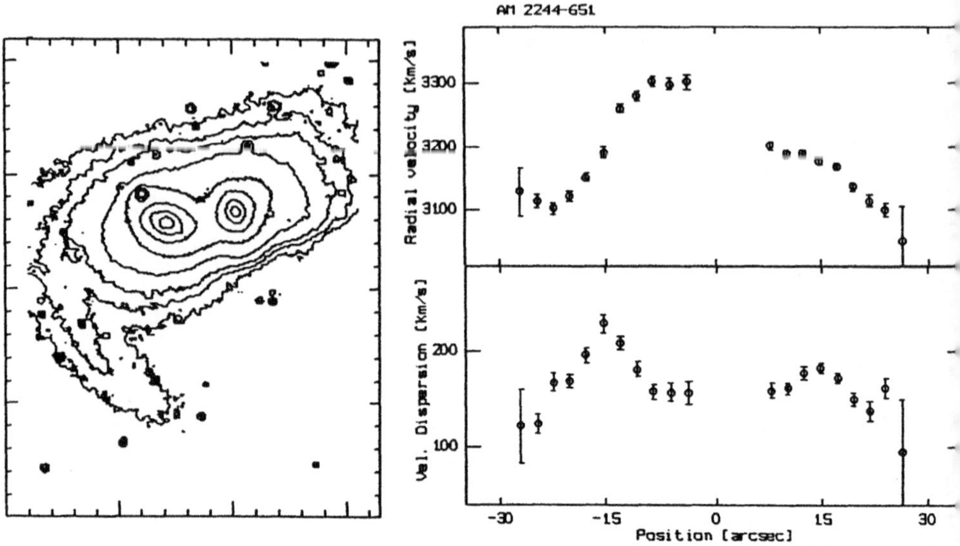

Figure 1. (left) Shows the peculiar morphology of AM 2244-651. The kinematical data of AM 2244-651 along the line connecting both galaxy centers are shown in **Figure 2** (right).

Similar relations between morphological and kinematical properties were also found fo other galaxy pairs not presented here (Madejsky, Bender and Möllenhoff, 1990; Madejsk and Bender, 1990). Unfortunately, there are many parameters which are not known priori for interacting galaxies and the different parameters may influence the ongoin interaction in a similar way, i.e. they often cannot be separated observationally. Im portant parameters are the relative velocities of the galaxies and the distance at closes approach. When these parameters provide the conditions for efficient interaction, th internal kinematics determine the strength of tidal interaction. As was shown with th aid of numerical simulations, the encounter of rotating galaxies on prograde orbits ca rapidly lead to merging (Barnes, 1990).

The author is grateful to R. Bender for providing his spectroscopy software.

References

Barnes, J.: 1990, *Dynamics and interactions of galaxies*, ed. R.Wielen, Springer-Verlag Heidelberg.

Bender, R.: 1990, *Astron. Astrophys.* **229**, 441.

Madejsky, R., Bender, R.: 1990, *Paired and interacting galaxies*, IAU n. **124**, in press.

Madejsky, R., Bender, R., Möllenhoff, C.: 1990, *Astron. Astrophys.*, in press.

MORPHOLOGY AND KINEMATICS OF SPIRAL GALAXIES WITHOUT BAR: THE IMPORTANCE OF THE CENTRAL MASS

G.C. Baiesi-Pillastrini
Department of Astronomy
University of Bologna, Italy

Using the central velocity gradient as a representative parameter of the central masses of spiral galaxies, a preliminary comparison has been made with the most important properties defining the Hubble classification scheme (Baiesi-Pillastrini, 1987,1988). Significant correlations with the bulge to disk ratio and the pitch angle of spiral galaxies have been found for 39 and 30 spirals, respectively. These preliminary results are an additional evidence that the light distribution reflects properties like central mass and density and the arm shapes strongly depend on the central mass concentration as expected from the density wave theory. Although it is desirable to confirm these results as soon as a larger sample of data would become available it appears that the central velocity gradient can be regarded as a classification parameter accounting for morphological and kinematical properties of spiral galaxies. As accurate photometric decomposition of the light distribution has become available for many galaxies, it is now possible to preliminary test the relation between the central velocity gradient and absolute integrated magnitude of the central part (bulge) of a spiral galaxy. Besides, a possible effect due to galaxy environment is investigated.

For 25 unbarred spiral galaxies the derived parameters used for the present statistical analysis are listed in table 1. Ho=100 km/s^{-1} Mpc^{-1} has been used for the distances and the central velocity gradient G is given in logarithmic scale from Baiesi-Pillastrini(1988). To convert the integrated magnitudes to the B–band, the relation (B–r)= 1.2 has been used for Kent's data and (B–V)= 0.8 for Kodaira's data. This sample represents the best match among galaxies with accurate and homogenous photometric decompositions.

The correlation between central velocity gradient vs. absolute magnitude of the bulge is apparent in figure 1. The relation is $\log G = -0.14\ \mathrm{M_{bulge}} - 0.22$ with a variance $\sigma(y/x) = \pm 0.2$ and the correlation coefficient r= 0.72. The scatter is probably due to large uncertainties in photometric decompositions: about 1.4 mag. In order to test the dependence of the sample on the environment, the above relationships has been studied for cluster or group galaxies and field galaxies. The correlation coefficient increases to r=0.77 for the first class of objects and decreases to r=0.46 for the second one. That is, the sample of cluster or group galaxies improves the relationships to the detriment of field galaxies.

As expected, a clear relationships between representative parameters of luminosity and mass of the central part of spiral galaxies is found. From figure 1 is also evident a clear correlation of both parameters with the Hubble type. This confirms the validity of

G. Longo et al. (eds.), *Morphological and Physical Classification of Galaxies*, 411–412.

the Hubble classification scheme in isolating different integral properties of spiral galaxi and provides further evidence that the visible matter reflects the mass distribution i their central regions. The different behavior of cluster or group galaxies with respec to the field ones, seems to confirm the recent results of Rubin et al. (1988). That i the dynamics and stellar content could depend on the density of the environment or b different for cluster and non cluster galaxies.

Table 1

NAME (1)	TYPE (2)	C (3)	DIST (4)	LOG G (5)	MBULGE (6)	REF (7)
I0342	6	1	2.3	1.57	-14.87	1
N0253	5	1	2.6	1.80	-17.16	1
N0300	7	1	1.0	1.59	-11.00	1
N0488	3	1	23.8	2.53	-19.14	1
N0598	6	1	0.5	1.53	-13.46	1
N1087	5	1	15.7	2.10	-14.78	2
N2683	3	2	3.1	2.48	-15.50	2
N2775	2	2	9.6	2.40	-18.37	1
N2782	1	2	25.5	2.52	-18.23	2
N2997	5	2	8.0	2.21	-17.13	1
N3031	2	1	1.0	2.41	-16.87	1
N3521	4	2	6.4	2.35	-17.19	2
N3893	5	1	10.4	2.00	-16.39	3
N3898	1	1	11.3	2.66	-18.44	1
N4321	4	1	15.1	2.38	-18.85	1
N4378	1	1	24.2	2.72	-19.48	3
N4698	1	1	9.2	2.41	-17.69	1
N4736	2	1	2.5	2.76	-17.61	1
N5005	4	1	10.5	2.22	-18.72	1
N5055	4	2	5.8	2.65	-17.80	1
N5248	4	2	11.1	2.30	-16.59	2
N5457	6	1	3.9	1.51	-15.89	1
N6946	6	2	3.5	1.64	-15.40	1
N7217	2	2	6.2	2.43	-17.70	1
N7331	4	2	11.0	2.31	-19.64	1

Col.(3) :
1-Cluster or group galaxies
2-Field galaxies
Col.(7) Photometric decomposition sources :
1-Simien, de Vaucouleur,1986,Ap.J.,302,564
2-Kent, 1985,Ap.J.Suppl., 59, 115
3-Kodaira et al.,1986, Ap.J.Suppl., 62,703

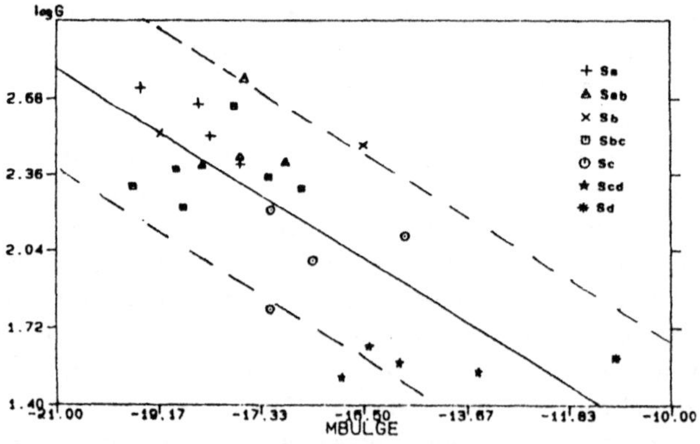

Figure 1. Correlation between central velocity gradient vs. absolute magnitude of the bulge. Regression line and the $\pm 2S(y/x)$ (dotted) lines are drawn. The sample is plotted with different symbols characterizing the morphological type.

Baiesi-Pillastrini, G.: 1987, *Astron. Astrophys.* **172**, 375.

Baiesi-Pillastrini, G.: 1988, *Astrophys. Lett. and Comm.* **27**, 27.

Rubin, V.C.,Whitmore, B.C.,Ford, W.K.jr.: 1988, *Astrophys. J.* **333**, 522.

HE CORRELATION BETWEEN VELOCITY GRADIENTS AND ARM 'LASSIFICATION IN SPIRAL GALAXIES

A. Biviano[1], M. Girardi[2], G. Giuricin[1,3], F. Mardirossian[1,3], M. Mezzetti[1,3]
[1]*Department of Astronomy, University of Trieste, Italy*
[2]*Scuola Internazionale Superiore di Studi Avanzati, Trieste, Italy*
[3]*C.I.R.A.C., Trieste, Italy*

. Introduction

Vhile looking for the fundamental parameters of spiral galaxies, Whitmore (1984, hereter W84), discovered a correlation between the arm structures and the gradients of the otation curve: the Outer Gradient (*OG*) was shown to increase with increasing DDO lass (van den Bergh 1960), for a sample of $\sim$ 30 galaxies. This result was in contradicon with the previous one by Kormendy and Norman (1979), but consistent with the rediction of the ***density wave theory*** (Lin and Shu 1964) that the higher is the fraction f the total mass in the disk (i.e. the lower the gradient of the rotation curve), the better efined and developed are the spiral arms (see, e.g., Toomre 1981, Carlberg and Freedian 1985). In this work, we have re-addressed this issue using a much larger sample f 94 galaxies, and well defined parameters for the gradients of the rotation curves and he arm morphology. Our results support W84, and hence, the density wave theory. A ecent work by Elmegreen and Elmegreen (1990; hereafter EE90), yielded similar results s ours, although for a much smaller sample of galaxies (41).

; Data-sample and results

1 order to investigate the correlation between arm class and velocity gradients, we ave chosen the arm class parameter (*AC*), as defined by Elmegreen and Elmegreen 1982; 1987), and two gradient parameters defined in the optical region where the arms evelop, i.e. the logarithmic gradient, *LG*, as defined by Persic and Salucci (1986; 988) and the outer gradient, *OG*, as defined by Rubin *et al.* (1982; see also W84). Iost of the references to the rotation curves can be found in Persic and Salucci (1988; 990), Whitmore, Forbes and Rubin (1988), W84, and Baiesi-Pillastrini, Palumbo, and 'ettolani (1983). The *AC*'s have been taken from Elmegreen and Elmegreen (1987).

We find a significant correlation between *AC* and *LG* at a 99.79 % significance level, hich appears even stronger when *OG* is used in lieu of *LG*: grand design galaxies have

'. Longo et al. (eds.), Morphological and Physical Classification of Galaxies, 413–414.

lower gradients than flocculent ones. This result is confirmed by the Rank Sum Te applied to the cumulative distributions of the LG (or OG) parameter for the two sub samples of galaxies with $AC \leq 4$ and $AC \geq 5$: the probability that the two subsampl have the same parent distribution is < 2 %. We use the Kendall's *partial* correlatic coefficient and the Multiple Regression Analysis, to show that the correlation betwee AC and the gradient is not induced by correlations of these two parameters with th morphological type, or the absolute blue magnitude, i.e. the correlation found is nc spurious. Finally, the correlation between AC and LG (or OG) appears to be higher i the subsample of barred galaxies, but not significantly.

Our result is consistent with those obtained by W84 and EE90, although we hav used different parameters for describing the arm structures and the gradients of the rota tion curves. The correlation between arm classes and velocity gradients can be explaine in the framework of the density wave theory, and suggests that the arm structure i not a transient feature, although it can be re-generated by several recurrent episodes c triggering and amplification of wave modes, as long as the disk retains its mass.

References

Baiesi-Pillastrini, G.C., Palumbo, G.G.C., Vettolani, G.: 1983, *Astron. Astrophys. Suppl. Ser.* **53**, 373.

Carlberg, R.G., Freedman, W.L.: 1985, *Astrophys. J.* **298**, 486.

Elmegreen, D.M., Elmegreen, B.G.: 1982, *Mon. Not. R. Astr. Soc.* **201**, 1021.

Elmegreen, D.M., Elmegreen, B.G.: 1987, *Astrophys. J.* **314**, 3.

Elmegreen, D.M., Elmegreen, B.G.: 1990, *IBM Research Division - Research Report* (**EE90**).

Kormendy, J., Norman, C.A.: 1979, *Astrophys. J.* **233**, 539.

Lin, C.C., Shu, F.H.: 1964, *Astrophys. J.* **140**, 646.

Persic, M., Salucci, P.: 1986, *Mon. Not. R. Astr. Soc.* **223**, 303.

Persic, M., Salucci, P.: 1988, *Mon. Not. R. Astr. Soc.* **234**, 131.

Persic, M., Salucci, P.: 1990, *Astrophys. J.* **355**, 44.

Rubin, V.C., Ford, W.K., Jr., Thonnard, N., Burstein, D.: 1982, *Astrophys. J.* **261**, 439.

Toomre, A.: 1981, in *The Structure and Evolution of Normal Galaxies,* eds. S.M. Fall, D. Lynden-Bell Cambridge Univ. Press, Cambridge, p. 111.

van den Bergh, S.: 1960, *Astrophys. J.* **131**, 215.

Whitmore, B.C.: 1984, *Astrophys. J.* **278**, 61 (**W84**).

Whitmore, B.C., Forbes, D.A., Rubin, V.C.: 1988, *Astrophys. J.* **333**, 542.

)PTICAL LINE/RADIO LUMINOSITY CORRELATION AND THE ΞFFECTS OF THE INTERSTELLAR MEDIUM

R. Morganti
Istituto di Radioastronomia, Bologna, Italy

1 Introduction

Starting with the qualitative work of Hine and Longair (1979), many results obtained from optical studies on early-type galaxies suggest a link between the emission at radio frequencies and the presence and the characteristics of the highly ionized gas in the host galaxy (Baum et al., 1988; Fosbury, 1989).

The great majority (if not all) of these optical studies of radio galaxies deal with samples of high power radio sources. However, when looking for correlations between the radio and optical properties, it is important to have complete sample of objects covering the largest possible range of radio luminosity. With this in mind, an investigation of the properties of the optical line-emitting gas in a sample of low luminosity radio galaxies (Bologna B2 sample) has been started. This paper presents some of the results obtained using both the B2 sample and the sample of high power radio galaxies (mainly 3C) observed by Baum et al. (1988). Combined with the recently published IR data (Impey et al., 1990), these data shed new light on how the interstellar medium (both "warm" gas and "cold" gas/dust) may affect the radio morphologies.

2 The new observations

The sample chosen for this study is the complete Bologna B2 sample, which consists of elliptical galaxies from the Bologna catalog of radio sources with $m_B < 15.4$ (Colla et al., 1975). These galaxies have been extensively studied at radio frequencies with the VLA (Fanti et al., 1987), and they were recently optically observed with the the 1m JKT (La Palma) using both broad band filters (R,B) and narrow band filters centered on the redshifted Hα line and on the nearby continuum.

Using these data, a comparison between the characteristics of the B2 objects and those of a sample of high power radio galaxies studied by Baum et al. (1988) was done. In order to avoid bias due to differences in the selection of the samples, the latter was restricted to the objects with z<0.2. In addition, a B2-selected sample was constructed

G. Longo et al. (eds.), Morphological and Physical Classification of Galaxies, 415–418.

excluding the compact objects (i.e. < 10 kpc), to be consistent with the criterion used b Baum et al. (1988) for selecting their sample. The objects excluded from the B2-selecte sample are circled in figure 1 and dashed in the histogram of figure 2.

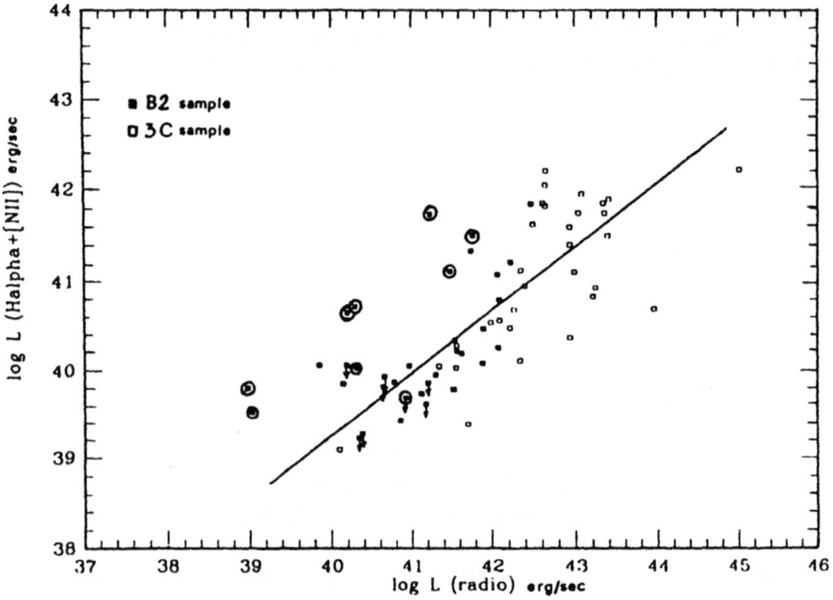

Figure 1.

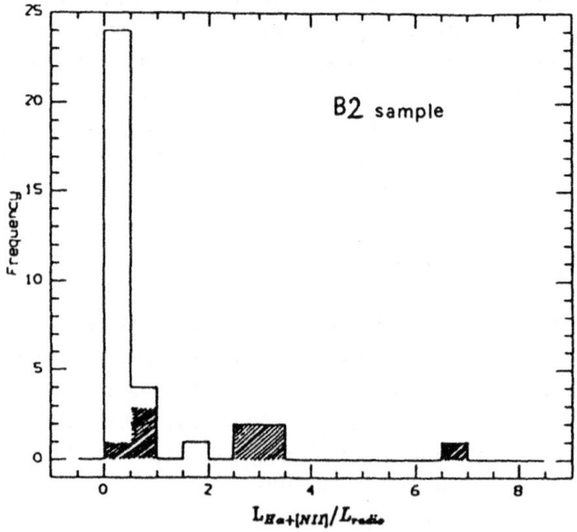

$L_{H\alpha+[NII]}/L_{radio}$

Figure 2.

ome of the preliminary results are summarized here:

- The existence of a correlation between the Hα+[NII] luminosity and the total radio luminosity (Baum et al., 1988; Saunders et al., 1989) is confirmed and extended, with the new data, over about six orders of magnitude in radio luminosity (this is for the sample that excludes the compact sources).
  Figure 1 shows the data from the two samples and the least-squares fit obtained using the Baum et al. (1988) data and the B2-selected sample. The relation found is

  $\log L_{H\alpha} = 0.73\pm0.09 \log L_{radio} + 10.0$

  in good agreement with that obtained by Baum et al. (1988) considering only the powerful radio sources.
  It is interesting to note that no correlation was found between the Hα+N[II] luminosity and the absolute blue magnitude.

- The majority of the compact objects excluded from the B2-selected sample appears to have an "Hα excess" for a given total radio luminosity. This is clear from the histogram in figure 2, which shows the distribution of the ratio between the Hα+[NII] and the radio luminosities: $L_{H\alpha+[NII]}/L_{radio} \sim 2.5$ for the compact objects (dashed in figure 2), whereas ≪1 for the other, more extended, radio sources.
  In addition, it is intriguing that all compact objects showing the Hα excess also have some other characteristics in common:

  1) they are all detected in the far- IR (by IRAS, Impey et al., 1990) and present a steep spectral index between 25 and 60 mμ that, according to de Grijp et al. (1985), classified these objects as "cold", i.e. starbust-like objects unlikely to contain an active nucleus;

  2) they also show peculiarities in their optical appearance (tidal tails, dust lanes etc..).

3 Discussion

There is a large body of observational evidence to support the idea that emission line gas in *powerful* radio galaxies is photoionized EUV/X-ray radiation from active nuclei (Robinson et al., 1987 and ref. therein).

If this is the case, the correlation between the radio and the optical line (Hα+[NII]) luminosities indicates that the radio plasma and the ionizing source are likely to both originate in central engine. Thus, from the fact that this relationship extends over a wide range of radio power, we would expect on the basis of continuity that photoionization is the main mechanism for producing the optical line emitting regions in low luminosity radio galaxies as well as in high luminosity ones.

Another possibility is that the observed correlation is in fact a secondary correlation induced by a third parameter, e.g. the characteristics of the ISM, or ionizing sources in addition to the nucleus. This is the most likely explanation for the different properties of the "compact" radio sources: the Hα excess observed in such objects might indicate that they have interstellar medium with different properties. If, for example, these objects have a somehow "richer" ISM, then this may intercept more light from the nucleus, an hence they will have larger emission line luminosities at given radio power. Alternativel there may be a higher rate of star formation associated with a richer ISM, which produc more ionizing photons. The presence of a richer ISM could also relate to the compa nature of radio emission, since the radio sources may be confined by the gas.

4 Conclusions

From the results obtained from narrow band images in the light of Hα+[NII] for a samp of low luminosity radio galaxies (Bologna B2 sample) it was found that an emissio line/radio luminosities correlation exists in good agreement with what observed for th powerful radio galaxies. This correlation can be due to a common origin of both th phenomena and, as already claimed for the powerful radio sources, the *main* source ionization is likely to be UV radiation from the nucleus. Some of the B2 objects sho an excess in the Hα luminosity for a given radio luminosity ($L_{H\alpha+[NII]} > L_{radio}$). Thi characteristic has been found to be always associated with compact radio sources (<1 kpc), which have steep IR spectral indices and peculiar optical morphologies. Such radi sources may be the key to understanding how other parameters (i.e. the amount of gas c the presence of an extra source of radiation) may influence the link between the nuclea activity and the optical line-emitting/radio luminosities.

To determine the relative importance of the various physical mechanisms that con tribute to the correlation between optical and radio properties, a detailed programme c spectroscopy of the low luminosity radio sources is required.

Acknowledgments

I would like to thank Marie-Helene Ulrich, Roberto Fanti, Clive Tadhunter and Tor Oosterloo for very useful discussions and suggestions.

References

Baum, S., Heckman, T., Bridle, A., van Breugel, W., Miley G.: 1988, *Astrophys. J. Suppl. Ser.* **68**, 643.

Colla, G., Fanti, R., Fanti, C., Gioia, I., Lari, C., Lequeux, J., Lucas, R., Ulrich, M.-H.: 1975, *Astron Astrophys. Suppl. Ser.* **20**, 1.

de Grijp, M.H.K., Miley, G.K., Lub, J., de Jong, T.: 1985, *Nature* **314**, 240.

Fanti, C., Fanti, R., de Ruiter, H.R., Parma, P.: 1987, *Astron. Astrophys. Suppl. Ser.* **69**, 57.

Fosbury, R.A.E.: 1989, in *Extranuclear Activity in Galaxies* (ESO Conference no. 32), p. 169.

Hine, R.G., Longair, M.S.: 1979, *Mon. Not. R. Astr. Soc.* **188**, 111.

Impey, C.D., Wynn-Williams, C.G., Becklin, E.E.: 1990, *Astrophys. J.* **356**, 62.

Robinson, A., Binnette L., Fosbury, R.A.E., Tadhunter, C.N.: 1987, *Mon. Not. R. Astr. Soc.* **227**, 97.

Saunders R., Baldwin, J.E., Rawlings, S., Warner, P.J., Miller, L.: 1989, *Mon. Not. R. Astr. Soc.* **238** 777.

'HE LUMINOSITY-DIAMETER RELATION FOR DISK GALAXIES: A EARCH FOR ENVIRONMENTAL EFFECTS

M. Girardi[1], A. Biviano[2], G. Giuricin[2,3], F. Mardirossian[2,3], M. Mezzetti[2,3]
[1]*Scuola Internazionale Superiore di Studi Avanzati, Trieste, Italy*
[2]*Dipartimento di Astronomia, Università di Trieste, Italy*
[3]*C.I.R.A.C., Trieste, Italy*

Introduction

ı the last few years, considerable effort has been devoted to outlining the possible ɔrrelations of the galaxy properties with local density. The tightness of the relationship etween luminosity and the absolute diameters of galaxies (hereafter referred to as L- relation) suggests that possible environmental effects should be detectable from the nalysis of its shape in different samples. However, Aguilar and White (1986) predicted ıat the tidal truncation should have a hardly detectable effect on the galaxy isophotal iameter. Peterson et al. (1979) found the Virgo disk galaxies appeared to have smaller iameters than the Hercules disk galaxies. Giuricin et al. (1985) found L-D relations ith shallower slopes in samples of spirals located in higher compactness groups. Giuricin t al. (1988) found differences in the L-D relation for disk galaxies in some clusters; but ıey did not find a dependence of the L-D relation on any cluster properties.

These results induced us to reconsider the environmental dependence of the L-D elation for disk galaxies. We collected data samples from the literature. We looked ɔr two basic parameters, the isophotal diameter D_{25} and the total blue magnitude B_T, oth corrected. These quantities were transformed to absolute quantities M_B and Δ_{25} ia the knowledge of the galaxy distances. Our data samples contain disk galaxies in he clusters by Aaronson et al. (1986), our neighbouring cluster Virgo (Binggeli et al., 985), two other nearby clusters, Fornax and Hydra (Lauberts and Valentijn, 1989); two eld samples (Davis and Seaquist, 1983 and Bottinelli et al., 1988). Furthermore, we onsidered the "Nearby Galaxy Catalog" (Tully, 1988; hereafter referred as "NBG").

Results of comparative analysis

Ve investigated the L-D relation in its logarithmic form M_B vs. $\log \Delta_{25}$. To each ata sample we fitted the straight regression line, obtained via the least squares method

. *Longo et al. (eds.), Morphological and Physical Classification of Galaxies,* 419–420.

applied to data affected by errors on both axes. The comparison of different lines wa given a statistical meaning via the use of two tests: the *Welch* test, that applies whe only two lines are involved, and the *Homogeneity* (or *Variance-ratio*) test, which applie in all other cases (see, e.g., Guest, 1961). These test yielded the probabilities that th differences in the intercepts and slopes of the compared fitting lines are significant.

First, we compared Aaronson's clusters. Then we compared Virgo different shell that is three subsamples of galaxies located at different projected distances from th cluster center; the same was performed to Hydra and Fornax clusters. We compared als all cluster and field samples together. We performed other three comparisons amon samples by NBG: clusters, field and groups and clusters together, subsamples of differen average galaxy densities. No significant differences have been found among all the severa relations, although they were chosen so as to span a large range of galaxy densities.

Our results showed that any possible environmental effect is not strong enough t affect significantly the L-D relation for disk galaxies. As regard the previous finding concerning the L-D relation for disk galaxies, we are led to say that any previous foun difference in the L-D relations can be ascribed to the poor sample, and/or statistic differences in the sample selection criteria, and/or to an incorrect choice of the statistica tool used in the comparison analyses.

Our result can be restated by saying that the environment has a negligible effect o the L-D relations, either because these are physically unaffected by the local density, o because the scatter in the relations is too large and the accuracy in the data is too low to allow us to detect any significant deviation from sample to sample. Lastly, we ma remark that our conclusion on disk galaxies is in line with the recent finding by Giurici et al. (1989) on elliptical galaxies, and the theoretical work by Aguilar and White (1986)

References

Aaronson, M., Bothun, G., Mould, J.R., Huchra, J., Schommer, R.A., and Cornell, M.E.: 1986, *Astrophys. J.* **302**, 536.

Aguilar, L.A., and White, S.D.M.: 1986, *Astrophys. J.* **307**, 97.

Binggeli, B., Sandage, A., and Tammann, G.A.: 1985, *Astron. J.* **90**, 1681.

Bottinelli, L., Gouguenheim, L., and Teerikorpi, P.: 1988, *Astron. Astrophys.* **196**, 17.

Davis, L.E., and Seaquist, E.R.: 1983, *Astrophys. J.* **53**, 269.

Giuricin, G., Mardirossian, F., and Mezzetti, M.: 1985, *Astron. Astrophys. Suppl. Ser.* **62**, 157.

Giuricin, G., Mardirossian, F., and Mezzetti, M.: 1988, *Astron. Astrophys. Suppl. Ser.* **72**, 151.

Giuricin, G., Mardirossian, M., Mezzetti, M., and Pisani, A.: 1989, *Astrophys. J.* **345**, 101.

Guest, P.G.: 1961, *Numerical Methods of Curve Fitting*, (Cambridge: Cambridge Univ. Press).

Lauberts, A., and Valentjin, E.: 1989, *The Surface Photometry Catalogue of the ESO-Uppsala Galaxies* (European Southern Obs.).

Peterson, B.M., Strom, S.E., and Strom, K.M.: 1979, *Astron. J.* **84**, 735.

Tully, R.B.: 1988, *Nearby Galaxy Catalog*, (Cambridge: Cambridge Univ. Press) – "NBG".

GLOBAL PROPERTIES IN EARLY–TYPE GALAXIES

A. Rifatto
Osservatorio Astronomico di Capodimonte, Naples, Italy

A subset of early–type galaxies from the catalogue of Knapp et al. (1989) was selected in order to derive some global properties. The selection criteria adopted were: a) detected fluxes at least in the 60 and 100 μm IRAS bands; b) distance known from the *Nearby Galaxies Catalog* by Tully (1988). The final sample was formed by 192 galaxies classified as E and S0. Dust temperature was derived assuming pure black–body radiation and that the dust emissivity is proportional to the frequency.

The infrared luminosity from 1 to 500 μm was computed by using both the 60 and 100 μm flux density, as described in the appendix to the catalogue of the IRAS survey (1985).

The total amount of warm dust in each galaxy was estimated on basis of the 60 and 100 μm IRAS flux densities and assuming a single dust temperature component.

The distribution of dust temperature, the distribution for the mass of dust, the distribution of B and infrared luminosity, the distribution of the ratio L_{IR}/L_B and the distribution of the ratio L_{IR}/M_{dust} were analysed for the E's, for the S0/SB0's and for both E's and S0/SB0's. The result is that there are not substantial differences in the histograms for E and S0/SB0 galaxies, so we can adopt the mean values of the whole sample.

Possible correlations among the previous derived quantities were also analysed. We expect a good correlation between the infrared luminosity and the mass of the dust, since both quantities are derived from the 100 μm flux density (Fig. 1a). On the other side the correlation between the mass of the dust and the total blue luminosity (Fig. 1b) is very weak. The scatter in both plot is a consequence of the dependence of the mass of the dust from the dust temperature.

Figure 1c shows the comparison between infrared and blue luminosity for the galaxies in the sample: again, the correlation (if any) is very weak.

The last two results do not agree with the conclusions of Young et al. (1989) who used a sample including also spiral galaxies.

G. Longo et al. (eds.), Morphological and Physical Classification of Galaxies, 421–422.

References

Lonsdale, C.J., Helou, G., Good, J.C., Rice, W.: 1985, *Catalogued Galaxies and Quasars Observed in th IRAS Survey*, Jet Propulsion Laboratory.

Knapp, G.R., Guhathakurta, P., Kim, D.W., Jura, M.: 1989, *Astrophys. J. Suppl. Ser.* **70**, 329.

Tully, B.: 1988, *Nearby Galaxy Catalog*, Cambridge University Press.

Young, J.S., Xie, S., Kenney, J.D.P., Rice, W.L.: 1989, *Astrophys. J. Suppl. Ser.* **70**, 699.

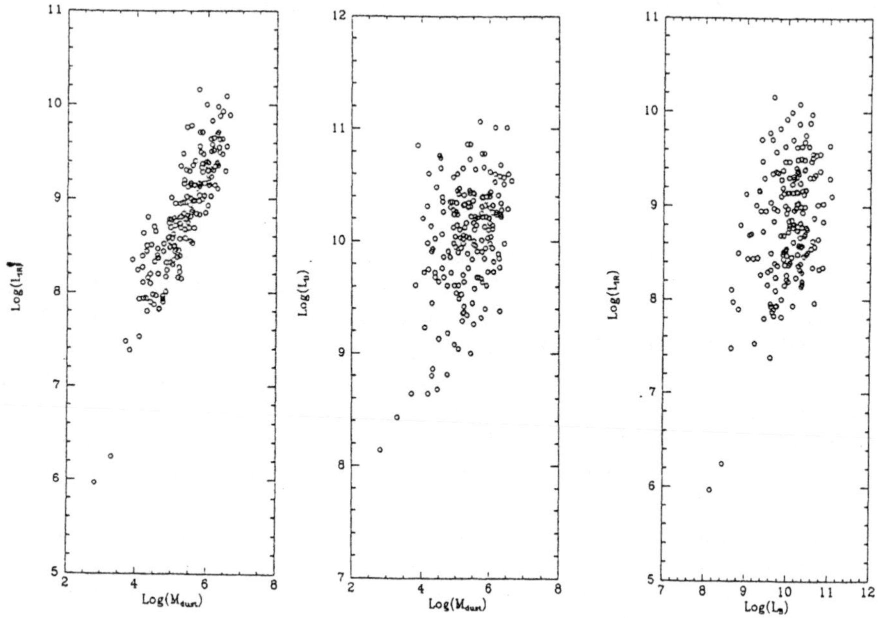

Figure 1.

MASSES AND MASS TO LIGHT RATIOS OF ELLIPTICAL GALAXIES

G. Busarello, G. Longo
Osservatorio Astronomico di Capodimonte, Naples, Italy

The knowledge of masses and mass-to-light ratios of galaxies is a clue to the understanding of their internal structure and dynamics, their formation and evolution processes, the composition of their stellar and gaseous contents, and the amount of dark matter. Many works have been devoted to the computation of the masses of galaxies, and, in particular, of ellipticals. The methods used are mainly based on applications of the virial theorem (e.g. Faber and Gallegher, 1979), thus implying the knowledge of some dynamical quantities (usually the central velocity dispersion) and some length scales (effective radii and core radii). In most cases only central (core) values of these quantities are used, as, for example, in King's method (King and Minkowsky, 1972; Richstone and Tremaine, 1986). Alternative methods use X ray emission from haloes, thus giving in principle mass to light ratios at large distances from the the centers (e.g. Trinchieri et al., 1986). Bacon et al. (1985) used a simple mass model (the same as that used in this work) and the virial theorem to derive expressions for the masses from i) σ_0 and V_{max} and/or ii) σ_0 only. They obtained the masses inside the effective radii for a large sample of galaxies, analyzing also their correlations with absolute magnitudes. Finally Bertin et al. (1987) used a set of self consistent anisotropic quasi spherical models to fit photometrical and kinematical data of some real galaxies, obtaining values of M and M/L directly from the fits. We computed masses and mass to light ratios of a sample of ellipticals by using a method based on the equilibrium (Jeans) equation for spheroidal systems and on the available kinematical data.

We start from the hydrodynamical equilibrium equation for a spheroidal system of eccentricity e :

$$\frac{1}{\rho(r)}\frac{d}{dr}\left[\rho(r)\sigma^2(r)\right] - \frac{V^2(r)}{r} = \frac{-4\pi G(1-e^2)^{\frac{1}{2}}}{r}\int_0^r \frac{da\ a^2\ \rho(a)}{(r^2-a^2e^2)^{\frac{1}{2}}} \qquad (1)$$

Equation (1) is an integro differential equation for r, which could be in principle be solved from the known functions $V(r)$ and $\sigma(r)$. To solve it, we made some assumptions on the density distribution by putting in equation (1) some particular functions for the density in the form $\rho(r) = \rho_0 f(r)$, and then checked for the radial variations of the 'constant' scale factors ρ_0. The best expression for the density resulted to be

$$\rho(r) = \rho_0 f_{MM}(r)$$

where $f_{MM}(r)$ is the function by Mellier and Mathez (1987), which is the deprojection of de Vaucouleur's law. This implies a constant mass-to-light ratio in the inner regions

G. Longo et al. (eds.), *Morphological and Physical Classification of Galaxies*, 423–424.

of the galaxies studied. The computed scale factor ρ_0 was finally used to compute th masses inside the semimajor effective axes. Rotation velocities and velocity dispersio profiles were taken from Busarello et al., 1990. These data result from the deprojectio (from line of sight integration) of observed rotation curves and velocity dispersion profile Table 1 shows the results of computations. Distances and absolute magnitudes were a taken from the Catalogue of Nearby Galaxies (Tully, 1988), implying a Hubble consta H_0=75 km sec^{-1} Mpc^{-1}. The details of the computations, selection criteria for the dat and kinematical data sources are given in Busarello and Longo, (1990).

Bacon, R., Monnet, G., Simien, F.: 1985, *Astron. Astrophys.* **152**, 315.
Bertin, G., Saglia, R.P., Stiavelli, M.: 1987, *Astrophys. J.* **330**, 78.
Busarello, G., Longo, G.: 1990, *Astron. Astrophys.*, submitted.
Busarello, G., Longo, G., Feoli, A.: 1990, *Astron. Astrophys.*, in press.
Faber, S.M., Gallagher, J.S.: 1979, *Ann. Rev. Astron. Astrophys.* **17**, 135.
King, I., Minkowsky, R.: 1972, IAU Symp. no. 44, p. 87.
Mellier, Y., Mathez, G.: 1987, *Astron. Astrophys.* **175**, 1.
Richstone, D.O., Tremaine, S.: 1986, *Astron. J.* **92**, 72.
Trinchieri, G., Fabbiano, G., Canizares, C.R.: 1986, *Astrophys. J.* **310**, 637.
Tully R.B.: 1988, *Nearby Galaxies Catalog*, (Cambridge University Press: Cambridge).

Table 1. Masses, mass to light ratios, and relative uncertainties.

(1) Ident.	(2) M ($10^{10}M_\odot$)	(3) M/L ($M_\odot/L_\odot$)	(4) $\frac{\Delta M}{M}$	(1) Ident.	(2) M ($10^{10}M_\odot$)	(3) M/L ($M_\odot/L_\odot$)	(4) $\frac{\Delta M}{M}$
A0559	118.56		0.13	N3904	8.51	10.55	0.16
A1350	19.50	2.22	0.19	N4125	37.70	14.91	0.22
A1847	63.62		0.15	N4278	4.47	18.17	0.14
I1459	34.69		0.03	N4291	8.74	9.78	0.05
I3370	11.62		0.04	N4365	19.64	16.68	0.10
I4296	58.16	5.32	0.14	N4374	20.57	11.75	0.07
N0221	0.04	3.61	0.07	N4387	1.21	7.84	0.11
N0584	9.09	6.66	0.18	N4406	21.68	10.30	0.09
N0596	4.90	5.49	0.02	N4472	41.11	10.54	0.10
N0636	3.67	6.57	0.14	N4473	7.44	8.57	0.16
N0720	13.56	13.10	0.35	N4478	1.58	5.05	0.12
N0821	14.54	16.43	0.04	N4486	35.81	10.84	0.10
N1052	11.35	18.54	0.08	N4551	1.38	8.48	0.10
N1395	15.63	15.66	0.15	N4564	5.27	13.77	0.16
N1407	30.27	16.22	0.09	N4621	16.18	14.66	0.07
N1531	0.85	9.06	0.14	N4636	15.89	11.65	0.33
N1549	8.67	14.69	0.21	N4649	21.71	8.51	0.51
N1600	85.51	8.11	0.19	N4660	3.71	9.70	0.18
N2300	21.99	14.56	0.08	N4697	29.11	8.57	0.19
N2778	3.90	8.78	0.16	N4742	0.78	1.28	0.17
N2974	19.68	15.10	0.07	N5266	41.56	17.37	0.15
N3087	7.28	5.29	0.13	N5322	27.02	8.03	0.14
N3091	38.91		0.13	N5638	5.42	6.65	0.10
N3108	8.67		0.50	N5745	21.17		0.08
N3379	2.63	6.32	0.12	N5845	2.45	6.41	0.16
N3557	35.81	6.59	0.05	N5846	29.69	13.99	0.08
N3607	19.60	16.34	0.09	N6251	22.86	5.45	0.12
N3610	3.86	2.70	0.21	N6703	6.53	5.34	0.10
N3613	20.02	11.03	0.17	N7562	30.75	7.53	0.11
N3640	9.53	6.37	0.11	N7619	35.74	5.42	0.15
N3818	4.68	11.15	0.15	N7785	35.47	7.50	0.11

·N THE PHOTOMETRIC SCALING LAWS FOR EARLY–TYPE ;ALAXIES

N. Caon[1], M. Capaccioli[2]
[1]*International School for Advanced Studies, Trieste, Italy*
[2]*Osservatorio Astronomico di Padova, Italy*

Ve have analysed the photometric data of the sample of 33 E and ordinary S0 galax-·s belonging to the Virgo cluster (according to the *Virgo Cluster Catalog* of Binggeli, andage and Tammann, 1987) studied by Caon, Capaccioli and Rampazzo (1990). The ample forms a homogeneous luminosity–limited set which is complete to $M_B = -17$ adopted $\Delta = 18.3$ Mpc). Deep Schmidt plates were coupled to unsaturated CCD expo-ıres in order to provide a *global* B–band mapping covering a surface brightness interval · 10 mag at state–of–art accuracy.

The effective photometric parameters R_e and μ_e were computed conforming to the riginal definition by de Vaucouleurs; in other words, they are not the scale parameters of ny interpolated fitting–law. Moreover, they do not depend too strongly on any arbitrary xtrapolation.

Figure 1 shows that R_e and μ_e are not simply related. In fact, galaxies with the ame total luminosity $L_T \propto I_e\, R_e{}^2$ are distributed over a rather wide range in μ_e and R_e: $\Delta\mu_e \simeq 2.5$ and, equivalently, $\Delta \log R_e \simeq 0.5$. This spread is not accounted for just y inclination effects, as inclination alone would imply intrinsic axis–ratios as small as : 0.1; furthermore, the position along the $M_B =$ const. sequence does not depend on llipticity. Thus, at least part of the spread must be a property intrinsic to early–type alaxies. The six brightest galaxies form a distinct group in the $(\log R_e,\ \mu_e)$ plane; hey are characterized by very large effective radii and faint surface brightnesses, and by ıminosity profiles with a marked upward concavity in a $r^{1/4}$ scale.

A least square fit of the relation $M_B = \alpha\mu_e + \beta \log R_e + \gamma$ to all but the six brightest bjects, combined with the Virial Theorem $\langle\sigma^2\rangle \propto \mathcal{M}_T/R_e$, gives:

$$L_T \propto I_e^{-0.76} \langle\sigma^2\rangle^{1.72} (\mathcal{M}/L)_T^{-1.7}$$

o be compared with Djorgovski and Davis' (1987) relation: $L_T \propto \sigma_c^{3.45} \langle I\rangle_e^{-0.86}$

:. Longo et al. (eds.), Morphological and Physical Classification of Galaxies, 425–426.

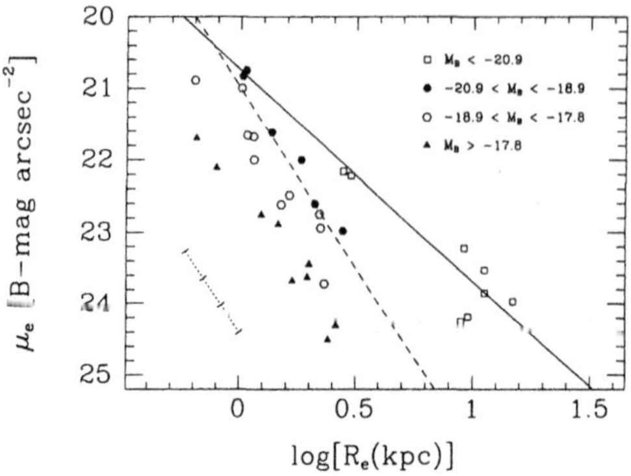

Figure 1. Distribution of 33 E and ordinary S0 Virgo galaxies in the ($\log R_e$, μ_e) plane; four luminosity classes are indicated by different symbols. The solid line is Hamabe and Kormendy's (1987) relation, properly rescaled. The drift induced by an error ΔB_T is shown by the dotted line (ticks are spaced by $\Delta B_T = 0.1$). The dashed line is the locus of points at $L_T \propto I_e R_e^2 = \text{const}$.

Our data seem to indicate the presence of a secondary maximum in the bright end o the luminosity function, produced by the same objects which lie well separated from th others in the ($\log R_e$, μ_e) plane. The occurrence of this secondary maximum is due to th fact that, for the brightest objects of the sample, our total luminosities are systematicall larger (up to $\simeq 0.5$ mag) than those reported in the literature.

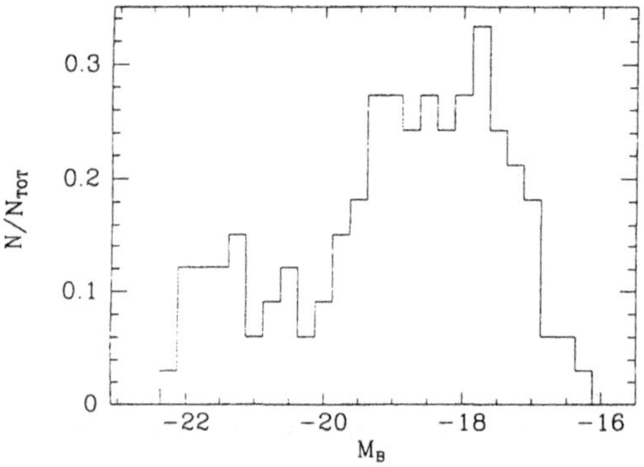

Figure 2. Luminosity function for the 33 E and S0 galaxies of the sample. Row data have been smoothed by a 1.0 mag wide running-window, moved in steps of 0.25 mag.

Binggeli, B., Sandage, A., Tammann, G., 1985, *Astron. J.* **90**, 1681.

Caon, N., Capaccioli, M., Rampazzo, R., 1990, *Astron. Astrophys.* **86**, 429.

Djorgovski, S., Davis, M., 1987, *Astrophys. J.* **313**, 59.

Hamabe, M., Kormendy, J., 1987. In *Structure and Dynamics of Elliptical Galaxies*, IAU Symp. no. 127 p. 379, ed. de Zeeuw, P. T., Reidel, Dordrecht.

PROPERTIES OF BRIGHTEST CLUSTER MEMBERS

S. Djorgovski[1], R. de Carvalho[1,2], I. Shlosman[1,3], J. Schombert[4]
[1]*Palomar Observatory, Caltech, USA*
[2]*Observatorio Nacional, CNPq, Rio de Janeiro, Brasil*
[3]*Dept. of Physics and Astronomy, Univ. of Kentucky, USA*
[4]*Dept. of Astronomy, Univ. of Michigan, Ann Arbor, USA*

Are the brightest cluster members (BCMs) somehow different from the "normal" ellipticals, or are their properties in accord with the high luminosity extension of the global properties and scaling laws of the family of ellipticals? We examined the behavior of BCMs in the parameter space defined by the luminosity (L), radius (R), surface brightness (μ), and velocity dispersion (σ), i.e., the so-called Fundamental Plane (FP) variables. We used the data from Malumuth and Kirshner, for both the core and the half-light parameters; the results on both radial scales were qualitatively the same.

There is an excellent $R - \mu$ correlation, which is much better than the equivalent relation for normal ellipticals. The scatter in this relation is reduced only very marginally with the introduction of σ as the "second parameter", in marked contrast with the normal E's, where the improvement in the bivariate fit is dramatic. In agreement with Malumuth and Kirshner, we find that the spread of σ at a given L or R is much smaller than for normal ellipticals. Thus, the bivariate scaling laws (FP solutions) for BCMs are different from those for normal ellipticals. We see a systematic trend, going from field ellipticals, to cluster ellipticals, to BCMs: the dimensionality of the data manifold (measured as the ratio of the variances corresponding to the first two eigenvectors) decreases. Whereas the field ellipticals are clearly a two-dimensional family, the BCMs are nearly one-dimensional: all quantities become better correlated. This variation suggests that the properties of ellipticals and their scaling relations may vary systematically in different environments, with clear consequences for measurements of peculiar velocities.

A possible explanation of these effects may be the dynamical evolution of BCMs in the homologous merger scenario (Ostriker, Tremaine, and others), in which: (1) the kinetic energy E_{kin} per unit mass, i.e., the velocity dispersion, should remain nearly constant: (2) mass and luminosity increase as the material is added; (3) as the orbital E_{kin} is converted into internal degrees of freedom (pressure support, σ), R increases while density and surface brightness decrease. Thus, one expects that: (1) there is a small dynamical range of σ at a fixed L, relative to the normal E's; (2) σ should be less important in the FP solutions (the $R - \sigma - \mu$ relation); (3) the $R - \mu$ relation becomes dominant, driven by the expansion of the merger products. All of these predictions are

G. Longo et al. (eds.), *Morphological and Physical Classification of Galaxies*, 427–428.

confirmed by the data, although alternative explanations may be possible. In any cas it appears that dynamical evolution can modify the FP.

An interesting possibility is that the $R - \mu$ relation may provide a new cosmologic tool, as it can be used to predict metric radii with a very small scatter; or alternativel as a test for the expansion of the universe using the predicted redshift dependence of μ

One long-standing mystery is the small scatter of BCM luminosities, as measure in apertures of 9.6 h^{-1} kpc radius (e.g., Gunn et al.) or 21.5 h^{-1} kpc radius (e.g Sandage et al.). This is the basis for the Hubble diagram as a cosmological test. Th observed scatter, corrected for the richness class and the Bautz-Morgan effect, is onl $\sim 0^m.30$ r.m.s., in all bands. However, measurements of the "total" mags (in practic measured at some very faint isophote, or extrapolated from a fit to the de Vaucouleur formula) indicate the r.m.s. scatter of 0.6–0.7 mag. This scatter must reflect the shap of the high-luminosity end of the luminosity function, and/or the scatter of characteristi luminosities from one cluster to another. Why are the smaller aperture mags so muc better behaved than the total mags? Why are the candles so standard?

We propose that this effect is a consequence of the excellent scaling relation betwee the radius and surface brightness. It then follows from the specific scaling relatio between the luminosity and surface brightness for BCMs, which is different that for th lower-rank ellipticals. Following several papers by Schneider, Hoessel, Gunn, and other and using our own fits to the published data, we arrive at the average scaling relation fo BCMs: $\mu_e = 3.1\log(r_e/\mathrm{kpc}) + \mu_0$, where the zero-point is 19.75 for the V band, or 19. for the Gunn r band ($H_0 = 100$). Assuming that galaxies follow the de Vaucouleurs formula, we obtain the scaling for the absolute magnitudes: $M_{tot} = -19\log(r_e/\mathrm{kpc}) + M_0$ where the zero-point is -20.2 for the V band, and –20.55 for the Gunn r band.

Since radius and luminosity are correlated, but are anticorrelated with surface brightness, and the profile in a fixed aperture (in kpc) gets shallower as the effective radiu increases, the luminosity in a fixed aperture must change more slowly than the tota luminosity: increase in the effective radius, which follows increase in the total luminosity is partly compensated by the lowering of the surface brightness. The couplin of radius and surface brightness for the higher-rank (normal) ellipticals has a differen slope, which makes the compensation less effective. Moreover, there is a larger scatter c surface brightness at a given R or L, which reflects the second parameter of the FP fo normal ellipticals. Thus, while BCMs have their luminosities ("candles") standardized the ordinary E's do not. A similar explanation was proposed previously by Morbey.

More details and complete references will be presented in a forthcoming journa paper, omitted here because of the space restrictions. SD acknowledges a travel gran from the AAS, and a partial support from the Alfred P. Sloan Foundation.

THE FUNDAMENTAL PLANE FOR EARLY TYPE GALAXIES

I. Jørgensen, P. Kjærgaard
Copenhagen University Observatory
Copenhagen, Denmark

Accurate CCD surface photometry in Johnson B and Gunn r has been carried out for 90 E and S0 galaxies in three clusters. Forty of the galaxies have also been observed in Gunn v. The observations are part of our investigation of the large-scale motions of galaxies both near the mass concentration of the "Great Attractor" suggested by Lynden-Bell et al. (1988) and in other directions. The clusters A3574 (Klemola 27) and S753 (the cluster around NGC 5419) are close (both angular and in distance) to the "Great Attractor". The third cluster, Coma, serves as standard and for comparison purposes. Here we present results from our study of the fundamental plane as a method for estimating distances to galaxies.

Surface photometry for Johnson B and Gunn r was determined by fitting ellipses to the isophotes of the galaxies. The effective radius, r_e, and the mean surface brightness, $<\mu>_e$, inside r_e were then determined by a fit to the de Vaucouleurs standard growth curve for a $r^{1/4}$-law. For Gunn v only central magnitudes have been determined. A radius r_n corresponding to the parameter D_n used by Dressler et al. (1987) and Lynden-Bell et al. (1988) has also been derived. For Johnson B r_n is the radius of a circle inside which $<\mu>= 20.75^m/arcsec^2$. For Gunn r $<\mu>= 19.75^m/arcsec^2$ is used.

The photometry was tied to the standard system by observations of standard stars. The photometry in Johnson B shows a very good agreement with results from Burstein et al. (1987) ("our"-"their"$= -0\overset{m}{.}005 \pm 0\overset{m}{.}04$ for magnitudes in circular apertures). The global parameters are in good agreement with data from Faber et al. (1989), e.g. for $\log r_n$ "our"-"their"$= 0.002 \pm 0.025$.

The fundamental plane has been investigated in the forms

$$\log r_e = \alpha \cdot \log\sigma + \beta \cdot <\mu>_e + constant \quad (1)$$

$$\log r_n = \alpha \cdot \log\sigma + constant \quad (2)$$

$$\log r_e = \alpha \cdot (v - r) + \beta \cdot <\mu>_e + constant \quad (3)$$

(Relation (1) and (2) only for Coma.) Data and relations are shown on the figures.

In conclusion one gets the formally best determination of distances to galaxies by using the relation $\log r_e - \log\sigma - <\mu>_e$, here an accuracy of 14% is reached. The fundamental plane in the form $\log r_e - (v - r) - <\mu>_e$ could be a possible way of determining

G. Longo et al. (eds.), *Morphological and Physical Classification of Galaxies*, 429–430.

distances to rich clusters but it needs further investigation to ensure that any systemat effects do not affect the results.

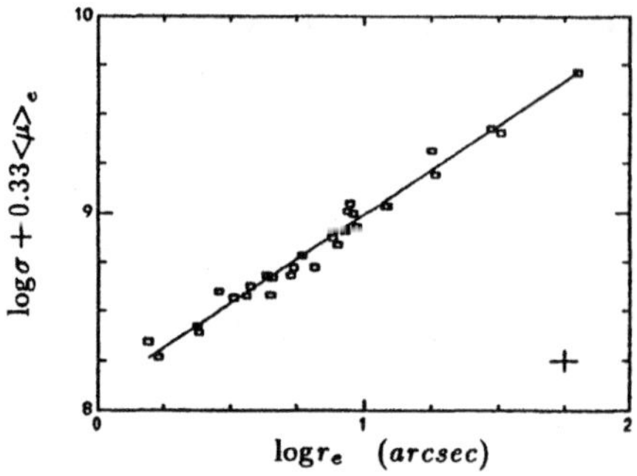

Figure 1. Data for Coma, Gunn r. $\log r_e = 1.11(\pm 0.10) \cdot \log\sigma + 0.37(\pm 0.01) \cdot <\mu>_e - 9.01(\pm 0.15)$. The scatter gives 14% uncertainty on distances to single galaxies. The coefficient to $\log\sigma$ is smaller than found by Djorgovski and Davis (1987), while we find agreement for the coefficient to $<\mu>_e$.

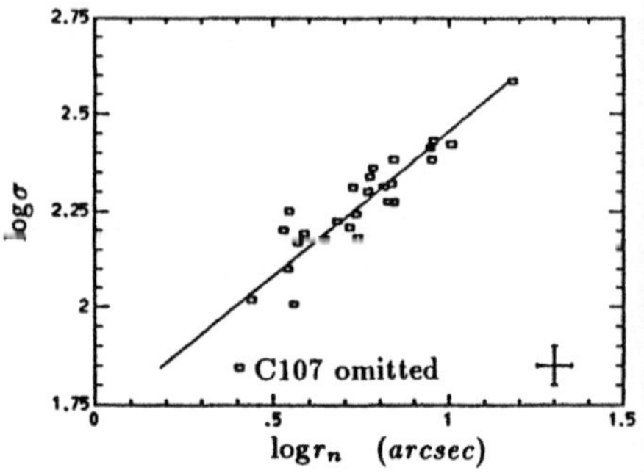

Figure 2. Data for Coma, Johnson B. $\log r_n = 1.33(\pm 0.13) \cdot \log\sigma - 2.27(\pm 0.29)$. The sca ter gives 18% uncertainty on distances. The r sults are in agreement with Dressler et al. (1987

The project is based on observations with the Danish 1.5m telescope at La Silla, Chil and is supported by The Danish Board for Astronomical Research and The Carlsber Foundation. We thank M.Franx for making the ellipse fitting program available to us.

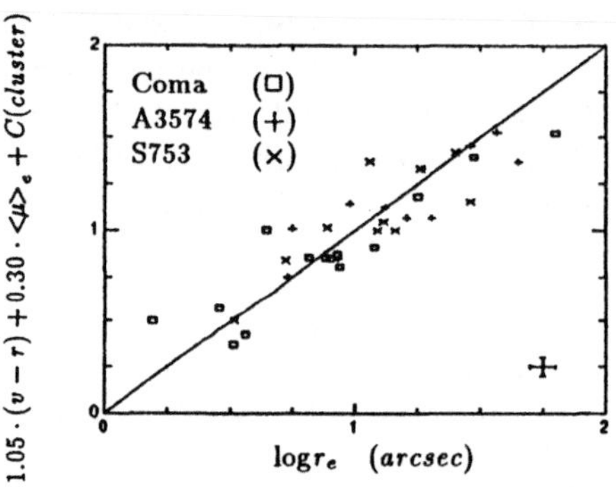

Figure 3. Data for all three clusters. r_e and $<\mu>_e$ are for Gunn r. $\log r_e = 1.05(\pm 0.25) \cdot (v - r) + 0.30(\pm 0.01) \cdot <\mu>_e + constant$. The scatter gives 33% uncertainty on distances to single galaxies. de Carvalho and Djorgovski (1989) have used the colours $(B - V)$, $(V - K)$ or $(U - R)$ for similar studies getting an accuracy of 25-35%. We find that replacing $\log\sigma$ with $(B - r)$ does not give a useful representation of the fundamental plane.

Burstein, D., et al.: 1987, *Astrophys. J. Suppl. Ser.* **64**, 601 (7Samurai, Paper III)

de Carvalho, R.R., Djorgovski, S.: 1989, *Astrophys. J.* **341**, L37

Djorgovski,S., Davis, M.: 1987, *Astrophys. J.* **313**, 59

Dressler, A. et al.: 1987, *Astrophys. J.* **313**, 42 (7Samurai, Paper I)

Faber, S.M. et al.: 1989, *Astrophys. J. Suppl. Ser.* **69**, 763 (7Samurai, Paper VI)

Lynden-Bell, D. et al.: 1988, *Astrophys. J.* **326**, 19 (7Samurai, Paper V)

LLIPTICAL GALAXIES AND THE THICKNESS OF THE UNDAMENTAL PLANE

F. Simien[1], P. Prugniel[2]
[1]*Observatoire de Lyon, France*
[2]*European Southern Observatory, Garching, Germany*

:e-analysis and comparison of the data presented in Dressler et al. (1978: 7S) and jorgovski and Davis (1987: DD) lead to the following generic representation of the ındamental plane: $10\log(\sigma) = (1+2\beta)M_t + < SB >_e + cst$, where M_t is the absolute ıagnitude and $< SB >_e$ then mean surface brightness (in mags) within the effective adius; we have adopted: $M/L \propto L^{\beta}$. The value of β clearly depends on the color: we nd $\beta_R = 0.13 \pm 0.03$ and $\beta_B = 0.23 \pm 0.03$. Figure 1 shows the B-calibrated relation for 2 galaxies in common between DD and 7S.

For the imposed condition $\beta = 0$, figure 2 shows that the residuals from the virial elation correlate with (U–V), with a slope of 7.6. Adopting this color correction yields new determination of the plane with a vanishing dependence of M/L on L. This result robably means that *a major part of the L-dependence of M/L is due to the dependence f metallicity on luminosity.*

The relatively slow rotation of an E galaxy may still represent a significant fraction f the total kinetic energy (Binney, 1978; Busarello et al., 1989), thus contributing to n increase in the thickness of the fundamental plane. And since the relatively faint Es re more likely to be isotropic than bright ones, this also biases the value of β. But ; is remarkable that for an oblate galaxy, the virial theorem implies that the distance o the fundamental plane (due to rotation) is independent from the projection angle. ndeed, the luminosity-weighted velocity dispersion σ and rotation v obey the relation: $\sigma_i^2 + v_i^2) = (\sigma_x^2 + v_x^2)(\sin^2 i + R(q)\cos^2 i)$, where the subscripts refer to inclined (i) or dge-on (x) observations (Bacon, Monnet and Simien, 1985 and references therein), and $R(q)$ is a function of the intrinsic axial ratio q. This implies that, with the v_i^2 term ncluded, the locus of a galaxy observed at different inclinations is almost parallel to he plane (Fig. 1). Thus, *it is possible to account for the rotational energy in a way ndependent from the viewing angle.* Unfortunately, the present sample appears too mited in size and quality to warrant the validity of the correction (Fig. 3). We note Fig. 4) a correlation between v/σ and $< SB >_e$, which is not unexpected, and which is

. Longo et al. (eds.), Morphological and Physical Classification of Galaxies, 431–432.

partly due to inclination effects: further study is in preparation.

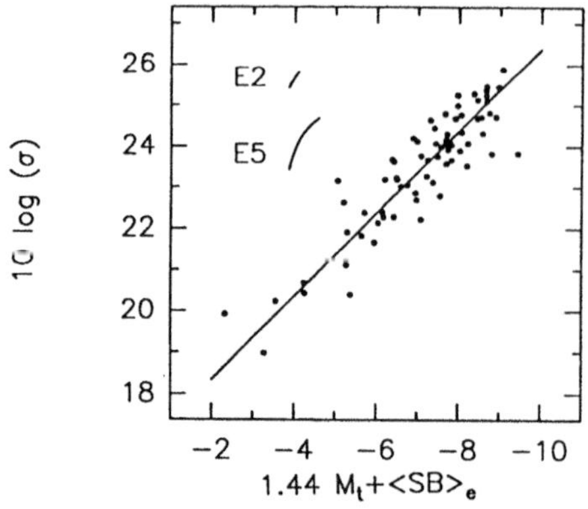

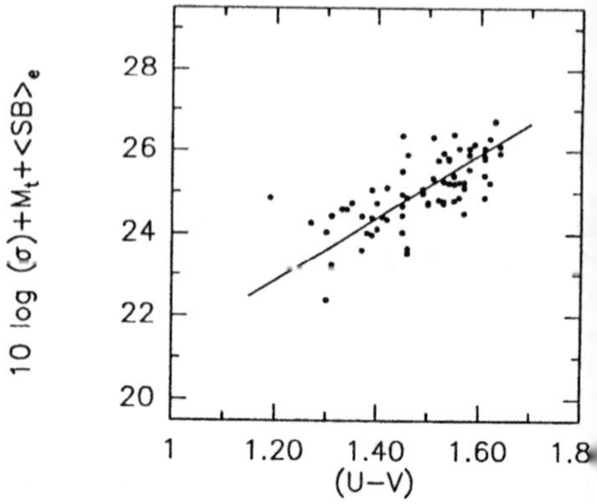

Figure 1. (left) The fundamental plane; for intrinsically E2 and E5 galaxies, locus from face-on to edge-on, when including the rotation, i.e. with $5\log(\sigma^2 + v^2)$ as the ordinate.
Figure 2. (right) Residual distances to the plane versus (U–V) ($r = -0.68$).

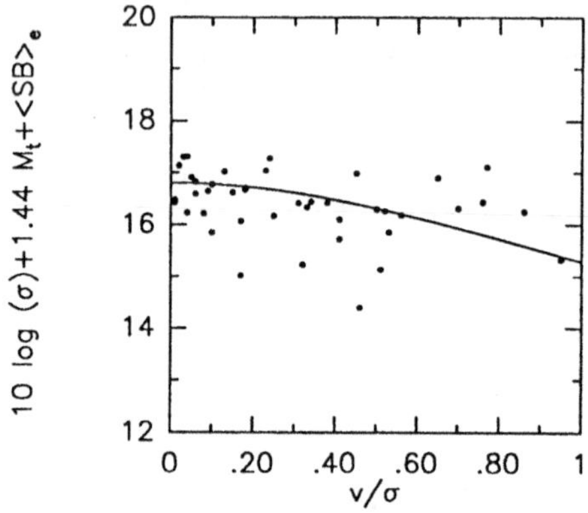

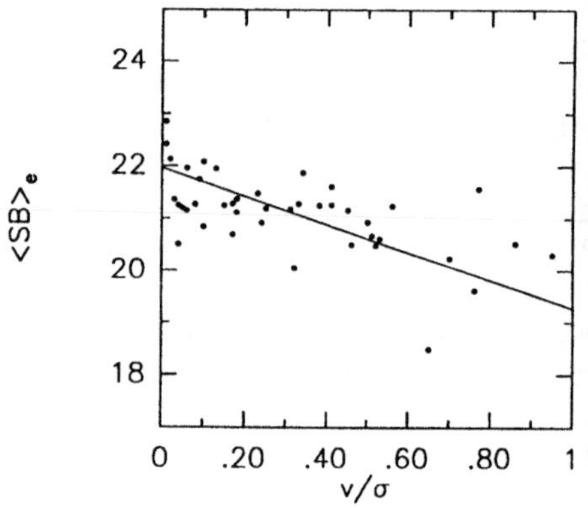

Figure 3. (left) Distances to the (uncorrected) fundamental plane as a function of the relative rotation. The curve shows the corrective term $5\log(1 + v^2/\sigma^2)$.
Figure 4. (right) The correlation between v/σ and $< SB >_e$ ($r = -0.58$).

References

Bacon, R., Monnet, G., Simien, F.: *Astron. Astrophys.* **152**, 315.

Binney, J.: 1978, *Mon. Not. R. Astr. Soc.* **183**, 501.

Busarello, G., Filippi, S., Ruffini, R.: 1989, *Astron. Astrophys.* **213**, 80.

Djorgovski, S., Davis, M.: 1987, *Astrophys. J.* **313**, 59.

Dressler, A., Lynden-Bell, D., Burstein, D., Davies, R.L., Faber, S.M., Terlevich, R.J., Wegner, G.: 1987, *Astrophys. J.* **313**, 42.

DISSIPATIONLESS FORMATION OF ELLIPTICAL GALAXIES

M. Stiavelli[1], P. Londrillo[2], A. Messina[3]
[1] *European Southern Observatory, Garching, Germany*
[2] *Astronomical Observatory, Bologna, Italy*
[3] *Department of Astronomy, University of Bologna, Italy*

Since the cursory work of van Albada (1982), dissipationless formation of elliptical galaxies has been the subject of several investigations, performed by exploring various sets of initial conditions and with different numerical techniques. It is now well established that dissipationless collapse, starting from initial conditions that are cold enough, produces a final state characterized by a luminosity profile well–fitted by the $R^{1/4}$ law of elliptical galaxies (de Vaucouleurs, 1948, 1953). What is much less clear are the details of this process and in particular: what is the influence of inhomogeneities in the initial conditions? how does the final shape relates to the initial conditions? what is the distribution function describing the final system? In order to give an answer to these questions we have performed a large set of N-body simulations by using two-different N-body codes, both based on a spherical harmonics expansion of the gravitational potential and already extensively described in the literature (Londrillo, Messina, 1990; Bertin, Stiavelli, 1989). The use of grids able to resolve the cores, of large numbers of particles (typically 8×10^4), and of small time steps (often $< t_{cr}/2000$) has allowed us to obtain a large *dynamic range*, necessary to distinguish the collapse dynamics from spurious numerical effects. A full account of our findings can be found in Londrillo, Messina, Stiavelli (1990). Our major results can be summarized as follows: i) inhomogeneities in the initial conditions play a crucial role in determining how well the density profile of the final system is fitted by the $R^{1/4}$ law, the presence of a tail in the initial density distribution helps in producing very realistic projected density profiles even for initial conditions which are not very cold (i.e. $2T/|W| \simeq 0.1$); ii) the shape of the final equilibrium cannot be easily related to the initial conditions, for systems that are very cold the final shape is reached only after many (i.e. > 20) half mass crossing times; iii) the final equilibrium system is characterized by positive temperature (Aguilar, Merritt, 1990).

This last result is illustrated in more detail in figure 1, where the histograms showing the velocity distribution for the most bound particles in two of our simulations are compared with the predictions of positive (*circles*, see Bertin, Stiavelli, 1984) and negative

G. Longo et al. (eds.), Morphological and Physical Classification of Galaxies, 433–434.

(*squares*, see Merritt, Tremaine, Johnstone, 1989) temperature distribution functions.

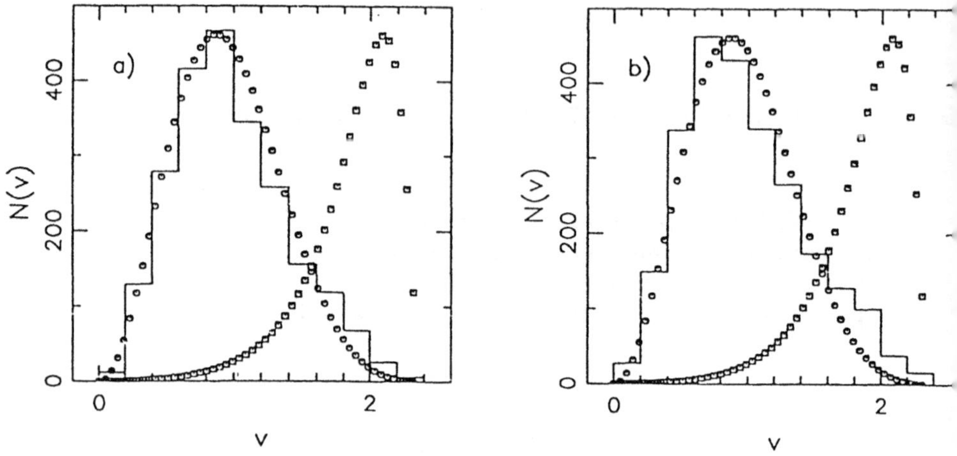

Figure 1

It is readily seen that, in contrast to the claim of Aguilar and Merritt (1990), w don't find any support for the negative temperature hypothesis. The end-products c dissipationless galaxy formation are characterized by larger anisotropy than the posi tive temperature models introduced by Bertin and Stiavelli (1984) but are neverthele positive temperature systems.

Acknowledgements

This work has been partially supported by Ministero dell'Università e delle Ricerca Scien tifica e Tecnologica, CNR, ESO, Università di Bologna and Scuola Normale Superiore d Pisa.

References

Aguilar, L., Merritt, D.: 1990, *Astrophys. J.* **354**, 33.

Bertin, G., Stiavelli, M.: 1984, *Astron. Astrophys.* **137**, 26.

Bertin, G., Stiavelli, M.: 1989, *Astrophys. J.* **338**, 723.

de Vaucouleurs, G.: 1948, *Ann. Astrophys.* **11**, 247.

de Vaucouleurs, G., 1953, *Mon. Not. R. Astr. Soc.* **113**, 134.

Londrillo, P., Messina, A.: 1990, *Mon. Not. R. Astr. Soc.* **242**, 595.

Londrillo, P., Messina, A., Stiavelli, M.: 1990, *Mon. Not. R. Astr. Soc.* submitted, and *ESO Preprir 728.*

Merritt, D., Tremaine, S., Johnstone, D.: 1989, *Mon. Not. R. Astr. Soc.* **236**, 829.

van Albada, T.S.: 1982, *Mon. Not. R. Astr. Soc.* **201**, 939.

THE TWO-COMPONENT COLOUR-STRUCTURE OF THE gE GALAXY NGC 4472

P. Surma[1], R. Bender[1] and J.-L. Nieto[2]
[1]*Landessternwarte Königstuhl, Heidelberg, Germany*
[2]*Observatoire Midi-Pyrénées, Toulouse, France and CFHT, Hawaii, USA*

From the observational point of view the 'missing link' in our understanding of the chemical and dynamical evolution of early-type galaxies still lies in the lack of a reliable data set of colour and metallicity distributions for a considerable number of objects. The problem in measuring colours mainly arises from the extremely small amplitude of the expected gradients and the difficulties to remove the sky background precisely. Typical colour-changes of the order of 0.01 mag/□" may be detected down to the brightness level of the sky only if the background determination is accurate to $\Delta F_{sky}/F_{sky} < 1\,\%$.

We used a focal reducer at the RC-focus of the 1.23 m telescope on Calar Alto (1.55 arcsec/pixel, field 10 × 15 arcmin²) and the 3.6 m Canada-France-Hawaii telescope on Mauna Kea (0.11 arcsec/pixel) with 22 μm GEC and 15 μm RCA CCDs respectively (filter system: Kron-Cousins B and R bands). High-resolution (FWHM= 0.8 arcsec) CFH data for the central regions are combined with the large-field focal reducer material which shows an accuracy of background removal of $0.1 - 0.3\,\%$ (cf. Surma et al., 1990 for details and extensive tests). The brightness profiles in B and R along the semi-major axis 'a' are derived by the isophote fitting procedure of Bender and Möllenhoff (1987). It has been modified for the low-S/N 'outer regions' of the galaxies and yields a *statistical* accuracy < 0.01 mag/□" throughout the brightness profiles.

Figure 1 shows the derived B-R colour profile of NGC 4472 ($1 < a < 180$ arcsec, 1 arcsec ≈ 0.1 kpc) reaching down 6.4 mag/□" to a surface brightness of $\mu_B = 23.9$ and $\mu_R = 22.4$ mag/□" (the letters 'B' and 'R' in figure 1 indicate the radial positions where $\mu_{galaxy} = \mu_{sky}$). Note that the error bars include both statistical *and* upper limits of the expected *systematical* errors. The total errors are < ±0.01 mag/□" for $a < 90$ arcsec and < ±0.035 mag/□" at $r < 180$ arcsec.

There is a clear-cut *two-component colour structure* in B-R. In the central region at $a < 10$ arcsec we find a steep, well-defined colour gradient of $\Delta(B - R) = -0.08$ mag/□" per decade in radius of inward reddening. The outer region of NGC 4472 shows a fairly constant B-R colour distribution. A comparison of our data with the scanning-CCD measurements of Boroson and Thompson (1987, BT) shows good coincidence within 15 arcsec. In the outer region, however, both data sets are consistent only when considering the comparably large background error of $1 - 2\,\%$ of BT.

G. Longo et al. (eds.), Morphological and Physical Classification of Galaxies, 435–436.

Standard theories of galaxy formation predict smooth colour gradients for elliptic galaxies, with dissipative models predicting higher colour gradients than non-dissipati ones. The anisotropic structure of Es argues in favour of dissipationless scenarios. Hov ever, dissipational processes during galaxy formation are necessary e.g. for the hig phase-space densities in Es (Kormendy, 1990), or counter-rotating, cold disks in the cor of Es (Bender 1990). NGC 4472 shows weak counter-streaming at $r < 5$ arcsec (Davie 1989) and the strong colour gradient towards the core of this galaxy reveals a central su system which is considerably redder than the outer main body. A similar two-componer colour structure has also been measured by Lachièze-Rey et al. (1985) for the cD galax NGC 6166. In fact, Rocca-Volmerange and Guideroni (1987) have argued on the basis IUE data for recent (about 5 Gyrs ago) star formation in ellipticals, which could explai centrally enhanced metallicity and reddening. An 'inhomogeneous' formation history fc NGC 4472 may additionally be indicated by the small, but significant colour variatior at large radii. One might speculate to attribute them to accretion events of objects low metallicities or even on-going low-mass star formation from cooling-flows.

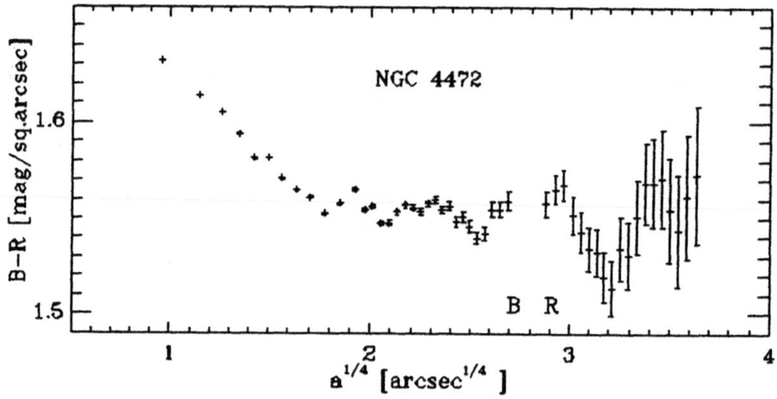

Figure 1.

References

Bender, R., Möllenhoff, C.: 1987, *Astron. Astrophys.* **177**, 71.

Bender, R.: 1990, in *Interactions and Dynamics of Galaxies*, ed. R. Wielen, Heidelberg.

Boroson, T.A., Thompson, I.B.: 1987, *Astron. J.* **93**, 33.

Davies, R.L.: 1989, in *Le monde de galaxies*.

Kormendy, J.: 1990, in *Interactions and Dynamics of Galaxies*, ed. R. Wielen, Heidelberg.

Lachièze-Rey, M., Vigroux, L. Souviron, J.: 1985, *Astron. Astrophys.* **150**, 62.

Rocca-Volmerange, B., Guideroni, B.: 1987, *Astron. Astrophys.* **175**, 15.

Surma, P., Seifert, W., Bender, R.: 1990, *Astron. Astrophys.*, in press.

'HE VELOCITY DISPERSION ANISOTROPY IN NGC 720

R. van der Marel[1], P. Cinzano[2]
[1]*Sterrewacht Leiden, Leiden, The Netherlands*
[2]*Department of Astronomy, University of Padova, Padova, Italy*

The model

ı Binney, Davies and Illingworth (1990; BDI) and van der Marel, Binney and Davies 1990) a model is presented that predicts the kinematics of an elliptical galaxy from its urface photometry. The model consists of the following steps :

- Assume axisymmetry and use Lucy's iterative algorithm to deproject for fixed inclination the 2–dimensional surface brightness Σ to a 3–dimensional luminosity density j.
- Find the 3–dimensional mass-density ρ by assuming a constant mass-to-light ratio.
- Find the potential Φ by solving the Poisson equation.
- Assume $f = f(E, L_z)$. The Jeans equations of stellar hydrostatic equilibrium are then a closed set, and can be solved for $\langle v_\phi^2 \rangle$ and $\sigma_R^2 \equiv \sigma_z^2$.
- Introduce an anisotropy parameter k that splits $\langle v_\phi^2 \rangle$ into streaming and dispersion, project the dynamical quantities back onto the sky, and compare with kinematical observations.

Application to NGC 720

n BDI this model was applied to the E4 elliptical galaxy NGC 720. The galaxy was ssumed oblate, since it rotates on its projected major axis. A good fit could not be btained. If the mass-to-light ratio is fixed to fit the minor-axis dispersion data, and k fixed to fit the observed major axis rotation, then the predicted major-axis dispersion xceeds the observations. This is illustrated in figure 1, in which the ratio $\mathcal{R}$ of the major o minor-axis RMS line-of-sight velocity $[\sqrt{\sigma^2 + v_{rot}^2}]$ is plotted. The predictions (solid ne) for inclination 90° (edge-on) clearly exceed the observations. Note that $\mathcal{R}$ neither epends on the assumed mass-to-light ratio (which influences both axes in the same

. Longo et al. (eds.), *Morphological and Physical Classification of Galaxies*, 437–438.

way) nor on the parameter k (since no splitting of $\langle v_\phi^2 \rangle$ into streaming and dispersio is required to evaluate $\mathcal{R}$). Therefore $\mathcal{R}$ only depends on the assumed inclination. the assumed inclination is decreased, then the galaxy has to be intrinsically flatter t project to the same axis ratio on the sky, which increases $\langle v_\phi^2 \rangle$. In BDI it is demonstrate that this increases the predicted ratio $\mathcal{R}$, making the fit to the data even worse. The conclude that $f \neq f(E, L_z)$ for NGC 720.

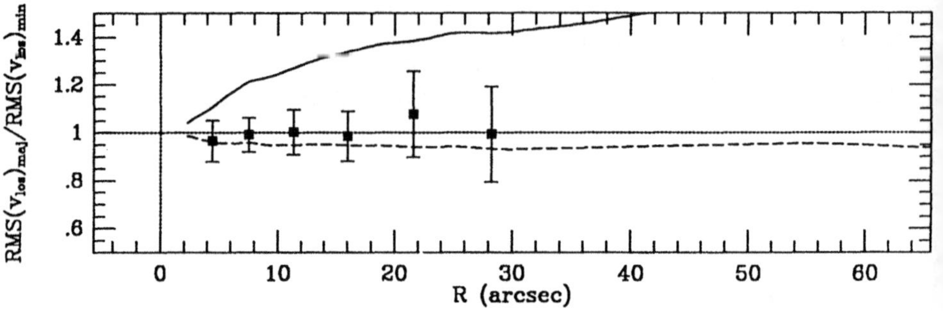

Figure 1. Predicted and observed ratio of the major to minor-axis RMS line-of-sight velocity $[\sqrt{\sigma^2 + v_{rot}^2}]$ as a function of distance from the center for NGC 720. The solid line is for $f = f(E, L_z)$, the dashed line is for the model described in section 3. The predictions were calculated assuming that the galaxy is seen edge-on.

3 An alternative model

Luminous elliptical galaxies are flattened by the anisotropy of their velocity dispersio tensor (Davies et al., 1983). For a galaxy with $f = f(E, L_z)$ this implies that $\langle v_\phi^2 \rangle$ exceed $\sigma_R^2 \equiv \sigma_z^2$. Since we see mainly v_ϕ on the major axis and mainly v_R on the minor axis this leads to models with $\mathcal{R} > 1$ (in agreement with figure 1). The kinematical data o NGC 720 suggest however that $\mathcal{R} \approx 1$. Fillmore (1986) studied several solutions of th Jeans equations for three-integral models. We tried to model NGC 720 with one of hi models, in which it is assumed that $\langle v_R v_z \rangle \equiv 0$ and $\langle v_\phi^2 \rangle \equiv \sigma_R^2$. The Jeans equations the form a closed set. Indeed such a model (Fig. 1; dashed line) provides a much better fi to the observations. In models of this type the velocity ellipsoid aligns with cylindrica coordinate axes. We are in the process of studying other three-integral models, in whic for example the velocity ellipsoid aligns with spherical coordinate axes.

References

Binney, J.J., Davies R.L., Illingworth, G.D.: 1990, *Astrophys. J.* **361**, 78.

Davies, R.L., Efstathiou, G., Fall, S.M., Illingworth, G.D., Schechter, P.L.: 1983, *Astrophys. J.* **266**, 41.

Fillmore, J.A.: 1986, *Astron. J.* **91**, 1097.

van der Marel, R.P., Binney, J.J., Davies, R.L.: 1990, *Mon. Not. R. Astr. Soc.* **245**, 582.

NGC 205 AND THE NATURE OF COMPACT NUCLEI IN DWARF GALAXIES

S. Djorgovski[1], O. Bendinelli[2], G. Parmeggiani[3], and F. Zavatti[2]
[1]*Palomar Observatory, Caltech, USA*
[2]*Dipartimento di Astronomia, Università di Bologna, Italy*
[3]*Osservatorio Astronomico di Bologna, Italy*

Compact nuclei are both a ubiquitous and a striking feature in low-surface brightness dwarf galaxies, sometimes containing a large fraction of the total light. Yet, their nature and origin are still a mystery. Spectroscopic and photometric observations suggest that the nuclei contain an intermediate-age stellar population. Their structural and dynamical parameters are still very poorly known, due to their compactness: essentially all are unresolved. The Local Group dwarf galaxy NGC 205 contains the nearest, albeit an unimpressive and relatively faint example. Even in this nearby case, the nucleus is almost unresolved, making the estimates of its structural and dynamical parameters difficult. The situation can be helped somewhat by seeing deconvolution.

Our data set consists of multiple CCD images taken at the Palomar 60-inch telescope in the Gunn-Thuan *gri* bands. The seeing was $\sim$ 1 arcsec, and the scale 0.246 arcsec/pixel. Stars for the measurements of PSFs were included in the frames. We use the profile by Kent (1987) to extend the radial range of our data. In the overlap region, there is an excellent agreement between Kent's and our profiles. The profiles were deconvolved using the Regularized Multi-Gaussian technique (Bendinelli, 1991; Bendinelli et al., 1990). We estimate that our final resolution limits are about 0.1 – 0.2 arcsec. In order to isolate the nucleus, we model the underlying galaxy light distribution as an exponential.

From our deconvolved profiles, we compute the structural and dynamical parameters for the nucleus alone, and for the underlying galaxy. We assumed the distance of 720 kpc, net extinction $A_V = 0.31^m$, and $(V - r) \simeq 0.1$. Our core radius of the nucleus should be regarded as an upper limit, the central surface brightness and the projected, volume, and phase-space densities should be regarded as lower limits, and the central relaxation time as an upper limit. For the nucleus, we obtain: apparent magnitude: Gunn r = 14.75, V= 14.75; absolute magnitude: $M_V = -9.53$; core radius: $r_c \leq 0.30$ arcsec = 1.05 pc; half-light radius: r_h = 0.4 arcsec = 1.4 pc; central surface brightness: $\mu_{0V} \leq 14.2$; projected luminosity density: $\Sigma_{0V} \geq 3.8 \times 10^4 L_\odot \mathrm{pc}^2$; volume luminosity density: $\rho_{0V} \geq 3.6 \times 10^4 L_\odot \mathrm{pc}^3$; phase-space density: $f_0 \geq 0.68$ (km/s)$^{-1}$ pc^{-2}; mass-to-light ratio: $(M/L)_{0V} \geq 0.8$; relaxation time: $t_{rc} \leq 8.0 \times 10^7$ yr. For the underlying galaxy, the corresponding quantities are: Gunn r = 8.0, V= 7.7; $M_V = -16.6$; r_c = 25 arcsec = 26 pc; r_h = 100 arcsec = 105 pc; μ_{0V} = 19.15; $\Sigma_{0V} = 800 L_\odot \mathrm{pc}^2$; $\rho_{0V} = 15 L_\odot \mathrm{pc}^3$; $f_0 = 2.6 \times 10^{-4}$ (km/s)$^{-1}$ pc^{-2}; $(M/L)_{0V} = 57$; $t_{rc} = 4.4 \times 10^9$ yr.

G. Longo et al. (eds.), *Morphological and Physical Classification of Galaxies*, 439–440.

We see a striking density contrast: by all measures, the nucleus is at least sever thousand times denser than the surrounding galaxy. This contrast and the difference i the relaxation times indicate that the nucleus is essentially an independent dynamic subsystem. The relatively short central relaxation time suggests that the nucleus ma be a subject to considerable dynamical evolution on the time scales of several Gyr, e.g the gravothermal instability (core collapse). On the whole, its estimated dynamic properties are similar to the most luminous globular clusters in our Galaxy.

The (M/L) ratios were evaluated using the velocity dispersions from Held et a (1990) and Carter and Sadler (1990). The low value for the nucleus is probably a lowe limit, since we do not fully resolve it; it is about right for the stellar population in i However, the (M/L) of the underlying galaxy is surprisingly high (even with a factor-of- error bar), indicating either that the simple King-Minkowski formula we used is a poo approximation here, and/or that there is a substantial amount of dark matter present.

Placement of the nucleus on the cooling diagram from the theories of galaxy forma tion indicates that it is a dissipatively formed system. The difference from the parameter of the underlying galaxy (which are typical for the late Hubble types) clearly shows tha if the nucleus formed from the material of the main body of the galaxy, the proces must have been highly dissipative. This is also suggested by the strong contrast in th phase-space densities (Carlberg, 1986). The gas may have been supplied by the stella evolution mass loss from the main body of the galaxy (Zinnecker et al., 1985), or b a returning wind infall (Silk, Wyse, and Shields, 1987). A dissipative origin is in ac cord with the earlier spectroscopic and photometric results. Tremaine (1976) propose that dwarf nuclei represent tidally accreted globular cluster remnants. In view of th discussion above, such purely dissipationless origin seems unlikely.

It is curious that the extreme core-halo structure of the nucleated dwarfs follow closely the unrestricted configuration of maximum gravothermodynamical entropy (Whi and Narayan, 1982). It is not clear how did dwarfs manage to achieve such a state, an why are they so different from the "power-law + core" profiles of normal ellipticals.

More details will be presented in a forthcoming paper. SD acknowledges a trave grant from the AAS, a partial support from the Alfred P. Sloan Foundation, and th cordial hospitality of Dipartimento di Astronomia, Università di Bologna.

References

Bendinelli, O.: 1990, *Astrophys. J.* **366**, 599.

Bendinelli, O., Parmeggiani, G., Zavatti, F., Djorgovski, S.: 1990, *Astron. J.* **99**, 774.

Carlberg, R.: 1986, *Astrophys. J.* **310**, 593.

Carter, D., Sadler, E.: 1990, *Mon. Not. R. Astr. Soc.* **245**, 12p.

Held, E., Mould, J., de Zeeuw, T.: 1990, *Astron. J.* **100**, 415.

Kent, S.: 1987, *Astron. J.* **94**, 306.

Silk, J., Wyse, R., Shields, G.: 1987, *Astrophys. J. Letters* **322**, L59.

Tremaine, S.: 1976, *Astrophys. J.* **203**, 306.

White, S., Narayan, R.: 1982, *Mon. Not. R. Astr. Soc.* **229**, 103.

Zinnecker, H., Cannon, R., Hawarden, T., MacGillivray, H.: 1985, in *The Virgo Cluster*, ESO Workshop eds. O.-G. Richter and B. Binggeli, p. 135.

IGHT AND COLOR DISTRIBUTION IN THE CORE OF M 32[1]

R. Michard[1], J.-L. Nieto[2]
[1]*Observatoire de la Côte d'Azur, Nice, France*
[2]*Observatoire Midi-Pyrénées, Toulouse, France*

We briefly report on results from high-resolution imaging of the central region of M 32 in he U, B, R and 1 μm bands. The resolutions (FWHM) range from 0.60″ at 1 μm to 0.92″ n the U band. All images show 'pointed' isophotes within the central 5″, the deviations rom ellipticity being small (maximum $\simeq$ 0.01) but significant (Fig. 1), suggesting that a nuclear disk is present within the central 15 pc. We find no color gradient in either U–B nor B–R at sub-arcsecond resolution (see figure 2; a small gradient is however suggested at longer wavelengths). An empirical model fit to the observations after convolution by he appropriate Point-Spread-Function yields an intrinsic half width at half maximum ('core' radius) of 0.32″, a central surface brightness B = 12.65, constant central colors (U–B = 0.56, B–R = 1.52), as well a a constant axis ratio ($c/a = 0.71$) and position angle throughout the central region ($r < 10''$). These results have to be considered along with the following observational constraints to build a consistent scenario of the formation process of M 32: the small anisotropy of the velocity field (Bender and Nieto, 1990), the need for gas dissipation to explain the very high central density (Kormendy, 1989; Nieto, 1990) and the presence of the nearby M 31 halo for gas supply. It is suggested that, in symbiosis with the protogalactic nearby M 31 supplying gas and strong tidal torques, M 32 has experienced several star formation bursts at an early epoch followed by efficient population mixing. A full account of this paper can be found in Michard and

[1]Based on observations made with the Canada-France-Hawaii telescope, operated by the National Research Council of Canada, the Centre National de la Recherche Scientifique of France and the University of Hawaii, USA

G. Longo et al. (eds.), *Morphological and Physical Classification of Galaxies*, 441–442.

Nieto (1991).

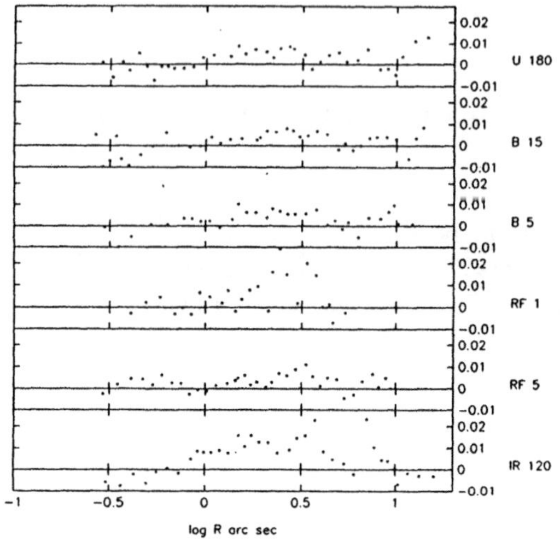

Figure 1. Profile of the *cos4θ* coefficient within the central 12″ for 6 high resolution CCD frames of M 32 all showing pointed isophotes between 1.2 and 5″.

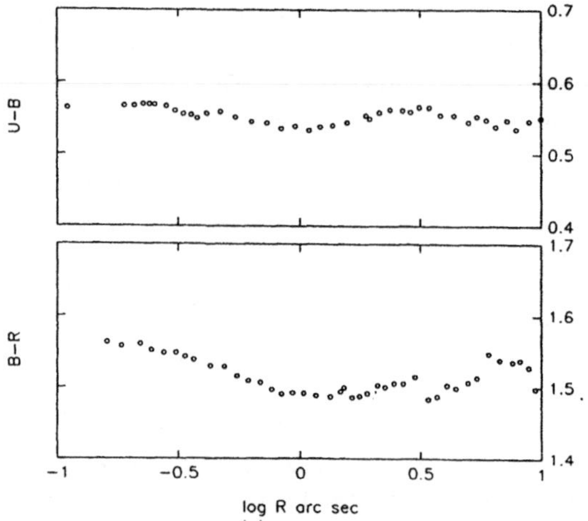

Figure 2. U–B and B–R profiles within the central 10″ at resolutions (FWHM) of 0.92 and 0.72″, respectively, showing no evidence for central color gradients.

References

Bender, R, Nieto, J.-L.: 1990, *Astron. Astrophys.* in press.

Kormendy, J.: 1989, *Astrophys. J. Letters* **342**, 63.

Michard, R., Nieto, J.-L.: 1991, *Astron. Astrophys. Letters* in press.

Nieto, J.-L., 1990, in *Dynamics and Interactions of Galaxies*, Ed. R. Wielen, p.258, Springer Verlag Berlin Heidelberg.

NGC 3384 AND ITS 'SATURN-LIKE APPEARANCE'

M. Ferrario, M. Capaccioli
Astronomy Department
University of Padova, Italy

NGC 3384 is a nearby galaxy unanimously classified as barred S0. It exhibits a morphological peculiarity first noticed by Pease (1920): *'a bright center* 40″ *diameter, on which is superimposed a very bright elongated nucleus* 19″ × 10″, *p* 45°, *crossed by a second bright nucleus* 40″ × 5″, *p* 130°, *presenting a Saturn-like appearance'*. This characteristic of NGC 3384 was rediscovered by Davoust et al. (1984; D84) who, analyzing high resolution CFHt plates, argued that the elongated *'second nucleus'* (hereafter third-component) is not a bar perpendicular to the major axis, but a small, inner *'polar ring'*. This interpretation rested on the fact that the elongated feature (1) protrudes from the bulge, indicating that it is not a local density enhancement within the bulge, (2) has ragged edges, not sharp ones as bars do, and (3) is offcentered with respect to the disk, a phenomenon not observed in other early-type barred galaxies.

We studied the inner regions of NGC 3384 with three CCD frames in the B and R bands, taken with the 2.2 m and NTT telescopes of ESO at La Silla and sky subtracted with the aid of a ESO Schmidt plate as in Capaccioli and Caon (1989). From the results of the standard photometric analysis (Fig. 1) we evict the presence of the third-component, visible in the orientation profile (where it gives raise to ~ 30° twist), and in the ellipticity profile (here amplified by the tendency of the bulge to become rounder with increasing radius). However, the NTT R-band frame, which has a 0.″8 seeing (FWHM), comparable with the 0.″7 of the CFHt plates of D84, does not show ragged edges of the feature, as D84 found. Either from the profiles of figure 1 or from isophotal maps, it is very difficult to obtain information about the photometric properties of the third-component, due to its very low luminosity. We disentangle it from the bright host galaxy by modeling the dominant components, bulge and disk, and by subtracting their sum from the original image (see e.g. Capaccioli et al., 1987). The models are built under the only assumption that isophotes of disk and bulge are axisymmetric and coaxial: since the third-component is very faint, this is a good first-order approximation. Each model computation requires three steps: a) masking the isophotes of the original image in the third-component region, b) fitting the remaining isophotes with ellipses, and c) creating an image with perfectly elliptical isophotes using the geometrical and brightness profiles obtained in the previous step. This procedure is clearly dependent on the chosen mask: we therefore produced several photometric models with different masks, to test the reliability of the resulting image. Our method obviously fails in the central regions. The resulting isophotes of the third-component are sufficiently regular to force an ellipse-fitting, so to measure its center and major-axis position angle (P.A.

G. Longo et al. (eds.), Morphological and Physical Classification of Galaxies, 443–444.

= 131°.5, only 8°.5 aways from the galaxy minor axis). We see (Fig. 2) that there is good agreement between the major-axis brightness profiles relative to different images of the third-component (deriving from various models of the galaxian main components at least for $r \lesssim 22''$, where the scatter is $\lesssim 0.1$ mag. Note that, at the most favorab location, the B-band surface brightness of third-component is yet 2 mag fainter tha the galaxy bulge+disk, and that the ratio of third-component luminosity to the tot luminosity inside $25''$ is $< 3\%$. The surface brightness profile is neither constant wit radius (as for bars of most S0's) nor exponential (as a disk), so we cannot discriminat between a bar and a disk.

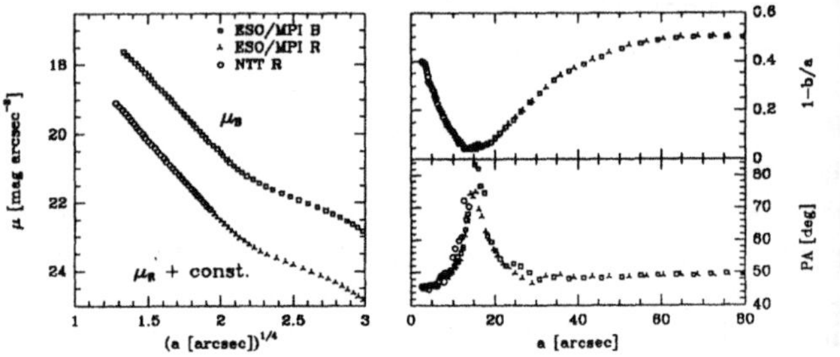

Figure 1. Brightness, ellipticity and orientation profiles of NGC 3384. The geometric profiles shown are obtained from both B and R frames, as no difference has been found in the two photometric bands.

We estimate the apparent length of the third-component to be $\simeq 50''$, in good agreemen with previous work (D84). The ratio of this value (corrected for the inclination $i = 26$ of the galaxy) to $D_{25} = 160''$ (Ferrario, 1990) is in the typical range for the bars of S0' (Elmergreen and Elmengreen, 1985).

We find the third-component to be concentric with the bulge within $0''.4 \simeq 0.02$ kpc and both components to be offcentered with respect to the disk: this effect is almos certainly due to asymmetries in this last component (Ferrario, 1990). The color of th third-component is very similar to that of the whole galaxy between $10''$ and $20''$, an tends to be bluer outside $20''$: this colour change, being in the outer regions of th structure, should be taken with care.

In summary, the aforementioned photometric evidences suggest that the structur of the inner region of NGC 3384 can be accounted for by the presence of a bar: the inne polar-ring hypothesis is not sufficiently supported by new evidences.

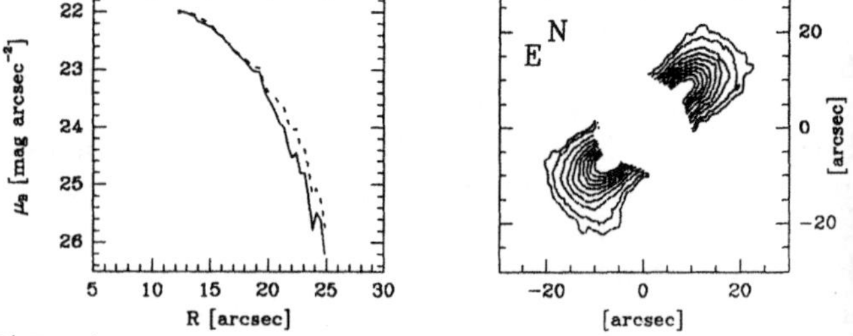

Figure 2. Isophote shape and surface brightness profile of the third-component.

Capaccioli, M., Caon, N.: 1989, 1st ESO/ST-ECF Data Analysis Workshop (Garching), P.J. Grosbøl, F Murtagh, R.H. Warmels eds., ESO Conferences and Workshop Procs. **31**, 107.

Capaccioli, M., Held, E.V., Nieto, J.-L.: 1987, *Astron. J.* **94**, 1519.

Davoust, E., Levievre, G., Maury, A., Nieto, J.-L.: 1984, *Mon. Not. R. Astr. Soc.* **209**, 503 (D84).

Elmergreen, B.G., Elmergreen, D. Meloy: 1985, *Astrophys. J.* **288**, 438.

Ferrario, M.: 1990, Thesis, University of Padova.

Pease, F.G.: 1920, *Astrophys. J.* **51**, 276 (Mount Wilson Contribution No. 186).

·IGALE OBSERVATIONS OF NGC 3198

R.L.M. Corradi[(1)], J. Boulesteix[(2)], A. Bosma[(2)],
M. Capaccioli[(3)], P. Amram[(2)], M. Marcelin[(2)]
[1] *Department of Astronomy, University of Padova, Italy*
[2] *Observatoire de Marseille, France*
[3] *Astronomical Observatory of Padova, Italy*

·canning Perot–Fabry observations (data–cubes of contiguous sub–ångstrom monochro-·natic images) allow us to isolate line–emission from nearby continuum using a 'local' ·ery narrow–band filter, without being troubled by the velocity blurring due to internal ·inematics.

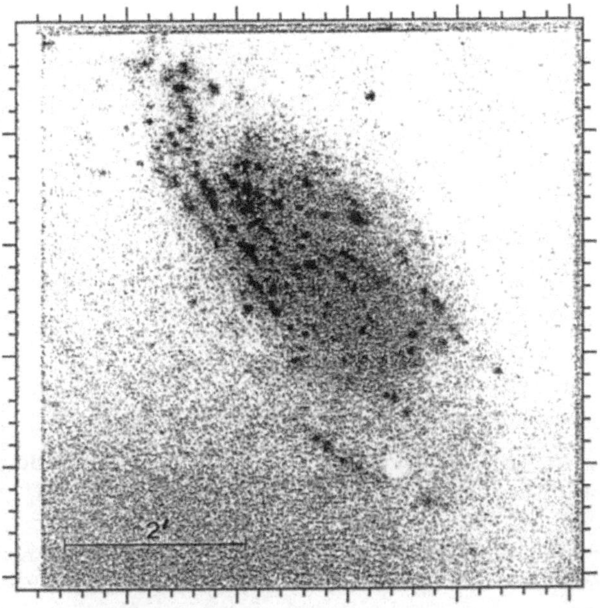

Figure 1. Reconstructed Hα image of NGC 3198. Observations were made at the 3.6–m telescope at CFH using the CIGALE instrument. North is at the bottom, West on the left.

For example, for each of the 256 × 256 picture elements a CIGALE (Boulesteix et ·l. 1983) observation gives a ∼10Å wide spectrum at ∼0.3Å resolution centered on the ·ine of interest. Working on these spectra it is then possible to remove completely the ·ontinuum and build 'true' monochromatic images by computing the integrated line–flux, ·i.e. the area under the line–profile). Furthermore, in the same time the radial velocity ·ield (with velocity resolution of 5–10 km s^{-1}) is available for the whole 6′ × 6′ field.

·. Longo et al. (eds.), *Morphological and Physical Classification of Galaxies*, 445–446.

Kinematics of an inner ringlike structure in NGC 3198

The analysis of the velocity field of NGC 3198 shows systematic deviations from the pur rotational axisymmetric model (Fig. 2). The major indications of non–circular motior are in a ringlike structure in the intermediate region of the optical disk of the galax (see the wiggles in the isovelocity contours in figure 2). This structure appears in CC images as a non–complete ring, and in the CIGALE Hα image (Fig. 1) as a string c HII regions from which the spiral arms emanate tangentially. The ring's diameter i 50″ – 2.5 kpc (if Δ=10 Mpc), and its apparent flattening $b/a = 0.23 = \cos 77°$; since th galaxy inclination is $i = 71°$, then the ring is either *non–circular* or *non–coplanar* to th disk, or both. In the case of coplanarity, the peculiar motions can be simply describe by a contraction with velocity ~20 km s^{-1} superimposed to the general rotation of th galaxy.

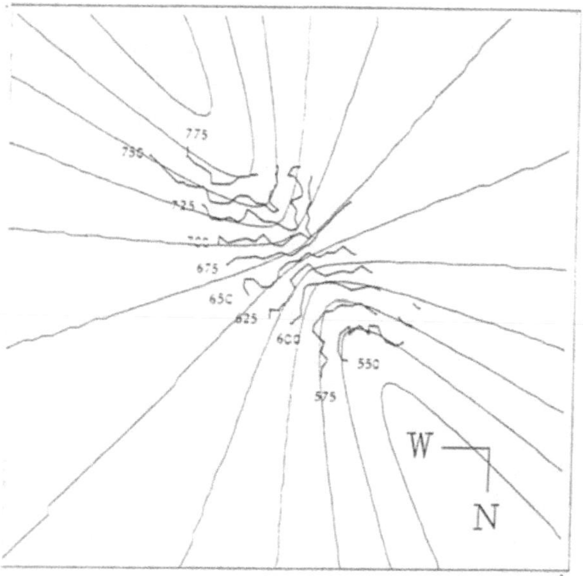

Figure 2. The velocity fiel of NGC 3198, compared to th best–fitting axisymmetric mode Heliocentric velocities are labelle

The 'bar' controversy: morphology *versus* dynamical modelling

NGC 3198 is classified as SB(rs)c in the RC2, but our CCD and CIGALE images do no show any clear evidence of a bar in the central region of the galaxy.

Another case of misclassification?

May be, but 1) the observed deviations of the velocity field from the pure axisymmetri model in the inner regions of the galaxy, and 2) the existence of the inner ringlike structure (very often present in barred galaxies), could be interpreted as possible indication of the presence of a bar lying in an unfavorable position to be easily detected.

Boulesteix, J., Georgelin, Y., Marcelin, M., Monnet, G.: 1983, in *S.P.I.E. Conf. Instr. Astr.* V, 445 37.

Corradi, R.L.M., Boulesteix, J., Bosma, A., Capaccioli, M., Amram, P., Marcelin, M.: 1990, *Astron Astrophys.* in press.

MAPPING THE STAR FORMATION IN NEARBY GALAXIES: THE MAGELLANIC IRREGULAR NGC 3109

F. Bresolin[1], M. Capaccioli[1], G. Piotto[1,2]
[1] *Department of Astronomy, University of Padova, Italy*
[2] *European Southern Observatory, Garching, Germany*

The SBm galaxy NGC 3109 (DDO 236) is located at the border of the Local Group, at a distance of $\sim$ 1.6 Mpc (estimated with Cepheids by Sandage and Carlson, 1988 = SC88). While comparable to the Small Magellanic Cloud for luminosity ($M_B = -16$), this galaxy is among the largest Magellanic systems known so far ($D = 14$ kpc).

SC88 and Demers et al. (1985) studied the stellar content of NGC 3109 with photographic material. We have used B and V CCD images taken with the 1.5-m Danish and 2.2-m telescopes of ESO at La Silla (Bresolin et al., 1990), forming a mosaic of $21' \times 3'$ (9 fields). Deep exposures under fair seeing conditions allowed us to measure stars down to V $\simeq$ 24, producing one of the most extended CCD samplings of the stellar content of a dwarf galaxy besides the Clouds (8250 stars). Zero-point errors are $\sim$ 0.03 mag in V and 0.05 mag in (B$-$V), and mean internal errors, estimated by artificial stars experiments, range from 0.02 mag at V = 19 to 0.5 mag at V = 23. The comparison of $\sim$ 200 stars in common with SC88 has revealed a systematic zero-point difference, our photometry being 0.3 mag brighter (placing NGC 3109 $\sim$ 15% closer).

Figure 1 shows the distribution of the brightest (V $\leq$ 22) blue ($-0.2 \leq$ (B$-$V) ≤ 0.4) and red ((B$-$V) $\geq$ 1.4) stars. Three main features stand out in the blue map, which are absent or barely visible in the red map:

- the central bar of NGC 3109, which stretches from SE to NW. de Vaucouleurs and Freeman (1972) described this feature in a study of the asymmetrical bar-disk configurations for SBm systems
- two spiral arms, leaving the southern end of the bar towards the east
- the clumpiness of the overall distribution of the blue stars.

With the aid of H_α CCD images obtained for the 6 central fields, we were able to identify several HII regions. These are found mainly along or near the bar, and are confined to a $\sim$ 300 pc thick strip, with a maximum density at a few minutes to the west of the optical center (an asymmetry common to late-type spirals and irregulars). The largest regions are $\sim$ 250 pc across; some extended loops are also found. The distribution of blue stars and stellar associations matches that of the HII regions rather well.

Individual c$-$m diagrams have been produced for the stars projecting onto each of the 30 most prominent HII regions of our catalog (note that crowding effects are enhanced by clustering, making photometry rather uncertain at faint magnitudes).

G. Longo et al. (eds.), Morphological and Physical Classification of Galaxies, 447–448.

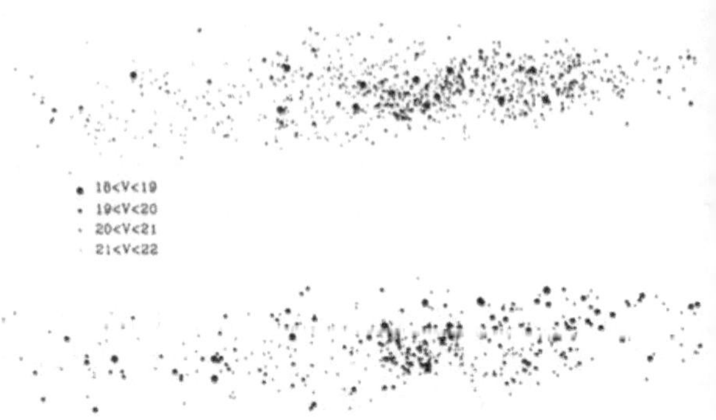

Figure 1. Distribution of blue (*top*) and red (*bottom*) stars over the image of NGC 3109 = DDO 236. North is at the top, east at the left; long axis is $\approx 21'$.

Ages of these star-forming regions were estimated by comparison with the theoretic isochrones computed by Bertelli et al. (1990); see figure 2.

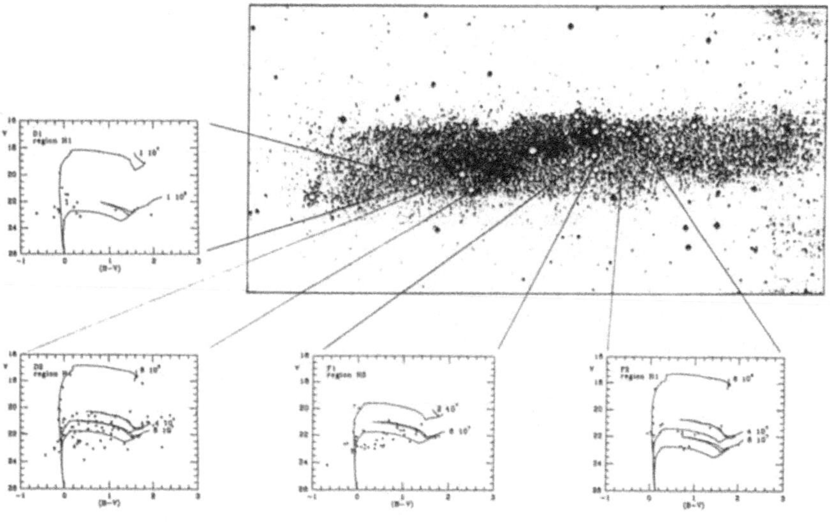

Figure 2. Distribution of prominent star-forming regions in NGC 3109 (white dots) and c−m diagrams for 4 of them. Superposed isochrones provide the reported ages.

Despite the uncertainties (internal absorption, metallicity, distance), this procedure give a fair estimate of the ages and age differences of the associations. The youngest star in NGC 3109 have ages of $\sim 6 \times 10^6$ yr and masses up to $\sim 30\ \mathcal{M}_\odot$. No appreciabl dependence of age on position in the galaxy is found: stars of 20-100$\times 10^6$ yr appear a over the disc. We conclude that NGC 3109 is still active in forming stars, possibly in rather regular manner in the last 10^8 yr.

Bresolin, F., Capaccioli, M., Piotto, G.: 1990, *The Messenger* **60**, 36.

Bertelli, P., Betto, R., Bressan, A., Chiosi, C., Nasi, E., Vallenari, A.: 1990, *Astron.Ap.Suppl.* **85**, 845.

Demers, S., Kunkel, W. E., Irwin, M. J.: 1985, *Astron. J.* **90**, 1967.

de Vaucouleurs, G., Freeman, K.C.: 1972, *Vistas Astron.* **14**, 163.

Sandage, A., Carlson, G.: 1988, *Astron. J.* **96**, 1599 (= SC88).

IM OBSERVATIONS OF THE MAGELLANIC CLOUDS FROM NTARCTICA

P. Calisse[1], G. Dall'Oglio[1], M.T. Di Bari[1], A. Iacoangeli[1], L. Martinis[4], P. Merluzzi[1], L. Piccirillo[1,2], L. Pizzo[1], L. Rossi[3], C. Santillo[1]
[1]*Department of Physics, University of Rome, "La Sapienza", Rome, Italy*
[2]*European Space Agency - ESTEC, Noordwijk, The Netherlands*
[3]*Istituto Superiore PT, Rome, Italy*
[4]ENEA - TIB, Frascati, Rome, Italy

)bservations and discussion

uring the austral summer 1987/88 and 1989/90, we performed two Antarctic expedi-ons with the main purpose of installing the first permanent Antarctic Infrared Obser-atory O.A.S.I. (Osservatorio Antartico Sub-millimetrico Infrarosso). At the same time, sing two small telescope (100 cm and 48 cm diameter[1,2]), we measured the continuum mission at millimetric wavelengths of the Large (LMC) and Small (SMC) Magellanic :louds. In both the expeditions, we detected a strong signal, both at 1mm and 2mm avelengths, coming from the LMC and SMC. These signals are very well spatially cor-elated with the 100 μm IRAS data. The resulting spectra on LMC and SMC indicate gray-body emission from cold dust at a temperature $T_d = 15K$ and a spectral index $= 1.1$. The longwavelength part of the spectrum shows an excess in the emission (Fig.) from the cool dust. A possible explanation of both the FIR and the mm measurements ould be the follow: since the mm data and the FIR data are well spatially correlated, is likely that the emitting source is defined in the same sky region, while the emitting ust is made of two different components. The first component, responsible of the FIR mission, is a dust at a temperature of 33 K and a spectral index $\alpha = 2$. The second omponent is a cold dust at a temperature of about 15 K and a spectral index $\alpha = 1.1$. 'ollowing Chini et al., we can also fit the longwavelength part of the spectrum with only he 100 μm IRAS data. In this case, we obtain a dust temperature of 17 K with $\alpha = 1.2$. n additional consideration on this model, following Stark could be done. In the LMC pectrum, the IRAS 60 μm point is contaminated by the contribution of fine structure adiation of neutral oxygen at 63 μm for an amount of 0.1 - 0.7 MJy srad^{-1}. After the ecessary correction, the 60 μm IRAS point is reduced in intensity, fitting better the 17

. Longo et al. (eds.), *Morphological and Physical Classification of Galaxies*, 449–450.

K model.

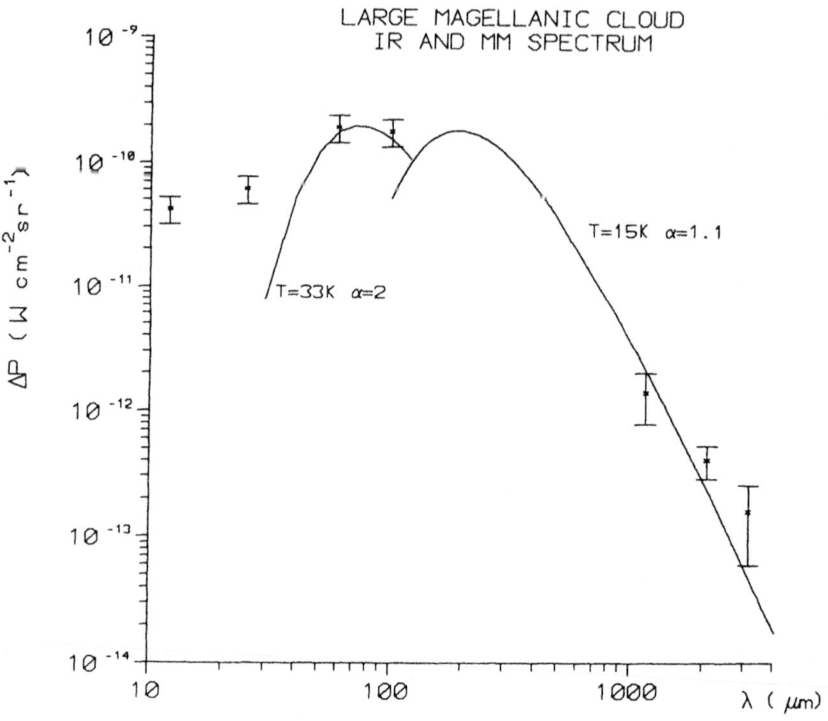

Figure 1

The correct determination of the dust model responsible of the entire IR, FIR, sub mm and mm spectrum could be done only with a direct measurements of the dus emission near its peak at about 350 μ m. At the same time, it would be desirabl to increase the angular resolution of the experiments from the present 1 degree to fe arcmins. This will be done starting from the 1990/91 Italian Antarctica Expedition. W are planning to use the OASI 2.6 m diameter telescope installed at Terra Nova Bay (Ros Sea - Antarctica) in order to repeat the mm measurements on the Magellanic Cloud with a better angular resolution ($\sim$ 5 arcmin).

This work was supported by the Programma Nazionale di Ricerche in Antartide an by the Istituto Superiore P.T. We would like to thank the team of ENEA for organizin the antarctic expeditions and the logistic support.

References

Andreani, P. et al.: 1990, *Astrophys. J.* **348**, 467.

Chini et al.: 1986, *Astron. Astrophys. Letters* **166**, L8.

Piccirillo, L.: 1990, submitted to *Rev. Sci. Instr.*.

Stark: 1990, *Astron. Astrophys. Letters* **230**, L25.

SEYFERT GALAXIES WITH CLOSE COMPANIONS. A METHOD FOR IDENTIFICATION

P. Rafanelli[1], **P. Marziani**[2]
[1]*Department of Astronomy, University of Padova, Italy*
[2]*International School for Advanced Studies, Trieste, Italy*

It has become evident in recent years that there is some relationship between galaxy interactions and the presence of nuclear activity. Dahari (1984), analyzing a redshift limited ($z < 0.030$) sample of Seyfert galaxies found a definite excess of physical and optical companions (38%) compared with a control sample of field galaxies (22%).

Motivated by these results we have undertaken a program of observations of close interacting Seyfert galaxies in order to study in detail the physical properties of these systems and of their interaction region. To this aim we have set up a method for identifying our targets by visual inspection of the POSS prints and applied this method to the Seyfert galaxies with $z < 0.1 (N_{S1} = 230,\ N_{S2} = 166)$ listed in the second edition of *A Catalogue of Quasars and Active Galactic Nuclei* (Veron-Cetty, Veron, 1985).

The only two parameters which can be measured on the POSS prints and used in a comparative way are:

- the separation D between pairs of galaxies, measured taking as unit the major axis d of the main component;
- an estimate of the difference in magnitude of the two galaxies.

In order to select the pairs of galaxies which are physical companions using both these parameters it is necessary to know their upper limits derived analyzing a sample of pairs of galaxies which:

- have been initially selected on the POSS prints;
- are physical companions;
- show the effects of a close interaction.

Such a sample is given in the *Atlas and Catalogue of Interacting Galaxies* by B. Vorontsov-Velyaminov (1959; hereafter V-V), which list 355 interacting galaxies defined as *objects connected by bridges of matter* and selected on the POSS prints. The following data can be easily extracted from the catalogue and from the enclosed atlas:

G. Longo et al. (eds.), Morphological and Physical Classification of Galaxies, 451–454.

- m_A = apparent magnitude of the main component of the system. The main com ponent is defined either as the brighter galaxy or as the central one if the syste is multiple.
- $\Delta m = m_{comp} - m_A$
- D = separation between the components of the system.

All companions of the 355 V–V objects result to be contained in a circle of radiu $3d$, centered on the main component of the system. Hence $D < 3d$ is the first conditio to be imposed in our attempt to identify interacting Seyfert galaxies.

The number of interacting V–V galaxies as a function of the magnitude of the principal component increases up to $m_A \sim 16$, and shows a rapid decrease for m_A : $16 - 17$, indicating that systems whose principal component is fainter than 16^m–17^m ar likely to be lost, either because $m_{comp} > (m_{lim})_{POSS}$ or the bridges of matter are faint than the level of the sky noise. Since 90% of the companions of the V–V galaxies hav $m_{comp} < m_A + 3$ we have taken $(\Delta m)_{max} = 3$.

Extending these results: $D < 3d$, $\Delta m < 3$ to all galaxies detectable on the POS prints we can affirm that:

- All galaxies which have a companion at a distance $D < 3d$ and not fainter tha $m_A + 3$ are quite likely components of a close and strongly interacting system;
- if $m_A < 16$ no selection effects are present.

This method of search has been applied to the Seyfert 1 (S1) and Seyfert 2 (S2) gala ies with $z < 0.1$ listed by Veron-Cetty and Veron (1985). It results that the percentage c interacting systems is 18% in the Seyfert–1 class and 27% in the Seyfert–2 class. It mus be remarked that all interacting Seyfert galaxies with $z < 0.03$ and $-45° < \delta < +90$ given by Dahari (1985) have been again identified using our procedure. In order to com pare our results with those of Dahari (1985) we have extracted the interacting Seyfer galaxies with $z < 0.03$ identified with our method. The percentages of interacting object that we have found (S1–21%; S2–29%) are lower than those found by Dahari (S1–27% S2–42%) and *very close* to the values found by us for $z < 0.1$.

A test for evaluating up to which redshift value the percentages of interacting Seyfer galaxies found with our method are reliable, can be based on the assumption that th expected frequency of interacting Seyfert galaxies should be almost the same for differen z -values. This is verified by S1 and S2 galaxies with $z < 0.05$ and $z < 0.02$ respectivel The frequency level $\sim$ 22% is almost the same for both classes of objects.

It is noteworthy that the apparent magnitude of S1(S2) galaxies at $z = 0.05$ (z = 0.02) derived assuming $M_{S1} = -21$ ($M_{S2} = -19$) and $H = 75$ km s^{-1} Mpc^{-1} is $\sim$16 namely the magnitude below which our method of selection begins to loose companions Anyway the observed excess, 30%–47%, of S2 galaxies for $0.02 < z < 0.05$ could be du to the fact that many S2 galaxies have been preferentially identified within spectroscopi surveys of interacting galaxies or of group of galaxies.

The preliminary results of our investigation on interacting Seyfert galaxies using ou method of identification show essentially that the frequency of interacting S1 and S2 i almost the same, $\sim$ 22%. This value is remarkably lower than the value $\sim$ 35% foun y Dahari (1985) and Fuentes-Williams and Stocke (1988), but still much larger than he percentage 2.5% of pairs of field galaxies having $D < 3d$ and $\Delta m < 3$, derived using he maps of Shane and Wirtanen (1967) and considering galaxies with $12 < m < 17$. t must be pointed out that the only condition $D < 3d$ would produce a percentage $\sim$ 5%, indicating that the restriction $\Delta m < 3$ reduces of a factor 10 the number of pairs f galaxies. The fact that our percentage 22% of interacting Seyfert galaxies is only factor 1.6 smaller than the percentage found by Dahari, who selected objects having) $< 3d$, strongly suggests that the condition $\Delta m < 3$ tends to isolate preferentially hysical companions. Concluding this means that Seyfert galaxies, as a class of objects, how a significant excess of physical companions, and then that interaction is more likely mong Seyfert than normal galaxies.

ℓeferences

)ahari, O.: 1984, *Astron. J.* **89**, 966.

)ahari, O.: 1985, *Astron. J.* **90**, 1772.

'uentes-Williams, T., Stoke, J.T.: 1988, *Astron. J.* **96**, 1235.

hane, C.D., Wirtanen, C.A.: 1967, *Publications of the Lick Observatory* vol. XXII, part 1.

'eron-Cetty, M.-P., Veron, P.: 1985, ESO Sci. Report n. 4.

'orontsov-Veliaminov, B.: 1959, *Atlas and Catalogue of Interacting Galaxies*, part I, Moscow University.

Hugo van Woerden while reviewing the posters.

LOBULAR CLUSTERS AS PROBES OF THE GALACTIC OTENTIAL

M. Stiavelli[1], G. Piotto[1,2], M. Capaccioli[2]
[1]*European Southern Observatory, Germany*
[2]*Department of Astronomy, University of Padova, Italy*

apaccioli, Ortolani and Piotto (1990) have shown that the slope of the mass function f galactic globular clusters (GCs) correlates with the cluster distance from the galactic lane. We report on the interpretation (Stiavelli et al., 1990) of this rather surprising esult as the cumulative effect of disk-shocking. The leading idea is that, during a assage through the disk, the cluster loses preferentially stars at large radii and, due o mass segregation, these are mainly low mass stars. As a start, we have tackled the roblem by a semi-analytical approach and modeled the disk-shocking in the impulse pproximation (Ostriker, Spitzer and Chevalier, 1972). We have taken into consideration ne model parameters describing the galactic mass-distribution and its effect on GC tidal adii, and GC structural characteristics (concentration, IMF slope, etc.), as well as those ccounting for observational selection effects and for some of the theoretical shortcuts. Ve find that, in addition to selection effects which may spoil the observed correlation, the esults are strongly dependent on i) the way tidal radius r_t changes with galactocentric istance R_g, and ii) the value of the surface mass density of the galactic disk in the solar eighbourhood $\sum_\odot$.

Tidal truncation is a well-known major effect of the gravitational field of the Galaxy King, 1962). A simple argument (Binney and Tremaine, 1987) shows that there is a elation between the tidal radius of a GC of mass M_{GC} and its galactocentric distance: $\propto (M_{GC}/M_g(< R_g))^{1/3} R_g$, where $M_g(< R_g)$ is the galactic mass interior to R_g. Being $M_g(< R_g)$ not known, we assume $r_t \propto R_g^a$. Two extreme models for the tidal force field, *a)* a central point source (which the bulge can be approximated to, for our purposes), nd *(b)* an extended mass distribution with $\rho \propto r^{-2}$ (consistent with a flat rotation curve) ive a $\alpha = 1$ and $2/3$ respectively. These limits are well consistent with the results of our alculations which imply $0.5 < \alpha < 1$ in order to reproduce the correlation observed by apaccioli et al. (1990). In fact, when α is outside this interval, the correlation gradually mears out.

A further constraint to α is offered by the plot of ($Log\ r_t - \frac{1}{3} Log\ M_{GC}$) vs $Log\ R_g$ or a sample of 131 globular clusters in the Galaxy (figure 1; data from Webbink, 1985;

. Longo et al. (eds.), Morphological and Physical Classification of Galaxies, 455–456.

Chernoff and Djorgovski, 1989).

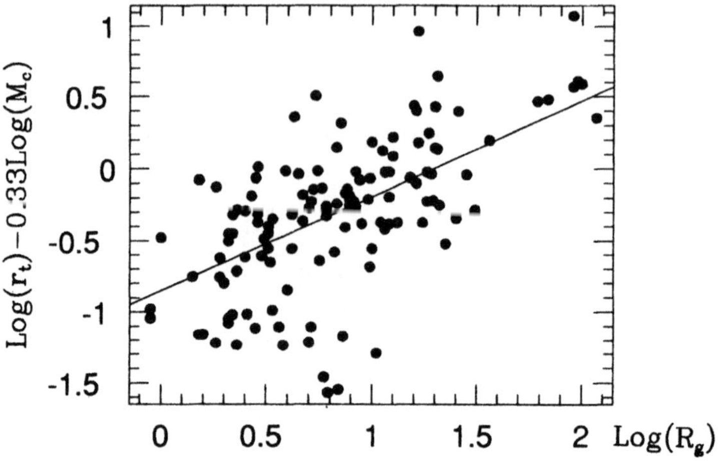

Figure 1.

The best fit value $\alpha = 0.66 \pm 0.08$ is in very good agreement with the value expecte for an extended mass distribution. (The relatively large scatter is mainly due to th observational inaccuracy in the determination of r_t and M_{GC}). Though not conclusiv this simple result is in good agreement with the results of the more sophisticated bu intrinsically different analysis by Innanen, Harris and Webbink (1983), who postulate the relation that we intend to test and derived constraints on the orbital distribution o globular clusters.

ii) The effect of disk-shocking is also strongly dependent on the surface mass densit $\sum(R)$ of the galactic disk, modeled by an exponential profile with scalelength of 3. kpc (Bahcall and Soneira, 1980). Our results agree better with the observations b Capaccioli et al. (1990) when $\sum_{\odot} \simeq 75 \pm 25$ $M_{\odot}pc^{-2}$. Indeed, for $\sum_{\odot} \leq 50$ $M_{\odot}pc^{-}$ only clusters at a distance from the plane $Z_g < 3$ kpc would be significantly affected. Fo $\sum_{\odot} \geq 100$ $M_{\odot}pc^{-2}$, instead, the effect would be too large. In order to derive a stronge constraint, detailed calculations and N-body simulations are presently in progress.

References

Bahcall, J.N., Soneira, R.M.: 1980, *Astrophys. J. Suppl. Ser.* **44**, 73.

Binney, J., Tremaine, S.: 1987, *Galactic Dynamics*, Princeton University Press.

Capaccioli, M., Ortolani, S., Piotto, G.: 1990, *Astron. Astrophys.* , in press.

Chernoff, D.F., Djorgovski, S.: 1989, *Astrophys. J.* **339**, 904.

Innanen, K.A., Harris, W.E., Webbink, R.F.: 1983, *Astron. J.* **88**, 338.

King, I.R.: 1962, *Astron. J.* **67**, 471.

Ostriker, J.P., Spitzer, L., Chevalier, R.A.: 1972, *Astrophys. J. Letters* **176**, L51.

Stiavelli, M., Piotto, G., Capaccioli, M., Ortolani, S.: 1990, to appear in *Formation and Evolution of Sta Clusters*, K. Janes ed., A.S.P. Conference Series, and ESO Preprint n. 727.

Webbink, R.F.: 1985, in *Dynamics of Star Clusters*, J. Goodman and P. Hut eds., Reidel (Dordrecht), p 541.

RELATION BETWEEN THE PROPERTIES OF GLOBULAR CLUSTER SYSTEM AND THE MASS DISTRIBUTION ON THE GALACTIC DISK

V.G. Surdin
Sternberg Astronomical Institute, Moscow, USSR

As it is was shown by Ostriker et al. (1972), one of the factors controlling the dynamical evolution of a globular cluster is the compressive gravitational shocks (or "tidal hocks") i.e., gravitational perturbations resulting from the passage of a clusters through he galactic disk. Actually, this is an important evolutionary factor, so far as it gives us the possibility to explain the relation between the mass concentration parameter $C = \log(r_t/r_c)$ and the galactocentric distance R of a globular cluster (Surdin, 1979). Therefore we may use this property of the globular cluster system as a tool for the determination of the mass distribution in the galactic disk.

The characteristic time for the destruction of a globular cluster under the action of compressive shock in the impulsive approximation (Ostriker et al., 1972) is

$$t_{sh} = \frac{3GMPV_z^2}{20r_h^3 g_m^2} \tag{1}$$

where $M, r_h,$ and P are respectively the mass, the spatial half-mass radius, and the orbital period of a cluster; V_z is Z-velocity when the cluster is approaching the galactic plane, and g_m is the maximum value of the gravitational Z-acceleration due to the galactic disk.

We must take into consideration that (Fall and Ross, 1977)

$$r_h \simeq 0.7 \cdot \sqrt{r_c r_t} \tag{2}$$

and that (Rastorguev and Surdin, 1978)

$$r_t = R_p \left(\frac{M}{(1+\nu)M_G(R_p)} \right)^{1/3} \tag{3a}$$

$$\nu = \frac{2e}{(1+e)^2 \ln\left(\frac{1+e}{1-e}\right)} \tag{3b}$$

where e is the eccentricity of the cluster orbit, and $M_G(R_p)$ is the mass of the Galaxy inside the perigalactical distance of the cluster orbit (R_p). Orbital period is connected with apogalactic distance of the orbit (R_a):

$$P \simeq \left[\frac{R_a^3}{GM_G(R_a)} \right]^{1/2}, \tag{4}$$

G. Longo et al. (eds.), *Morphological and Physical Classification of Galaxies*, 457–458.

and the gravitational acceleration depends on the surface density of the galactic disk (
near the perigalactic region of the cluster's orbit:

$$g_m = 2\pi G\sigma(R_p). \quad (5)$$

For exponential disk

$$\sigma(R_p) = \sigma_\circ \exp(R_\circ/h) \cdot \exp(-R_p/h) \quad (6)$$

where $\sigma_\circ$ is the surface density near the Sun, h is the radial scale, and $R_\circ$ is the distan to the galactic center.

For the Galaxy with constant rotational velocity (220 km/s), and values $R_\circ = 9$ kp $t_{sh} = 2 \cdot 10^{10}$ yrs, and $G_\circ = 54\ M_\odot\ \mathrm{pc}^{-2}$ (Gould, 1990) we have calculated from equatio (1)–(6) the conditions for low bound. The curves are drawn in figure 1 for different valu of the scale parameters (van der Kruit, 1987). Distribution of globular clusters (dots i figure 1; data by Chernoff and Djorgovski, 1989; Peterson and Reed, 1987) prove tha the value $h = 5$ kpc is more sufficient for the requirements of the dynamical evolution the clusters.

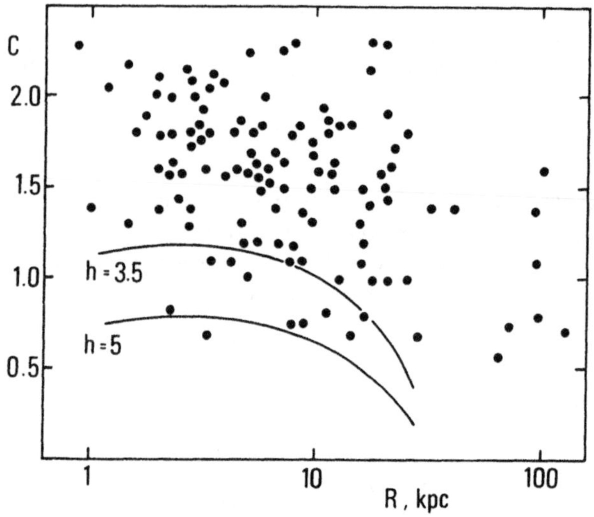

Figure 1.

Chernoff, D.F., Djorgovski, S.: 1989, *Astrophys. J.* **339**, 904.

Fall, S.M., Rees, M.J.: 1977, *Mon. Not. R. Astr. Soc.* **181**, 37P.

Gould, A.: 1990, *Mon. Not. R. Astr. Soc.* **244**, 25.

Ostriker, J.P., Spitzer, L., Chevalier, R.A.: 1973, *Astrophys. J. Letters* **176**, 451.

Peterson, C.J., Reed, B.C.: 1987, *Publ. Astron. Soc. Pac.* **99**, 20.

Rastorguev, A.S., Surdin, V.G.: 1978, *Astron. Tsirk.* **1016**, 3.

Surdin, V.G.: 1979, *Astron. Tsirk.* **1079**, 3.

van der Kruit, P.C.: 1987, in G. Gilmore and B. Carswell (eds.) *The Galaxy*, Reidel, Dordrecht, p. 27.

TOWARDS A PRECISE DEFINITION OF THE MORPHOLOGICAL STRUCTURES OF GALAXIES AS A BASIS FOR THEIR AUTOMATIC CLASSIFICATION

A. Accomazzi[1], D. Delfini[2], M. J. Kurtz[1], P. Mussio[2]
[1] *Harvard-Smithsonian Center for Astrophysics, Cambridge, MA, USA*
[2] *Department of Physics, University of Milano, Italy*

1 Introduction

We propose a new methodology to derive morphological types from digital images of galaxies in an automated fashion, in which "quantitative" methods, based on statistical model fitting techniques, are flanked by syntactic methods of pattern recognition (Kurtz et al., 1990).

The detection of important structures (arms, bars, rings, etc.) is based on a precise algebraic definition (Accomazzi et al., 1989; Delfini, 1990), which allows the design and implementation of a more complex automated system capable of classifying such structures (Kurtz et al., 1990). Automating the morphological classification guarantees objectivity and uniformity in the results. Furthermore, the contextual analysis of the results produced by this system on a training set of images allows the astronomical community to criticize and improve the definition itself.

2 Results

The definitions used in the extraction of morphological structures have been tested on a set of 20 CCD images providing encouraging results. An example is given in figure 1.

Preliminary results in the description and classification of these features have been obtained by exploiting a first set of elementary definitions (Accomazzi et al., 1989; see figure 2) which formalize the verbal description given by astronomers who collaborate on the experiment. Here the identification of arms and a nucleus in a nearly edge-on galaxy allows a preliminary classification of the object as a normal spiral galaxy.

We are currently studying ways to improve the current definitions in order to cope

G. Longo et al. (eds.), Morphological and Physical Classification of Galaxies, 459–460.

with the complexity of the morphological features in images of well-resolved galaxies.

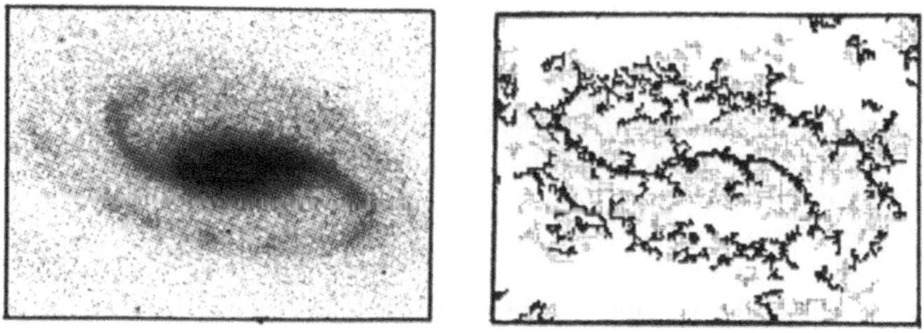

Figure 1. CGCG 157076. Left: CCD image taken with the 60″ telescope at the Loiano Observatory, Bologna. Resolution: 0.5″/pixel. The image is 130 ×170 pixels. Right: a bulge+disk model has been subtracted to the image and important morphological features have been detected. Dark pixels mark the ridges, while dotted pixels mark the valleys. Note how well the ridge structures match the arm patterns.

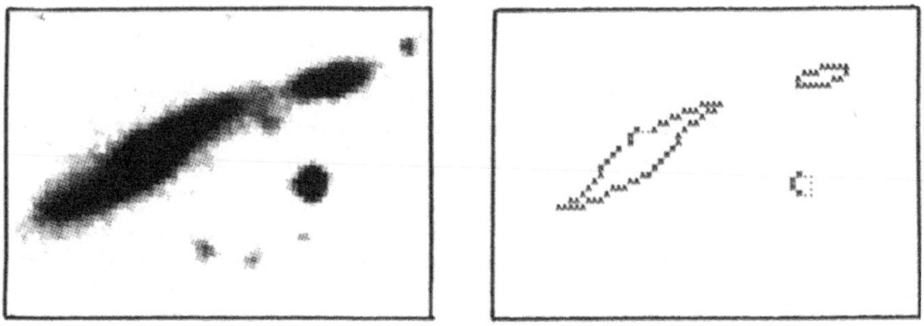

Figure 2. Subfield from an ESO Southern Sky survey Plate. The plate has been digitized at a resolution of 1.34″/pixel. The image shown here is 70 × 50 pixels. Right: some morphological features have been detected and classified: pixels marked 'A' denote the contours of possible arms, 'N' of possible nuclei/bulges. The remaining pixels are marked '.'.

Acknowledgements

We are greateful to Dr. Giuseppe Gavazzi for having supplied part of the CCD image under study. This work is partially funded by CNR contract 89.00215.12.

References

Accomazzi, A., Bordogna, G., Mussio, P., Rampini, A.: 1989, in *"Knowledge-Based Systems in Astronomy"*, A. Heck and F. Murtagh, eds., Springer-Verlag, 191.

Delfini D.: 1990, *Thesis*, Università degli Studi di Milano, Italy.

Kurtz, M.J., Mussio, P., Ossorio, P.G.: 1990, Pattern Recognition Letters, **11**, 507.

OURIER ANALYSIS OF BARRED GALAXIES: A CONVENIENT .LTERNATIVE TO THE COMPONENT DECOMPOSITION.

P. Dubath[1], B. J. Jarvis[2], L. Martinet[1], D. Pfenniger[1]
[1]*Observatoire de Genève, Sauverny, Switzerland*
[2]*European Southern Observatory, Santiago, Chile*

Difficulties to extract a bar component

ı the traditional decomposition approach, a barred galaxy is assumed to be the su- erimposition of three components, the bulge, the bar, and the disk. The bulge and ıe disk are assumed to be axisymmetric; since mass is positive they cannot contribute ıore to the galaxy than a component generated by a revolution of the bar minor axis. ubtracting this maximum axisymmetric component from the galaxy, the result can be ɔnsidered as the "minimum bar component" (Fig. 2). This result itself is not accept- ble as a bar component because of its strange "bow tie shape" (with zero value along s minor axis). Therefore, in order to get an elongated, convex bar an axisymmetric art has to be added to the minimum bar. However, it is impossible to decompose ıree **superimposed axisymmetric components** without introducing models and/or ssumptions.

The assumption that the bar contribution is negligible in the very center and in ıe outer part can be made. Fitting appropriate laws independently in both extreme ɛgions would allow an estimate of the bulge and disk contributions. Unfortunately, rolonging these laws into the barred intermediate region produces a profile higher than ıe minor axis profile, as illustrated in figure 1. Therefore, this does not permit to define n axisymmetric part for the bar.

In fact, as underlined in the following section, we have very little knowledge of ıe component behaviour in the overlapping region. No model or assumption with a heoretical or observational justification exist, which allow to define the contributions of

. Longo et al. (eds.), Morphological and Physical Classification of Galaxies, 461–464.

three components in the barred region.

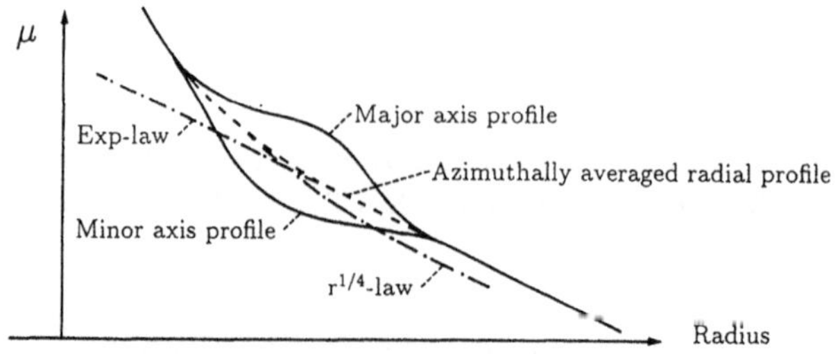

Figure 1. Schematic representation of the general behaviour of early type barred galaxy profiles.

2 The nature of components

Here, we argue about the usual representation of barred galaxies as the superimpositio of three systems. The idea that they are composed of a standard unbarred galaxy an of an additional bar component has probably no physical meaning, since the strong nor axisymmetric potential of a bar perturbs any underlying axisymmetric component. Fro orbit studies in barred galaxies, we know that no prograde circular orbits exist withi the bar region (see e.g. Contopoulos and Papayannopoulos, 1980; Athanassoula et al 1983), which is rather supported by elongated orbits, as confirmed by numerous N-bod simulations. Thus a physical bar component cannot be "bow tie" shaped. Concernin the bulge or any underlying axisymmetric component inside the bar region, the dynamic is necessarily different from the axisymmetric case, since the angular momentum integra J_z no longer exists. Knowing this, is there still a sense to try to define and to extrac one (or two) axisymmetric component in the bar region?

3 Alternative to the decomposition

As in previous works (Elmegreen and Elmegreen, 1985; Buta, 1986, 1987, and Ohta e al., 1990), we exploit the Fourier transform of the azimuthal profiles (intensity alon circles of radii r, lying in the plane of the galaxy and centered on the nucleus). Th intensity of the image may be written as:

$$I(\theta, r) = \sum_{m=0}^{\infty} I_m(r) \cos(m\theta + \varphi_m(r)) \quad (1$$

where $I_0(r)$ is the azimuthally averaged radial profile and represents the axisymmetri behaviour of the galaxy. The higher order terms describe the deviations from axisymme try. Fortunately, due to the symmetry of the galaxies, these higher order terms rapidl decrease, and only the first three or four *even* order terms contain non-negligible powe for a global representation of the galaxy. The phase of the Fourier transform componen is represented by $\varphi_m(r)$. In the barred region it is well defined and nearly constant bu becomes random in the axisymmetric regions, due to the almost circular isophotes. Th phase φ_2 represents the position angle of the bar.

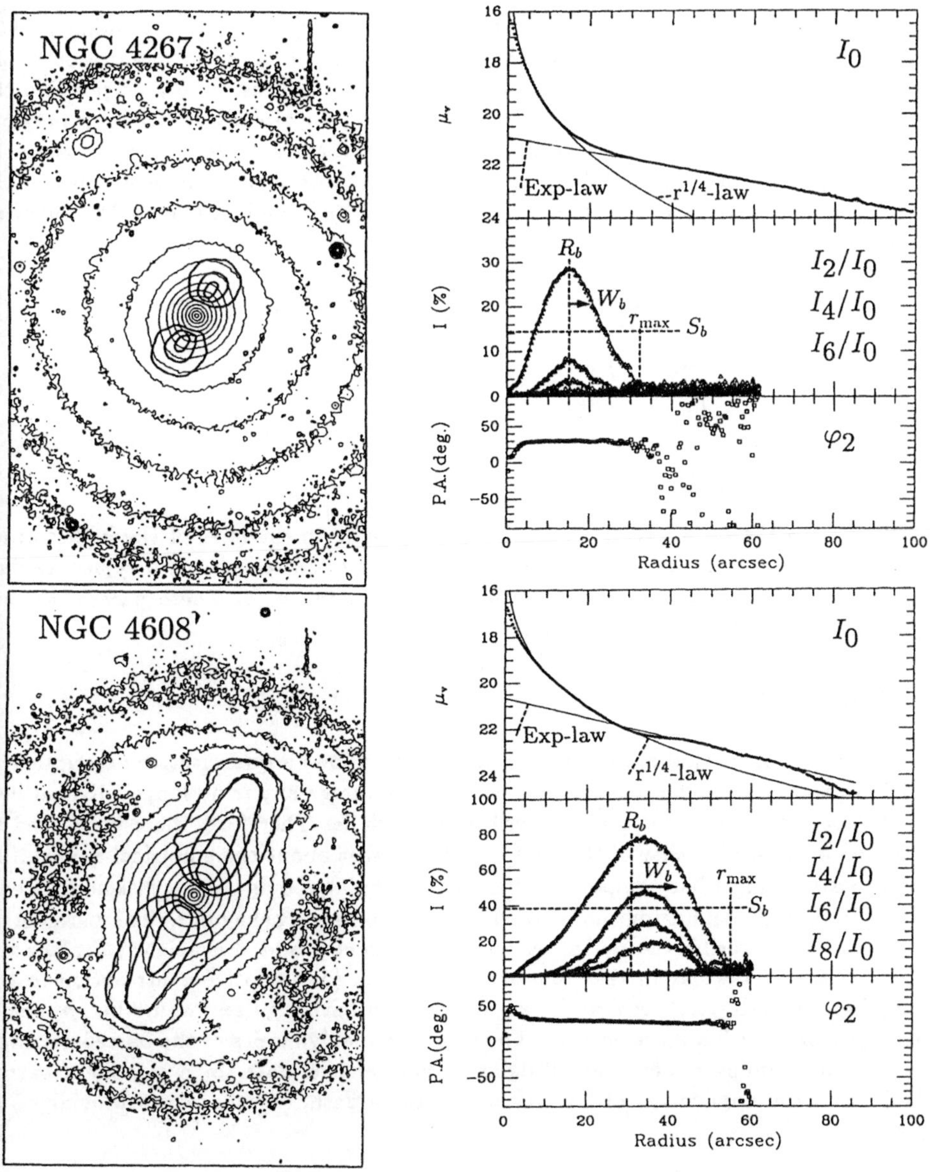

Figure 2. Left. Two examples of "minimum bar component" superposed to the isophotes of the galaxies, showing that minimum bars do not look like a physical independent component (Sect. 1). **Right.** Application to these galaxies of our Fourier method (Sect. 3). For each galaxy the figure shows: (top) the azimuthally averaged radial profile, (middle) the first 8 coefficients of the Fourier transform of the azimuthal profiles, I_1/I_0, I_2/I_0, ..., I_8/I_0, together with the parameters S_b, R_b and W_b, and (bottom) φ_2 which represents the position angle of the bar. The odd Fourier coefficients I_1/I_0, ..., I_7/I_0, (and I_8/I_0 in the case of NGC 4267) are not distinguishable on the middle panels, because they do not rise above the noise level.

As an illustration, two examples are presented in figure 2. These radial profil contain the same information as the 2D images and give a complete description of th galaxy. In a first general approach, one needs to quantify as simply as possible th dominants and global characteristics of the galaxy. As already noticed by previou authors (e.g. Ohta et al., 1990), $I_0(r)$ can be quite well represented by a $r^{1/4}$-law in th inner part and by an exponential-law in the outer part. We then propose to consider th ratio $I_2(r)/I_0(r)$ and, from the three first moments of this profile, to define the followin parameters:

$$S_b = \frac{1}{r_{\max}} \int_0^{r_{\max}} \frac{I_2(r)}{I_0(r)}\, dr, \qquad (2$$

$$R_b = \frac{1}{r_{\max}\, S_b} \int_0^{r_{\max}} r\, \frac{I_2(r)}{I_0(r)}\, dr, \qquad (3$$

$$W_b^2 = \frac{1}{r_{\max}\, S_b} \int_0^{r_{\max}} r^2\, \frac{I_2(r)}{I_0(r)}\, dr - R_b^2, \qquad (4$$

$r_{\max}$ being the radius where $I_2(r)$ becomes comparable to the level of the higher orde terms of the Fourier transform (Fig. 2). S_b measures the strength, R_b a typical radius and W_b the radial extent of the deviation from axisymmetry. These parameters ar represented in the two cases of figure 2.

4 Discussion

As for the decomposition approach, our alternative allows a quantitative description c a barred galaxy. Its advantage lies in the fact that the galaxy is no longer divided int model dependent components but described as a whole. Therefore, there is no need t argue about physical reality of the components, neither about the empirical ambiguitie linked to extracting a bar component. Instead of defining a bulge and a disk component we simply describe the azimuthally averaged axisymmetric behaviour. The parameter S_b, R_b and W_b describe the deviation from axisymmetry and conveniently replace th characteristics of the bar in the component approach (e.g. relative contribution to th galaxy light, length, etc.). A real decomposition of a barred galaxy would be physicall meaningful only with a more accurate knowledge of the dynamics of the barred galaxies

In a forthcoming paper, we will describe photometric data for a sample of barre galaxies and the application of our method to these data, together with an analysis o the results.

References

Athanassoula, E., Bienaymé, O., Martinet, L., Pfenniger, D.: 1983, *Astron. Astrophys.* **127**, 349.

Buta, R.: 1986, *Astrophys. J. Suppl. Ser.* **61**, 631.

Buta, R.: 1987, *Astrophys. J. Suppl. Ser.* **64**, 383.

Contopoulos, G., Papayannopoulos, T.: 1980, *Astron. Astrophys.* **92**, 33.

Elmegreen, B.G., Elmegreen, D.M.: 1985, *Astrophys. J.* **288**, 438.

Ohta, K., Hamabe, M., Wakamatsu, K.: 1990, *Astrophys. J.* **357**, 71.

IMPROVEMENT OF THE SIGNAL TO NOISE RATIO BY ADAPTIVE FILTERING

H. Lorenz, G.M. Richter
Zentralinstitut für Astrophysik der Akademie der Wissenschaften der DDR, Potsdam, Babelsberg

1 Introduction

For classification of galaxies a broad spectrum of spatial frequencies is relevant from the highest available resolution to the large very low surface brightness regions. Moreover, besides the visual recognition, quantitative evaluations by isophotes, profiles etc. are of growing importance. Particularly in the lower surface brightness regions such work is often hampered by noise and requires an increase of the signal to noise ratio (see e.g., figure 1a, b).

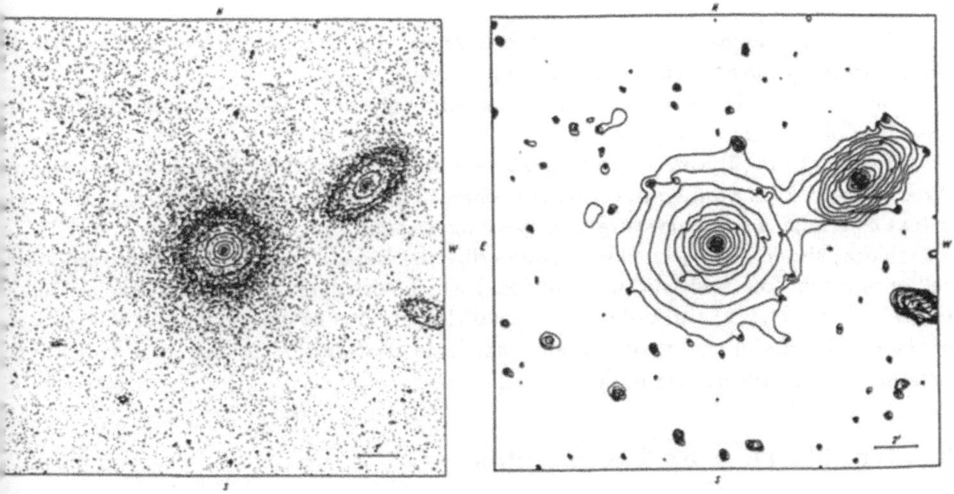

Figure 1a. Isophotes of NGC 3379 and its neighbours – **Figure 1b.** The same as figure 1a after adaptive filtering

G. Longo et al. (eds.), *Morphological and Physical Classification of Galaxies*, 465–468.

In principle this could be done by a classical filter, but this would result not only in degradation of the higher resolution features in the galaxy, but also in a smearing of t foreground stars in the low surface brightness regions. Hence, it would be useful to ha an adaptive filter, which recognizes at every point in the image the local spatial spectru (the resolution) of the signal and adapts its own impulse response so that the signal not degraded. In order to be unbiased it should be a linear filter (nonlinear filters as e. the median filter tend to introduce a systematic bias under some conditions).

2 Algorithm

We have developed an adaptive filter on the basis of the H-transform. This is an ortho onal transform defined by the following algorithm:
the image is divided into cells of 2×2 pixels. For every cell four values h_0, h_x, h_y, h_c, a computed by

$$h_t = \sum_{i=1}^{4} a_i p_i \qquad (1$$

where $t = 0, x, y, c$; p_i are the pixel values in the cell and a_i the corresponding values the following basis functions:

(2)

h_0 h_x h_y h_c

where black areas mean +1 and white areas –1. All values h_x, h_y, h_c together a the transform values (coefficients) of the first order.

The second order coefficients are obtained by the same algorithm, applied onto th h_0-values of the first order; etc..

It is seen from (2), that h_x and h_y are the components of the gradient in the respectiv direction and h_c is a measure of the curvature. The different orders represent the mea values of gradient and curvature over growing areas.

Hence, the h_x, h_y, h_c- coefficients describe the inner structure of the correspondin cell on a resolution level (spatial frequency) which corresponds to the size of this cell. for every type (x, y, c) and order the noise-distribution is known (e.g. from the statistic of the coefficients in the image), then coefficients exceeding a certain value indicate th presence of a significant signal in this cell.

3 Realization and examples

We developed up to now two realisations of an adaptive filter. In the first one, speciall adapted to slower and smaller computers, the nonsignificant coefficients are exclude and an inverse transform is made with the large significant coefficients only. Since thi responds to a nonequidistant raster of the resulting image, an interpolation by a two nensional Kalman-filter has to follow the inverse transform.

In the newer program the H-transform is used as a local operator, this means that transform is not made in disjunct cells but over every pixel of the image. The efficients are used only to recognize the signal, but the resulting image is constructed m the h_0-values of the different orders directly. This is by far more expensive than ordinary transform and the inverse transform and requires high computing power th large memories. However, it has the advantage of better homogeneity and isotropy the signal recognition.

Figure 1 shows an example made with the first program (from Capaccioli, Held, renz, Richter and Ziener: 1988, Astron. Nachr., **300**, 69).

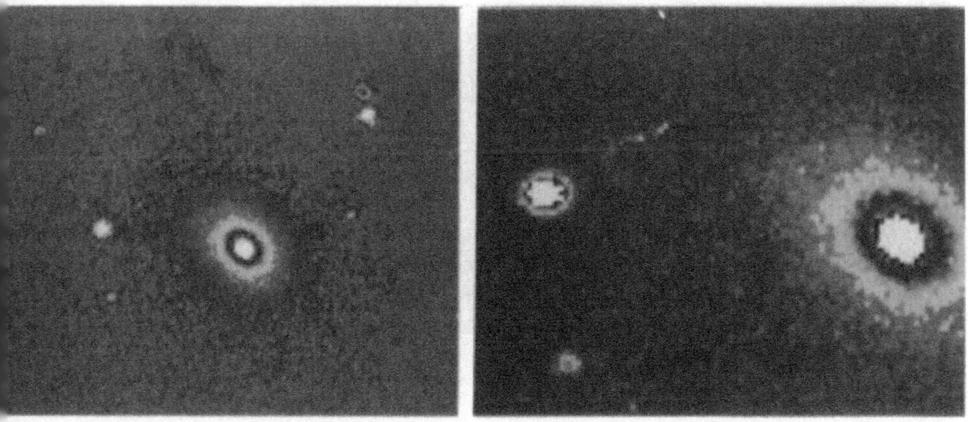

Figure 2. NGC 315 from a noisy plate and an enlarged part.

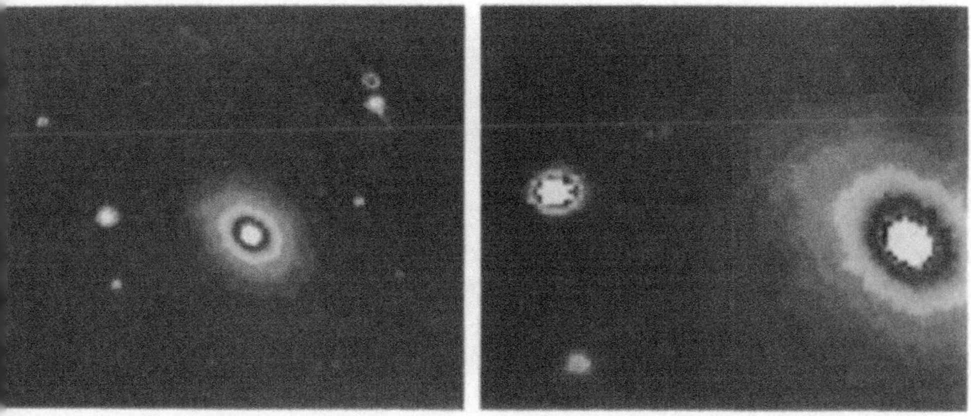

Figure 3. The same as Figure 2 after adaptive filter.

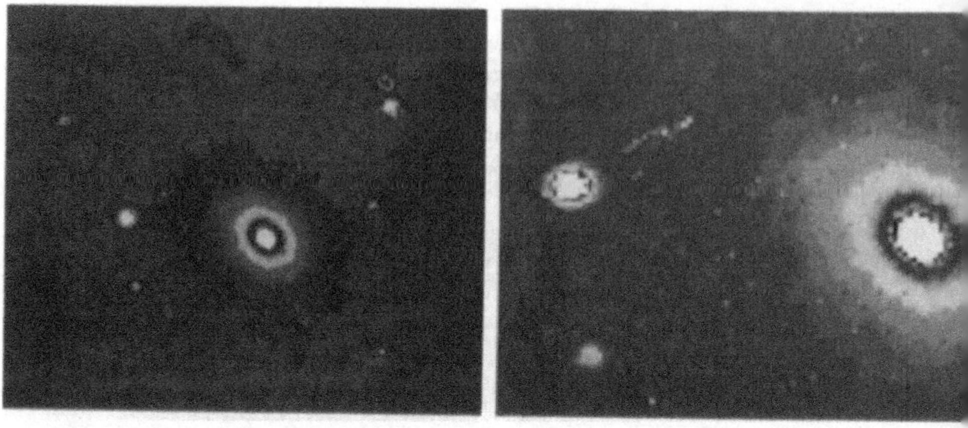

Figure 4. The same after elimination of the high frequency impulsive noise.

Figure 3 is the result of the newer program, applied to figure 2. However, som of the extreme noise spikes are misinterpreted as signal and appears as aberrant singl pixels. They can be recognized and eliminated, since their spectrum is too high frequen compared with the point spread function in the image. This is done by a filter simila to the one described above (Fig. 4).

bject index

33, 196.
85, 55, 61.
545, 62.
559, 424.
978, 54.
1350, 424.
1371, 54.
1413, 55.
1515+2146, 10.
1847, 424.
1927, 54.
2029, 55.
2670, 61.
M 2244-651, 409, 410.
ndromeda, 144.
rp 152, 223.
rp 155, 223.
rp 157, 223.
rp 220, 296.

arina, 164, 165, 166, 167, 169, 170, 171.
entaurus A, 27, 144, 146, 151.
Cen, 172.
GCG 157076, 460.
oma Cluster, 18, 19, 21, 70.

DO 154, 163.
DO 170, 163.
raco, 163, 164, 165, 166, 168, 169, 170, 171, 173.

234–21, 119.
406–30, 119.

568–6, 8.
ornax Cluster, 5, 41, 102, 163, 164, 166, 167, 168, 171, 173, 198, 261, 352, 353, 379, 419, 420.

R 8, 163.

I 1225+0146, 339.

342, 273, 209, 460.
1459, 77, 83, 193, 195, 198, 222, 223, 227, 228, 232, 366, 367, 424.
2006, 147, 151, 235.

IC 3370, 424.
IC 4296, 424.
IC 4767, 9.
I Zw 18, 339.

Leo II, 168.
Leo Triplet, 241.
Local Group, 172, 330, 353, 379, 439, 447.

M 49, 142, 143.
M 51, 7, 131, 270, 271, 296.
M 67, 88.
M 81, 7, 131, 132, 163, 241, 271.
M 82, 241, 273.
M 83, 208.
M 84, 196.
M 87, 27, 70, 71, 126, 142, 143, 151, 192, 193, 196, 201, 205, 381, 403.
M 92, 88.
M 100, 7, 132.
M 101, 239, 240, 241, 242, 243, 247, 248, 273.
M 104, 107.
Maffei II, 209.
Magellanic Cloud, Large (LMC), 256, 449.
Magellanic Cloud, Small (SMC), 256, 447, 449.
Magellanic Clouds, 241, 251, 252, 273.
Magellanic galaxies, 94.
Magellanic stream, 273.
Magellanic types, 18.
Malin 1, 8.
Milky Way, 85, 86, 87, 91, 92, 93, 94, 176, 241, 273, 278.
Mkn 348, 241.

NGC 45, 42, 51.
NGC 128, 9.
NGC 185, 339.
NGC 205, 172, 177, 261, 339, 439.
NGC 221, 69, 70, 71, 72, 172, 261, 339, 403, 441, 442, 424.
NCG 224, 42, 69, 70, 72, 86, 87, 89, 90, 91, 92, 158, 172, 240, 333, 403, 441.
NGC 253, 208, 460.
NGC 300, 460.

NGC 315, 467.
NGC 450, 42, 51.
NGC 484, 119.
NGC 488, 460.
NGC 520 , 223.
NGC 547, 119.
NGC 584, 396, 424.
NGC 596, 387, 388, 424.
NGC 598, 132, 134, 247, 333, 460.
NGC 628, 239, 242.
NGC 636, 388, 424.
NGC 681, 383.
NGC 708, 119.
NGC 720, 396, 424, 437, 438.
NGC 745, 119.
NGC 788, 383.
NGC 821, 389, 424.
NGC 891, 39, 40, 42, 96, 162, 239, 242.
NGC 936, 209, 213.
NGC 1023, 241.
NGC 1052, 119, 124, 146, 147, 192, 193, 194, 195, 198, 199, 234, 235, 242, 382, 383, 424.
NGC 1068, 207, 213, 215, 273, 383.
NGC 1073, 208.
NGC 1087, 460.
NGC 1097, 208, 211, 215, 273.
NGC 1172, 396.
NGC 1275, 193, 199, 202.
NGC 1291, 208, 214.
NGC 1300, 130, 208.
NGC 1316, 71, 73, 198, 199, 396.
NGC 1325, 154.
NGC 1326, 401.
NGC 1365, 208, 211.
NGC 1381, 107.
NGC 1395, 119, 424.
NGC 1399, 392, 396.
NGC 1404, 392.
NGC 1407, 396, 424.
NGC 1433, 210, 220, 396.
NGC 1453, 119, 396.
NGC 1530, 209.
NGC 1531, 424.
NGC 1533, 396.
NGC 1549, 387, 388, 392, 424.
NGC 1566, 7, 214.
NGC 1600, 24, 38, 78, 119, 424.
NGC 1800, 8.
NGC 1961, 242.
NGC 2217, 209.
NGC 2300, 424.
NGC 2403, 154.
NGC 2636, 404.
NGC 2683, 460.
NGC 2768, 119, 401, 402.
NGC 2775, 460.
NGC 2778, 424.
NGC 2782, 460.
NGC 2841, 138.
NGC 2974, 119, 146, 147, 424.
NGC 2997, 460.
NGC 3031, 460.
NGC 3059, 209.
NGC 3067, 241.
NGC 3077, 241, 271.
NGC 3081, 9, 210.
NGC 3087, 424.
NGC 3091, 424.
NGC 3108, 424.
NGC 3109, 447, 448.
NGC 3115, 72, 107, 108, 109, 110, 11 150, 232, 401.
NGC 3198, 154, 155, 445, 446.
NGC 3268, 393.
NGC 3359, 208, 242.
NGC 3377, 389.
NGC 3379, 70, 71, 107, 392, 424, 465.
NGC 3384, 387, 388, 443, 444.
NGC 3521, 460.
NGC 3557, 393, 424.
NGC 3607, 424.
NGC 3608, 366.
NGC 3610, 364, 424.
NGC 3613, 424.
NGC 3640, 424.
NGC 3641, 404.
NGC 3656 , 223.
NGC 3818, 424.
NGC 3862, 119.
NGC 3893, 460.
NGC 3894, 119.
NGC 3898, 460.
NGC 3904, 424.
NGC 3923, 386.
NGC 3928, 4.
NGC 3962, 119.
NGC 3992, 208.
NGC 3998, 124.
NGC 4027, 208, 214.
NGC 4105, 383.
NGC 4106, 383.

NGC 4111, 402.
NGC 4125, 100, 119, 424.
NGC 4151, 70, 208.
NGC 4179, 108.
NGC 4261, 28, 119.
NGC 4267, 463.
NGC 4278, 116, 119, 124, 146, 147, 222, 234, 235, 392, 424.
NGC 4291, 424.
NGC 4319, 220.
NGC 4321, 460.
NGC 4365, 29, 366, 367, 424.
NGC 4374, 119, 391, 392, 424.
NGC 4378, 460.
NGC 4382, 402.
NGC 4387, 424.
NGC 4406, 28, 29, 82, 118, 119, 231, 232, 366, 367, 424.
NGC 4452, 215.
NGC 4472 , 27, 78, 82, 142, 143, 242, 392, 424, 435.
NGC 4473, 424.
NGC 4478, 424.
NGC 4486, 118, 119, 381, 383, 392, 424.
NGC 4486B, 70, 403, 404.
NGC 4546, 209.
NGC 4551, 424.
NGC 4552, 119.
NGC 4564, 424.
NGC 4565, 42, 85, 94, 242, 243.
NGC 4589, 119.
NGC 4594, 72, 145, 162, 232.
NGC 4596, 213.
NGC 4621, 424.
NGC 4622, 137.
NGC 4627, 241.
NGC 4631, 241.
NGC 4636, 119, 392, 424.
NGC 4649, 424.
NGC 4656, 241.
NGC 4660, 424.
NGC 4694, 242.
NGC 4696, 118, 119, 193, 194, 199.
NGC 4697, 100, 107, 364, 383, 389, 424.
NGC 4698, 460.
NGC 4731, 208.
NGC 4736, 93, 208, 214, 460.
NGC 4742, 424.
NGC 4762, 215, 220, 402.
NGC 4782, 119.
NGC 4845, 116.
NGC 5005, 460.
NGC 5055, 138, 460.
NGC 5077, 119, 124, 126, 145, 146, 150, 151.
NGC 5090, 393, 394.
NGC 5091, 393.
NGC 5101, 208.
NGC 5102, 93, 401, 402.
NGC 5128, 235, 242.
NGC 5194, 216.
NGC 5194/5, 295.
NGC 5248, 460.
NGC 5252, 100, 101, 162.
NGC 5266, 124, 424.
NGC 5322, 74, 222, 223, 227, 366, 367, 424.
NGC 5383, 208, 211.
NGC 5419, 429.
NGC 5457, 460.
NGC 5477, 273, 274.
NGC 5635, 242.
NGC 5638, 424.
NGC 5666, 147, 404.
NGC 5745, 424.
NGC 5728, 208.
NGC 5813, 74, 83, 222.
NGC 5831, 383.
NGC 5845, 424.
NGC 5846, 119, 424.
NGC 6028, 10.
NGC 6034, 56.
NGC 6166, 55, 61, 436.
NGC 6221, 211.
NGC 6251, 70, 424.
NGC 6535 , 166.
NGC 6703, 424.
NGC 6868, 119, 396.
NGC 6876, 396.
NGC 6946, 209, 239, 240, 460.
NGC 6958, 198.
NGC 7020, 9, 10, 210.
NGC 7097, 119, 124, 145.
NGC 7098, 9.
NGC 7187, 9, 10, 210.
NGC 7217, 460.
NGC 7252, 75, 223, 224, 366.
NGC 7331, 460.
NGC 7457, 206.
NGC 7562, 396, 424.
NGC 7619, 424.
NGC 7626, 392, 396.

NGC 7702, 9.
NGC 7785, 424.
NGC 7814, 39, 40, 45, 96, 150, 162.

Sculptor, 164, 166, 167, 168, 170, 171, 173.

UGC 407, 403.
UGC 545, 403.
UGC 807, 42.
UGC 2748, 119.
UGC 3087, 403.
UGC 3654, 404.
UGC 4508, 404.
UGC 6132, 403.
UGC 7731, 403.
UGC 7778, 403.
UGC 9412, 403.
UGC 9799, 119.
UGC 10599, 403.
UGC 11546, 404.
UGC 11608, 403.
UGC 11668, 403.
UGC 11761, 403.
UGC 11763, 403.
UGC 12074, 403.
UGC 12389, 403.
UGC 12607, 403.
Ursa Minor, 163, 164, 165, 166, 168, 169, 170, 171, 173.

Virgo Cluster, 8, 17, 18, 19, 21, 27, 28, 29, 69, 102, 110, 143, 229, 261, 262, 263, 360, 367, 425, 426, 352, 353, 379, 419, 420.

GPSR Compliance

The European Union's (EU) General Product Safety Regulation (GPSR) is a set of rules that requires consumer products to be safe and our obligations to ensure this.

If you have any concerns about our products, you can contact us on ProductSafety@springernature.com

In case Publisher is established outside the EU, the EU authorized representative is:

Springer Nature Customer Service Center GmbH
Europaplatz 3
69115 Heidelberg, Germany

Batch number: 10816188

Printed by Printforce, the Netherlands